Handbook of Aquatic Food Products

Wiley Series in

INNOVATIONS IN FOOD SCIENCE AND TECHNOLOGY

Series Editors
Shaoping Nie
State Key Laboratory of Food Science and Resources
Nanchang University
China

Jianbo Xiao
Department of Analytical Chemistry and Food Science
Universidade de Vigo
Spain

Managing Editor
Jingjing Tian
College of Food Science and Technology
Nanjing Agricultural University
China

Handbook of Aquatic Food Products

Edited by

Haijin Mou
College of Food Science and Engineering
Ocean University of China
China

Chuang Zhang
Sanya Institute of Nanjing Agricultural University
College of Food Science and Technology
Nanjing Agricultural University
China

WILEY

IADNS

This edition first published 2027

Registered Office(s)
John Wiley & Sons, Inc., 111 River Street, Hoboken, NJ 07030, USA
John Wiley & Sons Ltd. New Era House, 8 Oldlands Way, Bognor Regis, West Sussex, PO22 9NQ, UK

For details of our global editorial offices, customer services, and more information about Wiley products visit us at www.wiley.com.

The manufacturer's authorized representative according to the EU General Product Safety Regulation is Wiley-VCH GmbH, Boschstr. 12, 69469 Weinheim, Germany, e-mail: Product_Safety@wiley.com.

Wiley also publishes its books in a variety of electronic formats and by print-on-demand. Some content that appears in standard print versions of this book may not be available in other formats.

Library of Congress Cataloging-in-Publication Data
Names: Mou, Haijin author | Zhang, Chuang author
Title: Handbook of aquatic food products / Haijin Mou, Chuang Zhang.
Description: Hoboken, NJ : Wiley, 2027. | Series: Innovations in food science and technology | Includes bibliographical references.
Identifiers: LCCN 2026010105 (print) | LCCN 2026010106 (ebook) | ISBN 9781394348909 hardback | ISBN 9781394348923 adobe pdf | ISBN 9781394348916 epub
Subjects: LCSH: Aquaculture industry | Aquatic resources | Aquaculture | Fishery products–Health aspects
Classification: LCC HD9450.5 .M68 2027 (print) | LCC HD9450.5 (ebook)
LC record available at https://lccn.loc.gov/2026010105
LC ebook record available at https://lccn.loc.gov/2026010106

Cover Design: Wiley
Cover Image: Xuanyu Han/Getty Images

Set in 9.5/12.5pt STIXTwoText by Straive, Pondicherry, India

Contents

List of Contributors

Rong Cao
Yellow Sea Fisheries Research Institute
Academy of Fishery Sciences
Shandong
China

Delvina
Taiwan Ocean University
Keelung
China

Carlo Esposito
Istituto di Scienze e Tecnologie
Chimiche "Giulio Natta"
Milan
Italy
and
Department of Pharmaceutical
Sciences
University of Milan
Milan
Italy

Yan Fan
College of Marine Life Sciences
Ocean University of China
Qingdao
China

Jerrell Felim
Taiwan Ocean University
Keelung
China

Hu Hou
State Key Laboratory of Marine Food
Processing & Safety Control
College of Food Science and
Engineering
Ocean University of China
Qingdao
China

Zwe-Ling Kong
Taiwan Ocean University
Keelung
China

Dongyu Li
Ocean University of China
Qingdao
China

Rui Li
College of Food Science and
Technology
Guangdong Ocean University
Zhanjiang
China

Zhuanzhuan Li
Yellow Sea Fisheries Research Institute
Chinese Academy of Fishery Sciences
Qingdao
China

Qingping Liang
College of Food Science and Engineering
Ocean University of China
Qingdao
China

Zhihong Liu
Yellow Sea Fisheries Research Institute
Chinese Academy of Fishery Sciences
Qingdao
China

N. M. Liyanage
University of Alberta
Edmonton
Canada

Peizhen Ma
Yellow Sea Fisheries Research Institute
Chinese Academy of Fishery Sciences
Qingdao
China

Haijin Mou
College of Food Science and Engineering
Ocean University of China
Qingdao
China

Sara Pozzo
Istituto di Scienze e Tecnologie
Chimiche "Giulio Natta"
Milan
Italy

Francesco Secundo
Istituto di Scienze e Tecnologie
Chimiche "Giulio Natta"
Milan
Italy

Shanshan Shui
College of Food Science and Pharmacy
Zhejiang Ocean University
Zhoushan
China

Yong Sun
Yellow Sea Fisheries Research Institute
Academy of Fishery Sciences
Shandong
China

Zhijun Tan
National Aquatic Product Quality
Inspection and Testing Center
Qingdao
China

Luying Tang
College of Food Science and Engineering
Ocean University of China
Qingdao
China

Chuanhai Tu
College of Food Science and Pharmacy
Zhejiang Ocean University
Zhoushan
Zhejiang
China

Jia Wang
Ocean University of China
Qingdao
China

Biao Wu
Yellow Sea Fisheries Research Institute
Chinese Academy of Fishery Sciences
Qingdao
China

Min Yang
Ocean University of China
Qingdao
China

Jeon You-Jin
Jeju National University
Jeju
South Korea

Bin Zhang
College of Food Science and Pharmacy
Zhejiang Ocean University
Zhoushan
China

Chuang Zhang
Sanya Institute of Nanjing Agricultural University
College of Food Science and Technology
Nanjing Agricultural University
Nanjing
China

Guanchao Zheng
National Aquatic Product Quality Inspection and Testing Center
Qingdao
China

Saiyi Zhong
College of Food Science and Technology
Guangdong Ocean University
Zhanjiang
China

Changliang Zhu
Ocean University of China
Qingdao
China

Lulu Zhu
College of Food Science and Engineering
Hainan University
Haikou
China

Series Preface

Food is fundamental to human health, social development, and quality of life. Since the beginning of the 21st century, food science and technology have been transformed by rapid advances in biotechnology, analytical science, information technology, advanced manufacturing, materials science, and nutrition. New food resources and ingredients are continually being identified. Innovative processing and preservation technologies are being developed while increasingly sophisticated approaches are being applied to food quality, authenticity, nutrition, and safety. At the same time, the global food system faces complex challenges arising from population growth, resource constraints, climate change, changing dietary patterns, and growing public expectations for safe, nutritious, and accessible foods produced through sustainable practices.

These developments have greatly expanded both the scope and the societal responsibilities of modern food science. Progress in the field increasingly depends on collaboration across disciplines and geographical boundaries as well as closer connections among fundamental research, technological innovation, industrial application, public health, and regulatory practice. There is consequently a pressing need for authoritative and systematically organized references that not only summarize established knowledge but also reflect emerging concepts, methods, technologies, and applications.

The *Innovations in Food Science and Technology* book series, developed in collaboration with the International Association of Dietetic Nutrition and Safety (IADNS) and Wiley, has been established in response to this need. IADNS is an international nonprofit academic organization committed to advancing and disseminating knowledge in food, nutrition, and safety. It seeks to strengthen cooperation among scientists worldwide and promote the responsible development of relevant industries. By bringing together the international academic network of IADNS and Wiley's experience in scholarly publishing, the series seeks to provide a durable platform for the exchange and integration of knowledge across the global food science community.

The series offers comprehensive and up-to-date coverage of the principal areas of modern food science and technology. Its volumes address food resources, processing, biotechnology, preservation and postharvest technology, analytical methods, authenticity, contaminants, functional foods, food enzymes, microorganisms, additives, and major classes of food components, including polysaccharides, polyphenols, proteins, peptides, lipids, and starch. Specialized subjects ranging from aquatic products and flavor science to the relationship between food and medicinal resources further extend the breadth of the series. Taken together, these volumes present a panoramic view of current progress while allowing each topic to be examined with the depth expected of a specialist reference work.

Each volume is led by recognized scholars and brings together contributions from experts with complementary backgrounds. The series emphasizes scientific rigor, an international perspective, systematic organization, and relevance to practical challenges. In addition to consolidating foundational principles, the books highlight recent discoveries, emerging technologies, evolving research directions and opportunities for application in food production, quality control, nutrition, safety management, and sustainable development. The series is intended to connect scientific understanding with technological practice and encourage productive dialogue among academia, industry, government agencies, and professional organizations.

The development of a series of this scale requires the close integration of academic leadership and professional publishing management. As Series Editors, Shaoping Nie and Jianbo Xiao shape the overall vision and academic framework of the series, define its principal subject areas, and guide the selection of individual volumes and their editors. Jingjing Tian serves as Publishing Director, supporting publication planning and coordinating the development and timely delivery of the volumes. Together, the three Series Editors promote international academic collaboration, engage leading researchers, and help ensure the scientific quality and global relevance of the series. Their complementary responsibilities enable the series to progress from an ambitious academic vision to a coordinated and sustainable international publishing program.

The series is designed for researchers and educators in food science, chemistry, nutrition, biotechnology, engineering, and related disciplines. It will also serve professionals working in food production, analysis, quality assurance, safety, and regulation as well as postgraduate students seeking advanced and reliable sources of knowledge. We hope that it will serve not only as a collection of reference works but also as a source of new ideas, interdisciplinary connections, and international cooperation.

We are deeply grateful to the volume editors, chapter authors, reviewers, advisers, and publishing professionals whose expertise and commitment make this series possible. We also thank IADNS and Wiley for their continuing support. We believe that modern food science and technology will play an increasingly important role in building healthier, safer, and more sustainable food systems. We hope that this series will contribute meaningfully to that shared endeavor.

August 2026

Shaoping Nie
Nanchang University

Jianbo Xiao
University of Vigo

Jingjing Tian
Nanjing Agricultural University

About the Series Editors

Dr. Shaoping Nie is a Distinguished Professor of Food Science and Food Nutrition at the State Key Laboratory of Food Science and Resources, Nanchang University. He is also Vice President of the Nanchang University, Associate Editor at the *Journal of Agricultural and Food Chemistry*, Founding Editor-in-Chief of Food Frontiers, Vice President of International Association of Dietetic Nutrition and Safety, and a Fellow of the Royal Society of Chemistry.

Dr. Nie's research interests are focused on the fields of food nutrition and food chemistry, food safety and analysis, especially focused on food complex carbohydrates (mainly bioactive polysaccharides and dietary fiber, especially on the structure, conformation, and bioactivities of dietary fiber and polysaccharides and the relationship of their structure and bioactivities), as well as on new product development of the precise regulation of food components for nutrition and health. He also focuses on developing bioactive ingredients for the functional foods and nutraceutical products, developing novel bioactive polysaccharides and dietary fiber from natural agricultural products, and exploring their applications in foods, medicinal, and pharmaceutical industries.

Due to his excellent research in bioactive polysaccharides and dietary fiber, he was selected as the Leading Talent of the National Special Support Program for High-Level Personnel in China and obtained supports from the National Natural Science Foundation of China for Distinguished Young Scholars and Excellent Young Scholars. Dr. Nie has published over 700 peer-reviewed scientific papers (over 500 papers in English, more than 300 papers as the first author, corresponding author, including co-corresponding author). Dr. Nie has authored and edited 8 books (3 books in English), participated in compiling 15 books, and obtained 45 authorized invention patents.

Furthermore, Dr. Nie has won some important awards: the Second-Class Award of the National Science and Technology Progress Award in 2023 (first recipient), the Second-Class Award of the National Science and Technology Progress Award in 2016 (third recipient), three First Prizes of the Natural Science of Jiangxi Province (two as the first recipient and one as the second recipient), two First Prizes of Scientific and Technological Progress of Jiangxi Province (one as the first recipient and one as the second recipient), one Second Prize of Natural Science Award of the Outstanding Scientific Research Achievement Award of the Ministry of Education (first recipient), etc. He was also awarded as a Highly Cited Researcher in Agricultural Sciences in 2018, 2019, 2020, 2022, and 2023 by Clarivate Analytics and as a Chinese Highly Cited Researcher in 2022, 2023, 2024, and 2025 by Elsevier.

Dr. Jianbo Xiao is currently a full professor in the Department of Analytical Chemistry and Food Science, Faculty of Science, University of Vigo (Spain), and a distinguished professor at Zhejiang A&F University (China). He received a PhD degree in nutrition from Okayama Prefectural University (Japan) in 2009. He worked as a postdoc supported by the AvH foundation at the University of Würzbug, Germany (April 2013–May 2015), and then worked as an assistant professor at the University of Macau (October 2015–December 2020). During his research career, Prof. Xiao has addressed several research topics focusing on dietary polyphenols, working mainly in the fields of food chemistry, nutrition, and phytochemicals. This interdisciplinary, cutting-edge background provides him outstanding versatility to succeed in attracting funds from competitive calls and to create his own research group in his career. He has accepted and published more than 500 peer-reviewed papers, including *Nature Reviews Drug Discovery, Nature Sensors, Nature Communication, Advanced Science, Journal of Advanced Research, Pharmacological Research, Food Chemistry*, and so on (Scopus citation: 41000, H-index 102). Dr. Xiao is the founding editor of *Food Frontiers, eFood, Food Safety and Health, and Future Postharvest and Food*; the former editor-in-chief of *World Journal of Diabetes and Human Nutrition & Metabolism*; the associate editor of *Journal of Advanced Research, Phytomedicine, Phytochemistry Reviews, Frontiers in Immunology, Quality Assurance and Safety of Crops & Foods, Discover Food, Current Protocols, Journal of Berry Research, and Phytomedicine Plus*; the section editor of *Current Neuropharmacology*; and the editorial board member of *Trends in Food Science & Technology, Critical Reviews in Food Science and Nutrition, Journal of Nutritional Biochemistry*, and so on.

Preface

The *Handbook of Aquatic Food Products* offers a comprehensive overview of the world's aquatic products. It focuses on the fundamentals of aquatic products, including the definitions, classifications, global value, and significance, as well as the specifics of major seafood categories such as fish, shrimp, shellfish, algae, and sea cucumbers. Each chapter introduces where these creatures live, what they're good for nutritionally, how they taste, and even how they're sold in the market. In addition, this handbook involves the development of aquatic processing technologies, the status and potential of high-value utilization of aquatic products, and by-products comprehensive utilization technologies. The handbook brings together knowledge from science, nutrition, and processing to give the readers a clear picture of the aquatic products industry.

This handbook is divided into 11 chapters in total. Chapter 1 is the overview of aquatic products. Chapters 2–7, respectively, introduce the types, distribution, nutritional value, flavor, processed products, and markets of fish, shrimp, shellfish, seaweed, marine microalgae, and sea cucumber. Chapter 8 briefly introduces other aquatic products, involving sea urchins, starfish, sea anemone, and jellyfish. Chapters 9 and 10 focus on the health benefits of aquatic products, including the functions and bioactivities of aquatic proteins, carbohydrates, lipids, and small molecules. At last, the quality and safety of aquatic products are introduced in Chapter 11. Equipped with insightful chapters and practical guidance, this handbook will serve as a valuable resource for industry professionals, food scientists, and university students, while also promoting a more sustainable diet and deepening our understanding of the complex relationship between human health, food security, and the environment.

Haijin Mou

1

Overview of Aquatic Products

Haijin Mou, Qingping Liang, and Luying Tang

College of Food Science and Engineering, Ocean University of China, Qingdao, China

1.1 Characteristics of Aquatic Products

Aquatic resources refer to the total of economic species and quantities of plants and animals in natural water bodies that have development and utilization value, also known as fishery resources. According to the type of water body, they can be divided into marine fishery resources and inland fishery resources. Broadly, aquatic products refer to aquatic economic plants and animals as well as their processed products, which include various foods, industrial goods, pharmaceuticals, and animal feed [1]. In recent years, simulated aquatic food products have rapidly developed worldwide. These products are mostly made by fishery enterprises using low-value fish, shrimp, or seaweed gel as raw materials and adding flavoring agents according to consumer preferences. They are also commonly included within the range of aquatic products [2].

1) *Wide Variety of Species.* Aquatic products encompass a broad range of species, including fish, shrimp, crabs, shellfish, algae, and other aquatic plants and animals [3]. This variety makes them an important and diverse resource in the global food market.
2) *Diverse Forms of Presentation.* While common food products typically appear in one or two forms, aquatic products are available in many different forms. These include fresh, live, chilled, frozen, rehydrated, salted, dried, and semi-processed forms, offering flexibility for consumers and catering to different preferences and storage needs [4].
3) *High Nutritional Value.* Aquatic products are rich in unique nutrients that other foods cannot easily substitute. Fish, shrimp, shellfish, and sea cucumbers, for example, are not only high in protein but also contain polysaccharides, eicosapentaenoic acid (EPA), docosahexaenoic acid (DHA), and other bioactive substances, making them indispensable sources of essential nutrition [5, 6].
4) *Significant Health Benefits.* The protein content of aquatic products is exceptionally high, and they are packed with important minerals such as calcium, iron,

zinc, selenium, phosphorus, potassium, and iodine. Additionally, they provide various vitamins and unsaturated fatty acids that are crucial for human health. These nutrients contribute to growth and development, wound healing, cancer cell inhibition, and the enhancement of the immune system [7].

5) *High Market Share*. Per capita consumption worldwide rose from 9.1 kg in 1961 to 20.7 kg in 2022, demonstrating the increasing demand and consumption of aquatic products globally [2].

1.2 Classification and Distribution of Aquatic Product Processing Raw Materials

Aquatic organisms live in environments that are vastly different from those of land-based species, which impart unique nutritional components to aquatic products. This uniqueness provides the foundation for the development of aquatic product processing [8]. However, factors such as the diversity of aquatic species, the seasonality of production activities, and the perishable nature of raw materials pose significant challenges for the processing of aquatic products [8, 9]. Aquatic product processing raw materials primarily refer to species that live in marine and inland water bodies, which have certain economic value and are available for use. The range of raw materials for aquatic product processing is extremely broad. Based on biological characteristics, these raw materials can be divided into two main categories: animal-based and plant-based materials [10].

1.2.1 Animal-Based Raw Materials

Common animals that can be used as raw materials for aquatic product processing belong to various animal phyla, including Chordata, Mollusca, Arthropoda, and Cnidaria.

1.2.1.1 Fish

Fish can be broadly categorized based on their morphological characteristics into two main classes: Cartilaginous fish (Chondrichthyes) and Bony fish (Osteichthyes) [11–13].

1.2.1.1.1 Cartilaginous Fish (Chondrichthyes) Cartilaginous fish are further divided into two subclasses: Elasmobranchii (sharks, rays, and skates) and Holocephali (chimaeras). Elasmobranchii are widely distributed across the Indian Ocean, Pacific Ocean, Atlantic Ocean, and the Southern Hemisphere, from the equator down to 55°S latitude, and in the Northern Hemisphere from the equator to 80°N latitude. Depending on their morphological features, Elasmobranchii can be divided into two orders: the Shark-like order and the Ray-like order.

Shark-like species (about 360) have streamlined bodies for fast swimming, lateral gill slits, and rough skin with placoid scales. They possess a strong sense of smell, and some deep-sea species can produce light. Ray-like species (over 430) have flattened, disc-shaped bodies with gill slits and mouths on the underside. Their pectoral fins are fused to the head, and they often feed on crustaceans or molluscs. Some rays have venomous spines or electric organs and usually live near the seafloor.

The subclass Holocephali, which includes the chimaeras, is much less diverse, with only around 30 species globally.

1.2.1.1.2 Bony Fish (Osteichthyes) Bony fish are divided into Sarcopterygii (lobe-finned fish) and Actinopterygii (ray-finned fish). Lobe-finned fish are mostly fossil species and have little relevance to the processing industry. Ray-finned fish, or Teleostei, contain the vast majority of living fish species and are economically most important. Key commercial groups include sturgeons, herrings, salmonids, eels, and carps, all of which play major roles in aquatic product processing.

1.2.1.2 Molluscs

Molluscs are primarily divided into bivalves and cephalopods, and they can be further categorized into various classes, including the Bivalvia, Gastropoda, Polyplacophora, Scaphopoda, and Cephalopoda. Among these, the classes with the highest economic value are the Bivalvia, Gastropoda, and certain species within the Cephalopoda class.

1.2.1.2.1 Bivalvia (Bivalves) Bivalves have two symmetrical shells and a laterally compressed body, lacking a distinct head and tentacles. They possess a shovel-shaped foot and plate-like gills, and they do not have a radula. About 15,000 species exist, mostly in marine environments. Economically important examples include Clams, Mussels, Scallops, Oysters, Geoduck, and Pearl oysters, which are nutrient-rich and widely used in fisheries, aquaculture, and export industries.

1.2.1.2.2 Gastropoda (Gastropods) Gastropods typically have a single spiral shell, though some species have reduced shells. They possess a well-developed head with tentacles and a broad, muscular foot for movement. Due to torsion, their internal organs are asymmetrical, and their gills may be modified into lungs. With about 75,000 species worldwide, gastropods inhabit marine, freshwater, and terrestrial environments. Important examples include Red snails, Land snails, and Abalone, many of which hold significant economic value.

1.2.1.2.3 Cephalopoda (Cephalopods) Cephalopods have a well-developed head, large eyes, and 8–10 arms or tentacles. They move by jet propulsion using a muscular mantle and funnel and possess a highly centralized nervous system that supports advanced behaviors. There are about 400 species, including squid, cuttlefish, octopus, and nautilus. Although species diversity is limited, some cephalopods—such as the Humboldt squid—reach extremely high annual catches of 6–7 million tons. Together with midwater fish and Antarctic krill, cephalopods are regarded as one of the world's major potential fishery resources.

1.2.1.3 Crustaceans

Crustaceans belong to the Arthropoda phylum and the Crustacea class, which includes species such as shrimp and crabs. The majority of crustaceans are aquatic, found in both freshwater and marine environments, although a few are terrestrial or semiterrestrial, and some even lead a parasitic lifestyle. There are over 30,000 known species of crustaceans. Based on morphological characteristics, they can be divided into two main subclasses: Malacostraca and Branchiopoda [14].

1.2.1.4 Echinoderms

Echinoderms are marine animals characterized by a lack of distinct head or body structures, with bodies that exhibit radial symmetry. They have a wide range of body shapes, including star-like, spherical, cylindrical, and tree-like branching forms. Echinoderms are abundant and diverse, with notable classes including Holothuroidea (sea cucumbers), Echinoidea (sea urchins), Crinoidea (sea lilies), and Ophiuroidea (brittle stars). Among them, species such as sea cucumbers from the Holothuroidea class and sea urchins from the Echinoidea class are of considerable economic importance due to their high market value [15].

1.2.1.5 Cnidarians

Cnidarians are among the most primitive multicellular animals, with over 9,000 known species. This group is divided into several classes, including Hydrozoa (hydra), Anthozoa (corals), Scyphozoa (true jellyfish), and Ctenophora (comb jellies). Of these, Scyphozoa, particularly the jellyfish, is highly valued in certain markets. The thick, gelatinous middle layer of the jellyfish bell is rich in water and mucilaginous substances. After processing, it is commonly used to make jellyfish skin, a delicacy in various cuisines, especially in East Asia [16].

1.2.1.6 Reptiles

Reptiles are terrestrial vertebrates capable of living on land throughout their adult lives, with embryos that also develop on land. Reptiles are cold-blooded, meaning their body temperature fluctuates with the environment. Among the most economically significant reptile species is the Chinese soft-shelled turtle (*Pelodiscus sinensis*), which is highly valued in food markets for its meat, considered a delicacy in many cultures [17].

1.2.2 Plant-Based Raw Materials

Algae are the main plant-based raw materials in aquatic products. They are simple autotrophic plants without differentiated roots, stems, or leaves, and most species are small in size. About 30,000 algal species are known worldwide. The most economically important groups are brown algae (e.g., kelp and wakame) and red algae (e.g., nori, Gracilaria, and Carrageenan-producing species). These algae are valued for their rich vitamins, minerals, and polysaccharides and are widely used in food, pharmaceuticals, and various industrial applications [18].

According to statistics, the global total fishery production (excluding algae) reached 185 million tons in 2024, with China's total aquaculture production amounting to 74.1 million tons, maintaining its position as the world's leading producer for several consecutive years. The representative species and production volumes from major marine regions are shown in Figure 1.1. In wild capture fisheries, the Pacific Ocean

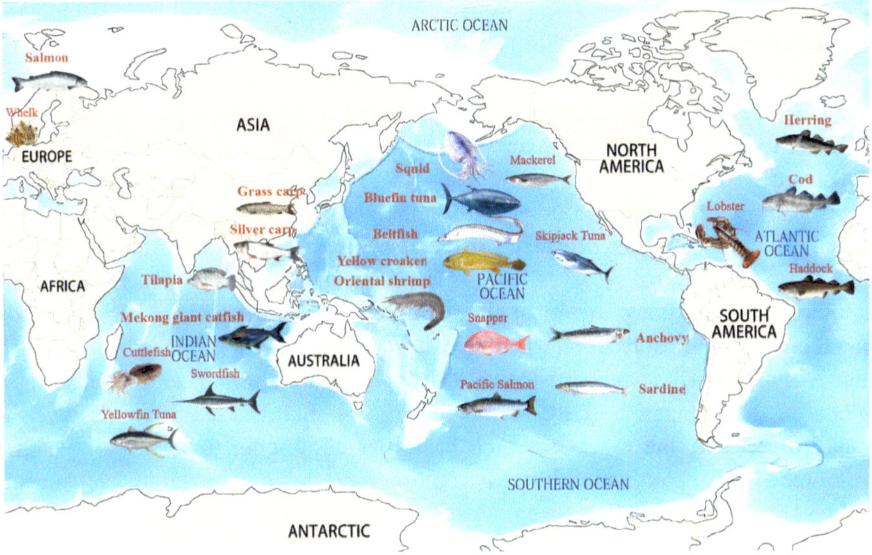

Figure 1.1 Distribution of major aquatic resources in the world.

accounts for nearly 60% of the global catch, with key species such as beltfish, yellow croaker, squid, and tuna from the Western Pacific, and anchovy and sardine from the Southeast Pacific. The Atlantic Ocean contributes around 21 million tons, with notable species including salmon, cod, and herring. The Indian Ocean's share is about 12 million tons, with important species such as tilapia and Mekong giant catfish [3, 19].

1.3 Importance of Marine Resource Development and Utilization

Since the beginning of the 21st century, global food security has become an increasingly urgent issue, drawing widespread social concern. With most of the land already developed and the availability of arable land, water resources, and environmental capacity severely limited, there is a growing need to integrate the vast oceans into the food production system. This shift aims to find food supply solutions with high-resource utilization efficiency. The ocean, as the birthplace of life on Earth, nurtures a vast array of plants, animals, and microorganisms. It is the largest biological treasure trove on the planet and a highly promising source of quality food. Marine biological resources are incredibly diverse, with approximately 200,000 known marine species worldwide—around 180,000 animal species and over 20,000 plant species. Data suggests that the ocean provides food equivalent to 135 billion tons of organic carbon annually for humanity. With the premise of maintaining ecological balance, it is estimated that the ocean can supply about 3 billion tons of aquatic products each year, enough to feed at least 30 billion people based on the food consumption requirements of an average adult. The food supplied by the ocean is roughly 1,000 times the agricultural output produced on land [3, 20].

In 2020, a report on sustainable food production published by *Nature* magazine in the United Kingdom predicted that by 2050, global ocean food production could increase to 21 million tons per year, with the potential to reach up to 44 million tons. This would contribute between 12% and 25% of the total increase in global meat production required for the projected 9 billion people by the middle of the century [20]. For example, **Antarctic krill**, a key component of the Antarctic ecosystem, has vast resource reserves, with estimates ranging between 650 million and 1 billion tons. It is one of the most abundant and successful species on Earth, making it one of the most resource-rich marine animals available for human consumption. Antarctic krill is also considered a critical potential protein source to address global food shortages. Initially, Antarctic krill was mainly used as an additive in aquaculture feed, but recent research has provided a better understanding of its resource status, ecological habits, distribution patterns, nutritional and biological values (BVs), and processing techniques. This knowledge has laid a solid foundation for the future development and improvement of processing methods, particularly in food production [21].

Currently, global consumption of aquatic food is growing rapidly, at twice the rate of the annual increase in the world's population. Aquatic products have become a significant solution to the global food security challenge. Around 88% of the global total fish production, about 171 million tons, was directly consumed by humans [3, 22]. The annual growth rate of fish available for human consumption has outpaced the overall growth rate of land-based meat products. As people's nutritional needs become more diverse, relying solely on land-based crops for nutrition is no longer sufficient to meet these demands. By focusing on the development of functional marine products, the industry can enhance the quality of life and overall health of the global population.

The development of marine aquatic products has given rise to various models of marine resource utilization, benefiting the sustainable growth of key industries such as mariculture, marine fishing, and seafood processing. **Mariculture** involves the cultivation of marine plants and animals in coastal shallow waters, tidal flats, and harbors, providing high-quality animal proteins for humans. This practice allows for the rapid production of economically valuable and often scarce marine species, including fish, shrimp, shellfish, algae, and marine delicacies, while overcoming the constraints of natural environmental conditions. **Marine fisheries**, on the other hand, use specialized fishing equipment to harvest marine aquatic products from seawater, providing essential animal proteins to society. Unlike mariculture, the marine fishing industry is limited by the sustainable capacity of nearshore fisheries and the maximum sustainable yield (MSY)—the largest quantity of aquatic products that can be sustainably harvested without depleting the resource base. **Seafood processing** emerged as a natural extension of mariculture and marine fisheries. It utilizes marine products such as fish, shrimp, shellfish, crabs, algae, and other marine plants as primary raw materials. Advanced processing technologies are then employed to enhance the economic value of these products. The seafood processing industry is characterized by its volatility, dependency on raw material availability, technological constraints, and high economic returns [3].

In addition, coastal waters can be transformed into artificial marine farms, serving as large-scale food production bases. Currently, many countries around the world are expanding their marine aquaculture operations into deeper and more distant waters, recognizing the immense potential of the ocean as a resource. The development of marine aquatic resources not only helps alleviate food security issues but also brings substantial economic benefits to residents living near coastal areas, creating additional employment opportunities. As a result, aquaculture, fishing, and processing industries are likely to expand under the framework of responsible resource utilization, generating more jobs and contributing to the optimization and upgrading of the marine industry structure. This will, in turn, improve economic efficiency and strengthen global food supply chains. The complexity of the marine environment, with its varying temperatures, salinity, and depths, creates distinct ecological zones. These habitat types, with their unique latitude, depth, and horizontal gradients, have a profound impact on the spatial and temporal distribution of marine species, including their survival and breeding patterns.

Marine organisms thrive in this specialized environment, leading to the development of distinct genomes and metabolic pathways, which differentiate them from terrestrial organisms. As a result, marine life produces unique natural compounds with distinct structures and bioactivity, making them an invaluable resource for discovering lead compounds and developing innovative medicines. Research on marine-derived pharmaceuticals began in the 1940s, gained momentum in the late 1960s and early 1970s, and received considerable academic attention from the 1980s onward, with a surge in interest in the mid-to-late 1990s. In recent years, there has been unprecedented global attention on marine resources. Initiatives such as the US Marine Biological Program, the EU's MAST Program, Japan's Marine Sapphire Program, the UK Marine Biological Development Program, and China's Blue Drug Bank have accelerated marine drug research. To date, scientists have identified nearly 30,000 compounds from marine organisms, with 13 of these developed into commercially available drugs. Over 40 compounds are currently in clinical and preclinical trials, and more than 1,400 compounds are undergoing pharmaceutical viability assessments. The rapid development of marine drug research underscores the ocean's enormous potential as a source of novel therapeutic agents, presenting vast opportunities for future healthcare advancements [23–25].

1.4 Nutritional Characteristics and Functional Values of Aquatic Products

In addition to providing essential nutrients for human health, aquatic products are also abundant in a variety of bioactive components, including omega-3 unsaturated fatty acids, high-quality proteins, polysaccharides, taurine, and sea cucumber saponins [26]. These components offer numerous health benefits, such as enhancing cognitive function and intelligence, preventing and combating cancer, strengthening the immune system, and regulating blood pressure, blood glucose, and blood lipids [26].

1.4.1 Protein

Aquatic products serve as a crucial source of high-quality protein, with the protein content generally ranging from 15% to 22%, and are distinguished by a well-balanced amino acid profile, a high BV, and excellent digestibility. The protein in aquatic products contains eight essential amino acids required by the human body, particularly featuring high concentrations of key amino acids such as lysine, threonine, and methionine. Compared to livestock and poultry meat, the protein in aquatic products has a higher BV, with most fish proteins reaching values of 80–90, indicating that the human body can utilize these proteins more efficiently, thereby helping to maintain nitrogen balance, meet protein synthesis requirements, as well as more effectively support the repair of tissues and the regulation of metabolic functions. Additionally, the protein content in aquatic products exhibits unique advantages in terms of both composition and content. For example, oysters are a rich source of protein, comprising 50.76–56.57% of their dry weight, and the essential amino acids within this protein content account for 12.20–14.15% of the total amino acids [27]. The protein content in Antarctic krill constitutes 15.40–16.31% of its fresh weight, and the essential amino acids within this protein comprise 42.32–45.28% of the total amino acids [27, 28]. The ratio of its essential amino acids to nonessential amino acids ranges from 73.37% to 82.74%, closely aligning with the ideal protein pattern recommended by the Food and Agriculture Organization (FAO)/WHO.

Based on the amino acid scoring model (AAS) proposed by the FAO/WHO in 1973, the protein nutritional values of various aquatic products, including fish, shrimp, and shellfish, were systematically evaluated. The results indicated that the AAS values for most fish species reached 100, a level comparable to that of poultry meat and higher than those of beef and milk. Compared to high-quality proteins such as those found in eggs and poultry, the protein content in aquatic products is more readily absorbed by the human body. For example, fish and shrimp possess delicate muscle fibers and low connective tissue content, resulting in a digestion and absorption rate exceeding 90%, which surpasses that of red meat and certain plant-based proteins [28]. Consequently, aquatic proteins are particularly well-suited for consumption by the elderly, children, and individuals with compromised gastrointestinal function. Additionally, common aquatic products such as sea bass, salmon, hairtail, prawns, and shellfish are characterized by their high-protein content, low-fat composition, and minimal calorie burden, which make them optimal choices for both weight management and enhancing nutritional density. Aquatic products are also a rich source of high-quality proteins with unique functional properties. For example, taurine found in fish and shellfish offers multiple health benefits, including blood pressure reduction, antioxidation, and bile secretion promotion. Furthermore, small-molecule bioactive peptides extracted from fish processing by-products have been demonstrated to exhibit functions such as antioxidation, antibacterial activity, and lipid regulation [27].

1.4.2 Lipid

The lipid content in aquatic products varies significantly based on the species and is generally lower compared to that in livestock and poultry. However, aquatic products possess exceptionally high nutritional value, particularly being rich in various unsaturated fatty acids, particularly omega-3 fatty acids such as EPA and DHA [29, 30]. These fatty acids play crucial roles in regulating blood lipids, reducing inflammation, preventing arteriosclerosis, and safeguarding the health of the cardiovascular and cerebrovascular systems. Additionally, DHA is a vital component for the development of the brain and retina in fetuses and infants. Depending on the type of aquatic product, high-fat fish such as salmon, mackerel, and tuna contain more than 10% lipid content, with the combined EPA+DHA content exceeding 2,000 mg per 100 g. In contrast, low-fat fish like cod and sea bass have a fat content below 5%, yet they still provide a certain amount of beneficial fatty acids. Moreover, the fatty acid compositions of the same species of fish vary as a result of differing growth patterns [29]. For example, the back muscle of naturally grown sweetfish exhibits higher levels of 18:1, 18:2, 20:4, and 22:6 fatty acids, whereas farmed sweetfish demonstrates elevated concentrations of 14:0, 16:0, 18:3, and 20:5 fatty acids. Compared to the fat content in livestock and poultry, fats within aquatic products exhibit a lower proportion of saturated fats and a higher proportion of unsaturated fats. Polyunsaturated fatty acids constitute 51.71% of the total fatty acids in squid, with EPA and DHA together accounting for 46.13% of the total fatty acids. Polyunsaturated fatty acids in oysters comprise 42.26–45.24% of the total fatty acids, with DHA accounting for 18.53–21.16% and EPA contributing 17.23–18.68% of the total fatty acids. Overall, lipids in aquatic products serve not only as an essential energy source but also play a pivotal role in modulating immune function, neurological health, and anti-inflammatory responses, thereby constituting an indispensable component of healthy fats in contemporary diets.

1.4.3 Carbohydrate

The overall carbohydrate content in aquatic products is relatively low; particularly, the fish and crustaceans typically contain less than 1% carbohydrates, which do not serve as a primary energy source. Scallops, clams, oysters, and other shellfish are rich in glycogen, which exhibits a structural similarity to human liver glycogen and can be rapidly metabolized into glucose, thereby providing immediate energy for the body. In contrast, seaweeds, including kelp, wakame, and nori, are abundant in functional polysaccharides such as fucoidan, algal polysaccharides, and alginate. Algal polysaccharide exhibits various biological activities, including antioxidation, anti-inflammation, antitumor effects, and antiviral properties [31]. It enhances the defense capabilities by modulating the activity and signaling pathways of immune cells. Alginates contribute to the delayed absorption of sugars and lipids while promoting the excretion of bile acids, thereby aiding in the regulation of blood sugar and lipid levels. Fucoidan is abundant in structural units such as fucose and sulfate groups, and it exhibits the functions of promoting the growth of beneficial bacteria as well as regulating the balance of intestinal microecology. Additionally, certain high-nutrition aquatic products, such as sea cucumbers and abalones, are abundant in glycoproteins, chondroitin sulfate, mucopolysaccharides, sea cucumber saponins, and other carbohydrate derivatives [32]. These components exhibit both nutritional and functional properties, demonstrating significant immune-enhancing, joint-lubricating, anti-inflammatory, analgesic, tissue-repairing, and antiaging effects. Consequently, they are extensively utilized in the formulation of nutritional health supplements and premium functional foods.

1.4.4 Vitamins and Minerals

Aquatic products serve as an excellent source of high-quality protein and healthy fats, and are also abundant in a variety of vitamins and minerals, which play a crucial role in regulating physiological functions, preventing nutritional deficiencies, and boosting immunity. The most representative aquatic sources of nutrients include fat-soluble vitamins A and D, which are predominantly found in deep-sea fish and cod liver oil. Aquatic products are also rich in water-soluble B vitamins, including vitamins B1 (thiamine), B2 (riboflavin), B6, and B12, which play essential roles in energy metabolism, nerve function regulation, and red blood cell production. Notably, vitamin B12, primarily derived from fish and shellfish among animal-based foods, is a nutrient that vegetarians are particularly prone to being deficient in. Additionally, certain aquatic products contain vitamins E and K, which possess antioxidant properties and blood-clotting functions, respectively [26]. In terms of mineral nutrition, aquatic products serve as a vital source for supplementing essential trace elements, including calcium, iron, zinc, iodine, and selenium. For example, oysters, clams, and other shellfish are rich in heme iron, which exhibits high bioavailability and thereby contributes to the prevention of iron deficiency anemia. Iodine is abundantly found in marine algae, such as kelp and nori, which serves as an essential element for the synthesis of thyroid hormones, and insufficient intake may result in goiter or metabolic disorders. Selenium, as a critical component of glutathione peroxidase, exhibits notable antioxidant and anti-cancer properties and is predominantly found in marine fish, shellfish, and shrimp.

1.4.5 Other Small-Molecule Active Compounds

In addition to the aforementioned nutrients, aquatic products are also abundant in various small-molecule bioactive compounds, which exhibit significant physiological regulatory functions and possess high nutritional value [26]. Among these, taurine is abundantly found in fish and shellfish and is known for its functions of lowering blood pressure, providing antioxidant effects, and protecting vision. Small-molecule active peptides possess the ability to reduce blood pressure, exhibit anti-inflammatory effects, and enhance immune function. Astaxanthin serves as a potent antioxidant and has been extensively applied in products designed for eye protection and antiaging. DHA derivatives, including neuroprotectin, exhibit neuroprotective and anti-inflammatory properties. Marine polyphenols derived from sources such as seaweed demonstrate potential in metabolic regulation and the prevention and management of chronic diseases. Overall, with the advancement of functional foods and health nutrition products, these natural active factors are emerging as one of the key directions for the high-value utilization of aquatic resources.

1.5 Development of Aquatic Processed Products and Technology

With the sustained growth in global aquatic resource production and the increasing diversification of consumer demands, the aquatic product processing industry is experiencing rapid development and has become a critical component in enhancing the value-added of fisheries. Currently, over half of the global aquatic products require varying degrees of processing, including decapitation, shelling, refrigeration, freezing, and preservation [33, 34]. Subsequently, through diverse processing technologies, these aquatic products are further transformed into processed products such as condiments, ready-to-eat meals, canned foods, and surimi-based products. Over the past two decades, as a result of the continuous optimization and adjustment of the industrial structure of aquatic product processing, the aquatic product processing system has gradually been established and perfected, which facilitated the development of diverse large-scale processing methods. Although frozen aquatic products still account for approximately 60% of the total output of processed aquatic products, the proportion of deeply processed products such as frozen processed items, surimi products, canned goods, fish oil, and algae-based derivatives that are obtained through various advanced processing technologies that alter the original characteristics of aquatic raw materials has now exceeded 50% (Table 1.1). Overall, the global aquatic product processing technology is currently experiencing a transformation and upgrading process toward intelligence, sustainability, and functionality. This evolution not only ensures food safety and caters to the demands of consumption upgrades but also promotes the efficient and sustainable utilization of marine biological resources.

Traditional processing techniques are associated with challenges, including low conversion rates of aquatic raw materials and limited added value of the resulting products. Consequently, this has prompted the transition of aquatic product processing technology from conventional methods to advanced technologies. Currently, several key advancements in aquatic product processing technologies encompass protein structuring technology, ultrahigh-pressure processing, supercritical fluid extraction, microencapsulation, ultrafine grinding, membrane separation, glass transition state storage, vacuum low-temperature frying, low-temperature plasma sterilization, and ohmic heating [33–37]. The development and application of these

Table 1.1 Overview of aquatic product processing technology.

Processed products types	Classification	Processing technology	Brief introduction to processing technology	Examples of aquatic processing products
Frozen aquatic product	Cooling technology	Air-cooling method	Place the aquatic products in the cooling air, and the heat of the aquatic products is carried away by the continuous circulation of the cooling air, thus cooling the aquatic products	Frozen fish fillets, frozen shrimp
		Water-cooling method	The method of cooling aquatic products by direct contact with cold water	
		Crushed ice cooling method	The method of directly contacting crushed ice with aquatic products, absorbing the heat of fusion, and turning into water, while cooling the aquatic products	
	Freezing technology	Air freezing method	The method of using low-temperature air as the medium to remove the heat from aquatic products, thereby freezing the aquatic products	
		Flat plate freezing or indirect contact freezing method	The method of freezing by contact between cold metal surfaces and aquatic products	
		Spray freezing or immersion freezing method	The method of obtaining freezing by contacting and exchanging heat between packaged or unpackaged aquatic products and liquid refrigerants or heat transfer media	
	Freezing new technologies	Magnetic field-assisted freezing	By applying a magnetic field when the material is being frozen, the magnetic field influences the growth process of ice crystals through vibration, thereby slowing down their growth rate	
		Ultrasonic-assisted freezing technology	The formation and distribution of ice crystals during the freezing process can be controlled by using ultrasonic waves to form fine and evenly distributed ice crystals, accelerate heat and mass transfer, and ultimately achieve rapid freezing	
		High-pressure auxiliary freezing technology	The technology of freezing products at low temperatures by applying high pressure	

Dried aquatic products	Drying technology	Convection drying	Using air as the drying medium, heat is transferred to the aquatic products through convection, causing the moisture in the aquatic products to evaporate due to heat and be removed, thereby achieving drying	Half-dried salt-cured saltwater fish, dried shrimp skin, dried sea cucumber, dried abalone, dried squid, dried kelp, dried seaweed
		Radiation drying	The method of drying aquatic products involves radiating heat to them through electromagnetic waves such as infrared rays and microwaves as heat sources	
		Freeze drying	Freeze the aquatic products first and then subject them to a higher vacuum degree	
Pickled aquatic products	Pickling processing technology	Dry brine method	The method of directly sprinkling an appropriate amount of solid salt on the surface of aquatic products for pickling	Salty fish product, fermented pickled products, such as natural fermented products, pickled fish, and drunken crabs, jellyfish products
		Brine salting method	The method of pickling aquatic products in salt	
		Mixed pickling method	The method combining dry salting and brine salting	
		Low-temperature salt curing method	It is divided into cooled salt curing method and frozen salt curing method	
Smoked aquatic products	Smoking processing method	Cold smoking method	The method involves long-term salting of raw aquatic products to slightly increase the salt content, and then drying them for a long time in a smoking chamber where the temperature is controlled below the thermal denaturation temperature of proteins	Smoked squid rings, smoked herring, smoked large yellow croaker
		Hot smoking method	The method involves briefly soaking the aquatic product raw materials in a seasoning liquid with an appropriate amount of salt, and then smoking them at a relatively high temperature near the heat source	
		Liquid smoking method	The method of using the smoke liquid obtained through processes such as condensation, sedimentation, and refinement of the smoke gas produced by traditional smoking methods as an additive to produce liquid-smoked foods can be divided into direct mixing method and surface addition method	

(*Continued*)

Table 1.1 (Continued)

Processed products types	Classification	Processing technology	Brief introduction to processing technology	Examples of aquatic processing products
Surimi and surimi products	Gel formation method	Heat-induced gel	As the temperature keeps rising, surimi gel is rapidly formed through the "gel deterioration zone"	Fish ball, fish tofu, octopus ball, simulated crab meat
		Acid-induced gel	By directly adding acid or acidifiers instead of high-temperature heating, proteins coagulate to form a gel	
		Fermented surimi gel	Adding microorganisms to surimi and subjecting it to fermentation can endow surimi with its characteristic flavor and texture	
		Ultrasonication	The technology that utilizes the vibration energy of ultrasound to generate powerful shear force and high temperature in the medium, altering the structure and function of substances and accelerating reaction rates	
		Ultrahigh-pressure processing	A food processing technology that uses high pressure to deactivate or denature biological macromolecular substances, such as enzymes and proteins, in food	
	Processing technology of surimi and surimi products	Frozen surimi processing technology	Selection of raw fish species and freshness → Pretreatment of raw fish → Meat collection → Rinsing → Fine filtration and dehydration → Mixing → Filling and weighing → Freezing and storage	
		Processing technology of surimi products	Thawing → Mashing or chopping → Shaping → Gelation → Heating → Cooling → Packaging and storage	

Canned aquatic products	Sterilization technology	Traditional thermal sterilization technology	The method of killing microorganisms by heating	Canned steamed turtle, seasoned pack, oiled can, soft canned fish and shrimp products, soft canned turtle with shellfish
		Ultrahigh-pressure sterilization technology	In a sealed container, water or other liquids are used as the pressure transmission medium to apply a pressure of 100–1,000 MPa to soft-packaged foods and other materials, thereby killing microorganisms	
		Other sterilization techniques	Cold sterilization technology; radiation sterilization; pulsed light sterilization; ultraviolet sterilization; high-pressure sterilization	
	Processing technology	Canned aquatic products	Raw material pretreatment → Canning → Exhausting → Sealing → Sterilization → Cooling → Low-temperature storage → Inspection → Packaging → Storage	
		Soft can	Raw material processing → Bagging → Sealing → Sterilization → Cooling → Drying → Inspection → Packaging	
Aquatic product seasonings	Processing technology	Fish sauce	The production process of fish sauce mainly relies on two key steps: salting and fermentation technologies. The salting technology is utilized to inhibit the action of putrefactive microorganisms, and the protease is used to hydrolyze the proteins in the fish body (fermentation process).	
		Shrimp paste	A thick paste made from small shrimps such as krill, brine shrimp, and phosphorescent shrimp, as well as shrimp processing by-products like shrimp heads. It is made through processes such as salting, mashing, and fermenting, and has a distinctive fresh and fragrant shrimp flavor	
		Oyster sauce	Traditional method: Concentrate and reduce juice → Add auxiliary materials and stir → Heat → Adjust color → Enhance flavor and aroma → Filter → Bottle → Pasteurize → Finished product	
			Enzymatic hydrolysis method: Oysters → Shell removal and washing → Grinding → pH adjustment → Enzymatic hydrolysis → Enzyme inactivation → Filtration of hydrolysate → Concentration → Blending and Filtration → Bottling → Sterilization → Finished product	

(*Continued*)

Table 1.1 (Continued)

Processed products types	Classification	Processing technology	Brief introduction to processing technology	Examples of aquatic processing products
Seaweed processing	Seaweed food processing technology	Lightly salted dried products	Raw material harvesting → Sun-drying → Impurity removal → Grading → Re-sun-drying	Dried kelp, dried seaweed, dried wakame
		Salted dried products	Raw material harvesting → Pickling → Sun-drying → Grading → Re-sun-drying	Salted dried kelp
		Seaweed salted products	Raw material collection → Water washing → Blanching → Cooling → Draining → Salt mixing → Salting → Draining again → Cutting and shaping → Inspection and packaging	Salted kelp, salted wakame
		Flavored seaweed products	Raw material selection → Sorting → Washing → Cutting into shreds → Steaming and boiling → Seasoning soaking → Drying → Packaging	Seasoned shredded kelp, seasoned wakame, seasoned seaweed slices, ready-to-eat seasoned kelp
		Seaweed seasoning	Pepper pretreatment → Seaweed pretreatment → Garlic pretreatment → Dissolution of stabilizer → Cooking → Blending → Bagging, sealing, and sterilization	Kelp chili sauce; south Korea
		Seaweed drink	Raw material cleaning → Chopping → Pretreatment → Colloid mill → Homogenization → Filling, degassing, and bottling → Packaging	Laminaria drink, wakame fermented beverage

novel food processing technologies can address the demand of consumers for high-quality and safe food products while extending the shelf life of aquatic processing-based foods.

1.6 Status and Potential of High-Value Utilization of Aquatic Products

The naturally active compounds derived from aquatic products are increasingly recognized as a critical source of raw materials for the global functional food and biopharmaceutical sectors [38]. Through processing technological innovation and development, by-products of aquatic products can be converted into high-value-added components, including collagen, small-molecule bioactive peptides, chitosan, polysaccharides, DHA/EPA, and others (Table 1.2). This transformation gradually establishes an industrial ecosystem that spans multiple sectors, such as food production, cosmetics manufacturing, agricultural inputs, biomaterials development, and biomedical applications. Additionally, the ongoing enhancement of consumer health consciousness has propelled the nutrition industry to evolve from fundamental nutrient intake toward more advanced areas such as precise nutrition, immune system support, and antiaging interventions.

1.6.1 Functional Foods

The high-value development of functional components in aquatic products has transitioned from traditional health supplements to encompass food, daily necessities, and precision products, establishing a closed-loop value chain that extends from natural raw materials to multifunctional end products [39]. The form of functional foods primarily includes functional beverages, beauty-oriented meal replacement products, collagen peptide gummies, brain health soft capsules, pressure-regulating fish peptide granules, marine oligosaccharide prebiotic powder, and other functional foods. Currently, small-molecule peptide products derived from fish by-products have been introduced into markets focused on blood pressure management, sports nutrition, and anti-fatigue treatment [40]. Omega-3 fatty acid products, with deep-sea fish oil and seaweed oil as key ingredients, are extensively utilized in areas such as cardiovascular protection, prenatal and infant nutrition, and brain health promotion. Chitosan and astaxanthin, which are extracted from shrimp and crab shells, have demonstrated distinct advantages in promoting intestinal health, enhancing immune function, and exhibiting antiaging properties. Components such as fucoidan and alginate, which are extracted from algae like kelp and wakame, are increasingly being utilized in the development of functional foods with anti-inflammatory and antiviral properties. Developed countries, including Japan, the United States, and Norway, have established a well-structured functional food industry system supported by regulatory oversight, standardized product specifications, and clinical validation, thereby facilitating the sustainable development of aquatic resources in a virtuous

Table 1.2 Overview of processing technology of bioactive components in aquatic products for their high-value utilization.

Classification	Processing technology types	Technology item and processing flow	
Oils and fats	Extraction process of oils and fats	Fish oil extraction process	Marine oil → Chopped → Extracted → Oil layer separation → Degumming → Deacidification → Decolorization → Deodorization → Fish oil
		Algal oil extraction process	Ultrasonic/microwave method Ultrasonic/microwave method Enzymolysis approach Slightly negative pressure method
		Antarctic krill oil extraction process	Organic solvent extraction process Supercritical CO_2 extraction method Enzymatic hydrolysis combined with organic solvent extraction method Expression method
	Separation and purification method	EPA and DHA	Low-temperature crystallization method Urea adduction method Silver nitrate column method/silver salt complexation method Supercritical CO_2 extraction method Vacuum distillation and molecular distillation method
	Microcapsule	Microcapsule granulation	Spraying method Air suspension Phase separation–coacervation Multiphase emulsion method Molecular embedding method Extrusion method
		Microcapsule encapsulation	Spray drying microencapsulation Vacuum drying microencapsulation under reduced pressure

Protein	Fish protein concentrate	Processing technology	Raw materials → Steaming → Pressing → Drying → De-greasing and deodorizing → Drying → Sterilization → Finished product
	Protamine	Preparation technology	Mature testicular tissue → Homogenization → Acid hydrolysis with sulfuric acid → Extraction → Filtration → Salting out of filtrate → Filtration of precipitate → Refinement → Washing → Drying → Crude product → Polyacrylamide → Gel filtration chromatography → Elution → Drying → Finished product
	Phycocyanin	Preparation technology	Cultivation of algal biomass → Drying → Extraction → Vacuum concentration → Spray drying
	Active peptide	Preparation technology	Protease hydrolysis method Acid hydrolysis method Alkaline hydrolysis method Directional synthesis method
		Desalination methods	Macromolecules: dialysis, ultrafiltration, nanofiltration Small molecules: electrodialysis, macroporous resin method, ion exchange resin method
		Separation and purification methods	Reversed-phase high-performance liquid chromatography Ion exchange chromatography Capillary electrophoresis Gel filtration chromatography
Polysaccharides	Chitosan	Preparation technology	Alkali fusion Sodium alcoholate method Enzymic method Microwave preparation method
	Seaweed dietary fiber	Extraction method	Crude separation method Chemical separation method Membrane separation method Chemical reagent and enzyme combined extraction method Fermentation method
	Chondroitin sulfate	Extraction method	Dilute alkali and concentrated salt extraction method Dilute alkali extraction method Alkaline hydrolysis-enzymatic hydrolysis extraction method

(*Continued*)

Table 1.2 (Continued)

Classification	Processing technology types	Technology item and processing flow	
Pigment	Astaxanthin	Extraction method	Alkali extraction Oil-soluble method Organic solvent method Supercritical CO_2 fluid extraction method
		Methods for improving extraction rate	Physical machining Acidizing Polymerization agent extraction
		Astaxanthin antioxidant measures	Low-temperature extraction (lower than 70 °C) Light-proof Antioxidant application Improve the vacuum degree Protease hydrolysis Microbial fermentation

cycle. Meanwhile, countries such as China and Thailand are also actively facilitating the transformation of aquatic products from table ingredients to functional raw materials.

1.6.2 Food Additives

Aquatic products and their by-products are abundant in diverse natural compounds that hold significant potential for application in food processing. These substances exhibit broad utility in functions such as thickening, water retention, antioxidation, preservation, and nutritional enhancement [40, 41]. For example, components such as alginate extracted from seaweeds like kelp and wakame, along with their residues, are extensively utilized in dairy products, jelly, meat products, and frozen foods. These substances contribute to gel formation, stabilize emulsions, and enhance texture, as well as play a role in promoting intestinal health regulation owing to their properties as soluble dietary fibers. EPA/DHA, astaxanthin, and other bioactive components are extensively utilized in infant formula, nutritionally fortified foods, and antioxidant systems to optimize fatty acid profiles and mitigate food oxidation. Surimi protein, fish maw, fish gelatin, and other components serve as texture modifiers, gelling agents, and foam stabilizers during processing, which are commonly utilized in a variety of products, including meat products, pastries, candies, and capsule shells.

1.6.3 Cosmetics Industry

High-value aquatic products have demonstrated significant potential for broad applications in the cosmetics industry [42]. Specifically, common raw materials, such as marine collagen, small-molecule peptides, chitosan, astaxanthin, and algal polysaccharides, exhibit exceptional antioxidant, moisturizing, repairing, and antiaging properties. Collagen derived from fish skin is extensively utilized in skincare products for applications such as firming and whitening. Meanwhile, astaxanthin, owing to its remarkable antioxidant properties, is incorporated into antiaging and sun protection formulations. Chitosan exhibits antibacterial and film-forming properties, making it suitable for applications in hair care products, acne prevention solutions, and degradable skincare films.

1.6.4 Agricultural Input Products

Aquatic by-products, such as fish bones, shrimp and crab shells, and seaweed residue, are abundant in protein, minerals, and bioactive polysaccharides. Through processes such as enzymatic hydrolysis, fermentation, or extraction, these materials can be transformed into high-value agricultural inputs, including fertilizers, eco-friendly pesticide substitutes, and seaweed-based mulching films [43]. Surimi liquid fertilizer is abundant in small-molecule peptides, which can effectively enhance root development and nutrient absorption. Chitosan possesses antibacterial properties and growth-promoting effects, making it widely applicable in biopesticides and foliar fertilizers. In addition, fucoidan and alginate derived from seaweed can be utilized to develop degradable seaweed mulching films, which possess functionalities such as moisture retention, temperature regulation, disease resistance, and root protection, as well as hold significant potential for application in ecological agriculture.

1.6.5 Biological Materials

Aquatic product by-products, such as gelatin and chitosan, can be utilized at high value as biological materials for capsule production. Gelatin, derived from clean and nonallergenic sources, is particularly suitable for applications in soft capsules and jelly-type health products, thereby effectively addressing market demands. Chitosan exhibits excellent biocompatibility and degradability, making it widely applicable in the development of controlled-release capsules, enteric-coated films, and edible films. These components are emerging as pivotal choices for green, natural, and functional capsule raw materials, thereby promoting the advancement of healthcare products and pharmaceutical packaging toward greater environmental sustainability and diversification.

1.6.6 Biomedicine Industry

Meanwhile, the research and development of drugs based on the "Blue Pharmacy" concept is gaining momentum. Certain polypeptides, lipids, sugars, and microbial metabolites have demonstrated significant antioxidant, antibacterial, antitumor, and neuroprotective properties, positioning them as crucial candidate molecules for future drug innovation [39]. Countries worldwide have placed unprecedented emphasis on aquatic product resources. Stimulated by initiatives such as the "Marine Biotechnology Program" in the United States, the "Marine Science and Technology (MAST) Program" in the European Union, the "Ocean Sapphire Program" in Japan, and the "Marine Biological Development Program" in the United Kingdom, the biopharmaceutical sector has experienced rapid growth and has emerged as a critical development area globally in the 21st century. Since the concept of "seeking medicine from the ocean" was officially endorsed by the state, China has made significant and fruitful advancements in the research and development of marine biomedicine. Notably, China has independently developed and successfully launched several marine-derived drugs, including Sodium Alginate Diester (also known as Poly Saccharide Sulphate, PSS), glycolate, Hailite, mannitol nicotinate, Doenkang, squalene, and Haikun Shenxi. Furthermore, a number of promising drugs such as 911, 916, 971, D-polyglycine, K-001, sea cucumber polysaccharides, and tetrodotoxin are currently undergoing clinical trials. With the advancement of processing technologies, such as the extraction and application of active ingredients from aquatic product by-products, the high-value utilization process has been significantly promoted. Globally, the high-value utilization of aquatic products is transitioning from "nutrient supply" to "health management," while expanding from "food processing" to "functional refinement."

1.7 By-products Processing of Aquatic Products and Their Comprehensive Utilization

With the sustained increase in global aquaculture production and the advancement of the aquatic product processing industry, the volume of by-products generated during processing has also been rising rapidly. According to FAO statistics, the current annual production of by-products from global aquatic product processing exceeds 60 million tons, representing 30%–60% of total raw material input. These by-products are diverse in nature and substantial in volume, encompassing items such as fish heads, bones, skins, internal organs, shrimp and crab shells, mollusc shells, and seaweed residues. They are rich in protein, lipids, minerals, and bioactive compounds, offering significant potential for resource utilization and value-added development. By-products from aquatic product processing were predominantly utilized as low-value feed for animals or discarded without further treatment, which not only led to a significant waste of resources but also imposed considerable environmental burdens. Over the past two decades, driven by the rapid advancement of food technology, bioengineering, and green processing techniques, aquatic product by-products have progressively been repurposed as high-value-added raw materials. These materials are now extensively utilized across various sectors, including food, health products, pharmaceuticals, agriculture, and environmental protection, thereby serving as a critical catalyst for the sustainable development of the economy.

Specifically, by-products, including fish heads, fish fillets, and fish bones, can be further processed through direct and advanced methods into a variety of value-added products, such as sausages, sauces, and snacks. Fish by-products serve as valuable sources for extracting collagen, small-molecule bioactive peptides, fish oil, calcium, and bile acids, which can be extensively applied in the development of functional foods, cosmetic and health products, as well as in animal nutrition. Fish offal serves as a crucial source for a variety of specific enzymes, including proteases and lipases. Additionally, collagen derived from fish skin is extensively utilized in the production of gelatin and cosmetic products. Crustaceans and bivalves are produced in large volumes, with their by-products potentially accounting for up to 70% of the total output. The chitin, chitosan, astaxanthin, and carotenoids that they are rich in can be utilized to develop antibacterial films, water treatment materials, nutritional supplements, and biomedical materials [43]. Seaweed serves as a critical resource for the extraction of alginate, agar, and carrageenan, and certain types of seaweed are abundant in vitamins, minerals, and plant-based proteins. Bioactive compounds, such as alginate and fucoidan, which are extracted from seaweed processing by-products, have extensive applications in health products, food, and pharmaceuticals [44]. Additionally, a wide range of seaweed-flavored products, including ice cream, candies, and beverages, are produced globally, among which the Asia-Pacific region represents the primary market, while markets in North America and Europe are experiencing gradual growth.

Globally, the comprehensive utilization of various aquatic product by-products is steadily increasing, progressively expanding from traditional applications in feed and fertilizer to diverse high-value-added domains, including functional foods, pharmaceuticals, cosmetics, and environmental protection materials [41]. In developed countries such as Japan, Norway, and the United States, a well-established system has been developed for the high-value utilization of aquatic by-products. The primary focus is on extracting bioactive compounds, including collagen, small-molecule peptides, fish oil, chitosan, and astaxanthin. These compounds are extensively utilized in the fields of nutrition and health, medical materials, and high-end cosmetics. In developing regions, such as Southeast Asian and South Asian countries, the utilization of by-products predominantly focuses on primary processing stages, including fish meal production, animal feed formulation, and organic fertilizer development. Limited by technological advancements, infrastructure capabilities, and market system maturity, the overall rate of comprehensive utilization remains relatively low, indicating substantial potential for further enhancement. As the largest producer and processor of aquatic products, China has recently been actively promoting the integrated development of by-products, an initiative that accelerates the establishment of an industrial chain focused on the deep processing of functional components and the comprehensive utilization of by-products. An increasing number of products are now specifically targeted toward the markets for functional foods, health products, and ready-to-eat meals, significantly enhancing resource efficiency and economic returns.

However, from a global perspective, the aquatic product processing industry still faces challenges such as inadequate utilization of by-products, uneven regional development, and low levels of product standardization. A significant amount of resources remains underdeveloped, thereby hindering the sustainable development of global fishery resources and the full realization of their added value. For example, primary processing continues to dominate as the mainstream approach, while the capacity for deep processing remains underdeveloped. In numerous developing countries, the processing of aquatic by-products remains predominantly focused on basic methods such as drying, crushing, and fermentation. These processes are utilized to produce primary products like fish meal, animal feed, and organic fertilizers, which generally exhibit low added value and limited market competitiveness. The extraction and refinement technologies for high-end applications such as functional protein peptides, collagen, chitosan, and marine active ingredients have not yet been widely adopted. The lack of core technical equipment and standardized production systems makes it impossible to extend the value chain from the raw material end to the consumer end. The extraction and refinement technologies for high-end applications, such as functional protein peptides, collagen, chitosan, and marine active ingredients, have yet to achieve widespread adoption. The absence of core technical equipment and standardized production systems hinders the extension of the value chain from the raw material stage to the consumer market. Furthermore, numerous countries have yet to establish comprehensive quality standards for by-product goods, component certification frameworks, and functional claim regulations, thereby constraining market access and impeding the development of international trade for related end products.

1.8 Current Issues and Solutions

The advantages of marine resources have contributed to the rapid development of the marine industry. However, despite the rapid market growth and technological advancements, the development of the marine food industry still faces several significant challenges [3, 20, 25].

1.8.1 Severe Water Pollution

The production processes in the marine food industry, including aquaculture and fishing, rely heavily on safe and clean water bodies. If water bodies are severely polluted, marine life is directly affected, which in turn destabilizes marine ecosystems and may even lead to irreversible environmental damage to the marine waters. Moreover, marine organisms living in polluted waters may contract diseases or become poisoned. When these contaminated products are harvested, processed, and consumed, they pose serious health risks to consumers. The main sources of water pollution are as follows.

1.8.1.1 Industrial Wastewater and Domestic Sewage

With the growth of the manufacturing industry, the discharge of solid waste, industrial effluents, domestic sewage, and substances like metals and pesticides into the ocean has significantly worsened marine pollution. This has led to substantial changes in the local marine environments. Many estuarine and coastal ecosystems have deteriorated, with coastal waters becoming increasingly eutrophic and frequent red tides occurring. These developments seriously threaten the survival environments of marine species.

1.8.1.2 Feed and Chemical Residues from Aquaculture

The rapid growth of the aquaculture industry has also placed considerable pressure on the ecological environment of coastal waters. In aquaculture and fishing industries, the feeding of marine organisms, along with the use of chemicals, antibiotics, and hormones, often leads to the accumulation of drug residues, environmental pollution, microbial contamination, and parasite infections. The environmental conditions in key aquaculture zones, such as estuaries, bays, and tidal wetlands, have become increasingly concerning, damaging the aquatic environments that marine life depends on. Additionally, the escape of farmed species could lead to the introduction of invasive species, disrupting the local ecological system and its species composition.

1.8.1.3 Overfishing and Overaquaculture

To maximize economic benefits, humans have been artificially farming marine species with high economic value. However, when the population of these species exceeds the natural carrying capacity of the ocean, it can reduce the dissolved oxygen content in the water, adversely affecting the metabolism of marine organisms. This disruption can lead to the degradation of coastal environments and the destruction of habitats for marine life.

1.8.2 Resource Depletion

The development of the marine industry relies heavily on the quantity, diversity, and quality of marine biological resources. On the one hand, the rich and diverse marine life supports the growth of the fishing industry, and the variety of marine species aids in the expansion of aquaculture, producing more marine products. On the other hand, the diversity of marine life contributes to the stability of marine ecosystems and enhances their ability to recover naturally. Therefore, the sustainable development and utilization of marine resources must be built upon a foundation of abundant and high-quality resources. Resource depletion, however, poses several challenges, which are reflected in the following aspects.

1.8.2.1 Decline in the Quantity and Diversity of Marine Life

Changes in ecological environments, destructive fishing practices, and global climate change have altered the structure of marine fisheries resources. Originally, marine ecosystems were densely structured across multiple trophic levels, but now certain trophic levels are being disrupted, leading to the collapse of some food chains. Higher trophic-level species, which are often economically valuable, have been overfished, causing their populations to decline. This, in turn, allows lower trophic-level species to proliferate, which negatively impacts the stability of marine ecosystems.

1.8.2.2 Limited Aquaculture Space

As human development of the ocean increases, various constructions such as port expansions, coastal tourism developments, beachfront housing, villas, entertainment facilities, and industrial projects that utilize seawater have emerged. These developments have significantly reduced the available space for aquaculture, with nearshore waters approaching saturation. Additionally, these construction projects can negatively impact the quality of seawater, further hindering the viability of marine aquaculture. This, in turn, poses a risk to the safety of seafood products. Furthermore, in an attempt to boost economic returns, people may introduce large numbers of juvenile fish into the aquaculture areas. However, excessive stocking of juvenile fish could overwhelm the ocean's ecological carrying capacity, potentially causing long-term damage to the marine environment.

1.8.3 Processing Technology Needs Improvement

1.8.3.1 Weak Theoretical and Technical Foundation

The foundational theoretical research on the processing and distribution of marine aquatic products remains at a relatively low level, with most scientific efforts focused on following up on existing studies. The primary issues include the lack of systematic research on the processing and nutritional characteristics of key nutrients such as proteins, fats, and polysaccharides in major marine raw materials like fish, shellfish, shrimp, and algae. Additionally, the mechanisms underlying quality changes in marine products during aquaculture, processing, and cold chain distribution are unclear. As the quality and safety of marine products gain increasing attention, the molecular basis for the generation and transformation of harmful substances remains poorly understood. There is also a lack of advanced detection technologies, such as trace and ultra-trace analysis. Furthermore, research on traceability systems, early warning mechanisms, and risk analysis is progressing slowly.

1.8.3.2 Slow Development of Mechanized and Intelligent Equipment

Compared to developed countries, developing nations still predominantly rely on labor-intensive methods for aquatic product processing, with a low level of mechanization. The main reason for this is the weak research and development in specialized processing equipment for marine aquatic products. The development of processing equipment is a highly interdisciplinary and capital-intensive field, involving not only the processing technologies of aquatic products but also mechanical manufacturing, material science, automation, and other areas. Developing such equipment requires prototyping and continuous improvements through pilot-scale experiments, leading to long development cycles. More investment is needed to accelerate research in this area.

1.8.3.3 Insufficient Technological Innovation in Processing Engineering

First, there is a lack of effective innovation, integration, and industrial demonstration of key technologies at the national level. Second, seafood processing enterprises have inadequate investment in technological innovation, and their ability to convert scientific achievements into practical applications is weak. Additionally, the capacity for cultivating high-level talent in the field of seafood processing is limited, and there are few innovation platforms available, which restricts the development of the marine product processing sector.

In the face of resource and environmental pressures, traditional marine food production models can no longer meet the development demands of the new era. At the same time, the full potential of the ocean in processing and production has yet to be fully realized. Therefore, the following recommendations are proposed to address the above challenges.

1.8.3.3.1 Strengthen the Protection of Marine Resources and the Environment The ocean is a fundamental component of the global life support system and is also a crucial resource reservoir and an important regulator of ecological environments. The entire industry should be encouraged to adopt a more comprehensive approach to sustainability, considering legal, social, and economic dimensions. First, continue advancing the legal framework for marine resource utilization. Building on existing environmental protection laws and regulations, there is a need to formulate or amend laws related to biodiversity protection and sustainable marine resource use. This will help preserve and select high-value marine species, enhancing both economic benefits and environmental sustainability. Efforts should focus on pollution control, tightening wastewater discharge standards, and upgrading treatment facilities for domestic and industrial wastewater in coastal areas. Additionally, noncompliant enterprises should be required to adhere to marine and environmental protection standards. Second, enhance the level of informatization in fisheries and aquaculture industries. By establishing comprehensive databases that link production, fishing vessel data, trade information, and other modules, a network can be built connecting different stages of the marine food industry supply chain. Third, increase supply diversification. It is essential to avoid over-reliance on traditional species like shrimp, salmon, and tuna. A diversified supply chain is crucial not only for improving living standards and ensuring food security but also for enhancing the nutritional value of special species such as molluscs, seaweeds, and small fish.

1.8.3.3.2 Accelerate Industrial Development Modern marine food enterprises should move away from exploitative practices that deplete marine biological resources and adopt sustainable aquaculture methods to ensure the continued development and utilization of marine resources. It is vital to attract both public and private investment into the marine food industry, stimulating its overall growth.

Brand development is essential, as it represents an inevitable trend toward marketization and commercialization. Strengthening the development of seafood processing brands will increase the international competitiveness of marine products. The focus should shift from primary processing to higher-value-added products, thereby enhancing the economic profitability of the industry. Production should be market-oriented, driven by consumer demand, and the supply-side structural reforms of the marine industry should be implemented to improve the industry chain and increase the proportion of processed marine products.

1.8.3.3.3 Integration of Technological Innovation Technological advancements and innovations in marine-related fields can drive and support marine resource development. Increased investment in Marine Science and Technology (MAST) is essential for harnessing relevant marine knowledge and biotechnologies to enhance aquaculture techniques and improve production lines in processing. Promoting technological innovation in the fishing industry will optimize its structure and facilitate industry upgrades. By improving the technological level of the industry, the development of sustainable, eco-friendly, and efficient marine food production systems will be accelerated.

References

1 United Nations Economic and Social Commission for Western Asia (UNESCWA). (2024). *Aquatic Resources - Definition*. UNESCWA Sustainable Development Glossary.

2 Arthur, R. (2024). Availability and importance of aquatic resources. In: *Review of the State of World Fishery Resources: Inland Fisheries*. Rome, FAO and Abidjan, AfDB.

3 Food and Agriculture Organization of the United Nations (FAO) (2024). *The State of World Fisheries and Aquaculture 2024: Blue Transformation in Action*. Rome: FAO.

4 Food and Agriculture Organization of the United Nations (FAO) (2022). Utilization and processing of fisheries and aquaculture production. In: *The State of World Fisheries and Aquaculture 2022*. Rome: FAO.

5 Sundaray, J.K., Rather, M.A., Ahmad, I., and Amin, A. (ed.) (2025). *Food Security, Nutrition and Sustainability Through Aquaculture Technologies*. Cham: Springer.
6 Zhang, T.T., Xu, J., Wang, Y.M., and Xue, C.H. (2019). Health benefits of dietary marine DHA/EPA-enriched glycerophospholipids. *Progress in Lipid Research* 75: 100997.
7 EFSA Dietetic Products, Nutrition, and Allergies (NDA) (2013). Scientific opinion on health benefits of seafood consumption in relation to methylmercury exposure. *EFSA Journal* 12 (7): 3761.
8 Pan, C., Wu, H., and Zhao, Y. (2022). Aquatic products' quality, processing and preserving: recent developments, trends and advances. *Applied Sciences* 12 (24): 12924.
9 Jiang, X. (2023). Overview of research progress on new processing technologies for aquatic products. *Highlights in Science, Engineering and Technology* 66: 124–134.
10 Shi, W., Yuan, C., Liu, S. et al. (2024). The aquatic product processing and by-product utilization. *Frontiers Media SA*. 8: 1498642.
11 Dash, B.P., Venkataraman, K., and Sivaperuman, C. (2024). Diversity of bivalve mollusks, their ecosystem services, and potential. In: *Bivalve Mollusks: Diversity and Conservation* (ed. C. Sivaperuman and K. Venkataraman), 1–25. Cham: Springer.
12 Jereb, P., Roper, C.F.E., Norman, M.D., and Finn, J.K. (ed.) (2014). *Cephalopods of the World: An Annotated and Illustrated Catalogue of Cephalopod Species Known to Date*, FAO Species Catalogue for Fishery Purposes No. 4. Rome: FAO.
13 González, Á.F. and Pierce, G.J. (2021). Advances in the study of cephalopod fisheries and ecosystems. *Fisheries Research* 242: 105975.
14 Myers, P., Espinosa, R., Parr, C. S., Jones, T., Hammond, G. S., and Dewey, T. A. (2025). Crustacea (shrimps, crabs, lobsters, water fleas, and relatives). *Animal Diversity Web* (University of Michigan).
15 (2010). Echinodermata. In: *Echinodermata*, 3–17. Vienna: Springer.
16 Leone, A., Lecci, R.M., Durante, M., and Meli, F. (2022). Jellyfish as food: a narrative review. *Foods* 11 (18): 2773.
17 Zhang, L., Chen, J., and Li, W. (2024). Global analysis of transcriptome and metabolomics revealed that fatty acid metabolism plays an important role in the growth of Chinese soft-shelled turtle. *Aquaculture Reports* 37: 101777.
18 Sharma, N. (2025). *Algae: Characteristics, Classes, Structure, Reproduction*. Microbe Notes: *Importance*.
19 United States Department of Agriculture, Foreign Agricultural Service (USDA-FAS). (2025). *China: 2025 China Fishery Products Report*. GAIN Report CH2025-0057, Washington, DC.
20 Costello, C., Cao, L., Gelcich, S. et al. (2020). The future of food from the sea. *Nature* 588 (7836): 95–100.
21 Meyer, B., Kawaguchi, S., Conry, J.A. et al. (2022). *Antarctic Krill and Its Fishery: Current Status and Challenges*. SCAR Antarctic Environments Portal.
22 Morales-Nin, B., Zimmermann, F., Merino, G. et al. (2024). Challenges on blue food provision. *Frontiers in Ocean Sustainability* 2: 1271783.
23 Zhang, Q., Lv, L., Wang, W. et al. (2024). Recent advances of bioactive marine natural products in drug discovery. *Journal of Ocean University of China* 23 (5): 1297–1318.

24 Banday, A.H., ul Azha, N., Farooq, R. et al. (2024). Exploring the potential of marine natural products in drug development: a comprehensive review. *Phytochemistry Letters* 59: 124–135.

25 Tigchelaar, M., Leape, J., Micheli, F. et al. (2022). The vital roles of blue foods in the global food system. *Global Food Security* 33: 100637.

26 Puri, M., Gupta, A., McKinnon, R.A., and Abraham, R.E. (2022). Marine bioactives: from energy to nutrition. *Trends in Biotechnology* 40 (3): 271–280.

27 do Nascimento Alves, R., dos Santos, J.R.S., Almeida, F.L.C. et al. (2024). Use of protein by-products obtained from aquatic organisms as bioactive compounds: a bibliometric review. *Food Reviews International* 40 (9): 2662–2681.

28 Zamborain-Mason, J., Viana, D., Nicholas, K. et al. (2023). A decision framework for selecting critically important nutrients from aquatic foods. *Current Environmental Health Reports* 10 (2): 172–183.

29 Ying, X., Li, X., Deng, S. et al. (2025). How lipids, as important endogenous nutrient components, affect the quality of aquatic products: an overview of lipid peroxidation and the interaction with proteins. *Comprehensive Reviews in Food Science and Food Safety* 24 (1): e70096.

30 Zheng, B., Wen, H., Zheng, B. et al. (2025). Current research on the processing, safety evaluation and functional properties of aquatic-derived lipids considering the sustainable processing of omega-3 polyunsaturated fatty acids: a critical review. *Critical Reviews in Food Science and Nutrition* 25: 1–19.

31 Mohan, K., Rajan, D.K., Ganesan, A.R. et al. (2023). Chitin, chitosan and chitooligosaccharides as potential growth promoters and immunostimulants in aquaculture: a comprehensive review. *International Journal of Biological Macromolecules* 251: 126285.

32 Xu, C., Zhang, R., and Wen, Z. (2018). Bioactive compounds and biological functions of sea cucumbers as potential functional foods. *Journal of Functional Foods* 49: 73–84.

33 Huang, X., Wang, H., and Tu, Z. (2023). A comprehensive review of the control and utilization of aquatic animal products by autolysis-based processes: mechanism, process, factors, and application. *Food Research International* 164: 112325.

34 Zeng, J., Song, Y., Cong, P. et al. (2024). A systematic review of drying in aquatic products. *Reviews in Aquaculture* 16 (1): 47–65.

35 Chen, L., Wang, Y., Zhu, C. et al. (2022). Effects of high-pressure processing on aquatic products with an emphasis on sensory evaluation. *International Journal of Food Science and Technology* 57 (11): 6980–6996.

36 Dong, H., Gai, Y., Fu, S., and Zhang, D. (2022). Application of biotechnology in specific spoilage organisms of aquatic products. *Frontiers in Bioengineering and Biotechnology* 10: 895283.

37 Yu, Y., Zhao, Y., Zhang, Y. et al. (2024). Recent advances in the effects of protein oxidation on aquatic products quality: mechanism and regulation. *International Journal of Food Science and Technology* 59 (3): 1968–1978.

38 Barrow, C. and Shahidi, F. (2007). *Marine Nutraceuticals and Functional Foods*. Boca Raton: CRC Press.

39 Zhang, C., Li, X., and Kim, S.K. (2012). Application of marine biomaterials for nutraceuticals and functional foods. *Food Science and Biotechnology* 21 (3): 625–631.

40 Zhang, X., Ma, S., and Liu, S.Q. (2025). Valorization of animal and aquatic product processing side-streams: upcycling into flavoring and functional ingredients. *Food Research International* 212: 116445.

41 Pandey, V.K. (2025). Exploring the significance of emerging blue food processing technologies for sustainable development. *Food Research International* 200: 115429.

42 Papadopoulou, S.N.A., Adamantidi, T., Kranas, D. et al. (2025). A comprehensive review on the valorization of bioactives from marine animal by-products for health-promoting, biofunctional cosmetics. *Marine Drugs* 23 (8): 299.

43 Gürdal, A.A. and Çetinkaya, T. (2024). Advancements in edible films for aquatic product preservation and packaging. *Reviews in Aquaculture* 16 (3): 997–1020.

44 Udayan, A., Pandey, A.K., Sirohi, R. et al. (2023). Production of microalgae with high lipid content and their potential as sources of nutraceuticals. *Phytochemistry Reviews* 22 (4): 833–860.

2

Fish

Saiyi Zhong and Rui Li

College of Food Science and Technology, Guangdong Ocean University, Zhanjiang, China

According to the latest assessment by the Food and Agriculture Organization of the United Nations (FAO) and industry reports, the current status of global fishery resources can be summarized as follows: 64.5% of fishery resources are within the biosustainable fishing level, but 35.5% of the populations are facing overfishing, and the proportion of overfishing is increasing at an average annual rate of 1%. In terms of production distribution, the total quota for wild white fish in 2025 is expected to be 6.82 million tons, a decrease of 100,000 tons compared to the previous year. This is mainly due to a sharp 20% reduction in the quota for Barents Sea cod (from 816,000 tons to 656,000 tons), significant production cuts in Norway and Russia, and Iceland, with 214,000 tons, becoming the largest exporter of cod. The expected production of Pacific salmon in 2025 is 766,000 tons, an increase of 45% compared to 2024, but a 30% decrease from the peak in 2023. The main reason is the continuous decline in the average water levels in Alaska and Canada. However, the catch in the Russian Far East has strongly rebounded (reaching 287,000 tons as of August, more than half of the annual forecast). Aquaculture has become the main source of supply. It is estimated that by 2025, the global production of Basa fish will reach 4.5 million tons (5.6%), tilapia 7.3 million tons (4.6%), and Atlantic salmon 3.12 million tons (3.9%), accounting for more than 50% of the total aquatic products. There are significant differences in regional sustainability. The Northeast Pacific (92.7% sustainable) and the Antarctic Ocean (100% sustainable) lead in scientific management, while the Mediterranean (35% sustainable) and the Southeast Pacific (46% sustainable) still face challenges such as fragmented governance and data deficiency. Countries like Peru are balancing resource conservation and fishery livelihoods through dynamic adjustments, such as restarting the fishing of giant squids and setting a 50,000-ton quota [1].

2.1 Species and Geographical Distribution

Around 73% of the Earth's surface is covered by oceans, and fish making up the majority of marine life, with their colorful patterns and beautiful fin tails. The distribution of fish is extremely wide, and they can survive in a variety of aquatic environments, from freshwater lakes to saltwater oceans, from mountain streams to deep ocean floors. More than 96% of the world's 34,800 named species of fish belong to the crown group of Actinopterygii, more than half of all vertebrates, and nearly 250 new species are discovered each year. Fish, as a nutritious and delicious food, is widely consumed worldwide. There were various ways to eat it, with rich nutritional value and many benefits for human health. Regular consumption of fish can bring various health benefits. Firstly, it helps to reduce the risk of cardiovascular disease, as the unsaturated fatty acids in fish can help regulate blood lipid levels. Secondly, the high-quality protein rich in fish meat helps to enhance immunity and improve the body's resistance. In addition, the nutrients in fish meat also help promote brain development and maintain neurological health. For children, moderate consumption of fish can improve intelligence and memory; for the elderly, fish can help prevent cognitive decline. Through diverse cooking methods, we can enjoy the deliciousness and nutrition brought by fish. To maintain health, it was recommended that people increase their intake of fish in moderation in their daily diet. The Green carp, grass carp, silver carp, and bighead carp all belong to the fish class and carp family and are traditional farmed fish species in China. Until today, it is still the main fish for freshwater fish farming in China. Because they were artificially farmed fish, they were called "the four famous domestic fishes" [2].

2.1.1 Marine Fish and Their Geographical Distribution

About 80% of the distribution of Marine fish is in the shallow continental shelf area, especially in the tropical and subtropical sea area of the Indo-Pacific Ocean. The main economic fish distributed in the cold zone and subcold zone are herring, cod, salmon, flounder, and mackerel. Sardine, saurel, and mackerel are found in the subtropical sea area. In tropical and subtropical sea areas, it distributes tuna and so on.

Marine fish account for about 70% of the total number of fish, including the common cod, salmon, tuna, and so on.

2.1.1.1 Cod

Cod can be divided into Atlantic cod, Pacific cod, and Greenland cod. Atlantic cod is large, with adults reaching more than 1.5 m in length and weighing up to 30 kg. The color is usually green, brown, or gray, with darker colors above and lighter colors on the belly. The iconic Atlantic cod, widely distributed in the cold waters of the North Atlantic Ocean, is managed into about 20 major North Atlantic stocks, and the living conditions of local cod populations on both sides of the North Atlantic vary greatly [3], which is exacerbated by ongoing anthropogenic climate change. The spawning cycle and juvenile survival rate of cod are sensitive to warm waters [4], mainly due to the physiological changes of their adult body, while the living conditions of embryos and larvae become more severe, which makes the reproductive success rate of Marine poikilothermal animals particularly low in the warming ocean [5]. At the same time, relevant studies have shown that the hemoglobin in cod is affected by natural selection [6]. Cod with a certain genotype of hemoglobin gene have stronger feeding behavior and higher success rate of hunting, and there is a significant correlation between the affinity of different genotypes of hemoglobin for oxygen in waters with different temperatures [7].

Pacific cod, which are found mainly in the North Pacific Ocean, including the Bering Sea, Gulf of Alaska, Sea of Japan, and other waters, are large, with adults usually reaching 50–80 cm long and reaching a maximum of 1.2 m. The body color is usually green, gray, or brown, with darker colors above and lighter colors on the belly. Some apparently contiguous Marine species have locally genetically distinct populations within their ranges, so careful consideration is needed when designating management units (MUs) for the assessment and management of Marine resources and for ethnogenetic studies. For coastal areas of the Sea of Japan, a boundary is established on the western coast of Honshu based on studies of geographic distribution and genetic variation. MU [8] to divide the Sea of Japan.

The Greenland cod is a fish adapted to the cold waters of the North Atlantic and is found mainly off the east coast of Greenland and Canada. Compared to Atlantic cod and Pacific cod, Greenland cod are small, with adults usually about 60–80 cm long, dark gray or brown, and lighter bellies. The fish usually live along coasts and in shallower waters, feed on small fish, crustaceans, and other Marine invertebrates.

Cod is a kind of Marine fish with high nutritional value, and its meat is rich in high-quality protein and low-fat content. According to the analysis, the protein content of cod meat can be as high as 18 g/100 g. In addition, cod is also a great source of omega-3 fatty acids, which play an important role in lowering cholesterol levels and preventing cardiovascular disease. When comparing the muscle composition of cod with other fish (silver carp, small yellow croaker, and moray eel), the results showed that cod had the highest water content and the lowest fat content. By comprehensive nutritional value and taste evaluation, cod muscle is superior to other fish in comparison. Cod is also a rich source of vitamins A and D, which are essential for maintaining vision and bone health. At the same time, cod contains minerals such as calcium, magnesium, and selenium, which have a positive effect on maintaining healthy bones and teeth. Cod is also rich in vitamin B12, which is important for the division of cells and the normal formation of blood, as well as the functional maintenance of the nervous system.

It is worth noting that cod is relatively low in mercury, which makes it a suitable seafood choice for all types of people. Cod also has positive effects, including reducing blood sugar, promoting blood circulation, relieving pain, and providing bactericidal and anti-inflammatory benefits. In terms of fatty acid analysis, gas chromatography (GC) analysis of five common cod products in the fisheries market showed that 13, 11, 13, 13, and 6 kinds of fatty acids were identified in Atlantic cod, Pacific cod, sable, silver cod, and oil fish, respectively. Nutritional assessments have shown that Atlantic cod, Pacific cod, sable, and codfish have healthier atherosclerosis and thrombosis indices [9], further confirming the nutritional value of these fish.

In terms of medicinal value, cod liver is large, and liver oil content is high, which is one of the important raw materials for manufacturing cod liver oil. Cod is rich in trace elements iron and phosphorus; these elements can promote the regeneration and metabolism of human red blood cells and improve hematopoietic function. Cod protein is mainly fibrin, loose meat, easy to digest and absorb, helping to regulate the spleen and stomach, and repair the damaged stomach mucosa. The trace elements potassium and magnesium in cod help to promote the metabolism of blood sugar in the human body and prevent blood sugar from rising. In addition, the omega-3 fatty acids in cod help lower blood cholesterol and triglyceride levels, thereby preventing the risk of heart disease and stroke. In terms of economic value, cod is called "gold in the sea" because of its high economic value. The processed products of cod, such as surimi, fish intestines, fish rolls, fish cakes, fish balls, crab feet, are important seafood foods. The heads and guts of cod fish can be processed into fish meal, fish oil, and other products, and the economic value of refined fish heads is higher. Cod is loved by consumers because of its delicious meat and fewer thorns, and has become a popular food fish. Cod can not only be eaten fresh, but also can be made into dry products, such as surimi, caviar, fillet, and so on.

2.1.1.2 Salmon

Salmon are widely distributed in fresh and Marine waters in the Northern Hemisphere. The main species of salmon include Atlantic salmon, Pacific salmon, and rainbow trout. The life cycle of salmon consists of both freshwater and marine phases, usually born in freshwater rivers, and the young fish migrate to the sea after living in freshwater for 1–3 years.

Adult Atlantic salmon are usually about 70–100 cm long and weigh between 4 and 12 kg, with some individuals reaching larger sizes. Adults are silver with a darker back, usually blue-green or gray. There are irregular black spots on the sides of the body and a larger abdomen. They are typically migratory fish, with streamlined bodies that help them move quickly through the water and are particularly suited to migrating long distances. It is widely distributed in the coasts and rivers of the North Atlantic Ocean, including the United States and Canada on the east coast of North America, and the rivers and seas of Britain, Norway, Iceland, and other places in Europe. In the ocean, it usually inhabits coastal and offshore areas and mainly stays in the water layer where the temperature is suitable. During the breeding season, return to the river where you were born to lay eggs. Choose a river with clear water and sufficient oxygen, usually a river with a gravel bottom. Juvenile Atlantic salmon have a limited ability to withstand high temperatures, and studies have shown that 23 °C is the upper limit for feeding, and behavioral changes occur at 22–24 °C, which may be a conservative baseline, but beyond this baseline, juvenile Atlantic salmon may be negatively affected [10]. In addition to this, other abiotic and biological factors, such as reduced habitat, low emissions, increased density, reduced food availability, disease, and high predation rates, can reduce the ability of young salmon to withstand high temperatures.

Pacific salmon are a key fish in the North Pacific, found along the west coast of North America and the east coast of Asia, and contain five main species: coho, sockeye, Chinook, pink, and king salmon. Each Pacific salmon species has unique characteristics and ecological habits, such as coho salmon, which are brightly colored during the breeding season, sockeye salmon, which are known for their bright red color, and king salmon, which are the largest and most fleshy species.

Rainbow trout are native to the Pacific coast of North America and the Kamchatka Peninsula, including parts of the Pacific coast of Canada, the United States, Mexico, and the rivers of Colombia. Now, rainbow trout has been spread across the Americas, Europe, Oceania, eastern Asia, and southern Africa and other about 20 countries, becoming the world's widely farmed varieties. Rainbow trout migrate small distances, and longer distances if they are anadromous or lake-based species. At high water temperatures, which can exceed the optimal range of many physiological processes in fish, stimulating fish to seek cool water refuges [11], it may become more common for summer-running steelhead trout to seek hot refuges when reentering their habitat using non-natal streams. At the same time, the optimal thermal habitat in some population birth areas may shrink over time [12].

Salmon is important in several ways. In terms of nutritional value, salmon is rich in high-quality protein, which is easy to digest and absorb, helping to maintain and enhance the body's immunity. In addition, salmon is rich in unsaturated fatty acids, especially omega-3 fatty acids (such as docosahexaenoic acid [DHA] and eicosapentaenoic acid [EPA]), which are important for heart health, brain function, and vision development. Salmon is also rich in many vitamins and minerals, such as vitamin D, vitamin B12, iron, zinc, and selenium, which play an important role in bone health, nervous system function, immune system health, and energy metabolism. At the same time, astaxanthin in salmon acts as a powerful antioxidant that can fight free radical damage, thereby protecting body cells and tissues. As a typical salmon, salmon is known as "treasure in water" because of its rich nutritional value [13]. Salmon is rich in polyunsaturated fatty acids such as EPA and DHA, which play an indispensable role in maintaining heart health, regulating blood lipid levels, and enhancing cognitive function. In addition, salmon is a great source of high-quality protein, many vitamins (such as VA, VD, VE, and VB), and minerals (such as zinc, selenium, copper, and manganese), which play an important role in maintaining human physiological function and strengthening the immune system.

In terms of medicinal value, salmon contains astaxanthin, which can mop up harmful free radicals in the body, combat oxidative damage to the brain and central nervous system, and help slow aging and prevent cancer. In addition, the rich unsaturated fatty acids in salmon oil promote the formation of cell membranes, reduce lipid and cholesterol levels, prevent cardiovascular disease, enhance brain function, and protect the brain from the effects of aging. At the same time, these unsaturated fatty acids can also inhibit the production of inflammatory substances, which provides a certain help for weight loss.

From an economic point of view, salmon has long been linked to human society and is one of the most valuable wild caught fisheries in the world. Salmon fishing contributes more than $2 billion to the economies of Russia, Japan, the United States, and Canada and directly creates 35,000 jobs. In addition, salmon fishing is relatively small, and many individuals, communities, and small businesses rely on it for their livelihoods, demonstrating its importance to local economies. Salmon also plays a vital role in the ecosystem. They provide food for a variety of predators and scavengers, including seals, whales, otters, bears, birds, and a variety of invertebrates. At the same time, salmon also transport essential nutrients from the Marine environment to freshwater and terrestrial habitats, promoting nutrient cycling in the ecosystem.

2.1.1.3 Grouper

The Grouper is classified within the family Serranidae (order Perciformes). It is a large and medium-sized Marine fish that lives in tropical and subtropical seas. It is widely distributed in the Atlantic Ocean, Indian Ocean, and Pacific Ocean. In China, it is mainly concentrated in the South China Sea and the southern part of the East China Sea. The groupers prefer to inhabit coral reefs, rocky reefs, and caves in stony and grandly sea areas, which are quiet, warm, and clear waters. They are solitary and reef fish. It is thick and laterally compressed, spindle-shaped or oval-shaped, with a body length usually ranging from 35 to 45 cm and a weight of 1.5 to 2.5 kg. Its body surface is covered with bright spots or bands, and its body color changes with the environment and its own health condition. It is darker black in weak light and lighter in strong light. Groupers are fierce by nature and feed on fish and crustaceans. They track their prey with their acute sense of smell. Most groupers are hermaphroditic, and the females reach sexual maturity first. Its embryonic development goes through stages such as the cleavage stage and the blastocyst stage, and the post-embryonic development is further divided into the larval stage, the juvenile stage, and the larval stage. Grouper is rich in nutrition, with tender and white meat, and is known as "sea chicken." It is rich in protein, unsaturated fatty acids, vitamin D, and trace elements such as potassium, magnesium, and selenium. Its protein helps the body's metabolism and tissue repair, unsaturated fatty acids can reduce the risk of cardiovascular diseases, and vitamin D promotes calcium absorption and maintains bone health. For this reason, grouper has become a precious economic fish, deeply loved by consumers, and in Cantonese wedding banquets, it even symbolizes "wealth, prosperity and good fortune."

2.1.1.4 Tuna

Tuna is an important economic fish, widely distributed in the world's oceans, belonging to the tuna family, which includes tuna, tuna, skipjack, and skipjack, a total of 14 species. Tunas are the most prominent branch of the group, encompassing eight major species, which are often collectively referred to as "true tuna." Specifically, the main species of the genus Tuna include Atlantic bluefin, Pacific bluefin, Southern bluefin, bigeye, albacore, yellowfin, blackfin, and albacore. Because of their delicious meat and high nutritional value, these species have become extremely important economic fish in the global fishery.

Tuna is a large pelagic food fish with significant commercial value and ecological importance. They mainly inhabit the pelagic waters of the ocean, showing a high degree of migration and adapting to a wide range of Marine environments. The natural distribution of tuna is wide and sparse, especially in tropical and temperate waters. In general, the distribution range of tuna is mainly concentrated between 45° north and south of the equator, including the Indian Ocean, the central Pacific, and the central Atlantic. As a tropical–subtropical oceanic fish, tuna has an extremely wide range of activities and can swim long distances. This characteristic allows tuna to play an important role in the global Marine ecosystem, being able to migrate between different seas. In the past, tuna have been recorded to swim from the coast of California to the coast of Japan, demonstrating their great migratory ability.

In the North Atlantic, the reproductive habits of albacore tuna are of particular concern. Studies have shown that this tuna adult fish usually selects the tropical area of the Sargasso Sea and its adjacent waters for spawning in the summer. This process is crucial for the reproduction and continuation of the population. Tuna entering the juvenile stage (1–4 years old) spend the winter in the mid-Atlantic, taking advantage of environmental conditions during this period to promote growth and development. As spring arrives, the young migrate north to feeding grounds in the Bay of Biscay and around the Northeast Atlantic. This migration behavior usually follows its preferred sea surface temperature (SST) range, and the tuna chooses the appropriate temperature conditions to maintain its physiological state during migration [14].

In the fall, young fish leave their previous feeding areas and return to their wintering areas. This phenomenon is closely linked to changes in SSTs, which gradually cool as the seasons change and may fall below their comfort zone, prompting young fish to migrate. Similarly, adult bluefin tuna from the Eastern Atlantic and Mediterranean also breed and spawn in the Mediterranean. Some young fish leave the Mediterranean during the summer for foraging migrations to the Bay of Biscay, while adults continue their spawning migrations, a process that is important for maintaining the sustainability of tuna stocks.

2.1.1.5 Shark

Sharks belong to the class Chondrichthyes and are important apex predators in Marine ecosystems. There are at least 440 known shark species, demonstrating the diversity and complexity of this group. In the classification system of sharks, the classification of their species is quite complex, mainly based on their physical characteristics, which are systematically classified, usually can be divided into eight orders, including *Hexanchiformes*, *Heterodontiformes*, *Orectolobiformes*, *Lamniformes*, *Carcharhiniformes*, *Squaliformes*, *Squatiniformes*, and *Pristiophoriformes*. Under these orders, they are subdivided into different families and genera, further enriching the classification system of sharks.

Common shark species include the great white shark, hammerhead shark, tiger shark, whale shark, basking shark, and bull shark (*Carcharhinus leucas*). These sharks are found in tropical, temperate, and polar oceans around the globe, showing their ability to adapt to different environments. Different species of sharks show some preferences in their living environments, with some species preferring warmer waters and others more adapted to colder waters. In addition, there are some shark species that prefer to live in shallow coastal waters or near the bottom of the sea. Great white sharks, for example, are commonly found off the coasts of

South Africa, Australia, California, and the eastern United States, demonstrating their dependence on specific ecological environments. Tiger sharks live mainly in warm tropical waters off the coasts of Hawaii, Central America, and the Caribbean, demonstrating their adaptability to warm waters. While hammerhead sharks prefer warm tropical waters, they can be found in a variety of environments, including shallow waters near coastlines and the deep seafloor. The physiological characteristics of sharks make them have unique adaptive mechanisms in the ocean. Because sharks lack swim bladders, they must constantly swim to maintain buoyancy and breathe, which makes their oxygen requirements relatively high. Despite their poor eyesight, sharks have extremely keen senses of smell, hearing, and touch, which allow them to hunt and survive effectively in complex Marine environments.

At present, many species and areas of the world of shark populations are experiencing rapid decline; this phenomenon has put many shark species facing a serious threat of extinction. The main factor contributing to this situation is overfishing [15], which includes both fishing for specific shark species and incidental fishing, where sharks are caught accidentally while fishing for other fish. The situation is particularly acute in many areas, so much so that current fishing rates have significantly exceeded the natural resilience of stocks. Other important factors affecting the decline of shark populations include habitat degradation, persecution of sharks, and the effects of climate change.

The relatively slow life history of the shark itself means that the process of population recovery is likely to be very slow, and even in areas where strong conservation strategies have been implemented, signs of recovery are rare. In response to this urgent problem, the need for large-scale conservation measures has been widely recognized worldwide. To this end, some coastal states have chosen to establish "shark sanctuaries" that prohibit commercial fishing and large-scale export of shark products, often covering their entire exclusive economic zones. It is worth noting that the area of shark sanctuaries worldwide is almost the same as that of Marine protected areas, which demonstrates the efforts and commitment of the international community in protecting shark populations [16].

2.1.2 Freshwater Fish and Their Geographical Distribution

Freshwater fish account for about 20% of the total fish, including common carp, crucian carp, grass carp, and tilapia. There are about 8,000 kinds of freshwater fish, distributed in rivers, lakes, and even groundwater layers around the world, mainly divided into daily food fish and ornamental fish.

2.1.2.1 Carp

East Asia is one of the regions with the richest diversity of freshwater fish in the world, forming a unique fish composition dominated by cyprinids. Among them, a group known as "East Asian carp" (about 154 species) is the most special; they occupy a large amount of ecological space in the monsoon area, showing a high degree of adaptation to river-lake ecosystems. These fish not only become an important object of freshwater aquaculture in East Asian countries but also cause serious ecological invasion in freshwater waters around the world.

Carp belongs to Cypriniformes, its family is Cyprinidae, and its genus is *Cyprinus*. Cyprinidae is the most widely distributed and diverse family in Cypriniformes, with more than 200 genera and more than 2,000 species of freshwater fish worldwide. Carp, as one of the important representatives of the family, has significant ecological and economic value. In China, the main economic fish in the carp family include black carp, grass carp, silver carp, and bighead carp, as well as carp and crucian carp, which are collectively known as the "four big fish." These fish are important objects for pond farming and are widely used in aquaculture to meet the market demand for aquatic products. Carp farming not only promotes the development of the local economy but also plays an important role in the aquatic ecosystem. Carp are native to Asia and are most common in the freshwater waters of China, in particular. However, with the promotion of human activities and aquaculture technology, the distribution range of carp has been significantly expanded and is now widely distributed around the world. In China, carp is found in almost all types of water bodies, except for a few areas in the Northwest Plateau, and can be found in paddy fields, ponds, ditches, streams, rivers, lakes, and swamps. Globally, carp are found in North America, South America, Europe, Oceania, and southern Africa. This wide distribution allows carp to survive and thrive in different hydrological environments; however, the introduction and spread of carp has also caused ecological problems in some areas. Carp are regarded as a notorious invasive species in some waters, causing certain negative impacts on the local ecosystem. Its competition with native fish and aquatic plants, its impact on water quality, and the destruction of ecological balance have aroused great concern of scientific and management departments.

Carp, as an important edible fish, has significant nutritional value and economic benefits. Its meat is rich in high-quality protein, which provides the body with essential amino acids to support growth and maintain body functions. In addition, carp is also rich in unsaturated fatty acids, which have a positive effect on cardiovascular health and lower cholesterol levels. These nutrients play an important role in promoting bone health, strengthening immunity, and improving energy metabolism. Therefore, carp is not only a delicious food, but also an important part of a healthy diet.

On the economic level, carp culture has shown remarkable economic benefits. Due to its fast growth rate, strong adaptability, and high yield, carp can bring considerable economic benefits to farmers in a short period of time and become the main farmed fish product in the northern region. This combination of high market demand and relatively low breeding costs has made carp breeding an important industry to promote rural economic development and thus improve the income level of farmers. In addition, with the increasing attention of consumers to healthy diets, the market potential of carp is also expanding, further providing a broad space for the development of the aquaculture industry.

According to the 2021 National Fishery Economic Statistical Bulletin, China's total import and export of aquatic products reached 9.55 million tons, with a total import and export value of US$39.95 billion. Among them, the export volume of aquatic products was 3.81 million tons, with an export value of US$21.93 billion; imports amounted to 5.75 million tons, and imports amounted to US$18.02 billion. This data shows that China has maintained a more active state in the seafood trade, with a trade surplus of US$3.90 billion, an increase of US$430 million from the same period last year. Carp and its products occupy a certain share in this import and export trade and have a certain competitiveness and importance in the international market.

2.1.2.2 Grass Carp

Grass carp as an important freshwater fish, is native to China, Russia, and Bulgaria, and mainly inhabits large river systems in East Asia. In their natural habitat, grass carp depend on sufficient turbulent conditions for successful spawning, a biological characteristic that gives them unique ecological needs in specific waters. In China, grass carp is mainly distributed in the Yangtze River, Pearl River, and Heilongjiang River basins, and the hydrological characteristics of these areas provide ideal habitat for the reproduction and growth of grass carp. Since the 20th century, grass carp has been introduced to 93 countries and regions in the world, including India, Hungary, and the United States, because of its rapid growth and effective control of aquatic vegetation. It is worth mentioning that Malaysia began to introduce grass carp in the 1980s, becoming one of the first countries to introduce the species from China. This introduction activity not only promoted the development of grass carp aquaculture in foreign countries but also made it an important biological tool for aquatic grass management and ecological restoration. When not involved in spawning, adult fish usually inhabit slow-moving or loose habitats, particularly those with abundant aquatic vegetation. This habitat makes grass carp an important consumer of aquatic plants, and its diet is mainly based on aquatic plants. Therefore, grass carp has shown good effects and economy in controlling aquatic vegetation and has been introduced to aquatic plant management in many areas of the United States.

However, grass carp's eating behavior on aquatic vegetation may have adverse effects on its ecosystem [17]. Stocking of grass carp may not only lead to direct destruction of aquatic plant communities but may also indirectly affect the animal communities associated with these vegetation as well as abiotic environmental conditions. In order to control the impact of grass carp in the ecosystem, several states in the United States allow diploid (fertile) grass carp to be stocked, although grass carp was used in these areas long before triploid grass carp became widely available. It is important to note that grass carp can escape and take root in natural waters during captivity, especially in the Mississippi River basin and the Great Lakes Region, which raises ongoing concerns about their ecological adaptability and potential ecological impacts. Grass carp, as a large freshwater fish native to East Asia, has been widely introduced to many countries in the world because of its ecological characteristics and economic value. In their habitat, grass carp not only plays the role of aquatic vegetation controller but also has an important impact on the structure and function of the ecosystem.

2.1.2.3 Tilapia

Tilapia belongs to the tropical crucian carp family, which belongs to the scientific classification of Teleoidea, Perciformes, and Cichlidae. This family of fish has attracted much attention because of its physiological characteristics and ecological adaptability, and has important economic value and ecological impact. Tilapia is famous for its fast growth cycle, strong stress resistance, and excellent feed/meat ratio, and its meat is delicious, which is widely used in aquaculture and related industries.

Tilapia is mainly native to Africa, especially the freshwater waters of the interior of Africa and the brackish water environment of the Atlantic coast of the Middle East. Its natural range extends northward to Israel and Jordan. Due to the remarkable adaptability and high growth rate of tilapia, the species began to be introduced to other parts of the world for artificial cultivation in the mid-20th century, especially in tropical and subtropical regions, where it has formed a relatively mature culture system.

Tilapia is a highly adaptable fish, able to survive in a wide range of waters and environmental conditions, showing strong ecological resilience. This adaptability is not only reflected in its ability to grow and reproduce, but also in its tolerance to different water quality and temperature conditions, providing favorable conditions for its cultivation on a global scale. Tilapia is one of the most important aquatic products because of its delicious meat and great market demand. With the introduction of tilapia, its spread around the world has aroused extensive ecological concern and research. Tilapia and carp are two of the most widely distributed exotic fish species in the world, and their invasion has had a significant impact on some ecosystems. Since tilapia was first introduced to southern China in 1957, more than seven species of tilapia have been introduced and farmed in China. So far, five tilapia species have been identified in rivers in South China, reflecting the competitive capacity of tilapia in terms of niche occupation and resource utilization. In the ecosystem, Nile tilapia often gains significant competitive advantages in the case of overlapping ecological niches due to its large size and rapid growth rate [18]. This advantage makes Nile tilapia and its hybrids the most widely distributed tilapia species in Southern China. In contrast, the red-bellied tilapia, despite its small size and relatively slow growth rate, also performs well in terms of adaptability. Red-bellied tilapia can tolerate lower water temperatures and has a higher reproductive rate and survival rate, thus forming stable natural populations in several rivers in South China [19].

The main nutrients of tilapia include high-quality protein, unsaturated fatty acids, minerals, vitamins, and retinol. It is rich in high-quality protein, about 19 g of protein/100 g of fish, especially the content of essential amino acids such as glutamate and glycine, is high, showing its high-quality and high-protein characteristics. Second, tilapia's fats are mainly polyunsaturated fatty acids, of which omega-3 fatty acids are particularly beneficial for cardiovascular health. In addition, tilapia is also rich in minerals such as potassium, magnesium, and phosphorus, which are important for maintaining good health. Tilapia contains high levels of vitamin B12, vitamin D, and folic acid, which can effectively boost immunity and protect the health of the nervous system. At the same time, tilapia has a high content of retinol, which helps prevent night blindness and vision loss, and maintains the normal function of the immune system. The content of taurine up to 210–220 mg/100 g can regulate blood lipids, reduce cholesterol and neutral fat, and enhance liver function.

In terms of medicinal value, tilapia's nutrients, such as omega-3 fatty acids and DHA, play an important role in brain development, which can improve memory and increase intelligence. At the same time, the unsaturated fatty acids in tilapia also help protect the cardiovascular system, lower blood pressure, and reduce thrombosis.

Tilapia is widely farmed worldwide and has become the third most internationally traded aquatic product after salmon and prawns in more than 150 countries and territories. According to the data, the total global production of tilapia in 2020 was about 6.10 million tons, making it the third-largest freshwater farmed fish in the world. China is the world's largest producer of tilapia, with the industry producing about 1.67 million tons in 2022, thanks to the improvement of farming technology and the growth of market demand, which has promoted the increase of farming area and farming volume. Tilapia is very popular in the European and American markets because of its delicious meat, fewer thorns, and high-protein content. The quality of artificial fish can meet the consumption habits of European and American consumers, with high economic value.

2.2 The Nutritional Compositions and Flavors of Fish

Green fish had a greenish color, a larger body size, and good meat quality. It contained many zinc elements and vitamins, such as potassium, phosphorus, selenium, magnesium, calcium, niacin, zinc, iron, and vitamin B2. It had the effects of clearing heat and dispelling fire, improving vision, strengthening the brain and tonifying deficiency, as well as antiaging, moisturizing the skin, supplementing trace elements, and anticancer effects. Its nutritional and economic value were both high. It was a traditional aquaculture variety in grassroots areas and a popular seafood among people, deeply loved by consumers [20]. Research has shown that the meat content was 71.08%, the protein content in muscle was 16.56%, the fat content was 14.08%, of which 70.73% was unsaturated fatty acids, the total amount of eighteen amino acids was 14.91%, the content of eight essential amino acids for the human body was 6.07%, the content of four umami amino acids was 5.62%, and there were also various trace elements [21]. Different cooking methods can affect the release of flavor compounds in green fish. For example, braised and steamed dishes were common cooking methods that could preserve the original flavor of green fish. The experimental results indicated that meat steamed for 8 minutes had a better taste and texture. The sensory evaluation score significantly increased to a maximum of 82.33 at 6 minutes. The content of umami and sweet amino acids significantly increased to a maximum of 1.68 g/kg after 6 minutes, while the inosine monophosphate (IMP) content was 1.91 g/kg, with a taste activity value (TAV) of 7.65, indicating that IMP had the greatest impact on taste. In addition, the total nitrogen content was 30.77 g/kg, which meant that a lot of protein degradation had occurred. According to the equivalent umami concentration (EUC) and its TAV, meat tastes best at 6–8 minutes. The longer the cooking time, the faster the protein degradation and the more flavor precursors [22].

Grass carp is one of the main freshwater aquaculture fish species in China, with tender and nutritious meat. Grass carp muscles contain various essential amino acids and polyunsaturated fatty acids, especially EPA and DHA, which have high levels. The moisture, crude protein, crude fat, and ash content of grass carp are 76.10%, 17.37%, 4.76%, and 1.07%, respectively. The essential amino acid content of the human body accounts for 39.32%, and the nonessential amino acid content accounts for 37.91%. Effective volatile compounds were detected from the back and belly meat of the fish, 46 and 45, respectively, with 37 of the same components. The key flavor components of grass carp back and belly meat are basically the same, mainly including nonanal, E-2-nonenal, E-2-decenal, octanal, 1-octen-3-ol, hexanal, decanal, E-2-octenal, 2,4-heptadienal, heptanal, and nonanol [23]. 1-Octen-3-ol was the main volatile flavor compound in grass carp muscle. The content of odor-producing 1-octen-3-ol, octanal, and (*E*,*E*)-2,4-decadienal in grass carp muscle cultured in low salt conditions was significantly lower than that in freshwater conditions. The hardness and chewiness of grass carp muscle cultured in low salt conditions were significantly higher than those cultured in freshwater conditions, indicating that low salt cultivation can improve the flavor of grass carp. In addition, ecological factors such as nitrite and nitrate in water bodies have a significant impact on the flavor quality of grass carp. High concentration nitrite stress can affect the growth performance of grass carp, accelerate the absorption of soil odor substances from the outside, and affect the synthesis of flavor metabolites such as amino acids and peptides in grass carp muscle tissue [24].

Silver carp is a freshwater fish rich in EPA and DHA, as well as rich in amino acids and essential trace elements for the human body. It was a highly nutritious food [25]. The content of EPA and DHA in the muscles of silver carp was considerable, accounting for 19.4–28.1% of the total fatty acids. The plankton ingested by silver carp is rich in linolenic acid, EPA, and DHA, which are metabolized and converted into EPA and DHA in the body of silver carp. Comparing the analysis results with seawater fish, the content was also comparable. Therefore, it can further prove that silver carp has good nutritional value and was a very valuable freshwater fish resource for processing and utilization [26]. The flavor of silver carp is mainly determined by the volatile compounds in its muscles. Research had shown that silver carp muscle contains various volatile flavor compounds, and 17 odor-active substances have been identified in silver carp meat, including (*E*,*Z*)-2,6-nonadienal, hexanal, 2,4-decadienal, octanal, nonanal, (*E*)-2-nonenal, and 1-octen-3-ol, which have odor characteristics such as cucumber, grass fragrance, fried oil, fat, fat, earthy smell, and earthy smell [27]. Among them, fat undergoes oxidative degradation to produce volatile odor substances such as aldehydes, alcohols, ketones, and hydrocarbons; Protein and peptide substances degrade to produce free amino acids, small peptides, and other flavor substances, or alter the composition of flavor substances.

Bighead carp is one of the freshwater fish species with the largest aquaculture yield in China, and it is also one of the most commonly used fish in daily cooking. Sixteen types of fatty acids were mainly detected in the abdominal meat of bighead carp, with a relative percentage of 69.77% of unsaturated fatty acids, including 28.80% of polyunsaturated fatty acids. The main unsaturated fatty acid component was oleic acid, with a content of 34.66%. DHA and EPA also have high levels [28]. The flavor of bighead carp was mainly determined by the volatile compounds in its muscles. Research has shown that the muscles of bighead carp contain various volatile flavor compounds. Through GC–mass spectrometry (MS) detection, seventeen odor-active substances were identified in the meat of bighead carp, including hexanal, heptanal, octanal, nonanal, 1-octen-3-ol, and key fishy substances such as *E,E*-2,4-decadienal [28]. These substances made important contributions to the overall flavor formation of bighead carp [29].

Tilapia is a fish species with strong adaptability, rapid reproduction, abundant nutrition, and high market demand. It occupies an important position in the aquaculture industry and also provides abundant food resources for consumers worldwide [30]. Visually, tilapia has a flattened body, a high back, a thick body, and an appearance similar to that of mandarin fish or sea bream. The body surface is covered with round scales, the head is small, the mouth is large, the lips are thick, the mouth has sharp and fine teeth, and there are no whiskers. The dorsal and anal fins have hard spines, and the fins are hard and sharp with black and white spots [30]. Tilapia had high nutritional value, with a crude protein content of 85.2% (dry basis) in muscle and a total hydrolyzed amino acid content of 84.14% in dry matter, of which essential amino acids account for 35.59%, or 42.30% of the total amino acid content. The amino acid value is 108, and the content of four flavor amino acids is abundant in free amino acids. Taurine was as high as 1.13%, accounting for 12.39% of the total free amino acids. The total amount of EPA and DHA accounts for 5.5% of fatty acids. Therefore, it was believed that red tilapia is a species with high nutritional and aquaculture value [31].

Salmon, as a popular cold-water migratory fish, not only enjoys a high reputation in the food industry but also represents a healthy diet. Its meat is delicious and nutritious, making it an excellent choice for both delicious dishes on the dining table and healthy food in a daily diet. Salmon meat is an important source of high-quality protein that is easily absorbed and utilized by the human body, playing a crucial role in maintaining bodily functions, promoting muscle growth and repair. Regular consumption of salmon helps maintain the balance of sodium and potassium in the body, which is crucial for maintaining normal physiological functions. Potassium helps regulate heart rhythm, while sodium participates in nerve conduction and muscle contraction. The balance between the two is crucial for maintaining physical health. In addition, salmon is also rich in various beneficial fatty acids for the human body, especially two omega-3 fatty acids, EPA and DHA. These fatty acids have a positive role in preventing cardiovascular disease, can reduce the cholesterol level in the blood, reduce the risk of atherosclerosis, and thus protect heart health. Meanwhile, they can also improve brain function, enhance memory and learn abilities, which is particularly important for the elderly. Salmon also contains abundant vitamin D, which is crucial for bone health. Vitamin D helps promote the absorption and utilization of calcium, maintaining normal bone growth and development. In addition, salmon also contains various minerals such as calcium, iron, zinc, magnesium, and phosphorus, which play an important role in maintaining normal metabolism and physiological functions of the body. For example, iron is an important component of hemoglobin and helps prevent anemia; Zinc is involved in regulating the activity of various enzymes and is crucial for the normal functioning of the immune system [32]. It was worth mentioning that the omega-3 fatty acids in salmon also have anti-inflammatory effects, which can reduce the body's inflammatory response and have a certain auxiliary effect on the prevention and treatment of heart disease, combined with heart failure, thrombosis, and other diseases. This discovery provides new ideas for the application of salmon in the medical field.

In summary, salmon is not only a delicious and tasty food, but also a highly nutritious and healthy food. Regular consumption of salmon can provide comprehensive nutritional support for the body, helping to maintain good health and vitality.

2.3 Fish Dishes

A cooking method for crispy bone fish, comprising the following steps: (1) removing the internal organs of freshwater fish and washing with clean water. (2) Wash the freshwater fish seasoning or auxiliary materials, place them in a stone or porcelain dish, and place them in the steaming tray of the high-pressure reactor. Control the pressure at 45–80 kPa and the temperature at 150–200 °C, steam for 35–60 minutes. The method of the present invention is simple, easy to operate, and can be cooked and eaten immediately, ensuring the deliciousness of the fish meat. It was suitable for cooking and consumption in canteens, restaurants, and homes, facilitating people's pursuit of high-quality food culture. Mandarin fish, commonly known as mandarin fish or seasonal flower fish, is a high-end freshwater fish that lives in the bottom layer of freshwater and feeds mainly on small fish and shrimp. There are many dishes cooked with mandarin fish, among which the most famous is undoubtedly the squirrel mandarin fish in Su Bang cuisine. Ingredients: a fish, two scallions, a piece of ginger, a tablespoon of cooking wine, three tablespoons of tomato sauce, a tablespoon of white vinegar, half a bowl of starch, a tablespoon of sugar, peas, and some dry starch. Method: first, wash the fish clean and remove the internal organs, cut off the fish head, horizontally slice the meat on both sides of the loach, cut off the fish bones, and use a diagonal cutting knife in both directions. Second, cut the scallions into chopped green onions, slice the ginger into slices, and clean the peas. Third, put chopped green onions, sliced ginger, and cooking wine into the loach, marinate for 20 minutes, then coat it with a thick layer of dry starch and set aside.

Boiled fish with pickled cabbage and chili was a famous dish in Sichuan cuisine. Generally, fish species such as green grass carp, grass carp, and fathead are used.

Wash and bone the fish, slice them, then mix the fish slices with salt and monosodium glutamate, and size them with cornstarch. Then, wash pickled Chinese cabbage, put a small amount of oil in the pot, fry ginger and garlic, then put pickled Chinese cabbage to fry, and add soup. First, boil pickled Chinese cabbage for 2 minutes, adjust the taste, then put the fish into the pot, and cook it for 2 minutes. Seafood sashimi refers to the practice of Hokkaido fishermen in Japan who, when supplying raw fish slices, often take some fish skins and stab them with bamboo sticks for easy identification due to the difficulty in distinguishing the type of fish slices after peeling. This bamboo stick and fish skin, which are pricked on the fish fillet, were originally called "sashimi." The raw materials used include fresh anchovies, salmon, tuna, Arctic scallops, bok choy, and white radish, cut evenly into thin strips. Place shredded radish under the plate and some garnish vegetables on top of the shredded radish. Cut the seafood into thin strips about 5 cm long and 1 cm thick, and spread them on shredded radish.

Sashimi, also known as Yusheng, was called fish chowder, chowder, or kuai in ancient times. It was a kind of food that is eaten with fresh fish and shellfish slices and seasonings. It originates in China and has a long history. Yusheng can be traced back to 823 BC. It was said that King Xuan of Zhou in the Western Zhou Dynasty specially ordered people to make "Caili" to reward the victorious soldiers. With the migration of Hakka people, Yusheng came to the south. Later spread to Japan, the Korean Peninsula, and other places, and it became very popular in Japan. Raw fish slices are easy to make, delicious to eat, and nutritious. From a nutritional perspective, sashimi has not undergone traditional cooking methods such as stir-frying, frying, or steaming, so the nutrients are not lost at all, making it a highly nutritious dish. However, from a hygiene perspective, if raw fish slices are not properly processed, they can become the root cause of people contracting infectious diseases. The present invention relates to a method for cooking cold water fish with one fish for two meals. After pretreatment of live fish, the fish head was made into steamed fish head soup, and the fish body was fried and grilled. The fish body was marinated with a special seasoning package before being fried and grilled on paper. After pickling, it was wrapped with food grade grilled fish paper. The food grade grilled fish paper, from inside to outside, consists of silicone oil paper, food grade polypropylene microporous filter membrane, and food grade coated cowhide Forest Stewardship Council (FSC)-coated paper. These three types of food grade paper are heated at 50–60 °C for 10–20 minutes in a baking pot coated with special seasoning and vegetable oil, and then placed in a refrigerated cabinet at a temperature of 10–20 °C for 35 minutes. After taking them out, fried and grilled, fried and grilled. The bottom of the baking pot did not need to be coated with vegetable oil; just pour hot oil with an oil temperature of 250–300 °C and fry. This invention innovatively uses cold water fish for two meals, with the fish head boiled in soup for nourishing and beauty purposes. The fish body was processed using high-temperature heating pretreatment and low-temperature refrigeration to preserve freshness and flavor. The grilled fish produced was crispy on the outside and tender on the inside, retaining the nutritional quality and taste of cold water fish, and was deeply loved by consumers.

2.4 Market Overview

The Green carp was a traditional farmed fish species in many parts of China, especially an important fishery resource in the middle and lower reaches of the Yangtze River, as well as in lakes and ponds where it is traditionally cultivated. The Green fish not only had a larger individual, fewer muscle spines, a delicious taste, tender meat, and high nutritional value, but also had a high market price and economic value for a long time. Its breeding prospects had always been promising. It was deeply favored by consumers, and the market was in short supply. The breeding benefits were considerable, and it had become one of the favorite breeding projects for farmers, with the highest price among the four famous domestic fishes [33].

The Crucian carp is an omnivorous fish species that mainly feeds on plants during its growth and development. It likes to live in groups and is widely distributed in various water bodies in China. The flesh of crucian carp was tender and had high nutritional value, rich in various minerals such as calcium, iron, and phosphorus. In addition, traditional Chinese medicine believed that crucian carp also had a good medicinal value, warming the stomach, removing dampness, and being beneficial for our bodies. It was one of the important edible fish species in China. With the gradual increase in demand for high-quality crucian carp in the market, the crucian carp produced under the traditional pond farming mode can no longer fully meet the market demand. It was necessary to guide farmers to transform their traditional farming mode and methods, strictly follow the requirements of ecological pollution-free farming technology, pay attention to strengthening water regulation in the breeding process, enhance effective prevention of diseases, and standardize the use of drugs.

In the global tilapia export trade market, China was the world's largest exporter of tilapia products. In recent years, the export volume and value of tilapia in China have shown a fluctuating upward trend, and the main products exported have gradually shifted from frozen tilapia to processed or preserved tilapia. Guangdong Province, as the main producer of tilapia in China, has always maintained its position as the largest province for tilapia exports. Tilapia products were mainly exported to countries such as the United States, Mexico, and Cote d'Ivoire. At the same time, there were also some problems in China's tilapia export trade, such as substandard product quality, overly concentrated export target markets, insufficient brand building, and low added value [34].

In 2024, the Chinese fish market showed a steady growth trend, with a national fish and aquaculture production exceeding 29 million tons. It was expected that this growth trend would continue until 2024. Freshwater aquaculture remains the main type of fish farming in China, accounting for over 90.0% of China's fish farming. In the category of marine aquaculture, large yellow croaker, oval-shaped pomfret, and other fish were the main aquaculture species. Guangdong and Fujian were the largest fish aquaculture production areas, with production accounting for 44.0% and 25.0%, respectively. Other major production areas included coastal provinces and cities such as Hainan, Guangxi, Shandong, Zhejiang, Liaoning, Jiangsu, Hebei, and Tianjin. Guangdong and Hubei were the largest freshwater aquaculture production areas for fish, with production accounting for 14.0% and 13.0%, respectively. Other major production areas included provinces and cities such as Jiangsu, Jiangxi, Hunan, Sichuan, Anhui, Guangxi, and Zhejiang, with coastal and Yangtze River provinces being the main ones. With the improvement of people's living standards and changes in consumption habits, the demand for fish products continues to increase, driving the continuous expansion of the fish market. Consumers' increasing demands for the quality, taste, nutritional value, and other aspects of fish products have prompted fish producers to continuously improve their production processes and enhance product quality (Report on in-depth research and investment direction of China's fish industry from 2024 to 2030, 2024).

In the future, the aquaculture industry will develop toward intelligence, greenness, and scale to improve production efficiency, reduce costs, and protect the environment. The application of biotechnology, innovation in aquaculture models, and progress in environmental protection technologies will all drive the development of the fish industry. For example, cultivating new varieties with specific excellent traits through gene editing technology, or developing more efficient and environmentally friendly feed and breeding equipment (Analysis of the Market Status and Development Trends of China's Aquaculture Industry in 2024: Industry Development toward Intelligence, Greenery, and Scale, 2024.)

With the continuous expansion of the fish market, more and more companies are entering this field, and market competition is becoming increasingly fierce. Enterprises need to continuously improve their competitiveness in order to establish themselves in the market. Brand has become one of the key factors for enterprises to stand out in market competition. Enterprises with well-known brands were more likely to gain the trust and recognition of consumers, thereby occupying a larger market share. The government had introduced a series of supportive policies to encourage and support the development of the fish industry. These policies, including fiscal subsidies, tax incentives, technical support, etc., provide strong guarantees for the development of the fish industry (In-depth investigation and development trend research report on the current situation of China's fish industry [2024–2031], 2024.)

In summary, the Chinese fish market will show a steady growth trend in 2024, but at the same time, it will also face challenges such as fierce market competition and constantly changing consumer demands. Enterprises need to constantly innovate and improve their competitiveness to adapt to market changes and developments.

2.5 Comprehensive Utilization of Fish By-products

During the processing of fish, a large number of by-products such as fish heads, bones, skins, scales, and internal organs are produced accordingly. According to relevant data from the Food and Agriculture Organization of the United Nations, taking herring, the fourth most produced fish in the world, as an example, the total annual catch reaches 2.16 million tons. During the production and processing, the by-products, such as fish fillets, fish heads, fish tails, and internal organs, exceed 1.3 million tons. These by-products were mainly used as animal feed in the past, and a large part of them were directly discarded, which not only caused a waste of resources but also brought environmental risks. However, in recent years, with the deepening of research, it has been found that fish by-products are rich in various nutrients such as protein, collagen, unsaturated fatty acids, and minerals, and have extremely high development and utilization value. Their comprehensive utilization and deep processing have gradually become research hotspots [35].

In the food industry, the enzymatic hydrolysis products of grouper heads can be used as natural antioxidants and raw materials for food processing. As shown in the research, the head of the pearl gentii grouper was used as the raw material and enzymically hydrolyzed by flavor protease under specific conditions (addition amount 3,000 U/g, pH 7.0, 45 °C, substrate mass concentration 300 mg/mL for 4.5 hours. The degree of hydrolysis of the crude enzymatic hydrolysis product obtained was 20.27% ± 1.28%. Among them, the protein mass fraction of the enzymatic hydrolysis product with a molecular mass < 3 ku was 13.18% ± 1.17%, and the antioxidant activity was significant. The IC50 values of the scavenging effects on 2,2-Diphenyl-1-picrylhydrazyl (DPPH) radicals, hydroxyl radicals, and superoxide anion radicals were 0.35, 4.60, and 0.46 mg/mL, respectively. It also has good solubility in the pH range of 4.0–8.0, with a nitrogen solubility index of 75.08–87.50%, thermal stability of 70.00–82.00%, and emulsifying activity of 23.52–37.45 m^2/g [36].

Similarly, the meat of pearl gentian grouper can be enzymatically hydrolyzed with flavor protease at an enzyme addition of 1050 U/g, pH 7.0, 53 °C, and a feed to water ratio of 1:35 for 5.5 hours to prepare bioactive peptides with strong antioxidant properties. The IC50 values of the enzymatic hydrolysis products and ultrafiltration components for the free radical scavenging ability of DPPH are between 0.63 and 0.88 mg/mL [37]. Hainan Huayan Collagen Technology Co., Ltd. innovatively utilizes by-products such as fish scales and fish skin, converting them into high-value-added small molecule peptides through biological enzymatic hydrolysis technology, to produce products like fish collagen peptides, which are widely used in the food industry. The products are exported to over 50 countries and regions. It is exported to over 50 countries and regions. In addition, fish by-products can be effectively and comprehensively utilized through a process involving enzymatic hydrolysis, peptide grading, and the Maillard reaction, they can be. For instance, after this treatment, the by-products of the dark striped Oriental pufferfish can significantly enhance the umami, volatile aroma, and overall acceptability of the hydrolyzed products, and reduce the bitterness and fishy smell.

In the fields of medicine and health products, fish by-products also show great potential. The collagen rich in fish skin and scales can be extracted and processed into collagen peptides, which have functions such as water retention, anti-wrinkle, and immunity enhancement, and are used in beauty, medical, and other products. Meanwhile, some active components extracted from fish by-products, such as polysaccharides and polypeptides, possess biological activities like antibacterial, antiviral, and antitumor properties, providing a new direction for new drug research and development. From an environmental protection perspective, the comprehensive utilization and deep processing of fish by-products can reduce waste emissions and lower environmental pollution. The previously discarded by-products have been fully utilized to achieve the recycling of resources, which is in line with the concept of sustainable development.

However, at present, the comprehensive utilization and deep processing of fish by-products still face many challenges. For instance, automatic classification is difficult, and by-products of different types and parts are mixed, increasing the difficulty of subsequent processing. It is prone to oxidation and spoilage. For instance, the by-products of herring have a short oxidation and spoilage period, which limits their storage and transportation. Moreover, the lack of large-scale by-product processing equipment hinders the industrial utilization process. In response to these problems, researchers have been actively exploring solutions. Professor Wu's team from Huazhong Agricultural University, leveraging their research expertise, conducted in-depth studies on the processing procedures and the characteristics of by-product raw materials of herring, cod, and salmon. They extended the oxidation and spoilage time of herring through natural plant impregnation treatment and produced high-quality surimi and protein using mechanical bone and meat separation and acid-base extraction methods. They developed a complete set of classification, stabilization, and high-value utilization technologies, and cooperated with enterprises to build production lines.

In the future, with the continuous advancement of technology, greater breakthroughs are expected in the comprehensive utilization and deep processing of fish by-products. On the one hand, new separation, extraction, and transformation technologies will continue to emerge, enhancing the utilization rate of by-products and the added value of products. On the other hand, interdisciplinary cooperation will become closer, with the integration of multiple disciplines such as food science, bioengineering, and materials science, providing the possibility for the development of more high-value-added products. At the same time, increasing investment in related technology research and development, cultivating professional talents, and improving policy support will further promote the development of this field, achieving efficient utilization of fish resources and sustainable development of the industry.

2.6 Conclusion

Fish are playing an increasingly crucial role in the global food supply system. With the continuous growth of the global population, the United Nations predicts that it will approach 10 billion by 2050. The pressure on land resources to meet human food demands is increasing day by day, and fish, as a high-quality food source, is becoming increasingly important. Fish is rich in high protein. The amino acid composition of its protein is close to human needs and has a high utilization rate. Moreover, most of its fats are unsaturated fatty acids. Omega-3 fatty acids, for instance, are particularly beneficial for heart health, reducing the risk of cardiovascular diseases and improving brain function. Meanwhile, fish contain A variety of vitamins (such as vitamin A, D, and B complex) and minerals (calcium, iodine, zinc, iron, selenium, etc.), providing essential nutrients for human growth and development as well as maintaining normal physiological functions. The per capita consumption of fish worldwide rose from 9.9 kg in the 1960s to 18.9 kg in 2010, and a preliminary estimate in 2012 further increased to 19.2 kg, fully demonstrating the growth trend of global demand for fish.

Fish have made outstanding contributions to meeting global protein demands. In 2022, the global output of the fishery and aquaculture industry reached 223.2 million tons, among which the output of aquaculture set a record at 130.9 million tons, surpassing that of the capture fishery for the first time. 89% of these aquatic products are consumed by humans, with an average per capita consumption of 20.7 kg. They provide nearly 20% of the animal protein for the world's 3.1 billion people. Especially in some coastal and island countries and regions, fish are the main source of protein. In addition, fish can not only be eaten directly but also have a wide range of industrial and medical uses in various parts. Cod liver oil is rich in vitamins A and D and is often used in the production of health products and medicines. It helps prevent night blindness, rickets, and other diseases. Fish maw is rich in collagen and is widely used in the fields of food, cosmetics, and medicine. It can be used to make healthy food, skin care products, and also as an auxiliary material for wound healing. Fish meal is a high-quality raw material for animal feed and is of great significance to the growth and development of livestock, poultry, and aquaculture animals.

In addition to its edible value, the ornamental fish industry is also booming worldwide. Goldfish, with their elegant postures and bright colors, have been cultivated for hundreds of years and have become an ornamental fish species with profound cultural heritage, deeply loved by the Chinese and East Asian cultural circles. Tropical fish such as the colorful guppy and angelfish are native to tropical and subtropical regions. Due to their brilliant colors and unique shapes, they are widely popular in the global aquarium market. The ornamental fish industry covers multiple links, such as aquaculture, trade, and the manufacturing of aquarium equipment, creating a large amount of economic value and job opportunities.

In the field of cooking, fish, due to its tender flesh, few bones, and abundant meat, can be cooked in various ways. Steaming can preserve the original flavor and nutrients of fish to the greatest extent. For instance, the traditional Chinese dish steamed sea bass has tender fish meat, and when paired with shredded scallions and ginger and steamed fish soy sauce, it exudes a refreshing fragrance. Stewing can infuse the nutrients of fish meat into the soup. For instance, the fish head and tofu soup has a rich and delicious milky white broth, which is rich in nutrients such as protein and calcium. Roasting endows fish with a unique, crispy, and fragrant texture. For instance, Brazilian grilled fish, after being marinated with special spices, is roasted over charcoal fire. The outer skin is golden and crispy, while the inside is tender and juicy. Making soup can fully release the freshness of fish. For instance, Thai Tom Yum Goong fish soup combines various flavors such as sour, spicy, sweet, and salty, giving it a unique taste. Different regions and cultures have given birth to numerous classic fish dishes. For instance, French Marseille fish soup, which is made from a variety of sea fish and shellfish and simmered with ingredients like tomatoes and Onions, is a representative of Mediterranean cuisine. Japanese sushi and sashimi have extremely high requirements for the freshness and quality of fish. The fish meat is sliced thinly and eaten with rice or directly, which reflects the ultimate pursuit of the original flavor of the ingredients.

With the improvement of living standards and the popularization of healthy eating habits, the global demand for fish as a healthy food continues to rise. In the ready-to-eat meal market, fish ready-to-eat meals are highly favored by consumers for their convenience and deliciousness. In the Chinese market, the pre-prepared pickled cabbage fish dish ingeniously combines tender fish meat with pickled cabbage. Consumers can enjoy restaurant-level delicacies simply by heating them up. The preprepared grilled fish retains the flavor of charcoal grilling, with a rich variety of side dishes, meeting the needs of family gatherings. The preprepared fish slices can be directly used for hot pot, stir-fried fish slices, etc., which is convenient and efficient. The preprepared boiled fish dish is spicy, fresh, and fragrant, recreating the classic flavor of Sichuan cuisine. Grilled eel preprepared dishes: The eel meat is tender and fresh, with a perfectly balanced sweet and salty taste, which is very popular among consumers.

In recent years, the scale of China's fish market has grown strongly, reaching over 29 million tons in 2023. This growth is attributed to multiple factors. From the demand side, consumers' understanding of healthy food has deepened, and their recognition of the nutritional value of fish has driven up consumer demand. On the supply side, aquaculture techniques have been continuously improved, such as the application of recirculating water aquaculture systems and precise feeding technologies, which have enhanced aquaculture efficiency and fish quality, while reducing aquaculture costs and risks. Meanwhile, the state policies strongly support the development of the fishery industry. Relevant laws and regulations such as the Fisheries Law have been promulgated and implemented to standardize fishery production and operation activities, ensure the sustainable utilization of fishery resources, and provide a solid legal foundation for the healthy development of the industry. In addition, the fishery subsidy policy encourages fishermen and aquaculture enterprises to expand production scale, improve aquaculture facilities, and adopt new technologies, promoting the modernization process of the fishery.

However, the future development of the fish industry will not be smooth sailing. On the one hand, market competition will become increasingly fierce. The global fishery and aquaculture industries are developing rapidly, and countries are increasing their investment to compete for market share. The rise of emerging fishing countries, relying on their cost and resource advantages, poses a challenge to traditional fishing powerhouses. On the other hand, environmental protection requirements are becoming increasingly strict. Overfishing has led to the depletion of some fish resources and the destruction of Marine ecosystems. For instance, the resources of Atlantic cod have decreased significantly due to overfishing. If the wastewater and waste residue produced by aquaculture are not properly treated, it will cause water pollution and eutrophication.

In the face of challenges, enterprises need to actively innovate. In terms of products, new fish processing technologies and product forms should be developed to increase the added value of products, such as creating functional fish products rich in specific nutrients. In the production process, more environmentally friendly and efficient breeding and processing technologies should be adopted to reduce the impact on the environment, such as promoting ecological breeding models and using microorganisms to purify the breeding water body. Meanwhile, the government will continue to strengthen supervision, strictly enforce regulations on the protection of fishery resources, and crack down on illegal fishing and behaviors that damage the fishery ecological environment. In terms of policy support, we will further increase investment in fishery science and technology innovation, promote the transformation and upgrading of the fishery industry, and help the fishery industry move steadily toward an environmentally friendly and sustainable direction, continuously contributing to global food supply.

References

1 China Fisheries Statistical Yearbook 2025 (2025).

2 Shiquan, Y. (1991). *The Great Dictionary of Chinese and Foreign Reputation*. The Great Dictionary of Chinese and Foreign Reputation.

3 Brander, K. (2005). Spawning and life history information for North Atlantic cod stocks. *ICES Cooperative Research Reports (CRR).*

4 Kjesbu, O. S., Sundby, S., Sandø, A. B. et al. (2022). Highly mixed impacts of near-future climate change on stock productivity proxies in the north East Atlantic. *Fish and Fisheries* 23 (3): 601–615.

5 Kjesbu, O.S., Alix, M., Sandø, A.B. et al. (2023). Latitudinally distinct stocks of Atlantic cod face fundamentally different biophysical challenges under on-going climate change. *Fish and Fisheries* 24: 297–320.

6 Halldórsdóttir, K. and Árnason, E. (2009). Multiple linked β and α globin genes in Atlantic cod: a PCR based strategy of genomic exploration. *Marine Genomics* 2: 169–181.

7 Ross, S.D., Behrens, J.W., Brander, K. et al. (2013). genotypes in cod (*Gadus morhua* L): their geographic distribution and physiological significance. *Comparative Biochemistry and Physiology A: Molecular & Integrative Physiology* 166 (1): 158–168.

8 Sakuma, K., Yoshikawa, A., Goto, T. et al. (2019). Delineating management units for Pacific cod (*Gadus macrocephalus*) in the Sea of Japan. *Estuarine, Coastal and Shelf Science* 229: 106401.

9 Feng, H., Huang, X., Tai, J. et al. (2020). Lipids analysis and rapid identification of cod products. *European Journal of Lipid Science and Technology* 122 (8): 1900444.

10 Mather, M.E., Parrish, D.L., Campbell, C.A. et al. (2008). Summer temperature variation and implications for juvenile Atlantic salmon. *Hydrobiologia* 603: 183–196.

11 Quinn, T.P. and Adams, D.J. (1996). Environmental changes affecting the migratory timing of American shad and sockeye salmon. *Ecology* 77 (4): 1151–1162.

12 Keefer, M.L., Peery, C.A., Bjornn, T.C. et al. (2004). Hydrosystem, dam, and reservoir passage rates of adult Chinook salmon and steelhead in the Columbia and Snake rivers. *Transactions of the American Fisheries Society* 133 (6): 1413–1439.

13 Murzina, S.A., Provotorov, D.S., Voronin, V.P. et al. (2023). Phospholipid composition of ringerlings of Atlantic salmon salmo salar during growth and development in aquaculture: the effect of different lighting and feeding regimes. *Doklady Biochemistry and Biophysics* 509: 51–55.

14 Dufour, F., Arrizabalaga, H., Irigoien, X., and Santiago, J. (2010). Climate impacts on albacore and bluefin tunas migrations phenology and spatial distribution. *Progress in Oceanography* 86: 283–290.

15 Dulvy, N.K., Fowler, S.L., Musick, J.A. et al. (2014). Extinction risk and conservation of the world's sharks and rays. *eLife* 3: e00590.

16 Ward-Paige, C.A. (2017). A global overview of shark sanctuary regulations and their impact on shark fisheries. *Marine Policy* 82: 87–97.

17 Whitledge, G.W., Chapman, D.C., Farver, J.R. et al. (2021). Identifying sources and year classes contributing to invasive grass carp in the Laurentian Great Lakes. *Journal of Great Lakes Research* 47: 14–28.

18 Gu, D.E., Mu, X.D., Xu, M. et al. (2016). Identification of wild tilapia species in the main rivers of south China using mitochondrial control region sequence and morphology. *Biochemical Systematics and Ecology* 65: 100–107.

19 Gu, D.E., Yu, F.D., Xu, M. et al. (2018). Temperature effects on the distribution of two invasive tilapia species (*Tilapia zillii* and *Oreochromis niloticus*) in the rivers of South China. *Journal of Freshwater Ecology* 33 (1): 521–534.

20 Hou, S.W. and Shi, Y.Y. (2023). Key points of green fish farming technology. *Guide to Getting Rich in Fishing Industry* 5: 29–31.

21 Cao, B.Y., Wang, L.P., and Wang, S.Y. (2004). Nutritional analysis and evaluation of ganlu blue fish muscle. *Fisheries Science* 23 (9): 34–35.

22 Wang, Y.W., Han, S., Huang, W.Z. et al. (2021). Changes of the flavor substances and protein degradation of black carp (*Mylopharyngodon piceus*) pickled products during steaming. *Journal of the Science of Food and Agriculture* 101 (10): 4033–4041.

23 Wang, X.F., Tu, X.H., Gua, G., and Dai, Z.Y. (2014). Nutritional evaluation and key flavor component analysis of grass carp. *Chinese Journal of Food Science* 14 (12): 182–189.

24 Zhang, J.M., Fu, B., Li, Y.C. et al. (2023). The effect of nitrite and nitrate treatment on growth performance, nutritional composition and flavor-associated metabolites of grass carp (*Ctenopharyngodon idella*). *Aquaculture* 562: 738–784.

25 Jin, Q.H. and Li, G.L. (1998). Research on the nutritional components of Chinese silver carp. *Food Science* 8: 41–43.

26 Zhang, D.M. and Yu, L.L. (1997). New nutritional evaluation of EPA and DHA in fatty acids of silver carp. *Aquatic Science and Technology Information* 24 (5): 195–198.

27 Liu, J.K. (2009). *The Flavor Characteristics of Silver Carp and the Influence of Thermal History on the Flavor of Silver Carp*. China: Huazhong Agricultural University.

28 Jiang, Q.X. (2015). *Research on the Hot Processing Characteristics and Mechanism of Bighead Carp Meat*. China: Jiangnan University.

29 Du, Y.Q., Fan, L.L., Ou, C.R. et al. (2022). Study on the correlation between volatile flavor components and bacterial population succession during the fermentation process of mackerel. *Science and Technology of Food Industry* 43 (20): 152–162.

30 Wang, Z.Y., Mou, X.D., and Wang, G.J. (2009). Overview and analysis of tilapia farming in China. *Aquatic Science and Technology Information* 4: 1–5.

31 Chen, Y.S. and Dai, C.J. (2003). Analysis of nutritional components in the muscle of red tilapia. *Journal of Fujian Normal University: Natural Science Edition* 19 (4): 62–66.

32 Luo, G. (2009). Overview of nutritional research on salmon. *Livestock and Feed Science* 5: 14.

33 Xuanlv, Q., Xuefei, L., and Xutao, M. (2024). Analysis of the technology, promotion, and application value of green ecological aquaculture of green fish in ponds. *Sustainability* 20: 105–107.

34 Yiyun, L. and Guohua, Z. (2024). Analysis of the current situation, problems, and countermeasures of China's Tilapia Export Trade. *National Circulation Economy* 6: 57–60.

35 Su, X.-N., Khan, M.F., Liu, D.-L. et al. (2025). Fabrication, modification, interaction mechanisms, and applications of fish gelatin: a comprehensive review. *International Journal of Biological Macromolecules* 288: 138723.

36 Fan, X., Xu, J., Lin, Z.A. et al. (2020). Preparation and functional properties of enzymatic hydrolysis products of Grouper head. *Journal of Guangdong Ocean University* 40 (4): 90–99.

37 Xu, J., Lin, Z.A., Li, Z. et al. (2020). The enzymatic preparation process of pearl gentian grouper meat peptides and the antioxidant activity of the enzymatic hydrolysis products were optimized by the response surface method. *Food Industry Technology* 41 (19): 205–211.

3

Shrimp

Bin Zhang, Shanshan Shui, and Chuanhai Tu

College of Food Science and Pharmacy, Zhejiang Ocean University, Zhoushan, China

3.1 Introduction

3.1.1 Global Significance and Economic Value of Shrimp

Shrimp, belonging to the order Decapoda within the class Malacostraca, are common crustacean species renowned for their rich nutritional profile, which includes high-quality protein, essential amino acids, minerals, unsaturated fatty acids, vitamins, astaxanthin, and antioxidants, among others. They are highly popular worldwide due to their excellent sensory qualities. The nutritional composition of shrimp varies depending on the aquaculture environment, but the basic composition of shrimp (on a wet weight basis) is water, crude protein, crude fat, and ash, making it a high-protein, low-fat, and low-calorie food source.

There are approximately 2,000 species of shrimp globally, but only about 400 species hold economic value. Most shrimp species are marine, with fewer freshwater varieties. Marine shrimp are primarily composed of species such as *Penaeus*, *Solenocera*, and *Acetes* shrimp, while river shrimp mainly include *Penaeus* and *Macrobrachium* shrimp. Among shrimp species, *Penaeus* shrimp have the highest production and economic value, which include the brown shrimp, white shrimp, and pink shrimp from the Gulf of Mexico, as well as the banana prawn, Chinese shrimp, and giant tiger prawn from the Western Pacific, with the vast majority of these species now being farmed. The *Acetes* shrimp, a group of small shrimp species, are not numerous in variety but form dense aggregations, making them one of the most important economic shrimp species in tropical shallow seas, especially in Southeast Asia, such as in Vietnam [1]. Currently, China is the world's largest base for *Penaeus* shrimp farming, with the highest production globally. According to the 2024 China Fisheries Statistical Yearbook, in 2023, the total freshwater aquaculture production in China was 34.14 million tons, of which shrimp production accounted for 4.4398 million tons, or 13.0% of the total production [2]. The total marine aquaculture production was 1.7663 million tons, with the top three contributing regions being Guangdong, Guangxi, and Shandong province. The total export value of China's aquatic products accounted for 20.1% of the national total export value of agricultural products in 2023. China's aquatic product exports represent 15.92% of the global total aquatic product exports, with the main export destinations being Japan, the United States, the European Union, and South Korea.

The shrimp industry is considered one of the most important industrial sectors due to its significant economic contribution and the high nutritional value of its products. In recent years, with the growing global awareness of healthy eating and the continuous increase in demand for high-nutritional-value seafood, the shrimp farming and processing industry has achieved remarkable development, becoming a key driving force for economic growth in many regions. The global shrimp market was worth approximately US$65 billion in 2023, demonstrating its strong economic potential and adaptability to diverse dietary cultures. According to a report by the global leading market research company Imarc Group, the global shrimp market size is projected to increase to US$105.4 billion by 2033, with a compound annual growth rate (CAGR) of 3.9% from 2025 to 2033. This production scale highlights the important position of shrimp in both domestic and international aquatic product markets, and also provides extensive employment opportunities in coastal and inland rural areas, contributing to regional economic stability and development. Moreover, the shrimp farming industry helps to achieve the United Nations Sustainable Development Goals (SDGs), including promoting food security, alleviating poverty, and promoting responsible consumption and production. Currently, shrimp have become the most valuable commodity in the global seafood trade, with a total global farmed shrimp production of 7.5 million tons, accounting for 62.7% of the total crustacean production [3]. This advantage is attributed to the continuous progress in aquaculture technology, such as genetic breeding, water quality management, and the improvement of environmentally friendly feed formulations, which have significantly increased production and farming efficiency.

Shrimp are typically sold in fresh form or processed into frozen, dried, canned, and snack foods. Fresh shrimp are highly favored because of their intact nutrition, superior flavor, and ease of digestion. In recent years, innovations in packaging technology, such as modified atmosphere packaging (MAP), and improvements in freezing technology, such as magnetic field freezing (MFF) and low-voltage electrostatic field (LVEF) assisted refrigeration, have further enhanced the quality and safety of shrimp products, meeting international standards and consumers' expectations for high-quality aquatic products. In addition, some studies have begun to experiment with high-pressure processing (HPP) technology and biodegradable active packaging materials, which improve shrimp muscle quality without the need for heating or the addition of preservatives, and achieve functional and green packaging, representing an important future direction for shrimp processing.

In addition to fresh shrimp, it is particularly noteworthy that shrimp processing greatly enhances their economic value and market importance, extends the shelf life of products, and broadens market supply in non-coastal areas. Through primary processing such as decapitation, shelling, and grading, as well as secondary processing to produce products like shrimp dumplings, shrimp paste, and shrimp sauce, not only is the shelf life extended and storage and transportation made more convenient, but the product value is also significantly increased. Moreover, different types of processing methods meet diverse consumer demands [4]. The industry of shrimp processing is a core growth point of the fisheries economy, effectively driving the development of the entire industry chain from farming and fishing to processing and export, and it has strategic significance for promoting employment and foreign exchange earnings in coastal areas. However, a large amount of by-products are generated during shrimp processing, including heads, tails, viscera, cephalothorax, and exoskeletons, which are often discarded without further treatment. This not only leads to the waste of biological resources but also causes environmental pollution, highlighting the urgency of developing sustainable waste management strategies. In fact, shrimp processing by-products are rich in micronutrients and bioactive components, such as chitin, chitosan, carotenoids, and biopeptides, which have antioxidant, anti-inflammatory, and antimicrobial properties, among others, and are highly beneficial to human health. Moreover, extracts from shrimp waste can be used for the development of sustainable and economical clean energy, such as astaxanthin extracts from shrimp shells that can serve as natural sensitizers for dye-sensitized solar cells (DSSCs) [5]. They provide important opportunities for the food and health industries to develop functional products such as fortified foods, dietary supplements, and natural preservatives, helping to achieve high-value utilization of by-products and transforming waste into high-value products, thus realizing a circular economy.

In summary, shrimp have significant global economic value due to their nutritional advantages, growing market demand, and key position in international trade. The continuous progress in farming technology, innovation in processing methods, and high-value utilization of by-products are constantly enhancing the sustainability and profitability of the shrimp industry, consolidating its important position in the global food system.

3.1.2 Chapter Objectives and Scope

The objective of this chapter is to provide a comprehensive and systematic overview of the shrimp industry, encompassing its global significance, resource distribution, nutritional profile, processing methods, and future prospects. Specifically, this chapter aims to:

1) *Highlight the Global Economic Value of Shrimp*. This section elucidates the substantial economic contributions of shrimp through its nutritional benefits, market demand, and role in international trade. It also emphasizes the importance of shrimp in achieving the United Nations SDGs, including promoting food security, alleviating poverty, and promoting responsible consumption and production.
2) *Examine the Shrimp Resources*. This section presents an analysis of global shrimp production, key producing regions, and the classification and distribution of major commercial shrimp species. It provides detailed information on the dominant species in various regions and their economic significance.
3) *Detail the Chemical Composition and Nutritional Profile*. This section explores the main components of shrimp, variations in composition among different

species, and the health benefits associated with key nutrients and bioactives. It also addresses the allergenicity of shrimp and its implications for consumer health.
4) *Discuss Shrimp Processing and Product Development*. This section describes the primary and value-added processing methods, innovations in packaging and preservation, and the significance of shrimp by-products in achieving sustainable resource utilization. It highlights the importance of processing in extending the shelf life of shrimp products and meeting diverse consumer demands.
5) *Address the Challenges and Future Directions*. This section identifies sustainability and disease issues in shrimp production, reviews technological innovations, and outlines future trends in the shrimp industry. It also emphasizes the need for sustainable waste management strategies and the potential for high-value utilization of shrimp by-products.

The scope of this chapter is to provide a detailed and authoritative account of the shrimp industry, emphasizing its global economic importance, nutritional benefits, and potential for sustainable development. It will serve as a valuable reference for researchers, industry professionals, and policymakers interested in understanding the current state and future prospects of the shrimp industry.

3.2 Overview of Shrimp Resources

3.2.1 Global Shrimp Production

Global shrimp production has undergone significant transformation over the past five decades, driven by technological innovations in aquaculture, shifts in consumer demand, and evolving trade dynamics. In 2023, global shrimp production was estimated at approximately 5.6 million metric tons (MMT), with aquaculture contributing over 60% of the total output, reflecting the sector's transition from wild-capture dependence to farm-based supply chains [6].

The Pacific white shrimp (*Litopenaeus vannamei*) remains the dominant species in global aquaculture, accounting for more than 80% of total farmed shrimp output due to its fast growth rate, adaptability to various farming systems, and resistance to common pathogens. Meanwhile, the black tiger shrimp (*Penaeus monodon*) has seen a resurgence, particularly in Asia, with production surpassing 550,000 metric tons in 2023 and projected to exceed 600,000 tons in 2024.

In terms of regional contribution, Asia continues to lead global production, although it experienced a slight contraction in 2023—its first in a decade—due to disease outbreaks, climate variability, and market saturation. China, India, Vietnam, Indonesia, and Thailand remain the top producers, collectively accounting for over 70% of global output. Latin America, led by Ecuador, has emerged as the second-largest producing region, with Ecuador alone contributing more than 1.49 MMT in 2023.

The industry faces several challenges, including price volatility, environmental concerns, and trade policy shifts. For instance, the US investigation into antidumping practices involving major exporters like India and Ecuador has introduced uncertainty into global trade flows. Additionally, climate phenomena such as El Niño have impacted production yields, particularly in Latin America.

Despite these challenges, the long-term outlook remains optimistic. Technological advancements in recirculating aquaculture systems (RAS), genetic improvement programs, and biosecurity protocols are expected to enhance productivity and sustainability. Global production is projected to grow at a CAGR of 4%–5% through 2030, with Asia and Latin America continuing to dominate supply [3].

3.2.2 Key Shrimp-Producing Regions

3.2.2.1 Asia-Pacific

Asia-Pacific remains the epicenter of global shrimp production, both in terms of volume and species diversity. China, the largest producer, has increasingly adopted intensive farming systems, including greenhouse-based RAS, to enable year-round production and mitigate environmental impacts [7]. However, estimates of China's output vary widely, ranging from 1.06 to 2.0 MMT annually, reflecting disparities in data collection and reporting methodologies.

India, the second-largest producer, has seen a contraction in *L. vannamei* production in recent years, partly due to disease outbreaks and market corrections. Nonetheless, its *P. monodon* output continues to expand, supported by improved broodstock and hatchery technologies.

Vietnam and Indonesia have similarly faced production declines in 2023 but are projected to recover in 2024, with Vietnam expected to achieve a 6% growth rate.

3.2.2.2 Latin America

Latin America has solidified its position as a major shrimp-producing region, with Ecuador at the forefront. Ecuador's industry has benefited from vertical integration, genetic selection programs, and strong export infrastructure. The country's shrimp exports have grown exponentially, particularly to China and the United States, making it the world's largest shrimp exporter by volume.

Other notable producers in the region include Brazil, Mexico, Venezuela, and Peru. While these countries have expanded production, their growth rates have slowed due to infrastructure limitations and environmental constraints. Nonetheless, the region's relatively low production costs and favorable climatic conditions continue to attract investment.

3.2.2.3 Middle East and Africa

Though smaller in scale, countries such as Saudi Arabia and Iran have emerged as significant producers, each contributing 60,000–70,000 MT of *L. vannamei* in 2023. These regions are leveraging advanced farming technologies and government support to develop export-oriented aquaculture sectors. Madagascar and Australia also contribute modest volumes of *P. monodon*, primarily targeting niche markets.

3.3 Classification and Distribution of Shrimp

3.3.1 Major Commercial Shrimp Species

Shrimp belong to the order Decapoda, infraorder Caridea, and Dendrobranchiata, with the latter including most of the commercially important species. The global shrimp market is dominated by a few key species, each with distinct biological traits, habitat preferences, and market values.

3.3.1.1 Pacific White Shrimp (*Litopenaeus vannamei*)

Native to the eastern Pacific coast, *L. vannamei* is the most widely farmed shrimp species globally [8]. Its popularity stems from its fast growth, high survival rate under intensive culture, and tolerance to a wide range of salinities and temperatures. The species is also amenable to selective breeding, leading to the development of specific pathogen-free (SPF) and high-growth lines.

3.3.1.2 Black Tiger Shrimp (*Penaeus monodon*)

Once the dominant farmed species, *P. monodon* was largely replaced by *L. vannamei* in the early 2000s due to disease susceptibility and slower growth. However, recent improvements in broodstock quality and hatchery practices have led to a resurgence, particularly in Asia [9]. The species is valued for its larger size and premium market price, especially in niche markets.

3.3.1.3 Indian White Shrimp (*Fenneropenaeus indicus*)

This species is native to the Indian Ocean and is commercially important in South Asia and the Middle East. It is moderately tolerant to environmental fluctuations and is often cultured in extensive or semi-intensive systems. Though less dominant than *L. vannamei*, it remains significant in regional markets.

3.3.1.4 Banana Shrimp (*Fenneropenaeus merguiensis*)

Found in the Indo-West Pacific, this species is commercially fished and farmed in countries like Australia and Thailand. It is valued for its sweet flavor and firm texture, but is less commonly farmed due to lower growth rates and limited breeding programs.

3.3.1.5 Kuruma Shrimp (*Marsupenaeus japonicus*)

Native to the Indo-Pacific, *Marsupenaeus japonicus* is highly valued in Japanese and Mediterranean markets. It is primarily cultured in extensive systems and is known for its large size and high market price. However, its farming is limited due to specific environmental requirements and slower growth.

3.3.2 Geographical Distribution of Shrimp Species

Shrimp species exhibit distinct biogeographical patterns, influenced by water temperature, salinity, depth, and substrate type. Their distribution spans tropical, subtropical, and temperate marine environments, with some species also found in brackish and freshwater systems.

3.3.2.1 Indo-Pacific Region

The Indo-Pacific is the most biodiverse region for shrimp, hosting the majority of commercial species. Countries such as Indonesia, the Philippines, Malaysia, and Thailand are home to a wide variety of *Penaeid* shrimp, including *P. monodon*, *Fenneropenaeus indicus*, and *Fenneropenaeus merguiensis*. The region's warm, nutrient-rich coastal waters provide ideal conditions for both wild fisheries and aquaculture.

3.3.2.2 Eastern Pacific

The eastern Pacific coast, particularly from Mexico to Peru, is the native range of *L. vannamei*. This region supports both capture fisheries and extensive aquaculture operations, especially in Ecuador, Mexico, and Honduras. The Humboldt Current enriches these waters, supporting high biological productivity.

3.3.2.3 Western Atlantic

The western Atlantic, including the Gulf of Mexico and the southeastern United States, supports significant wild shrimp fisheries. Species such as *Farfantepenaeus aztecus* (brown shrimp), *Farfantepenaeus duorarum* (pink shrimp), and *Litopenaeus setiferus* (white shrimp) are commercially exploited. These species are subject to strict management regimes to prevent overfishing and ensure sustainability [10].

3.3.2.4 Indian Ocean

The Indian Ocean is a key region for both wild capture and aquaculture. India, Sri Lanka, and Bangladesh exploit a variety of *Penaeid* shrimp, with *P. monodon* and *F. indicus* being the most prominent. The region's monsoon-driven coastal ecosystems provide seasonal breeding and nursery grounds for many species.

3.3.2.5 Red Sea and Arabian Gulf

These regions are emerging as aquaculture hubs, with countries like Saudi Arabia and Iran investing in large-scale shrimp farming operations. The warm, saline waters are suitable for *L. vannamei* culture, and the strategic location facilitates export to European and Asian markets.

3.3.2.6 Temperate and Cold-Water Regions

While most commercial shrimp are tropical, some cold-water species are economically important in temperate regions. For example, *Pandalus borealis* (northern shrimp) is fished in the North Atlantic and North Pacific and is highly valued in European and North American markets. These fisheries are typically seasonal and subject to strict quotas due to sensitivity to climate change and overexploitation.

3.4 Chemical Composition of Shrimp

The primary chemical constituents of shrimp are water and protein. Shrimp are rich in high-quality protein and contain very low levels of fat, predominantly unsaturated fatty acids, and are particularly abundant in beneficial omega-3 fatty acids

Table 3.1 Chemical composition and content of shrimp (per 100 g edible portion).

Chemical composition	Content (fresh weight)	Remarks
Moisture	75–80 g	Main components
Protein	16–22 g	High-quality complete protein, rich in essential amino acids
Fat	0.5–2.5 g	Rich in phospholipids and PUFA (EPA & DHA), it has a relatively high cholesterol content
Carbohydrate	< 1 g	It is mainly glycogen
Mineral	1–1.5 g	Rich in selenium, iodine, calcium, phosphorus, zinc, etc.
Vitamin	Trace	Rich in vitamin B12, vitamin E, and B vitamins
Astaxanthin	Variable	Powerful antioxidants exist in the shell and inside the body
Chitin	Mainly in the shell	Dietary fiber, the main component of inedible parts

such as eicosapentaenoic acid (EPA) and docosahexaenoic acid (DHA). Additionally, shrimp serve as a valuable source of essential minerals, including calcium, phosphorus, magnesium, selenium, and iodine, as well as vitamins such as vitamin A, vitamin E, and various B vitamins. The distinct umami taste of shrimp is primarily attributed to the presence of free amino acids and nucleotides, which contribute to its characteristic flavor profile. Notably, shrimp shells contain significant amounts of chitin and calcium, while astaxanthin, a key antioxidant compound, is responsible for the red coloration observed after cooking. The chemical composition and content of shrimp are shown in Table 3.1.

3.4.1 Main Components in Shrimp

3.4.1.1 Water

Water constitutes the primary component of shrimp muscle, typically representing 75%–80% of its total weight. This proportion may vary slightly based on factors such as the species, age, seasonal conditions, and habitat (marine or freshwater environments) of the shrimp. The substantial water content is primarily responsible for the tender and succulent texture characteristic of shrimp muscle.

3.4.1.2 Proteins

Protein represents the most significant solid constituent of shrimp muscle, present in remarkably high quantities (comprising over 80% of the dry weight and 16%–22% of the fresh weight). Shrimp protein is classified as a high-quality, complete protein, as it contains all the essential amino acids required by the human body. Notably, it is particularly rich in lysine, arginine, and glutamic acid. This amino acid profile contributes significantly to the distinctive flavor and palatability of shrimp muscle [11, 12].

Myofibrillar proteins (e.g., myosin and actin) are the primary proteins constituting shrimp muscle fibers and play a key role in determining the texture and gel-forming properties of shrimp muscle, such as the elasticity observed in shrimp paste and shrimp balls. Water-soluble proteins, such as sarcoplasmic proteins and certain flavor precursors, are readily dissolved in water and contribute to the development of distinctive flavors during the cooking process. The muscle also includes some enzymes (such as tyrosinase, which is the main cause of shrimp blackening) and allergens (such as protomyosin, which is the main allergen of shrimp).

3.4.1.3 Lipids

The total fat content of shrimp is relatively low, constituting approximately 1%–2.5% of its fresh weight; however, it plays a vital role in determining the flavor profile and nutritional value of the species [13].

3.4.1.3.1 Phospholipids It is the main component of shrimp fat and an important part of the cell membrane. The phospholipid fraction in shrimp is not a single entity but a complex mixture of several types. Phosphatidylcholine (PC) is often the most abundant phospholipid in shrimp, frequently constituting over 50% of the total phospholipid content. PC is a key structural component of cellular membranes. Its molecule consists of a choline head group, a phosphate group, and two fatty acid chains. In shrimp, these fatty acids are highly unsaturated. Phosphatidylethanolamine (PE) is the second most prevalent phospholipid, which plays a vital role in membrane fusion and curvature. Although the content of phosphatidylinositol (PI) is lower than that of PC and PE, PI, as the precursor substance of secondary messengers in cell signal transduction pathways, plays an extremely important role. Phosphatidylserine (PS) is another minor but essential component involved in apoptosis (programmed cell death) and other signaling events. Lysophospholipids are derivatives of phospholipids. Their levels can increase postharvest and are associated with flavor development and potential lipid degradation [14, 15].

3.4.1.3.2 Triglycerides It serves as an energy storage. A triglyceride molecule is composed of a glycerol backbone esterified to three fatty acid chains. Similar to their phospholipids, the fatty acids in shrimp triglycerides (TAGs) are characterized by a high degree of unsaturation. However, the profile can differ from that of membrane phospholipids. Saturated fatty acid palmitic acid (16:0) is often a major component, serving as a common product of de novo synthesis and a building block for other lipids. In addition, shrimp can accumulate polyunsaturated fatty acids (PUFAs) from their food, especially from phytoplankton and algae. Therefore, their triglycerides are rich in EPA (EPA-20:5n-3), DHA (DHA-22:6n-3), and arachidonic acid (ARA-20:4n-6) [16].

3.4.1.3.3 Cholesterol Shrimp has a relatively high cholesterol content (about 100–150 mg/100 g), which is a point to note. However, it is also rich in unsaturated fatty acids, so its impact on health needs to be comprehensively evaluated. Cholesterol is not uniformly distributed. The highest concentration is often found in the hepatopancreas (the digestive organ or "tomalley") and the cephalothorax (head), as these are sites of intense metabolic and hormonal activity. The muscle tissue itself contains a lower, but still significant, concentration integrated into its cellular membranes. It is important to note that while shrimp's cholesterol content is high, it is still lower than that found in organ meats from terrestrial animals (e.g., liver, kidney) and is comparable to, or slightly higher than, that in eggs [17].

3.4.1.3.4 Fatty Acids The fatty acid profile of shrimp is predominantly composed of unsaturated fatty acids, particularly PUFAs, such as EPA and DHA, which belong to the omega-3 family. These fatty acids are known to confer cardiovascular health benefits. While they play crucial physiological roles, they are also highly susceptible to lipid oxidation, which may result in spoilage and the development of off-flavors during shrimp storage [17].

3.4.1.4 Carbohydrates

The content of free carbohydrates in shrimp muscle is typically very low, generally less than 1%, and is predominantly present in the form of glycogen. Glycogen is a polysaccharide that serves as an energy storage molecule in animal organisms. Its degradation products, such as glucose, act as energy sources for biochemical reactions occurring in shrimp muscles postmortem, which can influence the spoilage process and pH changes in shrimp.

3.4.1.5 Minerals

Shrimp is an excellent source of various minerals, accounting for approximately 1%–1.5% of its fresh weight [18].

3.4.1.5.1 Calcium Calcium is more than a trace mineral in shrimp—it is a crucial macronutrient essential for their structural integrity and survival. Unlike vertebrates, which store calcium in an endoskeleton, shrimp use calcium to build and periodically renew their exoskeleton. This makes calcium metabolism a central process in their life cycle, accounting for 2%–10% or more of their dry body weight. The exoskeleton is the major calcium reservoir, comprising over 90% of the body's calcium, primarily as crystalline calcium carbonate. During pre-molt, shrimp form transient organs called gastroliths, which store calcium for rapid post-molt recalcification. The hepatopancreas also contributes to calcium metabolism and storage, though less significantly. Muscle tissue contains relatively little calcium. When consuming peeled shrimp meat, most calcium comes from surface shell fragments rather than the muscle itself.

3.4.1.5.2 Phosphorus Phosphorus is an essential macromineral that plays a fundamental and diverse role in the biology of shrimp. It is a critical component of all living cells, involved in energy transfer, structural formation, and genetic coding. In aquaculture, phosphorus is a key dietary consideration, not only for shrimp health but also for its environmental impact on culture systems. The total body phosphorus content is significant, typically ranging from 0.5% to 2.0% of the dry weight. Along with calcium, the exoskeleton is a major storage site for phosphorus in its mineralized form. The edible tail meat contains a substantial amount of phosphorus, largely incorporated into organic molecules like phospholipids in cell membranes, adenosine triphosphate (ATP), and other nucleoside phosphates. Hepatopancreas is rich in phosphorus due to its high content of nucleic acids (from rapid cell turnover), phospholipids, and energy molecules involved in digestion and metabolism. Phosphorus is present in all cells as part of nucleic acids (DNA/RNA) and membranes.

3.4.1.5.3 Selenium Selenium is an essential trace mineral for shrimp and humans. It accumulates in shrimp through their diet and aquatic environment, making them a significant dietary source. Selenium content in shrimp muscle ranges from 0.2 to 0.8 μg/g wet weight, varying with environmental and dietary selenium levels. Distribution within shrimp tissues is uneven, with higher concentrations in metabolically active organs and muscle. Selenium primarily exists in organic forms: selenomethionine, which nonspecifically incorporates into proteins for storage, and selenocysteine, which is specifically integrated into selenoproteins like glutathione peroxidase for biological activity. Inorganic forms (selenite/selenate), commonly used in aquaculture supplements, are present in relatively low concentrations.

3.4.1.5.4 Iodine Iodine (I), a vital trace element, is predominantly found in the marine environment. As marine organisms, shrimp accumulate iodine from seawater and their diet, making them a significant dietary source for humans. Iodine content in shrimp varies widely, typically ranging from 20 to over 100 μg/100 g of raw edible portion. A single serving can provide a substantial portion of or even exceed the adult Recommended Dietary Allowance (RDA) of 150 μg/day. Within shrimp, iodine is unevenly distributed: higher concentrations occur in the head (cephalothorax) and hepatopancreas due to their metabolic roles, compared to the muscle (tail meat). Iodine levels are influenced by species-specific accumulation, regional seawater concentration (e.g., higher in some North Atlantic waters), and dietary intake from prey such as algae and plankton. Processing methods also affect final content: boiling may leach iodine into cooking water, while including shells and hepatopancreas in pastes or powders can concentrate it.

Shrimp are also a rich source of zinc (Zn), crucial for immune function and enzyme activity; iron (Fe), primarily as heme-iron with high bioavailability for oxygen transport and energy production; copper (Cu), a cofactor for numerous metabolic and antioxidant enzymes; and magnesium (Mg), which is fundamental for neuromuscular function and as a catalyst in hundreds of biochemical reactions. This diverse mineral profile significantly contributes to the overall nutritional value of shrimp.

3.4.1.6 Vitamins

Shrimp is rich in various vitamins, especially water-soluble ones [19].

3.4.1.6.1 Vitamin B_{12} (Cobalamin) Vitamin B_{12}, an essential water-soluble vitamin with a complex cobalt-centered structure, is abundantly present in shrimp in highly bioavailable forms. As a rich dietary source, shrimp accumulate cobalamin through their diet—originating from bacterial synthesis—making them a potent source for human nutrition. In shrimp tissue, cobalamin occurs mainly in bioactive coenzyme forms: methylcobalamin, which functions in cytosolic methionine synthase reactions; 5′-deoxyadenosylcobalamin, active in mitochondrial methylmalonyl-CoA mutase reactions; and hydroxocobalamin, a stable precursor convertible to active forms. Cobalamin in shrimp remains relatively stable under typical cooking temperatures, though prolonged heating or boiling may lead to leaching into cooking water and potential degradation.

3.4.1.6.2 Vitamin E (α-Tocopherol) Vitamin E, the major lipid-soluble antioxidant in shrimp, is essential for protecting cellular integrity against oxidative damage—particularly critical given the high PUFA content in shrimp tissues. Unlike mammals, shrimp cannot synthesize tocopherol and must obtain it entirely from their diet, making feed composition a key factor for both shrimp health and product quality. Tocopherol content varies with dietary intake. Although fresh shrimp muscle contains moderate levels, these can be enhanced through targeted feed supplementation. α-Tocopherol is the most biologically active form; other forms (β-, γ-, δ-) are present but less potent. Tocopherol concentrates primarily in lipid-rich tissues: it is stored and redistributed via the hepatopancreas, incorporated into cellular phospholipid bilayers to protect PUFAs from oxidation, and is notably abundant in the ovaries and eggs.

3.4.1.6.3 Vitamin A Vitamin A, comprising fat-soluble retinoids and provitamin A carotenoids, is accumulated by shrimp through their marine diet, rendering them a nutritionally valuable source for humans. Content varies significantly by species, diet, and tissue, with moderate levels generally present and higher concentrations often found in the hepatopancreas. In shrimp, vitamin A is stored primarily as retinyl esters in the hepatopancreas and circulated as retinol. They also ingest carotenoids such as β-carotene, which can be converted to retinol, though with lower efficiency than in vertebrates. Shrimp acquire vitamin A mainly by consuming algae, plankton, and other marine organisms. In aquaculture, they obtain it through fortified feed.

3.4.1.6.4 Other B Vitamins Shrimp serve as a notable source of several essential water-soluble B-vitamins, including niacin (vitamin B_3), which functions as a precursor for NAD/NADP coenzymes critical in energy metabolism and redox reactions; pyridoxine (vitamin B_6), which acts as a cofactor for enzymes involved in amino acid metabolism, neurotransmitter synthesis, and hemoglobin production; and riboflavin (vitamin B_2), essential for electron transport in the mitochondrial respiratory chain via its derivatives flavin mononucleotide (FMN) and flavin adenine dinucleotide (FAD). The presence of these vitamins enhances the nutritional profile of shrimp, supporting human metabolic health through their roles in energy production, cellular function, and antioxidant protection.

3.4.1.7 Other Unique Bioactive Substances

3.4.1.7.1 Astaxanthin Astaxanthin is one of the most well-known characteristic compounds found in shrimp. As a powerful carotenoid antioxidant, it naturally occurs in microalgae and plankton, which are part of the shrimp's diet. Shrimp accumulate astaxanthin through the food chain and store it in their tissues, particularly in the shell and ovaries, imparting a red or orange-red pigmentation. The antioxidant activity of astaxanthin surpasses that of both vitamin E and β-carotene. Upon cooking, the pigment is released from protein complexes, resulting in the characteristic red coloration of cooked shrimp.

3.4.1.7.2 Chitin Chitin is the primary structural polysaccharide found in the exoskeletons of crustaceans, such as shrimp and crab shells. It is a polymer composed of *N*-acetylglucosamine units. Through chemical processing, chitin can be deacetylated to produce chitosan, which has a wide range of applications in the fields of food science, pharmaceuticals, and chemical engineering.

3.4.1.7.3 Free Amino Acids Glutamic acid, glycine, arginine, and other similar compounds are the primary contributors to the umami and sweet flavor profiles of shrimp meat.

3.4.1.7.4 Nucleotide-Related Compounds ATP and its degradation products (IMP, etc.) are also important flavor enhancers. ATP and its associated compounds form the fundamental energy currency and signaling network in all living cells, including those of shrimp. In shrimp, these molecules are central to postmortem biochemistry, directly determining freshness, flavor development, and shelf-life. Their rapid degradation is a primary indicator of quality loss in seafood products. In living shrimp, ATP is continuously synthesized and utilized to power muscular activity, neural function, and cellular maintenance. Upon death, aerobic respiration ceases, and cells transition to anaerobic pathways to generate ATP. The ATP degradation pathway involves several key intermediates and related compounds, including inosine (HxR), inorganic phosphate (P_i), and ammonia (NH_3).

3.4.2 Biochemical Composition and Variations

3.4.2.1 The Nutrient Differences Among Shrimp Species

The nutritional composition of shrimp is not uniform but exhibits significant inter-species variation influenced by biological, ecological, and environmental factors. These differences are observed across major nutrient classes. There are significant differences in the nutrient composition and content among various shrimp species, primarily manifested in the following aspects: protein content remains relatively stable (18%–22% wet weight), while lipid content and composition vary markedly—cold-water species (e.g., *P. borealis*) exhibit high hepatopancreatic lipid levels (5%–15%) rich in phospholipids and omega-3 PUFAs (with EPA+DHA accounting for over 30% of total fatty acids), whereas warm-water species (e.g., *L. vannamei*) show lower lipid levels that are heavily influenced by feed composition; cholesterol content generally ranges between 130 and 200 mg/100 g; among minerals, marine shrimp contain significantly higher iodine (20–100 μg/100 g) than freshwater species, and calcium levels in whole-shell shrimp can reach 500–1000 mg/100 g; vitamin B_{12} content is consistent (1–3 μg/100 g), and astaxanthin levels in wild marine shrimp are notably high (5–20 mg/100 g); chitin content is proportional to shell thickness. These variations are collectively influenced by species-specific traits, water temperature (higher omega-3 in cold-water species), feeding habits (carnivorous shrimp are richer in EPA/DHA), and farming practices (wild shrimp exhibit a superior fatty acid profile).

A study determined the muscle yield of eight common economic wild shrimp species (*Parapenaeopsis hardwickii*, *Exopalaemon carinicauda*, *Solenocera melantho*, *Palaemon gravieri*, *M. japonicus*, *Oratosquilla oratoria*, *Fenneropenaeus chinensis*, and *Solenocera crassicornis*) as well as the contents of conventional biochemical components, fatty acids, amino acids, nutritional minerals, and toxic heavy metal elements in their muscles. Research findings indicate that, in terms of crude biochemical components, *F. chinensis* and *S. crassicornis* possess excellent characteristics of high protein and low fat. In terms of fatty acid composition, *S. crassicornis* and *P. hardwickii* are rich in high-quality fatty acids, such as C20:4n6, C20:5n3, ΣPUFA, Σn-3PUFA, Σn-6PUFA, and ΣHUFA. Regarding amino acid composition, *F. chinensis* and *S. crassicornis* have relatively high contents of essential amino acids, nonessential amino acids, and total amino acids, and are rich in flavor amino acids. From the perspective of mineral composition, the majority of nutritional mineral elements in *E. carinicauda*, *S. crassicornis*, and *P. hardwickii* are relatively high, and the content of heavy metal elements in these eight types of shrimp is extremely low, all within the range of food safety standards [11].

3.4.2.2 Early Biochemical Changes after Death—Rigidity and Autolysis

After shrimp die, their blood circulation and respiration cease, and anaerobic glycolysis occurs in their bodies. The characteristic changes include the following aspects.

3.4.2.2.1 pH Value Drops Glycogen is broken down into lactic acid, causing the pH value in muscle tissue to rapidly decrease from the fresh 7.2–7.4 to 6.2–6.5 or even lower. The drop in pH value leads to protein denaturation and a reduction in water-holding capacity.

3.4.2.2.2 The Degradation of ATP and the Production of Umami Substances (IMP) ATP functions as a key energy carrier in biological systems. Following death, ATP undergoes a series of biochemical degradation steps: ATP → ADP → AMP → IMP (inosine monophosphate) → HxR (inosine) → Hx (hypoxanthine). IMP is a crucial flavor compound that contributes to a pronounced umami taste and enhances overall flavor. This stage represents the peak of flavor development in shrimp meat. Subsequently, IMP is further metabolized into HxR and Hx. The accumulation of hypoxanthine (Hx) is associated with the development of a bitter taste, which signifies the onset of quality deterioration. The concentration of Hx is commonly utilized as a chemical index for assessing the freshness of shrimp.

3.4.2.2.3 The Initial Degradation of Proteins The inherent cathepsins and calcium-activated enzymes in shrimp start to slowly break down myofibrillar proteins and connective tissue (collagen). This leads to the softening of muscle tissue and lays the foundation for subsequent spoilage.

3.4.2.3 Chemical Changes During Storage—Spoilage and Deterioration

Over time, microbial action and chemical oxidation have become dominant. The characteristic changes include the following aspects.

3.4.2.3.1 Lipid Oxidation This is one of the main chemical pathways for the deterioration of shrimp, especially due to their high content of unsaturated fatty acids. PUFAs react with oxygen to form hydroperoxides (primary oxidation products), which are highly unstable and further decompose into aldehydes (such as malondialdehyde), ketones, alcohols, and acids (secondary oxidation products). The low-molecular-weight aldehydes and ketones produced have a strong and unpleasant rancid taste. Oxidation products can react with proteins and amino acids, leading to color changes in shrimp meat. Oxidation causes protein cross-linking, resulting in textural changes such as hardening and drying. In addition, high temperatures, light, and the presence of metal ions (such as iron and copper) can greatly accelerate the oxidation process.

3.4.2.3.2 Protein Degradation and Spoilage It is mainly caused by the combined action of microorganisms and endogenous enzymes. Spoilage microorganisms (such as *Pseudomonas* and *Shewanella*) secrete a large amount of extracellular proteases, which break down large protein molecules into polypeptides and free amino acids. Free amino acids undergo a series of degradations under the action of microbial decarboxylases, deaminases, etc. Decarboxylation reactions produce biogenic amines, such as histamine, putrescine, and cadaverine. Excessive histamine can cause food safety problems (allergic poisoning). These amines have a strong fishy and ammonia smell. Deamination reactions produce volatile sulfur and nitrogen-containing compounds such as ammonia, hydrogen sulfide (H_2S), dimethyl sulfide (DMS), and trimethylamine (TMA). These volatile substances collectively constitute the typical fishy, ammonia, and foul smell of spoiled shrimp.

3.4.2.3.3 Melanosis The “black head” phenomenon is a very typical chemical change in shrimp, mainly caused by enzymatic browning. The heads and shells of shrimp contain rich phenolic substances (such as tyrosine) and polyphenol oxidase (PPO). After the shrimp die, the tissues are damaged, and PPO comes into contact with the substrate (tyrosine) and oxygen, catalyzing the oxidation of tyrosine to form melanin. Black or dark brown spots appear on the heads, joints, and tails of the shrimp. Melanosis itself does not directly represent spoilage (excessive microorganisms), but it greatly affects consumers’ purchasing desire and is considered a visual signal of quality deterioration. Usually, sulfites (such as sodium metabisulfite) are used to inhibit PPO activity and prevent melanosis.

3.4.2.3.4 Changes in Moisture and Texture Due to the pH value dropping close to the isoelectric point (pI) of proteins, the water-holding capacity of proteins is minimized, resulting in juice loss, weight loss, and nutrient loss. The continuous degradation of proteins causes the muscle tissue to lose its integrity, making the meat loose, soft, and mushy, and resulting in a loss of elasticity and overall texture softening.

3.4.2.4 Influence of External Factors on Chemical Changes

3.4.2.4.1 Temperature Temperature is the most critical factor influencing food preservation. Low temperatures ranging from 0 to 4 °C can markedly inhibit enzyme activity and microbial proliferation, thereby reducing the rate of chemical reactions. Freezing at temperatures below −18 °C effectively halts microbial activity and most chemical processes, with the exception of slow oxidation reactions.

3.4.2.4.2 Oxygen Exposure to oxygen will accelerate lipid oxidation and browning. Vacuum or MAP (replacing oxygen with CO_2 or N_2) can effectively delay these changes.

3.4.2.4.3 Processing Methods Heating denatures and coagulates proteins, inactivates enzymes, and kills microorganisms, but it also promotes the Maillard reaction to produce flavor and color, and may lead to the loss of certain nutrients (such as vitamins); drying reduces water activity, inhibits microbial and enzymatic activity, but may accelerate lipid oxidation; high salt or smoking components in pickling/smoking processing methods have preservative and antioxidant effects, but may introduce new chemical components.

3.5 Nutrition and Health Aspects of Shrimp

Shrimp serves as an excellent source of high-quality protein and contains a wide range of essential nutrients and bioactive compounds that play a vital role in promoting human health [20].

3.5.1 Nutrients and Bioactives in Shrimp

3.5.1.1 Essential Nutrient

3.5.1.1.1 High-Quality Protein Shrimp is recognized as an outstanding source of high-quality protein, characterized by a complete amino acid profile that includes all nine essential amino acids in proportions highly concordant with human physiological requirements. This results in superior bioavailability and a high-protein efficiency ratio (PER). On a fresh weight basis, shrimp muscle typically provides between 18 and 20 g of protein per 100 g, accompanied by minimal lipid content, positioning it as a premium high-protein, low-fat aquatic commodity. Beyond its nutritional composition, the protein in shrimp plays critical roles in tissue biosynthesis and repair, enzymatic and hormonal synthesis, and modulation of immune responses. Its high digestibility and excellent amino acid score further underscore its functional and metabolic importance in human nutrition.

3.5.1.1.2 Low Fat and Unique Fatty Acid Composition Shrimp is distinguished by its exceptionally low total lipid content, generally constituting less than 1% of its composition, while simultaneously delivering a nutritionally significant fatty acid profile rich in long-chain (LC)-PUFAs. It serves as a valuable dietary source of EPA and DHA—omega-3 fatty acids critically associated with cardiometabolic health, neurodevelopment, and modulation of inflammatory pathways. These bioactive lipids contribute to cardiovascular risk reduction, cognitive function support, and retinal integrity. Although shrimp contains dietary cholesterol (~100–200 mg/100 g), current evidence indicates that its impact on serum cholesterol is minimal in most individuals, particularly when compared to saturated and trans fatty acids. Furthermore, shrimp contain intrinsic bioactive compounds—including omega-3 PUFAs and astaxanthin—that may beneficially modulate lipid metabolism, such as through elevation of high-density lipoprotein (HDL) cholesterol. Thus, moderate consumption of shrimp is supported by nutritional science as both safe and beneficial within a balanced diet.

3.5.1.1.3 Vitamins Shrimp serves as a significant source of several essential vitamins, notably encompassing B-complex vitamins and fat-soluble vitamins. It is particularly rich in vitamin B_{12}, which functions as a cofactor in methionine synthase and methylmalonyl-CoA mutase—enzymes integral to DNA synthesis, erythropoiesis, and neurological integrity. Additionally, shrimp supplies vitamin E, a potent lipid-soluble antioxidant that protects PUFAs within cellular membranes from peroxidative damage. Vitamin A, present primarily as retinol and retinyl esters, contributes to visual cycle regulation, immune modulation, and cellular differentiation. Other B vitamins include niacin (B_3), essential in NAD/NADP coenzyme systems for energy metabolism; pantothenic acid (B_5), a component of coenzyme A; and pyridoxine (B_6), which serves as a cofactor in amino acid metabolism and neurotransmitter synthesis.

3.5.1.1.4 Minerals and Trace Elements Shrimp provides a notable spectrum of essential trace elements, with selenium being particularly abundant. As an integral component of selenoenzymes such as glutathione peroxidase, selenium plays a critical role in antioxidative defense, thyroid hormone metabolism, and immune regulation. Iodine, also richly present, is indispensable for the synthesis of thyroid hormones, which are central regulators of metabolic activity, growth, and development. Zinc acts as a cofactor for numerous enzymes involved in immune response, wound repair, DNA synthesis, and cellular proliferation. Phosphorus contributes structurally as a key constituent of hydroxyapatite in bone and teeth, and functionally in energy transfer through ATP. Copper facilitates iron metabolism, supports neurotransmitter synthesis, and aids in the formation of connective tissues. Although calcium is primarily deposited in the exoskeleton, consumption of small shell-on shrimp can serve as a dietary source of bioavailable calcium.

3.5.1.2 Bioactivator

3.5.1.2.1 Astaxanthin Astaxanthin is the most well-known and extensively researched. As a member of the carotenoid family, astaxanthin is classified as a keto-carotenoid. It occurs naturally in microalgae, and shrimp accumulate this compound by consuming the algae or plankton that have ingested them. This accumulation explains the red or pink coloration of shrimp meat, which appears blue when bound to proteins and turns red upon denaturation through heating. Astaxanthin possesses significantly stronger antioxidant activity compared to other carotenoids such as β-carotene and even vitamin E. It effectively neutralizes free radicals, suppresses lipid peroxidation, and protects cellular components—including cell membranes, mitochondria, DNA, and proteins—against oxidative damage. Furthermore, astaxanthin exhibits anti-inflammatory properties by modulating inflammatory pathways such as NF-κB, thereby inhibiting the production of pro-inflammatory cytokines and reducing chronic inflammation. Research suggests that astaxanthin may contribute to immune enhancement, protection against ultraviolet-induced skin damage, improved cardiovascular function, ocular health benefits (potentially due to its ability to cross both the blood–brain and blood–retinal barriers), and increased exercise performance [21].

3.5.1.2.2 Chitin and Chitosan Chitin and chitosan are primarily derived from shrimp shells and exoskeletons. Chitin, a nitrogen-containing polysaccharide, serves as a major structural component of shrimp shells. Chitosan is produced through the deacetylation of chitin and exhibits distinct physicochemical and biological properties. These compounds are not readily digested or absorbed by the human body; however, they possess unique biological activities. As dietary fiber agents, they can enhance intestinal motility and bind to lipids in the gastrointestinal tract, thereby reducing lipid absorption and lowering cholesterol levels. In the medical field, chitosan-based dressings are widely utilized for their antibacterial properties and ability to promote wound healing. Moreover, chitosan has been shown to activate immune cells such as macrophages, demonstrating significant immunomodulatory effects. In the medical field, chitosan-based dressings are utilized as antibacterial agents and wound healing materials, facilitating the repair and regeneration of damaged tissues. Moreover, chitosan has been shown to activate immune cells, such as macrophages, thereby exerting immunomodulatory effects.

3.5.1.2.3 Bioactive Peptides Bioactive peptides derived from shrimp proteins are released through enzymatic hydrolysis—either in vitro by specific proteases or in vivo during gastrointestinal digestion. These peptides, which are naturally encrypted within native protein sequences, exhibit multiple physiologically significant activities. Demonstrated functionalities include potent antioxidant capacity through free radical scavenging and metal chelation, antihypertensive effects via angiotensin-I-converting enzyme (ACE) inhibition, antimicrobial action against specific pathogens, and antiproliferative activity against certain cancer cell lines. Their structural characteristics, such as short sequence length and specific amino acid composition, contribute to their high bioavailability and functional efficacy, highlighting the potential of shrimp-derived peptides as valuable ingredients in nutraceuticals and functional foods.

3.5.2 Health Benefits and Allergenicity of Shrimp

3.5.2.1 Health Benefit

3.5.2.1.1 Cardiovascular Health Despite the relatively high cholesterol content in shrimp, numerous studies have shown that moderate consumption of shrimp has a net benefit for cardiovascular health. The omega-3 fatty acids (EPA and DHA) in shrimp have a significant effect in lowering triglycerides. Additionally, they increase the levels of HDL-cholesterol (C), which helps reverse cholesterol transport, taking cholesterol from the vessel walls back to the liver for metabolism and thus playing a role in preventing atherosclerosis. Chronic inflammation and oxidative stress are key factors in the development of atherosclerosis. Both astaxanthin and omega-3 fatty acids possess strong anti-inflammatory and antioxidant properties, which can inhibit the oxidation of low-density lipoprotein (LDL) (ox-LDL, a key factor in plaque formation) and reduce vascular inflammation. Shrimp is a good source of potassium, which helps relax the vessel walls by counteracting the effects of sodium, thereby contributing to lower blood pressure. Moreover, certain bioactive peptides derived from the hydrolysis of shrimp protein have been shown to have angiotensin-converting enzyme (ACE) inhibitory activity, similar to that of antihypertensive drugs. Furthermore, studies have indicated that moderate shrimp consumption within the context of an overall dietary pattern (such as the Mediterranean diet) is associated with a reduced risk of cardiovascular disease. The positive effects of beneficial components like omega-3 and astaxanthin far outweigh the potential negative impact of dietary cholesterol.

3.5.2.1.2 Brain and Nervous System Health DHA serves as a fundamental structural component of neuronal and retinal phospholipid membranes, essential for maintaining membrane fluidity, synaptic function, and signal transduction. Adequate DHA levels are correlated with a decreased risk of neurodegenerative disorders such as Alzheimer's disease, age-related cognitive decline, and macular degeneration. EPA contributes to neural health primarily through modulation of neuroinflammatory processes and support of mood regulation via eicosanoid metabolism and neurotransmitter pathways. Furthermore, astaxanthin demonstrates unique neuroprotective efficacy due to its ability to cross the blood–brain barrier. It provides potent antioxidant defense within the central nervous system, mitigating oxidative stress and inflammatory damage to neuronal tissues. The combined presence of these bioactive compounds in shrimp underscores its potential role in supporting cognitive function and neurological integrity.

3.5.2.1.3 Eye Health Shrimp contains several bioactive nutrients that contribute specifically to ocular structural integrity and visual performance. DHA is selectively concentrated in retinal cell membranes, where it supports photoreceptor function and synaptic transmission. Its presence is critical for maintaining retinal sensitivity and visual processing. Astaxanthin further enhances ocular protection due to its ability to permeate the blood–retinal barrier. It serves a dual role: filtering phototoxic blue light and attenuating photo-oxidative stress through potent free-radical neutralization. These functions help reduce the risk of retinal degeneration and cataract formation. By providing both DHA and astaxanthin, shrimp represents a functional food that supports visual acuity and may protect against light-induced and age-related ocular damage.

3.5.2.1.4 Anti-inflammatory and Immune Support Omega-3 fatty acids, including EPA and DHA, serve as precursors to anti-inflammatory mediators such as resolvins and protectins, which contribute to the alleviation of chronic inflammation. Astaxanthin further supports anti-inflammatory activity by inhibiting key inflammatory signaling pathways, such as NF-κB. Zinc and selenium are essential trace elements required for the proper functioning of the immune system. Zinc is vital for the development and activity of T cells, whereas selenium plays a critical role in the antioxidant enzyme system and provides protection against viral infections.

3.5.2.1.5 Thyroid Function and Metabolism Shrimp serves as a rich and bioavailable dietary source of iodine, a critical trace element required for thyroid hormonogenesis. Iodine is an essential constituent in the synthesis of thyroid hormones—triiodothyronine (T_3) and thyroxine (T_4)—which act as central regulators of metabolic homeostasis. These hormones modulate basal metabolic rate, macronutrient metabolism, thermogenesis, and overall energy expenditure, thereby influencing weight management and metabolic health. Adequate iodine intake is especially vital during early development, as thyroid hormones govern neuronal migration, myelination, and synaptogenesis in the fetal and infant brain. Iodine deficiency can impair neurodevelopment and lead to irreversible cognitive deficits. By providing substantial levels of bioaccessible iodine, shrimp supports not only the maintenance of normal thyroid function and metabolic regulation in adults but also contributes to neurological development during critical growth periods.

3.5.2.2 Allergenicity

3.5.2.2.1 Allergens Allergy to shrimp is mainly caused by specific proteins in muscle tissue, which are usually heat-stable and digestion-resistant, meaning that cooking and digestion cannot completely destroy their allergenicity. Tropomyosin is the main allergen in shrimp (and crustaceans), named Pen a 1 (*Penaeus aztecus*) or Lit v 1 (*L. vannamei*), etc. Tropomyosin is a highly conserved protein that also exists in other crustaceans (crabs, lobsters), mollusks (oysters, clams), and even arthropods (such as dust mites, cockroaches). These are also minor allergens, including arginine kinase, myosin light chain, and glycogen phosphorylase, but they may be major sensitizing proteins in some patients.

3.5.2.2.2 Immune Mechanism and Symptoms Shrimp allergy is a type I hypersensitivity reaction mediated by immunoglobulin E (IgE). After the first exposure to shrimp allergens, the body of an allergic individual produces specific

IgE antibodies, which bind to the FcεRI receptors on the surface of mast cells and basophils (sensitization stage). When shrimp is ingested again, the allergens cross-link with the IgE antibodies on the cell surface, causing cell degranulation and the release of histamine, leukotrienes, cytokines, and other mediators, triggering a series of symptoms. These usually occur within minutes to 2 hours after ingestion, with severity ranging from mild to life-threatening. On the skin, it can cause urticaria (hives), angioedema (especially around the eyes and lips, swelling), itching, and eczema. Additionally, it can cause nausea, vomiting, abdominal pain, and diarrhea. In the respiratory tract, it can cause nasal congestion, runny nose, laryngeal edema (resulting in hoarseness and breathing difficulties), and asthma attacks. It can also cause anaphylaxis, a potentially fatal severe reaction characterized by a sharp drop in blood pressure, breathing difficulties, and loss of consciousness, which requires immediate treatment with epinephrine.

3.5.2.2.3 Cross-Reaction People who are allergic to shrimp are usually also allergic to other crustaceans such as crab, lobster, and crayfish. Some shrimp-allergic individuals may also react to mollusks such as oysters, clams, and scallops because they contain similar tropomyosins. Cross-reactions with dust mites and cockroaches are very common. People allergic to dust mites may show false positive results in shrimp allergen tests, and vice versa. This is because the tropomyosins in their bodies have similar structures.

3.5.2.2.4 Management and Diagnosis Initial screening is conducted through skin prick tests (SPT) and serum-specific IgE tests (sIgE). Oral food challenge (OFC) is the gold standard for diagnosis, but it is risky and should be performed under medical supervision. Strict avoidance of shrimp and any food containing shrimp ingredients is currently the only effective management method. Patients need to carefully read food labels (such as seasonings, sauces, and soups which may contain shrimp powder). For those with a history of severe allergies, an epinephrine auto-injector should be carried at all times and used immediately in case of accidental ingestion.

In conclusion, shrimp is a highly nutritious food source that provides high-quality protein, beneficial fatty acids, and a rich array of vitamins and minerals. Its unique bioactive components, particularly the potent antioxidant astaxanthin, extend its value beyond basic nutrition, offering potential health-promoting and chronic disease-preventive properties. Moderate consumption of shrimp contributes positively to a balanced diet and overall health maintenance.

3.6 Processing and Products of Shrimp

Shrimp processing is a sophisticated segment of the seafood industry aimed at extending shelf-life, ensuring safety, meeting market specifications, and adding value. The transformation from raw material to final product involves a series of unit operations, each critical to determining the end quality (Table 3.2).

Table 3.2 Primary shrimp product types and processing technologies.

Product type	Core processing steps	Innovations
Whole raw chilled shrimp	• Rapid chilling with slurry ice • Modified atmosphere packaging	Real-time temperature sensors in packaging
IQF frozen shrimp	• Mechanized peeling/deveining • Sodium alginate glazing • Spiral freezing	Cryogenic LN_2 freezing for premium products
Breaded ready-to-eat	• Precooking • Pre-dusting with hydroxypropyl starch • Par-frying	Air-frying technology reduces fat by 30%
Fermented shrimp paste	• Salt mixing • Anaerobic fermentation • Sun-drying/Aging	Controlled-environment fermentation rooms
Shrimp surimi products	• Vacuum tumbling with TGase • Two-stage gelation • IQF freezing	Ultrasound-assisted protein restructuring
Antarctic krill oil	• Subcritical DME extraction • Molecular distillation • Astaxanthin microencapsulation	Enzymatic pre-treatment boosts the yield by 25%

3.6.1 Primary Product Forms and Their Processing Pathways

3.6.1.1 Whole Raw Chilled Shrimp

This form of shrimp is in its least processed state and is highly prized for its freshness. The processing typically involves several stages. During harvest and chilling, rapid cooling to 0–4° C is imperative to retard microbial growth and enzymatic activity. This is commonly achieved using slurry ice, which minimizes physical damage compared to flake ice [22]. In the packaging stage, MAP is increasingly employed. This technique uses specific gas mixtures to suppress spoilage bacteria such as *Pseudomonas* spp., extending shelf life by 25%–50% [23]. Throughout the logistics stage, maintaining an unbroken cold chain is essential. Temperature abuse during storage or transport can accelerate autolysis (due to proteolytic enzyme activity) and lead to melanosis, off-flavors, and texture degradation [24].

3.6.1.2 Frozen Shrimp Products

Frozen shrimp products dominate the international trade. Processing for frozen products is typically more extensive. Firstly, shrimp are sorted and graded by size (count per pound/kilogram) and quality. And then the shrimp are peeled and deveined manually or mechanically. Mechanized systems use water jets and rubber rollers, but may cause higher yield loss and physical damage.

Glazing is a critical step in preventing freezer burn, which can cause surface dehydration and lipid oxidation. In this process, shrimp are sprayed with a thin layer of water or sodium alginate solution, forming a protective ice shield that seals in moisture and preserves quality during frozen storage. Optimal glaze levels, typically 5%–15%, are a balance between protection and economic fraud prevention [25].

Freezing is typically carried out using individual quick freezing (IQF) in air-blast freezers, such as spiral freezers, which is the industry standard. This rapid freezing method promotes the formation of numerous small intracellular ice crystals, thereby minimizing mechanical damage to muscle tissue and helping preserve the product's texture after thawing [24]. While cryogenic freezing with liquid nitrogen offers even faster freezing rates, it is more expensive and generally reserved for high-value products.

Packaging plays a crucial role in maintaining product quality during frozen storage, with vacuum skin packaging or high-barrier laminated films commonly used to prevent oxidation and minimize moisture loss.

3.6.1.3 Heat-Processed (Ready-To-Eat) Products

Cooked shrimp are typically prepared using water-cooking or steam-cooking methods, heated to a core temperature of 70–75 °C to effectively inactivate pathogens and spoilage enzymes. Rapid cooling follows immediately to prevent overcooking and preserve texture [26]. This thermal treatment induces protein denaturation, resulting in the characteristic red color of the shell due to the transformation of astaxanthin, as well as an opaque appearance of the meat.

Breaded and battered shrimp are value-added products that involve coating, par-frying, and subsequent freezing. Key quality attributes include coating adhesion, crispness, and oil uptake during par-frying. Pre-dusting with gums and starches has been shown to enhance coating adherence and improve final product quality [27].

Canned shrimp are precooked, shelf-stable products made by cleaning, cooking, and sealing shrimp in cans or jars, followed by sterilization. This process significantly extends shelf life, making them a convenient pantry item. Canned shrimp are typically packed in water, brine, or oil, and may contain added salt or preservatives to enhance flavor and stability.

3.6.1.4 Dried Shrimp Products

Dried shrimp are a concentrated seafood product made by dehydrating fresh shrimp, renowned for their intense umami flavor and long shelf life. Commonly used as a seasoning or ingredient in various cuisines, they should be rehydrated in warm water for 10–30 minutes before use. The resulting soaking liquid is rich in flavor and can be incorporated into soups, stews, and sauces. Dried shrimp are versatile and can be added to soups, broths, braised dishes, and fried rice; enjoyed as a snack; or ground into a powder to season stir-fries, stews, and rice-based dishes.

Shrimp floss is a light, fluffy product created by cooking, shredding, and seasoning shrimp, followed by drying until it achieves an airy, fibrous texture similar to meat floss. It is often used as a savory topping or ingredient in snacks, rice dishes, and fillings.

Shrimp skin refers to very small shrimp, such as tiny marine shrimp or anchovies, that become thin and translucent when dried, resembling delicate skins. These are typically pale or reddish in color and are categorized into two main types: (1) sun-dried shrimp, which are raw shrimp dried directly under sunlight; this method requires dry, sunny conditions and produces a fresher-tasting product; and (2) boiled-dried shrimp, where shrimp are first boiled in salted water before drying, this more common method enhances preservation, flavor, and texture, resulting in a firmer, saltier product.

3.6.1.5 Mashed Shrimp Products

Shrimp surimi is a finely pulverized paste produced from shrimp meat, serving as a key base for a wide range of value-added seafood products such as shrimp balls, cakes, sausages, and analogs. The quality of shrimp surimi is primarily evaluated by its gel strength, texture, and water-holding capacity [28]. The standard processing workflow includes several critical steps: (1) Raw material preparation, frozen shrimp are thawed under controlled conditions, followed by removal of heads, shells, and digestive tracts; (2) Rinsing and dehydration, the meat is washed with cold water or dilute saline to remove impurities and soluble proteins, then dewatered via centrifugation or pressing; (3) Refining, the shrimp meat is ground into a fine paste; (4) Empty striking, an initial grinding step without additives to effectively break down muscle fibers; (5) Seasoning and mixing, salt, spices, cryoprotectants (e.g., sugar, sorbitol), and gel-enhancing agents such as tapioca starch, carrageenan, transglutaminase (TGase), or calcium chloride are added; (6) Grinding, further homogenization ensures a smooth, uniform paste; (7) Forming, the paste is shaped into desired forms such as balls, cakes, sausages, or sheets; (8) Gelation and cooking, the formed products are heated (typically with a two-stage process: setting at 40 °C followed by cooking at 90 °C) to develop a stable gel structure, then rapidly cooled to fix the texture; and finally (9) Freezing and packaging, products are quick-frozen and stored at −18 °C to extend shelf life.

Shrimp crackers are savory, textured patties made from coarsely chopped shrimp blended with binders (such as egg or starch), seasonings, and diced aromatics like garlic, ginger, or scallions. The mixture is shaped into patties and chilled for approximately 30 minutes to improve firmness before cooking. They are typically pan-fried for 3–5 minutes per side until golden and crisp, reaching an internal temperature of 63 °C (145 °F); baking or air-frying offer healthier alternatives. For storage, cooked shrimp crackers can be refrigerated for up to 2 days or frozen—either before or after cooking—for up to 2 months. Care should be taken not to overmix the batter, as this helps preserve a tender bite and desirable chunky texture.

3.6.1.6 Traditional Fermented Shrimp Products

These products are cultural staples, such as mắm tôm in Vietnam, belacan in Malaysia, Kapi in Thailand, and jeotgal in Korea, used as flavor bases or consumed directly. They offer nutritional benefits, including protein, calcium, omega-3s, vitamin B_{12}, and probiotics, but pose risks due to high sodium, histamine, allergens, and potential pathogen contamination [29].

Traditional production remains largely artisanal and labor-intensive, threatening sustainability due to overfishing of small crustaceans and declining production. Preservation efforts include United Nations Educational, Scientific and Cultural

Organization (UNESCO) recognition, geographical indications, and selective modernization, such as controlled drying environments and automated packaging, to maintain authenticity while improving safety, quality, and scalability amid rising global interest in fermented foods.

3.6.1.7 Antarctic Krill Oil

Antarctic krill oil, extracted from *Euphausia superba* in the Southern Ocean, is a highly valued marine ingredient renowned for its unique nutritional profile. It is particularly prized for its omega-3 fatty acids, EPA and DHA, which are naturally bound to phospholipids. This structural characteristic enhances their bioavailability and oxidative stability compared to triglyceride- or ethyl ester-bound omega-3s in fish oil. Additionally, krill oil contains the potent antioxidant astaxanthin, which imparts its distinctive red color and contributes to its stability and health benefits. The oil is widely used in nutraceuticals targeting cardiovascular, cognitive, and joint health, as well as in functional foods and pharmaceutical research.

Modern extraction methods aim to preserve bioactive compounds and include enzymatic or ultrasound-assisted extraction (for high phospholipid content), subcritical DME or supercritical fluid extraction (SFE) (to minimize oxidation), and enzymatic/ethanol processes with solvent recycling (for cost-efficiency) [30]. The industry is shifting toward solvent-free, enzyme-assisted, and eco-friendly technologies to maintain nutritional quality and reduce environmental impact.

Subcritical dimethyl ether (SDE) extraction is a highly efficient process that co-extracts oil and water directly from raw, undried krill, eliminating the need for pre-drying. This method removes approximately 85% of the moisture and produces high-quality krill oil with low peroxide values (1.46–2.14 mmol/kg), indicating minimal oxidation and excellent preservation of lipid integrity.

Enzymatic hydrolysis involves the use of proteases, and optionally chitinases, after cooking to break down krill proteins and facilitate oil release. This is followed by pH adjustment, centrifugation, and multi-stage ethanol extraction to recover lipids. The method offers high yields of valuable phospholipids and astaxanthin, operates at mild temperatures to preserve heat-sensitive compounds, and is considered environmentally friendly due to its reduced use of harsh chemicals and lower energy consumption.

Ultrasound-assisted ethanol extraction employs intermittent sonication to disrupt krill cell structures, significantly enhancing extraction efficiency. This method enables high recovery of phospholipids and astaxanthin in a single step, reduces processing time, and uses ethanol that can be recycled, making the process more sustainable and cost-effective.

Conventional solvent extraction typically uses solvents such as ethanol or hexane to recover krill oil. However, it faces several challenges, including potential fluoride contamination from the exoskeleton, high peroxide values indicating oxidation, the development of off-flavors, and thermal degradation of sensitive compounds like astaxanthin, which can compromise both quality and safety.

SFE utilizes supercritical CO_2, sometimes with ethanol as a cosolvent, to selectively extract lipids. Often performed in two stages, this method can separate neutral lipids from phospholipid-rich fractions, yielding oil of high purity and oxidative stability. While SFE is highly effective and solvent-free, it requires substantial capital investment and sophisticated equipment, limiting its accessibility despite its superior product quality.

3.6.2 Advanced Processing and Preservation Technologies

To mitigate nutrient loss and protein degradation associated with traditional thermal processing, the seafood industry is increasingly adopting advanced, nonthermal, and energy-efficient technologies:

Dense phase carbon dioxide (DPCD) is a nonthermal, eco-friendly method that uses pressurized CO_2 (typically below 50 MPa and 60 °C). Under optimal conditions (30 MPa, 55 °C, 60 minutes), DPCD enhances gel strength, water-holding capacity, and nutrient retention by promoting the formation of a finer, more uniform gel network in shrimp proteins, thereby improving texture and functional properties.

Ultrasound and Microwave treatments offer innovative approaches to protein modification and heating. Ultrasound induces structural changes in proteins, such as the conversion of α-helix to β-sheet conformations, improving solubility and gelation. Microwave heating enables rapid and uniform internal heating, reducing processing time. When used in combination, these technologies yield products with superior water retention and a denser, more stable microstructure.

Centrifugal dewatering systems are replacing conventional screw presses in industrial processing, offering up to 50% higher yield, improved recovery of fine protein particles, and reduced water consumption by minimizing the number of washing steps to one. These systems also feature clean-in-place (CIP) capabilities, enhancing hygiene and operational efficiency.

HPP applies pressures of 400–600 MPa to achieve non-thermal pasteurization, effectively inactivating pathogens and spoilage microorganisms. HPP can extend refrigerated shelf life and preserve sensory and nutritional quality, and it weakens the adhesion between the shell and muscle, facilitating easier shell removal [31].

Pulsed electric field (PEF) technology delivers short, high-voltage pulses to disrupt microbial and cellular membranes. It is used for microbial inactivation and to enhance the extraction of bioactive compounds from seafood by-products, improving process efficiency and valorization [24].

Ohmic heating generates heat internally by passing an electric current directly through the product, enabling rapid and uniform heating. This method significantly reduces processing time and minimizes thermal degradation, thereby preserving texture, color, and nutritional quality compared to conventional heating techniques.

Intelligent packaging integrates sensors, such as time-temperature indicators, pH sensors, and freshness monitors, to provide real-time data on product condition throughout the supply chain. This enhances traceability, ensures quality, improves safety, and helps reduce food waste by enabling more accurate shelf-life prediction [32].

3.7 Utilization of Shrimp By-products

The processing of shrimp for human consumption generates a substantial volume of waste, comprising heads, shells, and tails, which can account for 50%–60% of the whole shrimp weight. Historically considered a disposal problem, these by-products are now recognized as a rich source of valuable biomolecules. The modern paradigm shifts from "waste management" to "biomass valorization" through an integrated biorefinery approach, aiming to sequentially extract multiple high-value components, thereby enhancing economic viability and environmental sustainability [33].

3.7.1 Chitin and Chitosan Extraction from Shrimp Shells

Chitin, a β-(1→4)-linked *N*-acetyl-D-glucosamine polymer, is the main structural component of shrimp shells and the second most abundant natural biopolymer after cellulose [34]. Its deacetylated derivative, chitosan, is more commercially valuable due to its solubility in weak acids and cationic properties, which enhance its biological activities [35].

3.7.1.1 Raw Material Preparation

Shells are washed to remove residual meat and impurities, then dried (via sun, oven, or freeze-drying), and ground into powder to increase surface area for efficient chemical processing [36].

3.7.1.2 Step-by-Step Extraction Process

The conventional chemical extraction of chitin and chitosan from crustacean shells involves a sequential three-stage process, with an optional decolorization step and a final deacetylation step specifically for chitosan production.

Demineralization is the first step in chitin extraction and involves the removal of inorganic minerals, primarily calcium carbonate ($CaCO_3$), which accounts for 30%–50% of the dry weight of crustacean shells. This process is typically carried out using dilute hydrochloric acid (HCl, 0.5–1.5 mol/L) at temperatures between 25 and 50 °C for 1–24 hours [37, 38]. The reaction proceeds as follows: $CaCO_3\,(s) + 2\,HCl\,(aq) \rightarrow CaCl_2\,(aq) + H_2O\,(l) + CO_2\,(g)$. After treatment, the solid residue, consisting mainly of chitin and proteins, is thoroughly washed with distilled water to remove residual acid and soluble calcium salts until a neutral pH is reached [35].

Deproteinization is the second major step in chitin extraction, aimed at removing proteins, which constitute 20%–40% of the dry weight of the demineralized residue. This is achieved using sodium hydroxide (NaOH) at concentrations of 0.5–2 mol/L, typically at temperatures between 60 and 100 °C for 1–10 hours. The alkaline treatment hydrolyzes and solubilizes the proteins, leaving behind crude chitin. The resulting protein-free residue is then thoroughly washed with distilled water to remove residual alkali and dissolved proteins until a neutral pH is achieved. The final product is a white to off-white fibrous material, known as crude chitin [37].

Decolorization is an optional step in the chitin extraction process, used to remove pigments such as carotenoids and melanin from the crude chitin. This is typically accomplished using organic solvents like acetone or ethanol, or oxidizing agents such as hydrogen peroxide (H_2O_2) or sodium hypochlorite (NaClO). While this step helps produce a pure white chitin product, which is often preferred for commercial and pharmaceutical applications, it can increase processing costs and may lead to partial degradation of the chitin structure if not carefully controlled [37].

Deacetylation is the key step in chitosan production, where chitin is converted into chitosan through alkaline hydrolysis. This process involves treating chitin with concentrated sodium hydroxide (NaOH, 40%–50%) at elevated temperatures (90–120 °C) for 1–5 hours. The treatment removes acetyl groups from the chitin polymer, resulting in a deacetylated derivative known as chitosan. After reaction, the product is thoroughly washed to neutrality, dried, and ground into a fine powder. The degree of deacetylation (DDA) significantly influences chitosan's solubility, bioactivity, and functional properties, commonly determined using analytical methods such as Fourier-transform infrared spectroscopy (FTIR) or titration [39].

3.7.1.3 Green Extraction Technologies

Due to the environmental drawbacks of conventional chemical methods, such as high energy and water consumption, the generation of corrosive waste streams, and the use of harsh chemicals, greener, more sustainable alternatives are being developed for chitin and chitosan extraction.

Fermentation is a promising biological approach that utilizes proteolytic microorganisms, such as *Lactobacillus* spp., to simultaneously achieve demineralization and deproteinization. These microbes produce lactic acid, which dissolves inorganic minerals like calcium carbonate, while also secreting proteolytic enzymes that break down proteins in crustacean shell waste. This integrated process reduces chemical usage and environmental impact [40].

Enzymatic treatment employs specific proteolytic enzymes, such as Alcalase or pepsin, to selectively remove proteins under mild conditions. This method avoids the high temperatures and strong alkalis used in traditional deproteinization, thereby preserving the native structure and physicochemical properties of chitin. It also minimizes the formation of harmful by-products and is considered more eco-friendly [34].

Ionic liquids and deep eutectic solvents (DES) are emerging as innovative green solvents capable of efficiently dissolving and extracting chitin. These solvents operate under milder conditions compared to conventional methods, enabling high-purity chitin recovery with minimal structural degradation. Their low volatility, tunability, and potential for recyclability make them attractive for sustainable bioprocessing [41].

Together, these biotechnological and green chemistry approaches offer more sustainable pathways for valorizing shellfish waste into high-value chitin and chitosan.

3.7.1.4 Applications of Chitin and Chitosan

Chitin and its deacetylated derivative, chitosan, are biodegradable, nontoxic, and biocompatible biopolymers with diverse applications across multiple industries due to their unique functional properties.

Chitosan is widely used as a natural flocculant in wastewater treatment owing to its cationic nature and high chelating capacity. It effectively binds and removes heavy metal ions (e.g., Cu^{2+}, Pb^{2+}, Cd^{2+}) through electrostatic interactions and coordination bonding, and also aids in the removal of synthetic dyes and suspended solids, making it valuable for environmental remediation [42].

In agricultural applications, chitosan acts as a biopesticide, seed coating agent, and plant defense elicitor by stimulating the plant's natural immune system. It also serves as a matrix for controlled-release fertilizers, enhancing nutrient efficiency and reducing environmental leaching, thereby promoting sustainable crop production [43].

In the food industry, chitin and chitosan are widely used for their functional and bioactive properties. They serve as the basis for edible films and coatings that exhibit antimicrobial and antioxidant activities, helping to extend the shelf life of perishable foods such as fruits, vegetables, and seafood. Additionally, chitosan is employed as a natural preservative in various food products and acts as an effective juice clarifier by binding to suspended particles and facilitating their removal, thereby improving the clarity and stability of fruit juices [35].

In biomedicine, chitosan is utilized in wound dressings due to its hemostatic, antimicrobial, and tissue-healing properties. It is also a key component in drug delivery systems, enabling controlled release and targeted delivery of therapeutic agents. In cosmetics, chitosan functions as a moisturizing and thickening agent in creams, lotions, and hair care products, enhancing texture and skin hydration [44, 45] (Table 3.3).

Table 3.3 Shrimp by-product valorization pathways.

Target compound	Extraction method	Yield/purity	High-value applications
Chitosan	Chemical: Demineralization → Deacetylation	Shell-to-chitosan yield: 20%–25%	Antimicrobial food coatings Hemostatic wound dressings Heavy metal adsorbents
Astaxanthin	Supercritical CO_2 extraction + 10% ethanol cosolvent	92.55 μg/g from heads	Nutraceuticals (10× antioxidant potency vs. synthetic) Salmonid feed pigment
Bioactive peptides	Enzymatic hydrolysis → Ultrafiltration fractionation	Protein recovery ≥54%	Antihypertensive supplements Antioxidant functional foods
Calcium citrate malate	Neutralization of demineralization effluent with citric acid	95% of calcium recovery	High-bioavailability supplements (2.5× absorption vs. carbonate)

3.7.2 Protein and Nutraceutical Recovery from Shrimp By-products

Shrimp by-products (heads, shells, viscera) contain 30%–40% protein (dry basis) and valuable nutraceuticals, enabling high-value recovery beyond chitin. It is a rich source of bioactive peptides, amino acids, and the potent carotenoid antioxidant, astaxanthin [46].

3.7.2.1 Protein Hydrolysate and Bioactive Peptide Production

Enzymatic hydrolysis is the preferred method for producing protein hydrolysates and bioactive peptides from shrimp by-products, utilizing food-grade proteases such as papain, alcalase, and flavourzyme under controlled pH and temperature conditions. For instance, a combination of papain and flavourzyme (1:1.5 ratio) at 50 °C and pH 6.5 for 3 hours can achieve over 54% protein extraction. Following hydrolysis, enzymes are inactivated by heat treatment, and the resulting liquid hydrolysate is separated, concentrated, and dried into functional protein powders. This method operates under mild conditions, preserving the nutritional quality of proteins and generating bioactive peptides without significant degradation [30].

In contrast, alkaline or acid extraction involves treating shrimp by-products with dilute acid or alkali to solubilize proteins, followed by adjusting the pH to the pI of the target proteins to induce precipitation. Although this method is simple and cost-effective, it carries a higher risk of protein denaturation and may leave behind undesirable chemical residues, potentially affecting product safety and functionality [47].

The resulting protein hydrolysates can be further fractionated using techniques such as ultrafiltration to isolate low-molecular-weight peptides with specific bioactivities. These bioactive peptides exhibit a range of health-promoting properties, making them valuable for use in functional foods, dietary supplements, and nutraceuticals. Key bioactivities include: (1) Antioxidant activity, through scavenging free radicals and chelating pro-oxidative metal ions; (2) Antihypertensive effects, via inhibition of ACE, helping to regulate blood pressure; (3) Antimicrobial activity, showing efficacy against foodborne pathogens; and (4) Calcium-binding capacity, which enhances mineral solubility and bioavailability, supporting bone health [34, 45].

3.7.2.2 Nutraceutical Recovery

Shrimp by-products, particularly heads, shells, and viscera, are a rich source of valuable bioactive compounds with significant health benefits, collectively known as nutraceuticals. These include astaxanthin, omega-3 fatty acids, and essential minerals, which can be recovered and utilized in dietary supplements, functional foods, and pharmaceuticals.

Omega-3 fatty acids (EPA and DHA) are primarily found in the lipid fraction of shrimp heads and internal organs. These essential PUFAs play critical roles in brain development, cognitive function, cardiovascular health, and the reduction of chronic inflammation. Their high nutritional value makes them ideal ingredients for dietary supplements, fortified foods, and infant formulas [48].

Mineral recovery, especially calcium, offers another avenue for valorizing processing waste. The acidic effluent generated during the demineralization step of chitin extraction, rich in dissolved calcium from exoskeletons, can be neutralized and processed to precipitate high-bioavailability calcium salts such as calcium carbonate or calcium citrate malate. These purified minerals serve as sustainable, natural sources of calcium for use in dietary supplements and functional food products [49].

Astaxanthin, a red carotenoid pigment highly concentrated in the cephalothorax of shrimp, is one of the most potent natural antioxidants. It exhibits strong anti-inflammatory, immune-enhancing, and vision-supporting properties [50]. Two primary methods are used for its extraction: (1) Solvent extraction, employing acetone, ethanol, or ethyl acetate, often followed by saponification to convert astaxanthin esters into their free, more bioavailable form; and (2) SFE, which uses CO_2 with ethanol as a cosolvent. SFE is a low-temperature, solvent-free method that preserves astaxanthin's bioactivity and yields a high-purity product. However, due to its sensitivity to degradation by light, heat, and oxygen, leading to oxidation and isomerization, astaxanthin requires stabilization through techniques such as microencapsulation (e.g., spray-drying with maltodextrin) or storage in oil-based matrices [46]. As a high-value compound, astaxanthin is widely used as a pigment in aquaculture feed and as an active ingredient in human nutraceuticals and cosmetic formulations [51].

3.7.2.3 The Integrated Biorefinery Concept

The most efficient utilization of shrimp by-products often involves an integrated biorefinery approach, where multiple valuable components are extracted sequentially or simultaneously. The biorefinery concept emerges as a promising solution for the integrated exploitation of shrimp by the production of meat, chitin, chitosan, and astaxanthin [52].

First, recover the most labile and high-value compound, astaxanthin, via SFE or solvent extraction. Subsequently, extract proteins via enzymatic hydrolysis to produce hydrolysates and bioactive peptides. Then, the remaining solid is subjected to demineralization and deproteinization to isolate chitin. Finally, the chitin is deacetylated to produce chitosan.

This model ensures maximum resource utilization, transforms waste into a portfolio of valuable products, and exemplifies the principles of a circular bioeconomy, turning a disposal problem into a significant revenue stream for the shrimp processing industry [53].

3.8 Challenges and Future Prospects of Shrimp Industry

3.8.1 Sustainability and Disease Issues in Shrimp Production

3.8.1.1 Environmental Impacts and Barriers to Sustainability

With the continuous growth of the global population and the structural shift in dietary patterns toward high-protein diets, the negative impacts of high-density farming models on the ecological environment have become increasingly prominent, severely restricting the sustainable development of the industry. Intensive shrimp farming poses several challenges, including water pollution, habitat destruction, and a decline in biodiversity.

High-density shrimp farming relies on compound feeds [54]. Residual feed and excreta from farmed organisms lead to eutrophication of the water body. In addition, to prevent and treat shrimp diseases, excessive amounts of various chemical agents are used. These residual chemicals exacerbate pollution in water bodies and sediments. The total nitrogen and phosphorus in the discharge water from shrimp ponds often exceed the carrying capacity of natural water bodies. These pollutants spread to surrounding areas through the hydrological cycle, posing a threat to larger-scale ecosystems. High-density shrimp farming exerts a competitive exclusion effect on other local aquatic organisms. On the one hand, shrimp consume food resources, leaving other local organisms that rely on these resources facing food shortages. On the other hand, the activities of shrimp alter the environment of the water. These changes are detrimental to the survival and reproduction of local organisms. Long-term competitive exclusion leads to a decrease in biodiversity [55].

Shrimp farming is often conducted by reclaiming mangroves and coastal wetlands for ponds. This reduces mangrove areas and cuts the number of fish and shellfish relying on them. It also alters hydrological patterns and wetland ecological processes, threatening coastal ecosystem stability. Moreover, exotic shrimp species compete with local species for resources and ecological niches, and even spread pathogens. Exotic species may lack the restraint of natural predators, leading to their rapid proliferation. Current high-density shrimp farming is a "high-input, high-output, high-pollution" model. Exceeding its self-regulation and recovery abilities damages the ecosystem, as the model prioritizes short-term profits over long-term ecological capacity.

Over-high density and excessive feeding boost environmental load beyond water self-purification. Poor water quality causes shrimp stress, weak immunity, and disease. This leads to more chemical use, forming a vicious cycle that collapses the farming system and threatens farming sustainability. The "pollute first, then treat" model has long been common. Economic benefits over ecological protection, with remediation only starting when severe environmental issues arise. This end-of-pipe approach is costly, inefficient, and irreversible damage is hard to restore. A better way is the precautionary principle, recirculating systems to cut pollution at the source, and easing environmental pressure.

Regions differ greatly in ecological environments and socio-economics, leading current shrimp farming sustainability standards to poor adaptability. They often base their policies on specific regions: too lenient in fragile areas and too strict in high-capacity areas. Regional gaps in policy implementation, community participation, and farmers' technical capabilities also hinder unified standards. Thus, formulating flexible, region-specific standards based on local conditions is a key challenge for the industry.

The conflict between current shrimp farming models and ecological capacity limits the industry's sustainability. The need to take ecological capacity as core, build localized assessment models, abandon "pollute first, then treat," and set flexible regional sustainability standards. With policy supervision and multi-party participation, it can achieve green, sustainable development to support food security and marine protection.

3.8.1.2 Epidemiology and Management Challenges of Major Shrimp Diseases

Disease-related issues significantly hinder the growth of the global shrimp farming industry. Pathogens like white spot syndrome virus (WSSV), acute hepatopancreatic necrosis disease (AHPND), and infectious hematopoietic and subcutaneous necrosis virus (IHHNV) threaten aquaculture sustainability and efficiency [56]. Challenges in disease control, such as excessive antibiotic use, vaccine development difficulties, and limited management capacity of smaller farms, complicate prevention efforts [57].

WSSV, a double-stranded circular DNA virus characterized by a rod-shaped or elliptical structure, is accompanied by a triple-layered envelope. Outbreaks of white spot disease (WSD), the clinical condition induced by WSSV, are closely linked to various environmental factors. Notably, such outbreaks predominantly occur in water temperatures of 16, 25, 27, 28, and 30 °C, identified as critical triggering conditions. In addition to temperature, other factors significantly increase the risk of WSD outbreaks, including extreme salinity levels and fluctuations in pH within the range of 9.5–9.7. The disease is transmitted via two primary pathways: horizontally, through water-borne transmission, for instance, when shrimp ingest contaminated matter or come into contact with virus-laden water during feeding, and vertically, through infected oocytes. First detected in farmed shrimp in the Taiwan region of China and Japan in 1992, the virus has since spread globally, eventually becoming an endemic pathogen in nearly all countries engaged in commercial shrimp farming.

AHPND is a severe form of vibriosis caused by *Vibrio parahaemolyticus*, first identified in southern China in 2009 and quickly spread to major shrimp-producing areas in Latin America and Asia through global trade. The bacterium's core genome plasticity allows it to adapt to varying environments, promoting AHPND persistence. The potential transmission involves three steps: pathogenic strains migrate through shrimp trade and ocean currents, transfer pathogenic genes to local bacteria via plasmids, and eventually, local strains dominate the bacterial community.

IHHNV is the smallest known shrimp pathogen, measuring 20–22 nm in diameter. It is a non-enveloped virus with an icosahedral capsid and a 4.1 kb linear single-stranded DNA genome that includes three major open reading frames. Upon entering host cells, IHHNV disrupts cellular functions and gene expression by interacting with host proteins. First identified in 1981 in giant tiger prawns in Hawaii, IHHNV has since spread globally, infecting prawn populations on all continents. Transmission can occur horizontally through the ingestion of infected shrimp or vertically via ovarian tissue and oocytes.

Although antibiotics provide short-term benefits like reducing mortality and enhancing production efficiency, their use raises concerns about antibiotic-resistant bacteria, which pose a global health threat. Additionally, the high genetic variability of viral strains complicates research on disease mechanisms and shrimp immune responses, making vaccine development challenging. In intensive shrimp farming, coinfection with multiple viral pathogens can lead to similar clinical symptoms, complicating diagnosis and rendering single-virus vaccines ineffective. Small and medium-sized aquaculture operators often lack resources for disease prevention, relying on traditional methods without proper water quality control or monitoring. Many practitioners are local farmers with limited knowledge and skills, and financial constraints hinder disease management efforts. The demanding working conditions in the sector also deter skilled professionals, further complicating disease control.

3.8.1.3 Integrated Disease Management in Sustainable Shrimp Farming

The rapid development of the shrimp farming industry has significantly increased its scale, but with it comes a rise in disease outbreaks, leading to substantial economic losses and diminished market confidence. While traditional disease prevention methods primarily rely on chemical drugs and antibiotics, these have long-term drawbacks, including drug residues, environmental pollution, and the development of drug resistance. Such issues pose considerable challenges to the sustainable development of the shrimp farming industry.

Sustainable shrimp farming emphasizes the balance between ensuring healthy shrimp growth, enhancing production quality, and protecting the ecological environment. Disease prevention, being a critical component of this model, must address both the immediate health of shrimp and long-term environmental impacts. One promising solution is the integrated aquaculture model, which combines shrimp farming with the cultivation of other organisms. For example, in rice-shrimp coculture, the incidence of *Vibrio* bacteria is reduced by over 68%, and WSSV detection rates are significantly lower compared to monoculture ponds. This model fosters a mutually beneficial ecosystem, where rice and shrimp benefit from symbiotic relationships, leading to healthier shrimp and reduced reliance on harmful chemicals.

RAS represent another eco-friendly approach, utilizing modern technologies to purify and recirculate water, improving water quality, and adjusting it to the specific needs of shrimp. This helps stabilize the farming environment, promotes shrimp growth, and enhances their immunity to diseases.

Healthy shrimp seedlings are essential to prevent large-scale disease outbreaks, as infected fry can lead to high mortality rates, such as those caused by WSSV (over 90% mortality). Traditional quarantine methods, such as visual inspections and microscopy, are often insufficient, especially for detecting early-stage or latent infections. A more robust quarantine system that includes thorough laboratory testing and real-time monitoring of water quality is needed to ensure healthier seedlings.

Biological control strategies, such as the use of probiotics and plant-based drugs, are gaining attention for their potential to improve water quality and boost shrimp immunity. However, high costs and technical barriers, such as farmers' lack of skills and insufficient research and development, hinder their widespread use. Despite these challenges, these sustainable inputs show promise in disease control and could significantly reduce reliance on chemical treatments in the future [58].

In conclusion, the success of sustainable shrimp farming hinges on the collaborative prevention and control of diseases through integrated aquaculture, recirculating systems, and biological control. By focusing on ecological models and preventative strategies, the shrimp farming industry can reduce disease transmission risks and promote long-term sustainability. Overcoming current barriers to the adoption of sustainable practices will be crucial for the industry's future.

3.8.2 Technological Innovations and Future Trends in Shrimp Industry

3.8.2.1 Application and Breakthroughs of Intelligent Aquaculture Technology

With the continuous growth of global demand for shrimp products, traditional aquaculture models face numerous challenges. Internet of Things (IoT)-based aquaculture has become a field of increasing interest in the fishing industry. The application of intelligent technologies such as the IoT, big data analytics, and artificial intelligence (AI) has brought revolutionary changes to the shrimp industry. Through real-time monitoring, data-driven decision-making, and intelligent regulation, these technologies have significantly improved aquaculture efficiency, reduced resource waste, and minimized negative environmental impacts [59].

Modern IoT systems deploy multiparameter water quality sensors to wirelessly transmit data to a cloud platform. When any parameter exceeds the preset threshold, the system automatically issues an alarm and can link to and control equipment for intervention. IoT-driven precision feeding systems typically integrate underwater cameras, sonar, and other devices to observe the feeding behavior and aggregation distribution of shrimp, and even indirectly determine the amount of residual feed through sensors. Machine learning algorithms can be used to establish prediction models that take water temperature, dissolved oxygen, feeding quantity, etc., as input features and growth rate and body weight as output targets. Big data analytics is applied not only in the production end but also extended to the market end. By collecting and analyzing global shrimp market prices, a market demand prediction model is built to provide market demand early warnings for aquaculture enterprises and farmers.

AI technology, particularly image recognition and deep learning algorithms, provides a new method for early shrimp disease diagnosis by analyzing shrimp images captured by drones or underwater cameras. Büyükarıkan [60] developed a shrimp disease diagnosis system, with a diagnostic accuracy of 91%. Advanced AI algorithms, such as reinforcement learning, enable adaptive regulation of multiparameter coordination by allowing AI agents to continuously experiment and learn optimal strategies in simulated or real aquaculture environments. But the cost structure imposes significant economic pressure on small and medium-sized aquaculture farms. The substantial investment required has deterred many small and medium-sized farms. Small and medium-sized aquaculture farms face multiple applicability challenges in applying intelligent technologies.

The large amount of data generated by intelligent aquaculture faces multiple security threats; once leaked, it may lead to the loss of competitive advantages. Environmental data may expose sensitive information such as the farm's geographical location and ecological environment. Transaction data involves commercial secrets such as prices and customer information, which require strict protection. The improvement of benefits from technology integration depends on the design of a reasonable system architecture. This architecture not only ensures the real-time performance of the system but also makes full use of cloud computing resources.

Through the in-depth integration of IoT, big data, and AI, intelligent aquaculture technology is profoundly transforming the production methods of the traditional shrimp industry. In the future, it will be necessary to continue strengthening technological innovation, improving the industrial ecosystem, and promoting the transformation of the shrimp industry toward high-quality and sustainable development.

3.8.2.2 Innovation and Commercialization in Genetic Improvement Technology

Shrimp farming is a critical component of global aquaculture, underpinning food security and economic development. However, its sustainable expansion is constrained by persistent disease outbreaks (such as WSSV and AHPND), slow growth rates, and environmental sensitivity. Conventional breeding approaches, limited by long generation intervals and low heritability of complex traits, are insufficient to address these challenges, necessitating a shift toward molecular precision breeding.

Current innovations focus on two principal technologies: marker-assisted selection (MAS) and gene editing. MAS employs DNA markers linked to quantitative trait loci (QTLs) to enable early selection for traits such as disease resistance and growth rate. It has been successfully used to develop AHPND-resistant lines of *L. vannamei* and to enhance growth performance. Nevertheless, MAS is less effective for highly polygenic traits, prompting the adoption of genomic selection (GS), which improves prediction accuracy using genome-wide markers [61].

Conversely, gene editing—particularly CRISPR-Cas9—offers revolutionary potential by directly creating novel heritable variations. Studies have demonstrated its ability to reduce WSSV susceptibility by editing viral entry genes and to improve thermal tolerance via modification of heat shock protein genes. Despite promises, commercialization faces significant hurdles, including off-target effects, ethical concerns, and the absence of international regulatory frameworks.

Beyond technical barriers, systemic issues impede progress. Inadequate intellectual property protection discourages innovation due to easy replication of shrimp and weak enforcement mechanisms. Extending plant variety laws has proven insufficient; establishing a DNA fingerprint database is urgently needed to support legal verification. Furthermore, robust industry-academia-research collaboration is essential to bridge research and application. Integrated research and experimental development (R&D) platforms and clear IP-sharing agreements can facilitate the advancement of breeding programs. Finally, improving regional adaptability through genotype-by-environment evaluation and MAS-based selection is critical for deploying strains suited to specific ecological conditions.

In conclusion, while MAS provides immediate benefits and gene editing holds long-term promise, their success depends on synergistic integration within supportive frameworks. Strengthened IP regimes, effective partnerships, and rigorous regional testing are indispensable for achieving sustainable genetic improvement in penaeid shrimp.

3.8.2.3 Circular Economy: Future Trends and Pathways for the Shrimp Industry

The global shrimp industry has experienced rapid growth, with farmed shrimp production exceeding 5 million tons in 2024, accounting for over 12% of the global aquatic product consumption market [6]. However, traditional aquaculture practices face significant challenges, including excessive water consumption, environmental pollution, and low resource utilization efficiency. In Vietnam, for example, water loss in shrimp farms can reach up to 1,000 cubic meters, which represents over 20% of total water consumption. Furthermore, shrimp shells, which account for 35%–40% of the total weight of shrimp, remain underutilized, resulting in substantial resource waste.

The concept of a circular economy offers a sustainable solution to these issues by establishing a closed-loop system of "aquaculture-processing-waste utilization" [62]. This approach aims to minimize environmental impact while improving resource recycling and overall efficiency. For instance, the adoption of RAS in Vietnam has successfully reduced water consumption and water costs by 20%. Biological purification systems, such as mussel/microalgae/bacteria ecosystems, play an important role by removing pollutants and enhancing water quality, thus improving both environmental and aquaculture performance.

In addition, shrimp shell processing technologies, including bioenzymatic hydrolysis and ultrasonic assistance, have made significant strides in extracting high-value products such as chitin and chitosan. These innovations transform shrimp shells from waste into valuable raw materials, which can be used in industries like medicine, food preservation, and agriculture. However, large-scale adoption of these technologies still faces barriers, particularly the high initial investment costs and challenges faced by small-scale processors. Potential solutions to these challenges include providing subsidies, developing shared processing facilities, and establishing regional processing networks that lower entry barriers for smaller enterprises.

The circular economy model also benefits from green certification systems like the Aquaculture Stewardship Council (ASC) and Best Aquaculture Practices (BAP), which enhance market competitiveness and offer price premiums for certified products. Regional industrial clusters in countries like Ecuador and China demonstrate how resource sharing and economies of scale can improve both environmental and economic outcomes, as these clusters integrate aquaculture, processing, and waste treatment to create a more efficient and sustainable system. The future of the shrimp industry lies in the combination of technological innovation, market-driven demand, and supportive policies. Overcoming the economic and technical challenges associated with circular economy practices will be crucial in making these systems mainstream. By addressing these barriers, the shrimp industry can contribute significantly to global food security, environmental sustainability, and rural development, creating a more resilient and profitable industry for the future.

References

1 Cao, T.T., Le, H.A., and Eppe, G. (2025). Nutrient dynamics, environmental impacts, and feed efficiency in intensive whiteleg shrimp (*Litopenaeus vannamei*) farming on sandy soils in Ninh Thuan. *Vietnam. Aquaculture Reports* 44: 103050.

2 Fisheries Bureau of the Ministry of Agriculture and Rural Affairs (2024). *China Fisheries Statistical Yearbook 2024. China Agricultural Press.*

3 Food and Agriculture Organization of the United Nations. (2024). *The State of World Fisheries and Aquaculture 2024 Blue Transformation in Action*. Food and Agriculture Organization of the United Nations (FAO).

4 Hu, M.Y., Zhao, L., Sun, H.H. et al. (2024). Recent advances on shrimp paste: key flavor components and biochemical formation pathways, biogenic amine formation, microbial functions, and innovative process strategies. *Trends in Food Science & Technology* 152: 104694.

5 Milano, F., Mancarella, F., Giotta et al. (2025). Fast and green process to obtain astaxanthin-rich extracts from shrimp shell waste for dye sensitized solar cells. *Process Biochemistry* 156: 308–316.

6 Mandal, A. and Singh, P. (2025). Global scenario of shrimp industry: present status and future prospects. In: *Shrimp Culture Technology: Farming* (ed. P. Singh, A. Singh, A. Tyagi, and S. Benjakul), 1–23. Health Management and Quality Assurance.

7 NSouvi, K., Sun, C., Che, B., and Vodounon, A. (2024). Shrimp industry in China: overview of the trends in the production, imports and exports during the last two decades, challenges, and outlook. *Frontiers in Sustainable Food Systems* 7: 1287034.

8 Han, C., Li, L., Zhang, G. et al. (2021). Assessment of elemental profiling combined with chemometrics for authenticating the geographical origins of Pacific white shrimp (*Litopenaeus vannamei*). *Food Control* 130: 108231.

9 Ding, Y., Jiang, S., Jiang, S. et al. (2025). Metabolic response of black tiger shrimp (*Penaeus monodon*) to acute ammonia nitrogen stress. *Biology* 14 (5): 501.

10 Asche, F., Oglend, A., and Smith, M. (2022). Global markets and the commons: the role of imports in the U.S. wild-caught shrimp market. *Environmental Research Letters* 17 (4): 045023.

11 Wan, L., Peng, Y., Yu, H. et al. (2022). Comparing the muscle nutritional quality of eight common wild-caught economic shrimp species from the East China Sea. *Journal of Aquatic Food Product Technology* 31 (6): 549–564.

12 Yang, Z., Jiang, Q., Zhang, W. et al. (2024). Comparison of morphometric parameters, nutritional composition, and textural properties of seven *Crustaceans* species. *Fishes* 9 (4): 141.

13 Wang, L., Zang, M., Cheng, X. et al. (2024). Evaluation of changes in the lipid profiles of dried shrimps (*Penaeus vannamei*) during accelerated storage based on chemical and lipidomics analysis. *LWT-Food Science and Technology* 191: 115564.

14 Pascual-Silva, C., Alemán, A., Montero, M.P., and Gómez-Guillén, M.C. (2022). Extraction and characterization of Argentine red shrimp (*Pleoticus muelleri*) phospholipids as raw material for liposome production. *Food Chemistry* 374: 131766.

15 Ju, Z.Y., Forster, I., Dominy, W., and Lawrence, A. (2011). Classification and quantification of phospholipids and dietary effects on lipid composition in Pacific white shrimp *Litopenaeus vannamei*. *North American Journal of Aquaculture* 73 (2): 221–229.

16 Sun, H., Song, Y., Zhang, H. et al. (2020). Characterization of lipid composition in the muscle tissue of four shrimp species commonly consumed in China by UPLC-Triple TOF-MS/MS. *LWT-Food Science and Technology* 128: 109469.

17 Pires, D.R., De Morais, A.C.N., Coelho, C.C.S. et al. (2018). Nutritional composition, fatty acids and cholesterol levels in Atlantic white shrimp (*Litopenaeus schimitti*). *International Food Research Journal* 25 (1): 151–157.

18 Truong, H.H., Hines, B.M., Emerenciano, M.G. et al. (2023). Mineral nutrition in penaeid shrimp. *Reviews in Aquaculture* 15 (4): 1355–1373.

19 Nesara, K.M. and Paturi, A.P. (2018). Nutritional requirement of fresh water prawn and shrimps: a review. *Journal of Entomology and Zoology Studies* 6 (4): 1526–1532.

20 Chakraborty, K. (2025). *Recent Advances in Shrimp Nutrition and the Nutritional Significance of Shrimp to Human Health*, 313–339. Shrimp Culture Technology: Farming, Health Management and Quality Assurance.

21 Ren, F., Rao, C., Xiang, Q. et al. (2025). Production methods, biological activity and potential application prospects of astaxanthin. *Foods* 14 (12): 2103.

22 Li, Y., Lei, Y., Tan, Y. et al. (2022). Efficacy of freeze-chilled storage combined with tea polyphenol for controlling melanosis, quality deterioration, and spoilage bacterial growth of Pacific white shrimp (*Litopenaeus vannamei*). *Food Chemistry* 370: 130924.

23 Bono, G., Okpala, C.O.R., Alberio, G.R.A. et al. (2016). Toward shrimp consumption without chemicals: combined effects of freezing and modified atmosphere packaging (MAP) on some quality characteristics of Giant Red Shrimp (*Aristaeomorpha foliacea*) during storage. *Food Chemistry* 197: 581–588.

24 Kontominas, M.G., Badeka, A.V., Kosma, I.S., and Nathanailides, C.I. (2021). Innovative seafood preservation technologies: recent developments. *Animals* 11 (1): 92.

25 Shi, J., Lei, Y., Shen, H. et al. (2019). Effect of glazing and rosemary (*Rosmarinus officinalis*) extract on preservation of mud shrimp (*Solenocera melantho*) during frozen storage. *Food Chemistry* 272: 604–612.

26 Das, R., Mehta, N.K., Ngasotter, S. et al. (2023). Process optimization and evaluation of the effects of different time-temperature sous vide cooking on physicochemical, textural, and sensory characteristics of whiteleg shrimp (*Litopenaeus vannamei*). *Heliyon* 9 (6): 16438.

27 Pan, G., Ji, H., Liu, S., and He, X. (2015). Vacuum frying of breaded shrimps. *LWT-Food Science and Technology* 62 (1): 734–739.

28 Duppeti, H., Manjabhatta, S.N., and Kempaiah, B.B. (2025). Flavor profile and role of macromolecules in the flavor generation of shrimp meat and valorization of shrimp by-products as a source of flavor compounds: a review. *Critical Reviews in Food Science and Nutrition* 65 (1): 123–142.

29 Faithong, N., Benjakul, S., Phatcharat, S., and Binsan, W. (2010). Chemical composition and antioxidative activity of Thai traditional fermented shrimp and krill products. *Food Chemistry* 119 (1): 133–140.

30 Leiva-Portilla, D., Martínez, R., and Bernal, C. (2023). Valorization of shrimp (*Heterocarpus reedi*) processing waste via enzymatic hydrolysis: protein extractions, hydrolysates and antioxidant peptide fractions. *Biocatalysis and Agricultural Biotechnology* 48: 102625.

31 Khaliq, A., Chughtai, M.F.J., Mehmood, T. et al. (2021). High-pressure processing; principle, applications, impact, and future prospective. In: *Sustainable Food Processing and Engineering Challenges*, 75–108. Academic Press.

32 Zheng, D., Cao, S., Li, D. et al. (2024). Fabrication and characterization of chitosan/anthocyanin intelligent packaging film fortified by cellulose nanocrystal for shrimp preservation and visual freshness monitoring. *International Journal of Biological Macromolecules* 264: 130692.

33 Nag, M., Lahiri, D., Dey, A. et al. (2022). Seafood discards: a potent source of enzymes and biomacromolecules with nutritional and nutraceutical significance. *Frontiers in Nutrition* 9: 879929.

34 Younes, I., Hajji, S., Frachet, V. et al. (2014). Chitin extraction from shrimp shell using enzymatic treatment. Antitumor, antioxidant and antimicrobial activities of chitosan. *International Journal of Biological Macromolecules* 69: 489–498.

35 El-Araby, A., Janati, W., Ullah, R. et al. (2024). Antifungal efficacy of chitosan extracted from shrimp shell on strawberry (*Fragaria*× *ananassa*) postharvest spoilage fungi. *Heliyon* 10 (7): 29286.

36 Ait Hamdan, Y., Oudadesse, H., Elouali, S. et al. (2024). Exploring the potential of chitosan from royal shrimp waste for elaboration of chitosan/bioglass biocomposite: characterization and "in vitro" bioactivity. *International Journal of Biological Macromolecules* 278: 134909.

37 El Knidri, H., Belaabed, R., Addaou, A. et al. (2018). Extraction, chemical modification and characterization of chitin and chitosan. *International Journal of Biological Macromolecules* 120: 1181–1189.

38 Pakizeh, M., Moradi, A., and Ghassemi, T. (2021). Chemical extraction and modification of chitin and chitosan from shrimp shells. *European Polymer Journal* 159: 110709.

39 Andoko, A., Prasetya, R., Cakra, B.D. et al. (2025). Autoclave-assisted multi-objective optimization of the deacetylation process in chitosan production from shrimp shell waste via response surface methodology. *Results in Engineering* 27: 106353.

40 Rujirapong, C., Siripongvutikorn, S., Suyotha, W., and Cheirsilp, B. (2025). Possibility to utilize shrimp by-product as a fermented food and its health benefits on hypertension and cancer. *Applied Food Research* 5 (2): 101142.

41 Ma, J., Chu, D., Rakhimovna, V.N. et al. (2025). Efficient and green extraction of chitin from mantis shrimp shell using deep eutectic solvents at room temperature. *Bioresource Technology* 435: 132888.

42 Vázquez-Aldana, M., Sixto-Berrocal, A.M., Arcos-Casarrubias, J.A. et al. (2025). Biological and chemical synergy in chitin and chitosan production: the role of process sequencing in shrimp shell waste treatment. *International Journal of Biological Macromolecules* 306: 141247.

43 Charii, H., Boussetta, A., Ait Benhamou, A. et al. (2024). Exploring the potential of chitin and chitosan extracted from shrimp shell waste in enhancing urea-formaldehyde wood adhesives. *International Journal of Adhesion and Adhesives* 129: 103599.

44 Islam, M.A., Hasan, M.N., Evan, M.S.H. et al. (2025). Chitin nanofibers: recent advances in preparation and applications in biomedical and beyond. *RSC Advances* 15 (19): 14655–14690.

45 Hamdi, M., Sun, H., Pan, L. et al. (2025). Chitosan and its derivatives as potential biomaterials for biomedical and pharmaceutical applications: a comprehensive review on green extraction approaches, recent progresses, and perspectives. *European Polymer Journal* 229: 113882.

46 Liebana, C., de los Ángeles Pereira, N., Fernández-Gimenez, A.V. et al. (2025). Antioxidant and functional properties of bioactive peptide fractions derived from shrimp (*Pleoticus muelleri*) waste hydrolysates. *Biocatalysis and Agricultural Biotechnology* 67: 103639.

47 El-Beltagy, A.E. and El-Sayed, S.M. (2012). Functional and nutritional characteristics of protein recovered during isolation of chitin from shrimp waste. *Food and Bioproducts Processing* 90 (4): 633–638.

48 Gautam, A.R., Junarto, A.W., Benjakul, S. et al. (2024). Omega-3 and unsaturated fatty acid enrichment of shrimp oil: preparation, characterization, and its application in salad dressings. *LWT –Food Science and Technology* 212: 116993.

49 Sumi, K.R., Biswas, A., and Nusra, S.M. (2025). Assessment of nutritional quality of giant freshwater prawn (*Macrobrachium rosenbergii*) and black tiger shrimp (*Penaeus monodon*). *Food Chemistry Advances* 7: 100991.

50 Sun, L., Li, Y., Yang, A. et al. (2025). Astaxanthin: a comprehensive review of synthesis, biological activities and applications. *Food Chemistry* 488: 144847.

51 Patil, A.D., Kasabe, P.J., and Dandge, P.B. (2022). Pharmaceutical and nutraceutical potential of natural bioactive pigment: astaxanthin. *Natural Products and Bioprospecting* 12 (1): 25.

52 Moreno-Sader, K.A., Martinez-Consuegra, J.D., and González-Delgado, Á.D. (2021). Development of a biorefinery approach for shrimp processing in North-Colombia: process simulation and sustainability assessment. *Environmental Technology & Innovation* 22: 101461.

53 Yusoff, M.A., Mohammadi, P., Ahmad, F. et al. (2025). Valorizing shrimp waste through a circular economy approach in food processing. *Journal of Environmental Chemical Engineering* 13: 117704.

54 Emerenciano, M.G., Rombenso, A.N., Vieira, F.D.N. et al. (2022). Intensification of penaeid shrimp culture: an applied review of advances in production systems, nutrition and breeding. *Animals* 12 (3): 236.

55 Al Eissa, A., Chen, P., Brown, P.B., and Huang, J.Y. (2022). Effects of feed formula and farming system on the environmental performance of shrimp production chain from a life cycle perspective. *Journal of Industrial Ecology* 26 (6): 2006–2019.

56 Kumar, V., Roy, S., Sahu, A.K., and Das, B.K. (2025). Disease in shrimp aquaculture: diagnostic technique for sustainable management In: Das, B.K., Kumar, V. (eds). *Laboratory Techniques for Fish Disease Diagnosis*, Springer, Singapore, p. 147–173.

57 Xiong, J., Dai, W., and Li, C. (2016). Advances, challenges, and directions in shrimp disease control: the guidelines from an ecological perspective. *Applied Microbiology and Biotechnology* 100 (16): 6947–6954.

58 Seethalakshmi, P.S., Rajeev, R., Kiran, G.S., and Selvin, J. (2021). Shrimp disease management for sustainable aquaculture: innovations from nanotechnology and biotechnology. *Aquaculture International* 29 (4): 1591–1620.

59 Lim, L.W.K. (2024). Implementation of artificial intelligence in aquaculture and fisheries: deep learning, machine vision, big data, internet of things, robots and beyond. *Journal of Computational and Cognitive Engineering* 3 (2): 112–118.

60 Büyükarıkan, B. (2025). Robust shrimp disease detection using multi-model convolutional neural networks-based ensemble strategies. *Aquacultural Engineering* 111: 102616.

61 Lal, J., Vaishnav, A., Singh, S.K. et al. (2024). Biotechnological innovation in fish breeding: from marker assisted selection to genetic modification. *Discover Biotechnology* 1 (1): 7.

62 Alak, G., Atamanalp, M., Uçar, A., and Ozogul, F. (2025). Commercialization and pilot-scale production of fish protein hydrolysate: challenges and recent developments. *Fish Protein Hydrolysates* 315–341.

4

Shellfish

Zhihong Liu, Biao Wu, Peizhen Ma, and Zhuanzhuan Li

Yellow Sea Fisheries Research Institute, Chinese Academy of Fishery Sciences, Qingdao, China

As the second largest group in the animal kingdom after the phylum Arthropoda, shellfish hold a significant position in global ecosystems. Scientists have identified over 100,000 extant species and approximately 35,000 fossil species. Characterized by diverse morphologies, wide distribution, and high value, shellfish have maintained an inextricable relationship with human civilization since ancient times. From harvesting in intertidal zones by early societies to the vast global industrial chain today, shellfish have long been a crucial source of animal protein and an economic mainstay. According to the statistics from the Food and Agriculture Organization of the United Nations (FAO), global shellfish production in 2023 reached approximately 25.82 million tonnes, accounting for about 13.67% of the world's total fishery output. This included roughly 19.30 million tonnes from marine aquaculture and about 212,000 tonnes from freshwater aquaculture [1]. Owing to variations in species and environmental factors, shellfish exhibit remarkable diversity and uniqueness in their bioactive compounds. Additionally, filter-feeding shellfish remove suspended solids, nutrients, and unicellular algae, thereby enhancing water transparency and controlling eutrophication. Moreover, shellfish absorb carbon dioxide from seawater during growth to form their shells, giving them a carbon sink function. Consequently, shellfish deliver both economic and ecological benefits, presenting broad prospects for development and utilization.

4.1 Major Global Shellfish Species and Their Distribution

4.1.1 Principal Species of Shellfish Worldwide

The shellfish with significantly economic value primarily belong to the classes Bivalvia, Gastropoda, and Cephalopoda, encompassing nearly all species in the international seafood market. Among these, bivalves and gastropods are predominantly obtained through aquaculture, while cephalopods remain largely reliant on wild capture fisheries. Economically important bivalve groups include orders such as Ostreoida, Veneroida, Mytiloida, and Pectinoida, which dominate both wild harvest and aquaculture production. According to data from the FAO, bivalve aquaculture alone accounts for almost 50% of global mariculture production. Key species include oysters, clams, cockles, arkshells, scallops, mussels, and razor clams.

Oysters represent the world's most widely cultivated shellfish group and are a key focus of modern breeding technologies such as triploid breeding. In 2023, global oyster aquaculture production reached 7.50 million tonnes, with China, the United States, South Korea, and France as the primary producers. Oysters are distributed along coastlines worldwide, with 86 species registered in the World Register of Marine Species (WoRMS) database. China's oyster aquaculture output in 2023 amounted to 6.67 million tonnes, with *Magallana gigas*, *Magallana angulata*, and *Magallana hongkongensis* being the most commercially utilized species [2]. *Magallana ariakensis* and *Magallana sikamea* are cultivated on a relatively small scale. In China, *M. gigas*, also known as the Pacific oyster, is distributed in the marine areas north of the Yangtze River Estuary, and serves as the primary cultivated species in northern coastal regions. *M. angulata* is predominantly distributed in the intertidal and shallow subtidal zones of the East China Sea, constituting a major cultivated species in Fujian and Zhejiang provinces. *M. hongkongensis* is mainly found in Guangdong, Guangxi, and Hainan along the South China Sea, representing the principal cultivated species in Guangdong and Guangxi. In recent years, Chinese experts have developed superior new varieties characterized by rapid growth, high glycogen content, and diverse shell colors, which have gained significant market popularity [3]. France, the first European nation to engage in large-scale oyster farming, remains Europe's largest producer, consumer, and exporter of oysters. With dozens of varieties, French oysters are renowned for their delicate texture, smoothness, and plump flesh. In 2025, South Korea's Ministry of Oceans and Fisheries announced that the country had become the world's third-largest oyster exporter, after France and China. Representative species are shown in Fig. 4.1.

Figure 4.1 *Magallana hongkongensis.*

Clams, cockles, and arkshells are important farmed shellfish worldwide. Their production reached 5.90 million tonnes in 2023, accounting for 30.29% of the global total shellfish production according to FAO statistics [1]. Among them, the Manila clam, *Ruditapes philippinarum,* is a representative species that has become one of the most widely farmed shellfish worldwide. It has a wide suitable farming area and can be cultivated in ponds, tidal flats, and shallow seas. It is native to the coasts of the Pacific and Indian Oceans in Asia. In the 1930s, it was accidentally introduced from Japan to the west coast of North America. From the 1970s to the early 1980s, it was successively introduced to France, Spain, the United Kingdom, Italy, Australia, and other regions. The Manila clam is one of China's four major farmed shellfish and ranks first in production among single shellfish species. Cockles are the English common name for the bivalve family Cardiidae. The cockle, *Cerastoderma edule,* is one of Europe's most commonly consumed shellfish, with a wide distribution in the Atlantic Ocean, ranging from Norway in the north to Senegal in the south. They are abundantly distributed in the mud and sand of the intertidal zones and shallow waters. The cockle, *Acrosterigma elongata,* distributed in Southeast Asia, is generally about 13 cm in size, while the cockle, *Maoricardium pseudolima*, available in African markets, can reach 16–17 cm. The cockle commonly found in the Chinese market is *Clinocardium californiense*, which is generally 5–7 cm in size. The ark shell, *Anadara broughtonii*, a representative species of arkshells, is distributed along the northern Asia coasts of the Pacific Ocean. In China, it is distributed in the Bohai Sea, Yellow Sea, and East China Sea. It inhabits soft mud or sandy seabed from the intertidal zone to subtidal depths of 5–30 m. Its shell can reach a height of 8 cm, a length of 9 cm, and a width of 8 cm [4]. *Anadara broughtonii* is very popular in Japan, South Korea, and China. It is often consumed as sashimi and sushi in Japanese cuisine. Representative species are shown in Figs. 4.2, 4.3, and 4.4.

According to statistics from the FAO, global scallop aquaculture production reached 2.07 million tonnes in 2023 [1]. Scallops (order Pectinida) are medium to large marine bivalve mollusks characterized by well-developed adductor muscles, capable of short-distance movement through valve clapping. Most species attach to

Figure 4.2 *Ruditapes philippinarum.*

Figure 4.3 *Anadara broughtonii.*

Figure 4.4 *Scapharca subcrenata.*

substrates such as rocks and coral reefs using byssus, while a minority of species inhabit sandy or muddy seabeds. Commercially significant species include the eurythermal bay scallop (*Argopecten irradians*), cold-water Japanese scallop (*Patinopecten yessoensis*), warm-temperate Zhikong scallop (*Chlamys farreri*), warm-water noble scallop (*Chlamys nobilis*), and cold-water great Atlantic scallop (*Pecten maximus*). Among these, the bay scallop is relatively small, with a shell height of approximately 5–6 cm; the Zhikong scallop is medium-sized, reaching 6–8 cm in shell height; the Japanese scallop is comparatively larger, with adults typically measuring 8–15 cm and up to 20 cm; the noble scallop can attain a shell height of 7–8 cm within one year; originating from Brittany, France, the great Atlantic scallop benefits from unique geographical conditions and superior environmental factors, exhibiting a substantially larger overall size and plumper adductor muscles compared to common scallop species [4]. Representative species are shown in Fig. 4.5.

Figure 4.5 *Chlamys farreri.*

Species of the order Mytiloida are widely distributed and have strong environmental adaptability. Most attach their byssus to substrates such as rocks, docks, and ship bottoms and are also highly susceptible to becoming invasive species, as observed in cases like *Mytella strigata* and *Brachidontes pharaonis* [5, 6]. Owing to their palatability and rapid growth rate, mussels have become an important breeding and fishing target in many countries. According to FAO statistics, global mussel production reached 1.86 million tonnes in 2023, accounting for 9.52% of global shellfish aquaculture output. Commonly cultivated species include the blue mussel (*Mytilus edulis*), Chilean mussel (*Mytilus chilensis*), hard-shelled mussel (*Mytilus coruscus*), and green mussel (*Perna viridis*). The blue mussel exhibits particularly wide distribution across coastal waters globally, inhabiting regions including the North Atlantic, English Channel, Mediterranean Sea, Black Sea, Sea of Azov, and both western and eastern Pacific coastlines. Chile is the world's largest mussel exporter, with its mussel farms ranking among the highest-yielding aquaculture operations globally and producing approximately 300,000 tonnes annually.

Razor clams are a group of commercially significant aquaculture species, accounting for approximately 4.5% of global shellfish cultivation output, with most production in China. The primary cultivated species, the Chinese razor clam (*Sinonovacula constricta*), is widely distributed in intertidal zones and shallow coastal waters of China, Japan, and neighboring regions. It has a slightly flattened shape with a thin shell, typically measuring 6–8 cm in length [4]. Another economically valuable species is the giant razor shell (*Ensis siliqua*), which attains a larger size up to 21 cm in length, and is distributed from Norway to the Mediterranean, with Irish razor clams being particularly renowned [7]. Representative species are shown in Fig. 4.6.

Economically significant taxa within the class Gastropoda include orders such as Seguenziida, Neogastropoda, and Architaenioglossa. Most Seguenziida species inhabit the ocean, with only a minimal number adapted to freshwater or terrestrial environments. These species predominantly inhabit rocky areas from the lower intertidal zone to depths of 10 meters, though some extend to shallow waters as deep as

Figure 4.6 *Sinonovacula constricta.*

50 m. Their primary habitation mode of locomotion is crawling, and they are herbivorous, feeding mainly on benthic algae (e.g., diatoms and green algae) on rock surfaces. Among these, *Haliotis* species represent the most economically valuable gastropods, with global trade exceeding 2.87 billion dollars in 2023 [1]. Due to the severe depletion of wild stocks, over 90% of abalones in the market currently rely on aquaculture. Leading farms in regions such as South Africa, Australia, and China primarily cater to high-end Asian markets [8]. Neogastropoda exhibits high species richness, with a predominantly carnivorous diet. Their habitats span rocky intertidal zones, subtidal sandy or muddy seabeds, and coral reefs. Representative economically important species of Neogastropoda include the veined rapa whelk (*Rapana venosa*) and the areolated babylon (*Babylonia areolata*). Architaenioglossa is a group comprising both marine and freshwater species, with *Sinotaia purificata* as its representative economically significant species. Their habitats encompass marine intertidal zones, as well as freshwater rivers, lakes, and ponds. Freshwater species typically reside in shallow waters with sandy or muddy substrates, while marine species commonly inhabit intertidal mudflats and sandy shores. Representative species are shown in Fig. 4.7.

Figure 4.7 *Haliotis discus hannai.*

The groups of cephalopods encompass economically significant species such as squids, cuttlefish, and octopuses. With the decline of traditional commercial fish stocks, cephalopod catches have consistently increased, rendering them pivotal targets for both large-scale industrial fleets and small-scale artisanal fisheries. Although large-scale harvesting is concentrated in around four families, cephalopods have developed into a multibillion-dollar industry chain; the annual catch and trade value of just ten major commercial species already amounts to billions of US dollars [9].

4.1.2 Distribution Characteristics of Shellfish

The size of the shellfish's regional distribution range mainly depends on their adaptability to external environmental factors such as temperature and salinity. Eurythermal shellfish can adapt to a wide temperature range and are distributed in sea areas of several different climate types. For example, the Pacific oyster, *M. gigas*, spreads from the tropical Indian Ocean to the cold regions of Russia; the shipworm, *Teredo navalis*, is distributed almost worldwide. Euryhaline shellfish can live in waters with different salinities. The freshwater mussel, *Corbicula fluminea*, can grow and reproduce in brackish water, while the suminoe oyster, *M. ariakensis*, can survive in seawater with very low salinity. Stenothermal or stenohaline shellfish can adapt to a narrow temperature or salinity range, and their distribution range is relatively limited. For example, *Hyotissa* species, such as *Hyotissa sinensis*, only live in slightly higher-salinity seawater in the warm waters of the Indo-Western Pacific. Other species, such as *Protochaca jedoensis* and *Neptunea cumingii*, are only distributed in the temperate zone of the North Pacific.

The vertical distribution range of shellfish is remarkably extensive, with significant variations observed among different species. For instance, *Radix hookeri* inhabits the Qinghai-Tibet Plateau at elevations ranging from 3,900 to 5,470 m, whereas the so-called Dumbo octopus *Grimpoteuthis bathynectes* occurs at depths of up to 7,000 m.

The vertical distribution range of different individuals of the same species of shellfish can also be large, such as species in the families Poromyidae and Cuspidariidae, which can occupy seabed environments from tens to thousands of meters deep. Nautiluses are found in both deep and shallow marine habitats, predominantly at depths of 50 to 150 m. Most marine bivalves and gastropods inhabit intertidal zones and shallow seas. Species such as *Batillaris zonalis* and *Littorina brevicula* (Fig. 4.8) are typically found near the high-tide line, whereas *Astralium haematraga* and *Placiphorella stimpsoni* preferentially inhabit areas adjacent to the low-tide line. Numerous species reside in shallow seas, generally at depths ranging from several to tens of meters, including *Chlamys farreri* and *Hemifusus tuba*. Additionally, certain species such as *Siphonaria japonica* (Fig. 4.9) lead an amphibious lifestyle in the high-tide zone, adapted neither to prolonged submersion nor extended exposure to air.

Most economically important shellfish species inhabit intertidal and shallow sea areas, where their growth and development are predominantly governed by factors including tidal patterns, seawater physicochemical properties, substrate composition, and phytoplankton availability. These environmental parameters must remain within

Figure 4.8 *Littorina brevicula.*

Figure 4.9 *Siphonaria japonica.*

species-specific optimal ranges to support biological functions; when values approach or exceed critical thresholds, these factors become limiting factors that restrict organismal survival. Consequently, the quality of environmental conditions either facilitates or impedes the growth and survival of shellfish populations [4]. Suitable aquaculture waters can help improve production efficiency.

4.2 Nutritional and Functional Components of Shellfish

As a significant group of aquatic invertebrates, shellfish exhibit remarkable biodiversity and functional value regarding nutritional composition and health benefits. From a nutritional structure perspective, shellfish are generally characterized by high-protein content, low-fat, and rich mineral profiles. The protein content in their dry matter ranges from 47% to 83%, significantly higher than that of most terrestrial animals, making shellfish an ideal source of high-quality animal protein [10]. More importantly, shellfish proteins exhibit a comprehensive and balanced amino acid profile, with essential amino acids accounting for over one-third of total amino acids, which aligns with the ideal protein pattern recommended by the FAO/WHO [11]. Their notably high lysine content enables significant protein complementarity when combined with cereal-based foods, thereby enhancing the overall bioavailability of dietary protein. Flavor amino acids such as glutamic acid, aspartic acid, glycine, and alanine are particularly abundant in shellfish, which explains their characteristic umami and sweet flavors. Among these, glutamic acid, as the main umami substance, accounts for over 16% in some varieties. It is worth noting that shellfish are rich in taurine, a sulfur-containing amino acid. In particular, the content of taurine in the common oyster reaches 215.4 mg/100 g, accounting for 46.2% of the total free amino acids [12]. Taurine not only plays a critical role in retinal and brain development in infants and young children but also improves dyslipidemia, regulates blood glucose, and helps modulate blood pressure in middle-aged and elderly populations.

Regarding lipid composition, although shellfish have a relatively low crude fat content, typically 1%–15%, their fatty acid composition exhibits extremely high nutritional value. Marine shellfish, in particular, are rich in long-chain *n*-3 polyunsaturated fatty acids (PUFAs), notably eicosapentaenoic acid (EPA) and docosahexaenoic acid (DHA), which together account for 12%–34% of total fatty acids. These fatty acids play crucial roles in cardiovascular and cerebrovascular health, neural development, and anti-inflammatory responses. The DHA-to-EPA ratio in *Mimachlamys nobilis* approaches 2:1, indicating remarkable potential for cognitive enhancement and brain health [13]. In contrast, freshwater species such as *Cipangopaludina cathayensis* contain negligible amounts of EPA and DHA, highlighting substantial differences between marine and freshwater varieties. Furthermore, shellfish are high in phospholipids, which constitute approximately 14%–23% of total lipids, significantly higher than that of plant-based oils such as soybean oil. Existing research has demonstrated their efficacy in lowering blood lipids and improving lipid metabolism. Shellfish cholesterol content is generally low (0.21–0.35 mg/g), substantially lower than that of eggs (3.55 mg/g), aligning with modern dietary health requirements.

Glycogen is an unique energy storage substance in shellfish, and its distribution across different tissues is species-specific. For instance, in the pen shell (*Atrina pectinata*), it is predominantly stored in the adductor muscle (accounting for 28.7% of dry weight), whereas in the Pacific oyster (*M. gigas*), it is primarily concentrated in the gonad-visceral mass (27.13% of dry weight). The fluctuation in glycogen content is closely associated with the reproductive cycle, as it is extensively used during gametogenesis to support gonadal development. Another ingredient gaining significant attention is bioactive polysaccharides. Studies indicate that these polysaccharides possess diverse potential bioactivities, including immunomodulatory, antitumor, and antiviral effects, though the underlying mechanisms require further clinical validation. Additionally, glycosaminoglycans (GASs) such as chondroitin sulfate are abundant in abalone and exhibit anticoagulant and immunoregulatory activities [14].

Shellfish are a natural repository of mineral elements. Freshwater snails, such as river snails (*Margarya melanioides*) and Chinese mystery snails (*Cipangopaludina chinensis*), are rich in minerals such as calcium and iron, which highlights their potential as natural dietary supplements [15]. Marine bivalves, on the other hand, display species-specific elemental enrichment profiles: oysters are notably abundant in zinc, playing a critical role in immune regulation, reproductive health, and wound healing; clams such as hard clams (*Meretrix meretrix*) and razor clams (*S. constricta*) are rich in selenium, an essential component of various antioxidant enzymes that contributes significantly to mitigating oxidative stress and preventing cancer.

Shellfish are notably rich in various vitamins, particularly vitamin B_{12}. The tissues of oysters, blue mussels, and clams are excellent sources of niacin and vitamin B_{12}, while blue mussels, oysters, and scallops are high-quality sources of vitamin A [16]. Furthermore, alkaloids and terpenoids identified in shellfish extracts exhibit antimicrobial and antiviral potential and have traditionally been utilized in treating hepatitis and dysfunctional uterine bleeding.

However, the food safety of shellfish also requires close attention. Due to their filter-feeding behavior, shellfish are prone to accumulating heavy metals, among which cadmium (Cd) and arsenic (As) pose more prominent pollution risks. In the digestive gland of *M. edulis*, Cd accounts for up to 65% of the total body burden, whereas in scallops, the mantle and gills exhibit significantly higher heavy metal accumulation capacity than the adductor muscle. It is therefore advisable for consumers, particularly sensitive groups such as children and pregnant women, to prioritize consuming low-accumulation tissues like the adductor muscle.

4.3 Primary Processing Methods and Product Types of Shellfish

Shellfish processing refers to the procedure of transforming bivalve mollusks (e.g., oysters, clams, scallops, mussels, and razor clams) and gastropod mollusks (including abalone and snails) into food or nonfood products via physical, chemical, or biological methods. Conventional processing techniques primarily encompass the production of live shellfish products, frozen processing, dehydration, canning, and seasoning manufacturing. These methods aim to preserve freshness, extend product shelf life, enhance and enrich their flavor characteristics, increase added value, and improve consumption convenience. As processing technologies have evolved from traditional practices to modern high-value utilization, the shellfish industry has developed a diversified and refined industrial system.

Fresh bivalve products, as the most basic form of consumption, have consistently maintained significant importance. Taking abalone as an example, live products are transported directly to consumer markets via cold chain logistics. When prepared using traditional cooking methods such as blanching or charcoal grilling, they retain their delicate texture and fresh flavor to the greatest extent. Research indicates that a thermal processing protocol of 80 °C for 15 minutes can be applied to the industrial production of pickled pepper-infused fresh abalone, while steaming has been confirmed as the cooking method that best preserves the textural properties of abalone. However, fresh bivalves are prone to autolytic deterioration due to microbial and enzymatic activities and have a short storage period, which severely limits their sales radius [17].

Cryogenic processing technology has effectively extended the shelf life and sales radius of shellfish products. Traditional frozen abalones are typically marketed in two forms: whole abalones and shelled meat. Storage at −80 °C significantly slows quality deterioration. However, conventional freezing methods generate ice crystals that disrupt muscle fiber structures, leading to juice loss and textural softening upon thawing. To address this issue, liquid-nitrogen flash-freezing technology has been applied in abalone processing. The ultra-low temperature of −196 °C enables extremely rapid freezing, allowing the product to quickly pass through the maximum ice crystal formation zone, thereby limiting ice crystal size to below 10 μm and better preserving the structural integrity of shellfish tissues. Experimental results demonstrate that liquid nitrogen-frozen abalone maintains satisfactory sensory attributes even after three months of storage, with ozone treatment further enhancing product safety. Texture parameters of liquid nitrogen-treated abalone show an approximately 30% improvement compared to conventionally frozen products, though repeated freeze-thaw cycles still exacerbate protein denaturation. To mitigate this, pretreatment by immersion in composite salt solutions has been adopted. The synergistic effect of calcium chloride and sodium chloride enhances the thermal stability of muscle proteins, increasing water-holding capacity by over 15% after thawing. Furthermore, high-pressure-assisted freezing technology shows promising potential: treatment at 200 MPa alters water phase transition curves and promotes the formation of unique amorphous ice.

Drying methods yield distinctive shellfish products through diverse methodologies, with each exhibiting unique characteristics and application scenarios. The technical spectrum includes natural sun-drying, hot-air drying, cold-air drying, vacuum freeze-drying, and microwave vacuum drying. Natural sun-drying relies on solar radiation and wind to achieve gradual dehydration, forming a characteristic "sugary core" texture and intense flavor profile. Hot-air drying triples the efficiency of natural drying while ensuring controllable hygiene conditions, albeit with risks of protein denaturation and quality degradation. Cold-air drying maximizes the retention of heat-sensitive components and preserves original quality effectively, yet suffers from prolonged processing durations. Vacuum freeze-drying optimally preserves nutritional constituents and original morphology, though accompanied by substantial equipment and operational costs [18]. Microwave vacuum drying, conversely, balances efficiency and quality to a considerable extent, emerging as a focal point in contemporary research and applications. Common dried products such as dried clams, dried mussels, and dried abalones effectively inhibit microbial proliferation through dehydration, facilitating storage and transportation while imparting unique organoleptic properties that garner significant market appeal.

Canned foods have enabled long-term preservation of shellfish products. As an important form of industrialized shellfish processing, canning includes various types such as canned clams in tomato sauce, smoked oysters, and abalones in soy sauce. These products rely on high-temperature sterilization and hermetic sealing to achieve extended shelf life, yet technical challenges, including melanosis and texture deterioration, persist in practical production. Research indicates that pretreatment with a composite solution of 0.25% citric acid and 0.02% ascorbic acid effectively inhibits hydrogen-sulfide-induced can corrosion and improves the color stability of canned abalones. Sterilization techniques have evolved from conventional high-temperature methods (121 °C) to stepped sterilization (70–80 °C for 25–35 minutes), which, combined with ozone-based microbial control technology, enhances product texture resilience by 20%. Innovative packaging solutions, such as stand-up pouches integrated with microwave sterilization, reduce the required sterilization time at 121 °C to 8 minutes.

Seasoning processing focuses on the advanced utilization of cooking broths and enzymatic hydrolysates, with representative products including oyster sauce and hoisin sauce. The application of a composite enzymatic hydrolysis method to oyster cooking broth significantly enhances the umami and flavor complexity of oyster sauce. The successful development of low-salt oyster sauce via multi-strain fermentation technology not only aligns with the health-conscious consumption trend but also expands the application scope of shellfish in the condiment industry [19].

With the acceleration of the rhythm of modern lifestyles and the upgrading of consumption demands, the development of novel shellfish products has increasingly focused on the ready-to-eat and prepared-foods sectors. The application of ultrahigh-pressure processing technology (e.g., 600 MPa for 20 minutes) in fresh abalone processing achieves sterilization and enzyme inactivation under non-thermal conditions, effectively preserving the products' original texture and flavor. Liquid-smoked abalones, utilizing modern smoking techniques, improve flavor stability and safety while also balancing convenience and high quality. In the field of prepared dishes, research has demonstrated that preprocessed abalones subjected to vacuum low-temperature heating at 70 °C for 1 hour maintain optimal textural properties upon reheating. Composite prepared foods such as "*Buddha Jumps Over the Wall*" effectively preserve the original broth flavor and ingredient texture through steam reheating, meeting consumers' dual demands for convenience and quality. Functional products development has emerged as a research hotspot in the high-value utilization of shellfish in recent years, as exemplified by oyster peptide beverages derived from enzymatically hydrolyzed oyster extracts or oral liquid products developed from abalone-derived polysaccharides.

High-value utilization represents an important direction for the extension of the shellfish processing industrial chain and the sustainable development of resources. Abundant bioactive compounds in shellfish, such as antioxidant peptides, antihypertensive peptides, and immunomodulatory polysaccharides, are being extracted and purified via modern biotechnological processes such as enzymatic hydrolysis, ultrafiltration, and chromatographic separation [20]. These components are now widely applied in health foods and pharmaceuticals, substantially enhancing the added value and market potential of shellfish products. Research indicates that sulfated polysaccharides derived from shellfish exhibit inhibitory activity against COVID-19, with efficacy up to 3.3 times that of conventional polysaccharides, thus demonstrating significant medicinal potential. Shells constitute over 70% of shellfish by-products and are primarily composed of calcium carbonate (approximately 94.3%). Through calcination and other treatments, they can be transformed into CaO catalysts or $CaCO_3$ fillers for applications in biodiesel synthesis and polymer composites, thereby facilitating waste-resource recycling. Protein components in shells, such as conchiolin, can be hydrolyzed for use as cosmetic ingredients [21]. The blue polyene pigments present in shells also show potential applications in the food industry. Furthermore, shells possess excellent adsorption properties, making them suitable for environmental management applications. For instance, research has demonstrated their effectiveness in removing hydrogen sulfide from seawater, thus offering a cost-effective and high-performance solution for water purification.

In summary, shellfish processing technology continues to evolve toward diversification, value addition, and refinement. The optimization of traditional processing techniques and the development of new products are mutually reinforcing, not only enriching product variety and improving market adaptability but also significantly enhancing resource utilization efficiency and reducing environmental pollution risk through the comprehensive utilization of by-products. Future research should further focus on elucidating the structure–activity relationships of bioactive compounds in shellfish, developing large-scale extraction and purification processes, and promoting interdisciplinary and cross-sectoral integrated applications, thereby comprehensively advancing the shellfish processing industry toward high-quality and sustainable development paths.

4.4 High-Value Shellfish Products and Applications

Amidst global population growth and rising protein demand, the efficient utilization of shellfish resources has emerged as a critical pathway to address food security challenges and advance the marine economy. According to the 2024 FAO report "*The State of World Fisheries and Aquaculture*," the annual global production of shellfish (including cephalopods) reaches nearly 30 million metric tons, serving as a core pillar of the fishery economies in Southeast Asia, Europe, the Americas, and other regions. Nevertheless, under traditional industrial models—where species such as oysters, scallops, mussels, squid, octopus, and cuttlefish are predominantly sold as live products or as minimally processed products (e.g., dried or frozen goods)—the resource utilization rate remains below 40%. This results in the underutilization of their bioactive compounds, functional components, and by-products, leaving substantial high-value potential untapped.

High-value shellfish products refer to those derived from shellfish bodies and their by-products through advanced bioprocessing, separation, and purification technologies, which confer high economic value and distinctive functional properties. These products are widely applied in diverse sectors such as food, pharmaceuticals, cosmetics, and novel environmental protection materials. Key categories encompass functional foods (e.g., shellfish-derived bioactive peptides, calcium supplements, and functional condiments), cosmetic ingredients (e.g., mussel polysaccharides and abalone polysaccharides), healthcare and medicinal products (e.g., antiaging formulations and immunomodulators), and environmentally friendly materials (e.g., heavy metal adsorbents and water purification agents).

4.4.1 Shellfish-Derived Bioactive Peptide Products

Shellfish-derived bioactive peptides represent a pivotal research direction in the high-value utilization of marine bivalves, with current studies focusing on the precise extraction of active components and their functional applications [22]. At the technological level, traditional single-enzyme hydrolysis processes have been progressively enhanced through integrated techniques such as "multi-enzyme synergistic hydrolysis coupled with membrane separation" and "supercritical CO_2 extraction." These advanced methodologies facilitate the efficient enrichment and purification of low-molecular-weight bioactive peptides and high-purity polysaccharides, significantly improving both the yield and bioactivity of functional components. In terms of research priorities, the industry has shifted from general nutritional supplementation to targeted functional applications, for instance, developing ACE-inhibitory peptides for cardiovascular protection, developing allergen-specific removal technologies for infant nutrition, and optimizing branched-chain amino acid profiles for athletic populations, thereby addressing the specific health needs of diverse consumer groups.

Previous studies have confirmed that oyster peptides possess a range of physiological functions, including antitumor effects, antioxidant activity, blood glucose-lowering activity, antiaging properties, inhibition of angiotensin-converting enzyme (ACE) activity, and promotion of growth and development [19]. In practical applications, Japan has utilized the Pacific oyster (*M. gigas*) to extract ACE-inhibitory peptides with a molecular weight of less than 3 kDa, developing a functional oral liquid that has become a benchmark for functional aquatic beverages in Japan. New Zealand has addressed the need for hypoallergenic infant nutrition by using muscle protein from *Octopus vulgaris*, adopting ultrafiltration membrane purification technology to remove allergens. The resulting hypoallergenic peptide powder has significantly reduced allergy incidence. In the United States, protein from *Todarodes pacificus* has been processed via low-temperature extrusion to produce energy bars for athletes. Rich in branched-chain amino acids, these products effectively delay muscle fatigue. Currently, Europe, Japan, and New Zealand serve as the core R&D regions for functional foods based on shellfish-derived bioactive peptides. China is also experiencing rapid development in the integration of industry, academia, and research in this field.

4.4.2 Shellfish Polysaccharides and Their Derivatives

The rich diversity of shellfish species and their substantial biomass provide a solid foundation for the development of bioactive compounds derived from shellfish. Differences in taxonomy, ecological habits, and environmental conditions lead to the structural complexity and diverse molecular structures of shellfish polysaccharides. These polysaccharides are natural macromolecules characterized by complex architectures and a wide range of biological activities, primarily composed of amino-containing acidic mucopolysaccharides [23]. Research has shown that shellfish-derived polysaccharides include neutral sugars, sulfated polysaccharides, GAGs, and other heterogeneous glycans. Shellfish polysaccharides commonly exhibit antioxidant and anti-hemolytic activities; some also demonstrate antibacterial properties, immunomodulatory effects, and inhibitory effects on cancer cell proliferation. They show promising therapeutic potential in managing various acute and chronic diseases, offering benefits such as fatigue reduction, regulation of blood glucose and lipid levels (commonly referred to as "three highs"), hepatoprotection, thrombolysis, and antiaging effects. Notably, certain shellfish polysaccharides possess antiviral activity, including activity against influenza virus. For example, a sulfated polysaccharide with a molecular weight of 750 kDa, isolated from pearl oysters and purified via ion-exchange chromatography, exhibits significant antioxidant and neuroprotective activities. Moreover, these polysaccharides display moisturizing, skin-whitening, and skin-repairing capabilities, with favorable safety profiles. Hyaluronic acid, a representative acidic mucopolysaccharide, demonstrates excellent hydration capacity, effectively shielding against ultraviolet radiation and bacterial invasion while penetrating into deeper skin layers to improve skin condition. Similar functional properties have been observed in polysaccharides from pearl oysters and oysters.

The demonstrated efficacy of shellfish polysaccharides underscores their potential as valuable sources for functional foods, specialized dietary formulations, and novel pharmaceuticals, highlighting broad prospects for development and application. Although research on plant polysaccharides—particularly their bioactivity and use in functional foods and clinical therapeutics—is more advanced, shellfish polysaccharides exhibit significant similarities in biological activities and pharmacological effects, suggesting comparable therapeutic potential. These compounds show promising applications in healthcare products, cosmetics, and related industries. In the food sector, shellfish polysaccharides serve not only as functional additives but also as antibacterial preservatives of natural origin. Nevertheless, due to their complex structural features, studies on certain shellfish polysaccharides remain confined to preclinical animal experiments, limiting their translation into practical applications.

4.4.3 Mineral Supplement Products

Shellfish shells and soft tissues are rich in mineral elements, with calcium content in shells exceeding 95%. They also contain trace elements such as magnesium, zinc, iron, and phosphorus, as well as ultra-trace elements such as copper, manganese, and chromium [24]. These components make them excellent raw materials for high-quality mineral supplement products. Currently, shellfish-based mineral products are primarily categorized into three types: the first category comprises organic calcium supplements, including calcium citrate (with a higher absorption rate) produced through the conversion of shell-derived calcium carbonate, calcium lactate, known for its high bioavailability and low gastrointestinal irritation, and amino acid-chelated calcium, which forms a five-membered ring structure for sustained release and absorption. The second category consists of compound mineral supplements containing multiple trace elements such as zinc, iron, and phosphorus. These products are particularly suitable for individuals with osteoporosis and children in the growth phases. The third category is functional food additives, such as the use of oyster shell powder to fortify bread, which can not only improve the rheological properties of the dough but also increase the calcium content. Nanoscale oyster shell powder can also be used to fortify calcium in milk tablets.

4.4.4 Eco-friendly Application Products

Driven by the "Dual Carbon" goals, shell-based new materials have emerged as an innovative direction for the valorization of shellfish, exhibiting significant potential in environmental protection. These materials enable synergistic resource recycling and environmental conservation through high-value utilization [25]. Specifically, high-value products and applications of shellfish waste in the environmental sector primarily fall into three categories: (1) Flame-retardant materials. Eco-friendly fire-resistant coatings and materials can be developed using extracts from shellfish shells. For instance, self-healing fire-retardant coatings based on phosphorylated polysaccharides and catechol compounds derived from shells exhibit excellent fire-resistant properties. (2) Bio-adsorption materials. Calcined shell powder demonstrates strong adsorption capacity for heavy metal ions such as Pb^{2+}, Cd^{2+}, and Cu^{2+}, with removal rates reaching up to 97% for Pb^{2+} and Cd^{2+}. Hydrophobically modified shell-based materials can also effectively adsorb organic dyes from textile wastewater and marine oil spills. (3) Soil amendments. Oyster shells, rich in calcium and other essential crop nutrients, form unique porous structures and abundant CaO after calcination. When processed into soil conditioners, they effectively mitigate soil acidification, immobilize heavy metal activity, and contribute to soil pH regulation, calcium supplementation, and overall soil remediation.

4.5 Challenges and Solutions

4.5.1 Challenges

4.5.1.1 The Unclear Mechanisms of Active Substances' Efficacy Constrain the Development of High-Value Products

Shellfish, such as oysters, are rich in bioactive compounds, including polysaccharides and taurine, which have demonstrated potential antioxidant and immune-enhancing properties. However, research on their specific mechanisms of action remains substantially underdeveloped. Current studies on shellfish-derived active components remain predominantly focused on extraction techniques, while critical aspects—such as the efficacy and mechanisms of bioactive peptides, the regulatory pathways of oyster polysaccharides on immune signaling pathways (e.g., NF-κB and MAPK), and the molecular targets underlying taurine's anti-fatigue effects—remain poorly elucidated. This knowledge gap has led to the promotion of shellfish-based functional products primarily as general nutritional supplements, lacking scientific substantiation of the "component–mechanism–efficacy" relationship, thereby hindering the development of high-value products [26].

4.5.1.2 Underdeveloped Processing Industrialization Leading to Nutritional and Value-Added Losses

The global shellfish processing industry exhibits a three-tier technological divergence: Northern European countries lead in value-added processing innovations, the EU emphasizes standardized production, while developing nations predominantly engage in primary processing. Over 80% of shellfish in Southeast Asian countries are still processed using traditional methods, remaining at a rudimentary and extensive stage of production. The majority of processing enterprises are small-scale, workshop-style factories that lack the technical capacity for product research and development. First, mechanization levels are inadequate, with preprocessing stages, including impurity separation, dead shellfish sorting, and cleaning/grading, heavily reliant on manual labor. Manual sorting efficiency is only one-third that of mechanical methods and shell damage rates exceed 15%. Second, value-added processing is limited: products primarily consist of frozen cooked meat and dried goods, with the value addition merely one-fifth that of deep-processed products. Third, traditional methods result in nutritional degradation: hot-air drying causes over 60% loss of EPA/DHA; salting leads to 40%–70% loss of B vitamins; and high salinity levels (15%–20% salt content) contradict contemporary health-conscious dietary trends.

4.5.1.3 By-products and Circulation Bottlenecks Constrain Industrial Efficiency

Global shellfish processing exhibits low resource utilization efficiency, with byproducts remaining largely untapped. The annual output of shellfish-processing byproducts worldwide exceeds 12 million metric tons, of which shells, viscera, and mantles account for 60%–70% of the raw material weight. Currently, these residues are predominantly disposed of in landfills or discarded, resulting not only in resource wastage but also in adverse environmental impacts. In terms of circulation, the survival of live shellfish depends on specific environmental conditions. Fluctuations in temperature, humidity, and oxygen concentration during transport can readily cause mortality or reduced vitality, leading to rapid deterioration of texture and freshness. Certain species necessitate "water-based oxygenated" transport, which consumes three times more energy than anhydrous survival technologies and incurs loss rates of 15%–25%.

4.5.2 Solution Approaches

4.5.2.1 Strengthen Fundamental Research on Active Substances and Elucidate Efficacy Mechanisms

Establish a scientific and systematic evaluation framework for typical bioactive compounds derived from shellfish, such as polysaccharides, protein-derived peptides, lipids, and minerals, to clarify their mechanisms of action and assess their development potential. Utilizing techniques such as Western blot and RT-PCR, the research will investigate the regulatory effects of shellfish-derived bioactive components on immune and metabolic signaling pathways. Develop a "component–mechanism–efficacy" database to provide a scientific foundation for the functional development of related products.

4.5.2.2 Advancement of Processing Technologies and Product Innovation

Technological innovation should balance advanced capabilities with accessibility, focusing on "mechanization, deep processing, and value-added enhancement." First, dedicated processing equipment must be developed, with emphasis on breakthroughs in pretreatment machinery (e.g., integrated shellfish cleaning and grading systems and shell-meat separation equipment) and deep-processing equipment (such as ultra-high-pressure sterilizers and freeze-drying systems). Second, product diversification should be pursued through the development of demographically targeted products, functional foods, and cross-industry applications. The adoption of enzymatic hydrolysis techniques and innovative product development will further elevate added value.

4.5.2.3 Optimization of High-Value Utilization of By-products and Production Circulation

Promoting resource recycling and reducing circulation costs are essential to improving industrial efficiency. First, strengthening comprehensive utilization of byproducts: for instance, processing shells into nano-calcium carbonate through high-temperature sterilization and pulverization (with 30% higher absorption rate than conventional calcium carbonate), extracting EPA/DHA from viscera via supercritical CO_2 extraction, and preparing collagen through enzymatic hydrolysis of mantles. Second, optimizing circulation models by categorizing shellfish based on transport tolerance—prioritizing air freight or short-distance cold chain for perishable varieties while adopting more cost-effective land transport for durable species. Processed products adopt an "integrated cold chain + direct e-commerce" approach to advance the "origin-to-consumer" model, shortening circulation cycles (2–3 days from processing plants to end consumers). Concurrently, regional branding is strengthened to enhance product premiumization.

References

1 FAO. 2025. Fishery and Aquaculture Statistics – Yearbook 2023. FAO Yearbook of Fishery and Aquaculture Statistics. Rome . . https://doi.org/https://doi.org/10.4060/cd6788en.

2 Ministry of Agriculture and Rural Affairs Bureau of Fisheries, National Fisheries Technology Extension Center, China Society of Fisheries (2024). *China Fishery Statistical Yearbook*, 2024. Beijing: China Agriculture Press.

3 Jiang, K., Chen, C., Jiang, G., et al. Genetic improvement of oysters: current status, challenges, and prospects. *Revies in Aquaculture* 2024, 16, 796–817. https://doi.org/https://doi.org/10.1111/raq.12868

4 Wang, Z. and Zheng, X. (2024). *Marine Shellfish Culture and Breeding*. Beijing: Science China Press.

5 Ma, P. Li, H., Liu, Y., et al. First confirmed occurrence of the invasive mussel *Mytella strigata* (Hanley, 1843) in Guangdong and Hainan, China, and its rapid spread in Indo-West Pacific regions. *BioInvasions Records* 2022, 11(4): 947–963. https://doi.org/https://doi.org/10.3391/bir.2022.11.4.1

6 Battiata, M.; Curatolo, T.; Naser, M.D.; Sarà, G.; Lo Brutto, S. The state of the art of the global distribution of the invasive mytilid species *Brachidontes pharaonis* (P. Fischer, 1870). *Diversity* 2024, 16, 381. https://doi.org/https://doi.org/10.3390/d16070381

7 Diaz, A.G., Seijo, C.L., Gaspar, M.B. et al. (2011). *Razor Clams: Biology Aquaculture and Fisheries*. Santiago de Compostela: Consellería do Mar, Xunta de Galicía.

8 Cook, P.A. Worldwide abalone production: an update. *New Zealand Journal of Marine and Freshwater Research* 2025, 59, 4–10. https://doi.org/https://doi.org/10.1080/00288330.2023.2261869

9 Ryotaro, L. (2024). Sustaining cephalopod fisheries in a changing marine environment. *Journal of Aquaculture Research & Development* 15: 937. https://doi.org/10.35248/2155-9546.24.15.937.

10 Li, Y., Sun, C., Liu, C. et al. (2012). Comparison of macronutrient components of several marine and freshwater shellfish. *Journal of Shanghai Ocean University* 21 (2): 297–303.

11 Li, N. (2011). *The Amino Acids, Fatty Acids and Heavy Metals Content Analysis of the Shellfish and Its Product Quality Evaluation*. Hebei: Agricultural University of Hebei.

12 Liu, Y., Gong, X., Xu, Y. et al. (2014). Determination and comparative analysis of flavor-enhancing nucleotides and amino acids in three common shellfish from offshore. *Journal of Fishery Sciences of China* 21 (2): 351–360.

13 Liu, S., Li, D., Gao, J. et al. (2009). The lipid components and nutritional evaluation of three seashells. *Acta Nutrimenta Sinica* 31 (4): 414–416.

14 Ulagesan, S., Krishnan, S., Nam, T.J. et al. (2022). A review of bioactive compounds in oyster shell and tissues [J]. *Frontiers in Bioengineering and Biotechnology* 10: 913839.

15 Venugopal, V. and Gopakumar, K. (2017). Shellfish: nutritive value, health benefits, and consumer safety. *Comprehensive Reviews in Food Science and Food Safety* 16 (6): 1219–1242.

16 Zhang, A., Shao, S., Zhao, K. et al. (2016). Nutritionalcom position and edible safety analysis of two estuarine bivalve clams, *Meretrixm eretrix* and *Mactra veneriformis* in Shuangtaizi Estuary. *Journal of Fisheries of China* 40 (9): 1497–1504.

17 Chen, S., Yang, S., Liu, X. et al. (2019). Recent progress in comprehensive processing and utilization of abalone and its byproducts. *Meat Research* 33 (10): 76–81.

18 Gao, J., Shen, J., Zhang, C. et al. (2015). Effects of vacuum freeze-drying on the quality of the oyster *Ostrea rivularis*. *Modern Food Science and Technology* 31 (4): 253–257.

19 Zhao, Q., Wei, X., Sun, J. et al. (2021). Research progress on comprehensive utilization of oyster resources. *China Food Additives* 7: 150–159.

20 Naik, A.S. and Hayes, M. (2019). Bioprocessing of mussel by-products for value added ingredients. *Trends in Food Science & Technology* 92: 111–121.

21 Asaoka, S., Yamamoto, T., Kondo, S. et al. (2009). Removal of hydrogen sulfide using crushed oyster shell from pore water to remediate organically enriched coastal marine sediments. *Bioresource Technology* 100 (18): 4127–4132.

22 CAI K., CAI S., WU J., et al. Functional and bioactive properties of marine shellfish derived peptides: a review. *Journal of Fisheries Research*, 2016, 38(2) : 157-164.

23 Nilesh, N., Didem, D., Seda, C. et al. (2024). Polysaccharides from shell waste of shellfish and their applications in the cosmeceutical industry: a review. *International Journal of Biological Macromolecules* 265: 131119.

24 Cao, M., Ding, X., Xu, L. et al. (2021). Progress of the utilization of oyster shell resource. *Journal of Jimei University (Natural Science)* 26 (5): 390–397.

25 Zhao Y., Wu Y. Li Y. et al. Research progress of recycling and high-value utilization of protein from aquatic by-products. *Journal of Fisheries of China*, 2021, 45(11): 1943-1953.

26 Jiang, C. (2012). Major problems and solutions of shellfish processing in China. *Fishery Information and Strategy* 27 (2): 87–93.

5

Seaweed

Min Yang and Jia Wang

Ocean University of China, Qingdao, China

5.1 Introduction

Seaweeds are marine resources that occur in suitable marine habitats in all climatic regions of the globe. They are able to photosynthesize to form huge algae forests that provide shelter and food for other creatures. There are many kinds of seaweeds, and to date, tens of thousands of different seaweeds have been identified. The seaweeds mainly include brown seaweed, red seaweed, and green seaweed.

Seaweeds have long been a staple in diets around the world, particularly in Asia. As a primary producer in the marine ecological food chain/web, seaweeds are known for their umami taste, which can also enhance other tastes and flavors. Due to the unique environmental factors of the ocean, there are various bioactive components with specific functions present in marine organisms, such as vitamins, fatty acids, carotenoids, polysaccharides, polyphenols, minerals, and pigments. These compounds exhibit significant physiological activities in the human body and are the basic components of marine functional foods, such as regulating metabolism, boosting immunity, and antioxidation. Among the thousands of compounds found in seaweed, not only do they fully encompass all the various chemical structure types present in terrestrial plants, but there are also some special chemical structure substances that are not found in terrestrial plants. Thus, it can be seen that the diversity of seaweeds and their bioactive compounds structures far exceeds that of terrestrial plants. The rich and diverse resource conditions provide the necessary material basis for the continuous development of seaweed functional foods.

In recent decades, the seaweed industry has evolved from harvesting wild seaweed to large-scale artificial cultivation of seaweed. The aquaculture industry has produced a large amount of *Laminaria*, *Gelidium amansii*, and other large seaweeds. Based on this, various seaweed polysaccharides such as alginate, carrageenan, and agar, as well as a variety of seaweed active substances, have been prepared, becoming an important part of the marine biotechnology industry. Various seaweed-derived bioproducts have been widely applied in functional foods, health products, and other health-related products. With the advancement and development of new technologies, people have conducted in-depth research on the biochemical composition and functional characteristics of seaweed, which has led to a wider application of seaweed. According to statistics, seaweed in food applications held a significant market share from 2017 to 2021, and its use for human consumption has increased globally. It is reported that the seaweed market was valued at $15.01 billion in 2021 and is expected to reach $24.92 billion by 2028. In this part, the resources, nutrition, processing, and food applications of seaweed will be discussed.

5.2 Species and Distribution of Seaweed

5.2.1 Species of Seaweed

5.2.1.1 Brown Seaweed

Brown seaweeds (Phaeophyceae) are multicellular algae with a high degree of evolution that are found in cold water of continental coastal waters and are uncommon in freshwater. It is the largest seaweed, with some species reaching up to 35–45 m in length and a wide range of shapes. Fucoxanthin is the main component of brown seaweed and is significantly higher in content than other pigments, including chlorophyll *a* and chlorophyll *c*, which gives these seaweeds their brown color.

Ascophyllum, *Laminaria*, *Saccharina*, *Macrocystis*, *Nereocystis*, and *Sargassum* are the most prevalent genera. *Sargassum* represents a genus among the most expansive and taxonomically rich within the class of brown seaweeds, encompassing approximately 400 recognized species that exhibit diverse morphological characteristics. In addition, as shown in Fig. 5.1, dominated by species like kombu (*Laminaria japonica*) and wakame (*Undaria pinnatifida*), brown seaweed rich in iodine and alginate [1]. They account for a significant portion of the edible seaweed market, particularly in East Asian cuisines.

Figure 5.1 Two typical brown seaweeds of *Laminaria japonica* (a) and *Undaria pinnatifida* (b).

Figure 5.2 Two typical red seaweeds of *Corallina officinalis* (a) and *Pyropia yezoensis* (b).

5.2.1.2 Red Seaweed

Red seaweeds (Rhodophyta) are widely distributed, and most of them grow in the deep sea. *Chondrus*, *Porphyra*, *Pyropia*, and *Corallina* are some of the most common red seaweed genera (Fig. 5.2). Red seaweeds are rich in phycoerythrin and phycocyanin, which give them red color. The walls are made of carrageenan and cellulose agar. These two long-chain polysaccharides are often used in industrial production. *Corallina officinalis* is an important red seaweed whose cell surface secretes calcium carbonate to form tough and calcified cell walls that are important for building coral reefs [2]. In addition, nori *(Pyropia yezoensis)* is a widely edible red seaweed in Japan, England, etc.

5.2.1.3 Green Seaweed

Green seaweed (Chlorophyta) can be found in both seawater and freshwater, which appear green due to the presence of chlorophyll *a* and chlorophyll *b*. All green seaweeds are classified in a common class, called Ulvophyceae, which is a very diverse group including about 1927 species [3]. Common green seaweed species belong to *Ulva*, *Cladophora*, *Enteromorpha*, and *Chaetomorpha* (Fig. 5.3). Four major species of green seaweed belonging to the genus *Ulva* include *Ulva lactuca*, *Ulva prolifera*, and *Ulva linza*. Among them, *U. lactuca,* also known as sea lettuce, is commonly grown on rocks and found naturally in sublittoral waters around the world. *U. lactuca* is the softest seaweed with the least sticky, elastic, crispy, cartilaginous, and cohesive properties. In addition, *Caulerpa* spp. was found to be the most abundant green seaweed species, including *Caulerpa taxifolia* and *Cladophora prolifera*. Green seaweed has a variety of apparent forms that are closely related to its ability to produce bioactive compounds. For example, *Ulva* and *Cladophora* are filamentous seaweeds, growing in long chains or filaments. Their increased surface area benefits light absorption and nutrient absorption, which contribute to the accumulation of active substances such as polysaccharides and vitamins [4].

Figure 5.3 Two typical green seaweeds of *Ulva lactuca* (a) and *Enteromorpha* (b).

5.2.2 Distribution of Seaweed

5.2.2.1 Environmental Influences on Seaweed Communities

The distribution of seaweed generally includes three aspects: horizontal distribution, vertical distribution, and spatiotemporal variation. Horizontal distribution refers to the distribution with latitude (regional distribution), and vertical distribution is the distribution along the tidal zone. Tidal zones are generally divided into three parts, which are distributed in different areas horizontally, including the supratidal zone, the intertidal zone, and the subtidal zone. Green seaweed, brown seaweed, and red seaweed are generally distributed in the intertidal zone, tidal or upper subtidal zone, and subtidal zone, respectively. The fixed tide level of seaweed is mainly related to the types and proportions of pigments it contains. Different pigments require different wavelengths of light. With the changes in light intensity and light quality, the distribution of seaweed is also affected. Generally, in darker areas or the deep sea, phycocyanin can absorb blue and green light more effectively than chlorophyll. Therefore, the areas near the low tide line and deep sea are mostly inhabited by red seaweed. While the habitats of green seaweed, which only contain chlorophyll and carotenoids, are mostly located near the sunny shallow waters.

Temperature is the primary determinant of seaweed distribution [5]. Tropical species like *Eucheuma* thrive in waters above 20°C, dominating regions such as the South China Sea and Indonesian archipelago. Temperate zones (10–20°C) host commercially valuable species like *L. japonica* and *Pyropia*. In addition, sea temperature is also an environmental factor that causes the seasonal change of seaweed. Seaweed increases and decreases with the change of season and latitude. For example, in the northern coastal areas of China, the peak period of large seaweed (*Enteromorpha*) is mainly in summer. In winter and spring, wakame (*U. pinnatifida*) and some green algae dominate. In addition, the topography, substrate, humidity, salinity, wind waves, ocean currents, pollutants, animal feeding, and competition among algae in the ocean are factors that affect the growth and distribution of seaweed.

5.2.2.2 Global Patterns of Seaweed Distribution

Wild seaweed resources are distributed all over the world. For decades, seaweed has been widely cultivated in 61 countries and territories, contributing over 50% of global marine and coastal aquaculture production. *Laminaria, Eucheuma/ Kappaphycus, Gracilaria, Undaria,* and *Porphyra* are the five main cultured genera, accounting for about 95% of the world's cultivated seaweed production [6].

According to statistics, there are currently 6,495 species of seaweed such as brown seaweed, red seaweed and green seaweed in the world, among which there are 1,485 species of brown seaweed, 4,100 species of red seaweed, and 910 species of green seaweed. Among them, brown seaweeds are dominant in cold temperate waters. There are abundant wild resources in North Atlantic areas, such as Norway, Ireland, the United Kingdom, Iceland, Canada, and the United States, and South Pacific areas, such as Chile and Peru. Red seaweeds are distributed in almost all latitude zones, while green seaweeds have the highest degree of evolution in tropical waters. The *Laminaria* of the *Phaeophyta* is mainly distributed in the Russian Far East, Japan, China, North Korea, Norway, Ireland, the United Kingdom, France, and other places. The *Macrocystis* is mainly distributed in Chile, Argentina, and parts of the United States and Mexico. The *Ascophyllum* is mainly distributed in Ireland, the United Kingdom, Iceland, Norway, Canada, and other places. *Gracilaria* of the Rhodophyta is distributed in global seas. In the Southern Hemisphere, they are mainly found in Argentina, Chile, Brazil, South Africa, and Australia, while in the Northern Hemisphere, they are mainly distributed in Japan, China, India, Malaysia, the Philippines, and other countries.

Seaweed can be produced on a large scale industrially through artificial breeding. Research reports from the Food and Agriculture Organization (FAO) show that global seaweed production has experienced rapid growth over the past two decades. In 2021, the global total production of seaweed was approximately 36.3 million tons, almost tripling compared to 11.8 million tons in 2001. From 2017 to 2021, the average annual growth rate of seaweed production was 1.86%. Approximately 97% of seaweed production comes from aquaculture, reaching 35.1 million tons in 2021. Most of the seaweed supplied globally is produced by the four major producers in Asia: China (60%), Indonesia (25%), South Korea (5%), and Philippines (4%). Meanwhile, the production of seaweed harvested from the wild has fluctuated over the past 20 years, but has generally held steady at around 900 thousand tons.

5.3 Nutritional Value of Seaweed

Seaweed is an important biological resource in the ocean, rich in various unique nutrients and bioactive substances (Table 5.1), and has extremely high nutritional value and a wide range of health benefits. From traditional consumption to modern medicine and functional foods, seaweed has great potential to contribute constructively to people's health, food security, and become an essential component of a healthy diet.

Table 5.1 The main distribution and nutritional components of seaweed.

Seaweed species		Distribution	Water content (%)	Carbohydrates (% dw)	Proteins (% dw)	Ash (% dw)	Fatty acids (% dw)
Green seaweed	*Cladophora* sp.	Atlantic coast, Scandinavia, Mediterranean, Adriatic, Baltic	/	34.8–39.9	13.9–18.4	20.7–36.5	/
	Ulva intestinalis	Russia, South Korea, Japan, China, Philippines, Malaysia, Australia, Singapore, Indonesia, Arabian Sea, Atlantic Ocean, Indian Ocean	About 89.6	7.3–41.5	0.9–19.5	15.2–32.1	0.1–10.5
	Ulva lactuca		75.6–96.6	30.8–70.9	9.3–25.5	1.4–70.2	0.1–10.5
	Caulerpa racemosa	/	95.2–98.1	33.0–41.1	17.8–23.78	7.0–19.1	Under 11.9
Red seaweed	*Ahnfeltia plicata*	/	About 64.1	29.3–44.9	4.43–20.1	About 23.2	/
	Brogniartella byssoides	/	/	About 24.1	About 15.8	About 41.9	/
	Ceramium sp.	Tropical, subtropical, and temperate sea areas	83.6–89.9	18.7–38.9	14–34.43	11.35–32.9	/
	Chondrus crispus	North Atlantic coast, Baltic Sea, Iceland	/	55.6±68.5	11.3–21.0	21–27.2	1.0–3.1
	Cystoclonium purpureum	/	85.71–93.1	19.88–35.6	17.2–29.46	38.6±0.3	/
	Delesseria sanguinea	Coastal area of China	/	About 25.9	About 18.3	About 31.2	/
	Dilsea carnosa	/	/	44.9–50.5	About 15.2	About 24.0	/

	Furcellaria lumbricalis	North Atlantic, Baltic Sea	76.9–77.4	29.4–61.4	14.5–17.1	About 32.9	/
	Rhodomela confervoides	Yellow Sea	79.8–83.8	32.3–47.6	14.8–25.5	About 32.2	/
	Gracilaria corticata	Japan, North Korea, South Vietnam, Philippines, South Pacific Coast of North America	94.8–96.3	About 43	16.1–28.7	8.2–23.1	0.1–22.6
Brown seaweed	*Ascophyllum nodosum*	Atlantic Ocean, Norwegian, Icelandic coast	/	About 31.7	About 5.9	About 20.2	/
	Chorda filum	Yellow Sea, Bohai Sea, Kuril Islands, Japan Sea coast	/	25.9–32.5	About 6.3	About 39.0	/
	Desmarestia aculeata	India	/	20.0–40.2	About 11.5	About 25.4	/
	Fucus serratus	North Pacific, North Atlantic, Baltic Sea	/	26.6–29.8	About 7.1	About 20.3	/
	Fucus vesiculosus	Cold sea area	/	24.1–29.1	About 7.1	About 24.4	/
	Halidrys siliquosa	/	/	17.0–30.4	About 7.9	16.9–18.3	/
	Laminaria digitata	West Coast of Europe	About 94.3	47.7–56.1	6.6–16.1	16.7–19.6	/
	Saccharina latissima	Nordic coasts of Spain and Galicia	/	54.3–57.1	4.4–12.9	11.5–52.2	/
	Sphacelaria cirrosa	Paracel Islands	/	25.6–27.8	About 12.0	About 28.8	/
	Sargassum ilicifolium	Temperate seas	94.4-96.2	About 32.9	8.9–43.87	5.33–29.9	0.1–25.5

5.3.1 Main Nutritional Constituents

5.3.1.1 Seaweed Polysaccharides

Seaweed polysaccharides are the most abundant active components in seaweed, generally accounting for 50% to 76% of the dry weight [7]. These polysaccharides are not only the main organic components of the seaweed, but their specific composition and structure also vary significantly among different types of seaweed, and are influenced by species characteristics, habitat, and growth stage.

According to their biological sources and functional properties, seaweed polysaccharides can be divided into three categories: extracellular (intercellular matrix) polysaccharides, structural (cell wall) polysaccharides, and intracellular storage polysaccharides. Extracellular polysaccharides are distributed in the intercellular spaces (mucilage layer) and participate in adhesion and water retention; structural polysaccharides form the microfiber framework of the cell wall and confer mechanical strength, while storage polysaccharides are stored within the cell in an energy reserve form. Among them, extracellular (intercellular matrix) polysaccharides are represented by alginate, agar, and carrageenan. These polysaccharides are mostly linear macromolecules and often carry sulfate groups and other functional groups. Alginate is a linear anionic polysaccharide composed of β-D-mannuronic acid (M) and α-L-guluronic acid (G), with high water affinity and gel properties, and is widely present in the cell walls and intercellular substances of brown seaweed; agar (mainly composed of β-D-galactose and 3,6-anhydro-α-L-galactose) and carrageenan (consisting of various sulfated β-1,3/1,4-galactose) in red seaweed are the main gelatinous polysaccharides in the intercellular layer. Carrageenan can be classified into three types (κ, ι, and λ) based on the position and quantity of sulfate groups. Brown seaweed also contains another important sulfated polysaccharide—fucoidan, whose main chain is composed of α-1,3/1,4-linked fucose residues and is decorated with varying degrees of sulfate groups and other carbohydrate side chains. The ulvan of green seaweed is mainly composed of sulfated rhamnose and uronic acids. The structure contains L-rhamnose-3-sulfate (Rha3S), D-glucuronic acid (GlcUA), L-iduronic acid (IdoA), and D-xylose (Xyl), which provide a certain degree of flexibility and conformational diversity to the molecule. In addition, common cell wall polysaccharides such as cellulose and hemicellulose (such as xylan and mannan) are also widely present in various types of seaweed, and these polysaccharides are highly diverse in monosaccharide composition, glycosidic bond connection mode, branching degree, and esterification/sulfation modifications. In terms of intracellular storage polysaccharides, brown seaweed mainly exists in the form of laminarin (β-1,3-glucan) and mannitol. For example, in brown seaweeds of the *Saccharina* and *Laminaria*, the content of laminarin can be as high as over 30% of the dry weight. Red seaweed and green seaweed mainly store starch, and the content of these polysaccharides is closely related to the intensity of photosynthesis and nutritional status.

Overall, seaweed polysaccharides exhibit remarkable diversity in both structure and function. From the complex cell wall matrix to the mucopolysaccharides that can regulate the intercellular environment, and to the energy-storing polymers, all these types of polysaccharides together constitute the material basis of the seaweed cells. They not only participate in the construction of cell walls and intercellular mucilage layers but also store energy and influence cellular metabolism. Due to their unique functional group modifications, monosaccharide compositions, and molecular weight, seaweed polysaccharides possess rich physicochemical properties and bioactivities, thus attracting extensive attention in the fields of food, medicine, and materials.

5.3.1.2 High-Quality Protein

The protein content of seaweed can reach up to 50% of its dry weight (DW), depending on factors such as growth conditions, harvesting time, and seaweed species. Generally, the protein content in red seaweed is generally higher (20% to 47% of DW), followed by green seaweed with about 9% to 26% of DW, and finally brown algae, about 3% to 15% [8].

Most seaweeds contain more essential amino acids (except lysine and threonine) than the human requirements. Moreover, compared with common plant protein sources such as soybeans, the digestibility and absorption rate of some seaweed proteins are higher. For instance, the protein content in *Porphyra* is approximately 20% to 30%, and its amino acid score (AAS) is close to the ideal protein model recommended by the FAO, with a reasonable proportion of essential amino acids. Compared to animal protein sources such as meat and fish, seaweed proteins are cholesterol-free and have extremely low-fat content, making them a low-fat and high-quality protein resource. Seaweed proteins also contain bioactive peptides and lectins, which have various functional activities such as antioxidant, anti-inflammatory, blood pressure-lowering, and antibacterial properties [9]. In addition, there are several structurally unique amino acids in seaweed, such as taurine, which reduces cholesterol and strengthens liver function, and laminine, which has a blood pressure-lowering effect.

5.3.1.3 Unsaturated Fatty Acids

The lipid content of brown, red, and green seaweed was 0.1%–11.5%, 0.4%–12%, and 0.2%–15% of their dry weight, respectively [10]. It is rich in various unsaturated fatty acids (UFAs), with polyunsaturated fatty acids (PUFAs) being the predominant type. Overall, PUFAs in the total fatty acids (TFAs) of seaweeds account for approximately 30%–50%, while the monounsaturated fatty acids (MUFAs) account for about 10%–20%. Among PUFAs, eicosapentaenoic acid (EPA, 20:5 n-3) in red seaweed (such as *Porphyra*) can account for 40%–50% of the TFAs, while EPA in brown seaweed (such as *Undaria*) can reach 15%–19%. Docosahexaenoic acid (DHA, 22:6 n-3) content is relatively low, generally within the range of 1%–8%. Other ω-3 fatty acids, such as alpha-linolenic acid (ALA, 18:3 n-3) and stearidonic acid (SDA, 18:4 n-3), can also reach tens to hundreds of micrograms per gram of dry weight in kelp leaves. Among ω-6 fatty acids, linoleic acid (LA, 18:2 n-6) and arachidonic acid (ARA, 20:4 n-6) each account for 2%–3% and 0.4%–0.7% of the TFAs, respectively. The ω-6 to ω-3 ratio in algae is generally low, typically between 0.5 and 4.4, with an average of approximately 1:1. This fatty acid composition can alleviate inflammation and prevent cardiovascular diseases [11]. In summary, the content of MUFAs and PUFAs in seaweed is significantly higher than that of saturated fatty acids, making them a good source of healthy dietary fats.

5.3.1.4 Minerals and Vitamins

Seaweed are natural treasure trove of minerals and trace elements, containing a variety of minerals and vitamins beneficial to human health. Typical minerals include iodine, potassium, calcium, magnesium, iron, and zinc. Among them, the Na concentration ranged from 0.5% in *Ulva lactuca* (Chlorophyta) to 15.8% in *Dermonema virens* (Rhodophyta). The concentration of K in *Hypnea spinella* (Rhodophyta) and *Dictyota dichotoma* (Phaeophyceae) ranges from 0.2% to 24.7%. Moreover, every 100 g of fresh *Ulva lactuca* contains about 2200 mg of Ca. Therefore, eating 8 g of *Ulva lactuca* can meet 37% of the calcium reference nutrient intake (RNI) of an adult male. In contrast, the same portion of cheddar cheese provides only 5% RNI [12]. Besides, the bioavailability of iodine in *Ascophyllum nodosum* (Phaeophyceae) and *Chondrus crispus* is 4.6 and 2.1 mg/45 g, respectively. This means that eating 1.5 or 3 mg of seaweed can meet the daily recommended dietary iodine intake [13].

It is reported that seaweed is rich in various vitamins, including vitamin A precursors (β-carotene), vitamin C, vitamin E, vitamin K, and various B vitamins (such as vitamin B_1, B_2, niacin, folic acid, and the rare vitamin B_{12}). Especially vitamin B_{12}, seaweed is one of the few plant sources of B_{12}. These minerals and vitamins work together to help support thyroid function, enhance the immune system, and provide antioxidant protection.

5.3.1.5 Other Bioactive Components

Polyphenolic compounds are widely present in brown seaweed, mainly formed by the polymerization of phenyl structural units, including structures such as eckol, phloroglucinol, and dieckol, and they exhibit a high degree of structural diversity. These substances are mainly found in the cell wall and intercellular matrix, originating from the polyphenolic pathway during biosynthesis. Particularly, brown seaweed contains high levels of phlorotannins, which are complex polymers composed of oligomers of phloroglucinol (1,3,5-trihydroxybenzene). Red or green seaweed contains more phenolic acids and flavonoids [14].

Carotenoids are widely present in seaweed; and fucoxanthin is the representative substance. Fucoxanthin is a unique oxidized carotenoid with an enol ketone structure. It is the main pigment unique to brown seaweed and gives them an orange-brown color. Furthermore, during the metabolic process, seaweed can also synthesize various enzymes, including amylase, protease, pectinase, and other hydrolases, and they are mainly present in the extracellular or intracellular fluids. These diverse bioactive substances are an important part of the biochemical composition of seaweed. The unique molecular characteristics and species specificity that they show are valuable for developing the function of seaweed resources.

5.3.2 Health Benefits

5.3.2.1 Maintain Metabolic Balance

The iodine in seaweed can maintain the balance of thyroid function. Iodine is a necessary raw material for the synthesis of thyroid hormones (T_3, T_4), and thyroid hormones are the main regulators of the body's basal metabolic rate. Adequate intake can effectively prevent thyroid function decline caused by iodine deficiency, avoiding a decrease in metabolic rate, fatigue, and abnormal weight gain. Polysaccharides, dietary fiber, and special fatty acids in seaweed are closely related to human blood glucose and lipid homeostasis. These components maintain metabolic balance by regulating weight, blood glucose, and blood lipid levels. The alginate found in seaweed is a key component for weight loss. When alginate enters the stomach, it reacts with stomach acid to form a highly viscous stomach gel. This gel is like a fine "network" that wraps around the food, significantly slowing down the gastric emptying rate. As the gastric emptying time prolongs, the digestion and absorption process of food in the gastrointestinal tract also slows down. This means that the body's absorption of calories decreases, thereby achieving the effect of controlling calorie intake. Fucoxanthin can activate the lipolytic enzymes in fat cells, such as hormone-sensitive lipase (HSL), prompting the breakdown of triglycerides into glycerol and fatty acids, which are then released into the bloodstream and subsequently oxidized and utilized by other tissues of the body, thereby accelerating the metabolic breakdown of fat. Several animal experiments have shown that the body fat percentage and liver fat content of the fucoidan-supplemented experimental group were significantly lower than those of the control group [15].

5.3.2.2 Protection of Cardiovascular Health

Multiple components in seaweed, such as polysaccharides, fucoxanthin, and fatty acids, play a crucial role in maintaining cardiovascular health. These components mainly exert their effects on maintaining cardiovascular health by regulating blood glucose, blood lipids, and anti-thrombosis.

Seaweed polysaccharides exhibit a unique mechanism in regulating blood glucose. Especially, the polysaccharides inhibit the activity of α-amylase, which is the key enzyme to degrade starch into glucose. Polysaccharides bind to the enzyme, altering enzyme's spatial structure, reducing its catalytic efficiency, and dramatically slowing the breakdown of starch into glucose. This effectively controls the increase in postprandial blood glucose, preventing sharp blood glucose spikes. In the animal model experiments of diabetes, the fasting blood glucose levels of diabetic mice that were given seaweed polysaccharides for a long time were significantly reduced, and their glucose tolerance was significantly improved [16]. Further studies have found that ulvan and carotenoids may exert hypoglycemic effects by enhancing insulin sensitivity.

Fucoidan exhibits a significant hypolipidemic bioactivity. It can combine with cholesterol in the intestines to form insoluble complexes, hindering the absorption of cholesterol and promoting its excretion through feces at the same time. In addition, fucoidan can also stimulate the liver to synthesize more bile acids. The synthesis of bile acids requires the consumption of cholesterol, which prompts the liver to take in more cholesterol from the blood for the synthesis of bile acids, thereby reducing the cholesterol content in the blood. Animal experiments have shown that fucoidan can also reduce the deposition of fat in the liver and alleviate the degree of fatty liver, further demonstrating its important role in regulating lipid metabolism. Moreover, ω-3 PUFAs in seaweed can inhibit fatty acid synthesis and promote fatty acid oxidation, especially EPA and DHA, thereby reducing triglyceride levels in the blood. These fatty acids also increase high-density lipoprotein (HDL) levels and decrease low-density lipoprotein (LDL) levels [17].

Fucoidan also has anticoagulant and antithrombotic bioactivities to prevent cardiovascular disease. It lowers the risk of thrombus by preventing platelets from aggregating and decreasing blood viscosity. Fucoidan also has antioxidant and anti-inflammatory properties that can reduce vascular endothelial cell damage and inhibit the inflammatory response, consequently reducing the risk of cardiovascular disease. Furthermore, in vitro studies have shown that seaweed enzymes (including various hydrolases and transferases) can degrade polysaccharides, generate oligosaccharides, and improve the structure of the gut microbiota, thereby promoting the absorption of nutrients and maintaining the function of the intestinal mucosal barrier. It is beneficial for improving overall metabolic health and digestive function [18].

5.3.2.3 Immunoregulation

The active components such as polysaccharides and polyphenols in seaweed have demonstrated immunoregulation effects at different levels. Seaweed polysaccharides mainly exhibit immunomodulatory activity by activating immune cells. Macrophages are an important line of defense in the immune system because they can phagocytose and remove invading pathogens. Seaweed polysaccharides activate signal pathways and enhance macrophage activity through binding to macrophage surfaces' receptors. This improves macrophages' ability to phagocytose pathogens. Seaweed polysaccharides also promote the secretion of cytokines by macrophages, including tumor necrosis factor-α (TNF-α) and interleukin-1 (IL-1) [19]. These cytokines play a crucial regulatory role in the immune system, which can activate other immune cells, such as T lymphocytes and B lymphocytes, thereby further enhancing the immune response.

Free radicals can be produced during the normal metabolic process of the human body, which have strong oxidizing properties. When free radicals accumulate to a certain level, they will cause damage to cells and tissues, thereby affecting the normal function of the immune system. The antioxidant components in seaweed, such as vitamin C, vitamin E, carotenoids, and polyphenols, can react with free radicals, converting them into stable substances, thereby eliminating free radicals and reducing the damage to the immune system caused by oxidative stress. Epidemiological studies have found that immune indicators are generally better in people who consume seaweed long-term than in those who do not. For instance, in a study targeting the elderly, the experimental group consumed a certain amount of seaweed every day. The results showed that their immune cell activity was enhanced, serum immunoglobulin levels were increased, the frequency of catching colds and flu was reduced, and the course of the disease was shortened. These results prove seaweed nutrient components significantly enhance immunity.

5.3.2.4 Maintaining Intestinal Health

The dietary fiber in seaweed plays an irreplaceable role in maintaining intestinal health. It is significant in regulating the balance of intestinal flora and preventing intestinal diseases. Seaweed dietary fiber includes both soluble and insoluble fiber. Insoluble fiber increases the volume of feces, promotes intestinal peristalsis, and allows feces to be discharged more smoothly. Soluble dietary fiber plays a vital role in regulating intestinal flora balance. It can be fermented and utilized by beneficial bacteria in the intestinal tract, providing them with rich nutrition and promoting their growth and reproduction. *Bifidobacterium* and *Lactobacillus* are two types of beneficial bacteria in the intestinal tract that can inhibit the growth of harmful bacteria and maintain the balance of intestinal microecology. Soluble dietary fiber can provide the energy and nutrients needed for the growth of *Bifidobacterium* and *Lactobacillus*, increasing their quantity and enhancing their activity. When beneficial bacteria are dominant in the intestines, they can produce short-chain fatty acids (SCFAs), such as acetic acid, propionic acid, and butyric acid. These SCFAs can not only provide energy for intestinal cells, promote their growth and repair, but also regulate the pH value of the intestine, inhibit the growth of harmful bacteria, and enhance the immunity of the intestine. Studies have shown that regular consumption of seaweed rich in dietary fiber can increase the number of beneficial bacteria in the intestines, improve the structure of the intestinal flora, and reduce the risk of intestinal diseases. In a study on intestinal diseases, it was found that adding soluble dietary fiber extracted from seaweed to the daily diet of mice could significantly improve the intestinal flora of the mice and alleviate the symptoms of intestinal diseases [20].

5.3.2.5 Beauty and Skin Care

Seaweeds have a lot of natural ingredients to nourish the skin. These ingredients include vitamins, minerals, and polysaccharides. These ingredients can protect skin barrier, maintain skin's moisture balance and structural integrity through multiple mechanisms, and thus are widely used in skin care products. Vitamin C can keep the skin bright and elastic through scavenging free radicals generated by external stimuli, inhibiting melanin production, and participating in collagen synthesis. Vitamin E captures lipid peroxidation free radicals, maintains membrane integrity, and improves microcirculation in the skin and cell membranes. Minerals such as calcium can strengthen the skin barrier and prevent transdermal water loss; iron ensures that skin cells receive sufficient oxygen; zinc is involved in the metabolism and repair of keratinocytes and has a regulatory effect on acne-prone skin [21]. Seaweed polysaccharides (such as fucoidan and alginate) form a moisturizing film on the skin surface, significantly increasing skin moisture content and reducing transdermal water loss. Meanwhile, their antioxidant and anti-inflammatory activities can alleviate irritation and redness, promoting the recovery of skin barrier function. In addition, seaweed polysaccharides promote keratinocyte renewal, accelerate skin metabolism, and improve skin texture, making it more delicate and smoother. In the future, applying these active ingredients from seaweed to functional skin care formulas is expected to provide a solid scientific basis and innovative direction for natural, gentle, and highly effective skin care products.

5.4 Cultivation and Harvesting Methods of Seaweed

5.4.1 Cultivation Methods

The main seaweed culture method is open water systems. According to the different marine areas for cultivation, it can be divided into shallow sea and intertidal aquaculture. Depending on the different cultivation methods, the open water systems can be divided into floating raft aquaculture, rock-based aquaculture, and bottom-seeding aquaculture.

Shallow sea aquaculture usually uses vertical or flat culture to cultivate seaweed such as *Laminaria* and *U. pinnatifida*, and realizes low-cost production through floating cages or raft frames. The **intertidal aquaculture** adopts integrated multi-trophic aquaculture (IMTA). For example, the Sanggou Bay model in China combines fish, shellfish, and seaweed to form a carbon and nitrogen material circulation system. The annual output exceeds 240,000 tons, effectively reducing environmental load. However, precise control of density is necessary to prevent deterioration of water quality.

Floating raft aquaculture is widely used in seaweed cultivation. It involves suspending macroalgae seedlings on the sea surface using rafts or wooden boats. The rafts provide a stable platform for the growth of seaweed, allowing it to access sunlight, nutrients, and carbon dioxide effectively. For example, in the cultivation of *Laminaria*, floating raft systems are commonly employed. The *Laminaria* seedlings are attached to ropes, which are then hung from the rafts. This method enables easy management and monitoring of the algae, facilitating the adjustment of growth conditions as needed. Additionally, floating raft cultivation can be carried out in both shallow sea and intertidal zone areas, providing flexibility in location selection.

Rock-based aquaculture involves attaching seaweed seedlings to rocky substrates until they reach full growth. This method mimics the natural growth environment of seaweed, which often attaches to rocks in the wild. By using artificial rocks or natural rock formations, farmers can create suitable habitats for seaweed cultivation. Rock-based aquaculture is a relatively simple and cost-effective method, especially in areas with abundant rocky substrates. It also promotes the growth of a diverse range of seaweed species, contributing to the overall biodiversity of the marine ecosystem. However, it may be limited by the availability of suitable rocky areas and the need to protect the natural rock formations from damage.

Bottom-seeding aquaculture is a farming method that directly seeds the seeds or seedlings of seaweed onto the seabed. This method is applicable to some seaweed varieties that can naturally attach and grow on the seabed, such as kelp. The advantages of bottom-seeding aquaculture are low cost, minimal impact on the environment, and the ability to utilize the natural resources of the seabed. When conducting bottom-seeding aquaculture, it is necessary to select an appropriate seabed environment to ensure that the seaweeds have sufficient light and nutrients. It is also necessary to pay attention to preventing the invasion of other organisms to ensure the growth and yield of seaweed.

5.4.2 Harvesting Techniques

The traditional harvesting method is dominated by manual harvesting, and its technical core depends on the labor-intensive operation mode. Fishermen usually use simple tools such as sickles and rakes to harvest intertidal seaweed during low tide, or complete the harvest through diving in shallow water areas. This method has the advantages of low equipment investment and intuitive operation, but there are significant limitations: the harvesting efficiency is strictly limited by tidal cycles and weather conditions, the returns are relatively low, and intense physical labor increases the risk of work accidents. More seriously, a root-uprooted harvesting method that seriously damages the regenerative tissue at the base of the seaweed. This extensive harvesting not only restricts the large-scale development of the industry but also leads to the imbalance of the offshore ecosystem. In some production areas, ecological problems such as the degradation of seaweed beds and changes in the structure of benthic communities have already emerged.

In recent years, the industrial-scale seaweed harvesting technology has accelerated its transformation from traditional mechanized and extensive modes to an intelligent and ecological one. By integrating automation equipment, bioengineering, and clean energy technology, the harvesting efficiency and environmental compatibility have been significantly improved. In the field of mechanization, the intelligent harvesting system based on satellite remote sensing and AI image recognition can accurately locate the distribution of seaweed. At the same time, the flotation technology in seaweed harvesting was improved and adapted to the preprocessing of macroalgae. The degradable sodium borate silica beads were used to adsorb the fragments of *Sargassum*. Combined with the gas–liquid two-phase flow enrichment device, the energy consumption of harvesting in the intertidal zone was reduced by more than 60%.

Although it is still necessary to overcome the technical bottlenecks such as the corrosion resistance of flotation materials and shortage of efficient and universal harvesting equipment. The promotion of the EU's blue economic policy and the cross-disciplinary innovation are accelerating the transition of seaweed harvesting from a resource-consuming model to a precise and sustainable one.

5.4.3 Postharvest Handling

Rapid cooling technology rapidly inhibits enzyme activity and microbial growth through low-temperature treatment, thereby prolonging the shelf life of seaweed and retaining its bioactive components. For example, freeze-drying (−80 °C) and cryogenic storage (−40 °C) are widely used in laboratory and industrial scenarios to maintain the physical and chemical properties of macroalgae. The harvesting-processing continuum technology significantly improves macroalgae processing efficiency and reduces energy consumption by integrating harvesting, pretreatment, and deep processing. The seamless operation chain of seawater cleaning, boiling water short-term sterilization (3–5 minutes), −40 °C frozen storage, and rapid fermentation (completed in 3 days) not only integrates traditional multistage treatment into continuous production, but also minimizes the risk of microbial contamination by reducing intermediate exposure.

5.4.4 Processing Methods of Seaweed for Food Application

5.4.4.1 Drying

Traditional sun-drying and air-drying are the most fundamental and widely used drying methods in the field of seaweed processing. The process flow includes four core steps: fresh harvesting, flat drying, storage, and quality certification. The core advantages of this method lie in its zero energy consumption and low cost, which make it particularly suitable for small-scale production in tropical regions. However, its limitations are also significant: rainy and cloudy weather will prolong the drying period and increase the risk of microbial contamination. Long-term exposure to ultraviolet rays will cause the degradation of photosensitive components such as chlorophyll and carotenoids (such as phaeoplysin), and the fucoidan is prone to generating brown pigments, which need to be decolorized through formaldehyde and hydrochloric acid treatment, potentially introducing chemical residues. Nevertheless, this method performs exceptionally well in preserving polyphenolic substances.

Modern drying technology is centered on spray drying and transforms liquid seaweed extract into powder by atomization (Table 5.2). Spray drying technology is characterized by its speed and efficiency, and it can effectively preserve the bioactive components and nutrients in the seaweed extract. For example, in the extraction process of astaxanthin, spray drying has been proven to be one of the effective methods to optimize the downstream processing process [22]. The

Table 5.2 The advantages and disadvantages of drying methods.

Methods	Advantages	Disadvantages
Solar	Sustainable and no energy consumption	Dependence on the weather
Spray	Fast and economical method, suitable for seaweed production for human consumption	Degradation in the quality and operational cost
Freeze	Highly energy intensive	Applicable for small scale
Oven drying	Less energy intensive	Suitable for small scale
Crossflow air-drying	Economic and fast drying	Energy cost
Incinerator	Seaweed biomass burning can be avoided	High cost and complicated

operating conditions such as feed rate, gas flow rate and inlet temperature can be optimized by experiments to improve the recovery rate and quality of the product. Despite the high cost of spray drying, it has broad application prospects in macroalgae drying, especially in the case of efficient drying and maintaining high-value-added product quality.

5.4.4.2 Salting

The typical process flow of modern salted seaweed includes harvesting, blanching, cooling, salting (with a salt concentration of 10–30%), pickling, and water control. Among these steps, the control of blanching temperature and time is crucial for retaining chlorophyll and texture. It was found that the microbial activity was closely related to the storage conditions during the salting process: 30% salt concentration combined with 4 °C low temperature could effectively inhibit bacteria that produce alginate lyases (such as *Dyella* and *Halomonas*), while promoting the enrichment of salt-tolerant archaea without degrading ability, thereby preventing the softening or rotting of the algae [23]. At room temperature (25 °C), even with a 20% salt concentration, microbial degradation would accelerate, and significant spoilage would occur within 30 days. The effects of salting on nutritional components show two sides: although it causes partial loss of iodine, protein, and dietary fiber in fresh algae (such as the decrease of iodine content by more than 80% after salting), it can better retain antioxidant active ingredients compared with the drying process. For example, the contents of chlorophyll and carotenoid in salted kelp are significantly higher than those in sun-dried ones.

5.4.4.3 Extraction

The extraction process of seaweed begins with the pretreatment of biomass, which is a crucial step to ensure the efficiency of subsequent procedures. Pretreatment mainly includes cleaning, drying, and crushing. The purpose is to remove attached pollutants (such as sand, salt, and epibiont) and increase the solvent contact area through mechanical crushing. The size of the crushed seaweed particles needs to be controlled within a reasonable range to balance extraction efficiency and subsequent separation difficulty.

The traditional extraction technology is represented by solvent extraction, Soxhlet extraction, and hot alkali extraction, and its core depends on the solubility of the target compound in a specific solvent. For example, the hot alkali method is widely used in the extraction of brown algae polysaccharides by converting insoluble alginate into soluble form by pH adjustment. However, these methods generally have defects of high energy consumption, large solvent consumption, and longtime consumption, and high temperature may lead to the degradation of polyphenols.

The development of modern technology focuses on efficiency improvement and sustainability. Among them, the more widely used extraction methods are shown in Table 5.3. The core of seaweed separation technology lies in selecting appropriate methods based on the physicochemical properties of the target components (such as polysaccharides, polyphenols, and proteins), in order to meet the functional, safety, and economic requirements of the food industry. Therefore, in the food industry,

Table 5.3 The extraction methods widely used in seaweed processing industry.

Extraction methods	Description	Advantages	Disadvantages
Conventional extraction	Depending on a particular solvent's solubility of the target compound. Extraction is also carried out by adding acid or base.	The method has been tested.	High energy consumption, large solvent consumption, and longtime consumption may result in the degradation of bioactive compounds.
Microwave-assisted extraction (MAE)	Using electromagnetic radiation, temperature, time, seaweed/water ratio, and other conditions to maximize yield.	Short extraction time with relatively small amount of solvent.	High heat causes the degradation of polyphenols.
Ultrasound-assisted extraction (UAE)	Ultrasound disrupts the cell structure, resulting in a high extraction rate at low temperatures. It is suitable for the extraction of heat-sensitive substances.	Short extraction time, low temperatures, and relatively little solvent.	Degradation and changes in the structure of polysaccharides.
Supercritical fluid extraction (SFE)	Using CO_2 as the solvent, the target substances can be selectively extracted under high-pressure conditions. It is suitable for the purification of lipophilic components.	The solvent is easy to remove from the extract.	High-pressure treatment may have a negative impact on the compound.
Pressurized solvent extraction (PSE)	PSE as pressurized liquid/fluid extraction (PLE/PFE), accelerated solvent extraction (ASE), high-pressure solvent extraction (HPSE).	The method is similar to Soxhlet extraction, but the solvent has high extraction properties.	The combination of high temperature and pressure to increase the solubility and diffusion of the solvent is a challenging task.
Enzyme-assisted extraction (EAE)	Utilizing enzymes catalytic to enhance the efficiency of extraction.	Does not use organic solvents, is environmentally friendly, and is safe.	The extraction efficiency is influenced by the properties of the enzyme.

Table 5.4 The commonly used separation and purification methods.

Methods	Description	Advantages	Disadvantages
Ethanol fractional precipitation	Precipitate target components (such as polysaccharides) step-by-step through different ethanol concentrations.	Low-cost, easy to operate, and suitable for preliminary separation.	Poor selectivity, low recovery rate (50–60%), and repeated purification may lead to activity loss.
Ion exchange + Size-exclusion chromatography	Combine ion exchange (based on charge differences) and size-exclusion chromatography (based on molecular size differences).	High purity (>90%), recovery rate increased to over 85%, suitable for purifying active components such as polysaccharides from brown algae and rock algae.	Requires optimization of buffer pH and ionic strength, and the equipment cost is relatively high.
Membrane separation (UF/NF)	Separation is carried out by ultrafiltration (UF) or nanofiltration (NF) based on molecular weight.	No chemical reagents are needed and low energy consumption.	Membrane fouling is a prominent issue, and frequent cleaning increases maintenance costs.
Solvent extraction (Ionic liquids)	Use specific ionic liquids to separate brown algal glycans in one step through specific intermolecular binding.	High selectivity, significantly better purity than ethanol method (recovery rate 70%), and can couple the extraction–purification process.	High cost for ionic liquid recovery, and strict detection is required for possible residues.
SFE + Molecular distillation	Use supercritical CO_2 extraction combined with molecular distillation to extract heat-sensitive components (such as EPA/DHA).	Avoid oxidative loss, retain nutritional activity, and have no solvent residue.	High equipment investment and high energy consumption for large-scale production.
Microfluidics + affinity chromatography	Combine microfluidic technology (inertial migration sorting) with affinity chromatography (ligand–receptor binding) to enrich microalgal proteins or polypeptides.	High specificity, efficient sorting of active cell metabolites, suitable for the development of functional foods (such as fortified nutritional beverages).	Limited throughput, difficult to meet large-scale production requirements.

separation technology needs to take into account high purity, low residue, and large-scale feasibility. Table 5.4 lists the commonly used methods, including ethanol fractional precipitation, ion exchange and size-exclusion chromatography, membrane separation, solvent extraction, supercritical fluid extraction (SFE), molecular distillation, microfluidics, and affinity chromatography.

5.4.4.4 Fermentation

Seaweed is rich in polysaccharides, phenolic compounds, γ-aminobutyric acid (GABA), and dietary fiber, but its natural complex cell wall structure and low bioavailability of some bioactive compounds limit its direct application. At present, there are two main types of seaweed fermentation, including bacterial fermentation and fungal fermentation. For example, through the fermentation dominated by lactic acid bacteria (LAB), the seaweed components can undergo significant biotransformation: LAB secretes extracellular enzymes (such as cellulase and alginate lyase) to degrade the polysaccharide network in the seaweed cell wall, releasing the internal active substances [24]. *L. japonica* fermented by *Aspergillus oryzae* can remove unpleasant odors and enhance antioxidant activity. The comprehensive utilization of seaweed fermentation by-products is a key link to improving the economic efficiency of the industrial chain. The fiber components that are not completely degraded in the fermentation residue can be converted into protein feeds rich in small peptides and free amino acids by solid-state fermentation. Although various microorganisms can be used for the fermentation of seaweed, research on them has been scarce to date. Further studies on the physicochemical and biological properties of the final products produced under different fermentation conditions are needed to provide strong evidence that fermented seaweed can be used as a product with high nutritional value.

5.4.4.5 Puffing

The core of seaweed puffing technology is to destroy the structure of seaweed cell walls by means of the instantaneous changes of pressure, temperature, and other parameters. Compressional puffing is a typical process. Practice shows that high pressure and high temperature can effectively deconstruct the polysaccharide-protein complex in seaweed. In addition, the sensory score results showed that the odor of seaweed was significantly reduced after puffing treatment, which indicated that the technology also had deodorization function.

5.4.4.6 Non-heat Treatment

In the field of non-heat sterilization, high-pressure, ultrasonic, plasma, and pulsed electric technologies are reshaping the seaweed preservation system. These technologies accurately inactivate microorganisms through physical field effects, while avoiding the destruction of active ingredients such as seaweed polysaccharides and pigments by traditional thermal processing. For example, high-pressure treatment can maintain the antioxidant capacity of brown seaweed, and pulsed electric field can simultaneously improve the protein extraction efficiency [25]. Multi-technology combination (such as ultrasonic-assisted plasma) further breaks through the limitations of a single mode and achieves synergy between sterilization and quality control.

5.5 The Application of Seaweed in the Food Industry

Seaweed is widely used in the food industry. For instance, it is added to many processed meals as a taste or nutritional booster and is most popular in traditional dishes like sushi, soups, and salads. Hydrocolloids such as carrageenan, agar, and alginate, which are derived from seaweed, possess emulsifying, stabilizing, and thickening properties and are important components in the food industry [26, 27].

5.5.1 The Application of Seaweed

5.5.1.1 Food Additives

Seaweed can be used to extract a variety of functional components, which have a wide application prospect in the food industry. Seaweed can not only be directly used to make seaweed foods, such as seaweed sausages, yogurt, bread, and algae jelly but also enhances the flavors, texture, taste, and appearance, as well as extending shelf life of the products. The extensive use of seaweed-derived hydrocolloids (such as alginate, propylene glycol alginate [PGA], and carrageenan) has been substantiated in numerous studies as essential food additives with diverse functional roles, including viscosity modulation, emulsification, and texture enhancement.

5.5.1.1.1 Gelling Agents and Thickeners The gel-forming seaweed polysaccharides can be used as gelling agents and thickening agents in food. Agar is a polysaccharide extracted from seaweed algae. It can be used as a gelling agent in soft candies and as a thickening agent in yogurt, cheese, jam, and canned foods. Sodium alginate is a polysaccharide extracted from brown algae. Studies have found that sodium alginate has a unique tissue improvement effect in baked foods such as bread, cake, biscuit, and egg roll, especially in some high-fiber health bread and cake, which can increase the elongation, toughness, elasticity, and texture of bread, and reduce the slime when slicing. Leveraging the thermally irreversible gelation properties of alginate-calcium ion interactions, this technology significantly enhances meat products' texture and functional performance. When applied to meat enemas, it improves product texture and slicing consistency. In meatballs, it markedly increases brittleness and elasticity. PGA is a derivative of alginate, which can play an acid-resistant thickening role in ketchup and other low pH sauces. Four seaweed extracts—sodium alginate, carrageenan, alginate propylene glycol ester, and agar—were combined to replace alum in the production of aluminum-frcc potato noodles, thereby enhancing food safety.

5.5.1.1.2 Stabilizers and Emulsifiers Alginate is widely applied in the production of biscuits and egg rolls, effectively reducing product breakage, enhancing surface smoothness, and ensuring superior moisture resistance. Alginate can endow ice cream and other cold beverages with a silky-smooth appearance and a smooth taste. PGA can be used as a stabilizer of beer foam, and the general dosage is 40–120 ppm. In addition, adding 0.1%–0.2% PGA to soy protein drinks can eliminate the grit taste of soy globulin and increase the stability of the product.

Carrageenan, characterized by its anionic polyelectrolyte structure, forms stable complexes with positively charged proteins through electrostatic interactions [18]. This unique binding mechanism enables carrageenan to function as a multifunctional food additive, serving as a stabilizer (preventing phase separation), emulsifier (enhancing droplet dispersion), and clarifying agent (precipitating suspended particles) in various food systems, including chocolate milk, ice cream, soymilk, ricotta cheese, canned whipped cream, infant formula, nutritional supplements, low-fat sandwich meats, and condensed milk.

5.5.1.1.3 Water-Retaining Seaweed polysaccharides are extremely important and widely used natural food water-retaining agents, playing a crucial role in the food industry. Algal polysaccharides, especially rich in hydrophilic structures such as L-fucoside and sulfate groups, can be used to synthesize water-retaining agents. As a surface coating for barbecue and fried meat products, it effectively reduces cooking-induced juice loss while maintaining tenderness and juiciness.

5.5.1.1.4 Extend Shelf Life Seaweed polysaccharides can form hydrophilic gel networks or high-viscosity solutions, locking in the moisture within food, reducing the water activity (Aw), inhibiting the growth of moisture-dependent microorganisms and enzymatic reactions, thereby effectively delaying food spoilage and deterioration in quality.

Alginate and carrageenan make the cake form porous swelling tissue, fine and uniform pores, rich elasticity, soft endoplasmic lubrication, and good dry resistance, which can appropriately extend the shelf life. An edible coating made by mixing brown algae polyphenols, chitosan, and sodium alginate can be applied to the surface of fresh fruits and vegetables. This coating can inhibit the growth of mold, yeast, and bacteria, and delay the spoilage of fruits and vegetables. Brown algal polyphenols and fucoxanthin have antioxidant properties and can be applied in foods rich in unsaturated fats to delay the off-flavor caused by fat oxidation.

5.5.1.1.5 Marine Flavor Seaweed can be used in foods that offer a special, rich aroma and flavor, which helps to enhance consumers' acceptance. Incorporating seaweed into processed foods can endow them with aroma and a unique Marine flavor, and sometimes it can also produce astringency or bitterness. At present, more than 200 volatile molecules have been identified in seaweed, including hydrocarbons, ketones, aldehydes, alcohols, halogens or sulfur-containing compounds, acids, esters, furans, and phenols [28]. Therefore, seaweed has the most significant potential to influence the perception of sensory information.

5.5.1.2 Functional Foods

The nutritional potential of seaweed is directly related to its biochemical composition and biological activity properties, and these properties vary greatly among

Table 5.5 The health benefits and applications in food industry of the main bioactive compounds of seaweed.

Types of seaweed	Examples of seaweed	Bioactive compounds	Application in the food industry	Health benefits
Brown algae	*Saccharina japonica*	Alginate	Stabilizer, gelling agent, water-retaining agent, thickening agent, functional food	Antioxidation, antitumor, anti-inflammatory, antiviral
	Fucus vesiculosus	Fucoidan		Anticoagulant and antithrombotic, antivirus, antitumor, immunomodulation, hypolipidemic, antioxidant, anticomplementary
	Eisenia bicyclis	Laminaran		Antiviral, anticoagulant, antioxidant, antimicrobial
	Undaria pinnatifida	Fucoxanthin		Anti-obesity, antioxidation, antitumor, anti-inflammatory, antiviral
	Ecklonia cava	Phlorotannins		Antioxidant, anticancer, antidiabetic, antivirus, hypotensive, matrix metalloproteinase and hyaluronidase inhibition, radioprotective, anti-allergic
Red algae	*Gigartinaceae*	Carrageenans	Thickeners, gelling agents, suspending agents, emulsifiers, and stabilizers	Antioxidant, antiviral, anticancer, anticoagulant, hypolipidemic, immunoregulation
	Gelidiaceae	Agar		Antioxidation, antitumor, anti-inflammatory, whitening effects, anti-obesity, and antidiabetic
Green algae	*Ulva* sp.	Ulvan	Functional foods	Immunoregulation, antiviral, anticancer, antioxidant, hypolipidemic
	Caulerpa racemose	Flavonoids	Preservative	Antioxidant, antimicrobial, cardioprotective, hyperglycemic, anti-inflammatory, anti-allergic

different species. The polysaccharides (such as alginate and algal polysaccharides), proteins (such as phycocyanin), polyphenols (such as xanthine), carotenoids (such as fucoxanthin), and PUFAs (such as EPA) extracted from seaweeds are key components of seaweeds and can be applied in functional foods. The main bioactive compounds of seaweed and their important bioactivity are summarized in Table 5.5.

5.5.2 Types of Seaweed Products

5.5.2.1 Food and Food Additives

Seaweed has a long history as a food source in Asia, particularly in Korea, Japan, and China. The well-known edible seaweeds include *L. japonica*, *Coccoloba uvifera*, *Caulerpa racemosa*, *Ulva pertusa Kjellman*, *Gracilaria verrucosa*, *Nori nei*, and *U. pinnatifida*. The diversity of seaweed makes its global application possible. For instance, nori is the main ingredient for making sushi, and kelp is often made into kelp soup, which is deeply loved by both Japanese and Korean people. Seaweed extracts can be used to make dietary supplements or functional foods.

Seaweed extracts can be used as emulsifiers, thickeners, gelling agents, water-retaining agents, etc. to enhance the sensory properties of food. They can also be added as nutritional fortifiers to foods such as bread, cookies, and energy bars to increase the content of dietary fiber and minerals.

5.5.2.2 Agriculture and Animal Feed

Seaweed fermentation products can be used as organic fertilizers, providing crops with trace elements, natural growth hormones (such as cytokinins and auxins), organic matter, and beneficial microorganisms; they also can improve soil structure, enhance plant stress resistance (resistance to drought, cold, and diseases), and increase yield and quality. Seaweed powder can be added as feed to livestock, poultry, and aquaculture feed, providing minerals (especially iodine and selenium), vitamins, and essential amino acids for animals, thereby enhancing their immunity. Studies have shown that feeding chickens with *U. lactuca* can increase meat production and breast meat percentage, while reducing abdominal fat content [29].

5.5.2.3 Pharmaceuticals and Biological Products

The active components extracted from seaweed, such as polysaccharides, terpenoids, and alkaloids, possess bioactivities including antitumor, antiviral, antibacterial, antioxidant, lipid-lowering, hypoglycemic, and anticoagulant properties, and can be used to develop drugs for treating various diseases. For instance, the drugs such as Haikun Shenxi Capsules for kidney failure and anticoagulant drugs like PGA sodium sulfate (PSS) are all seaweed polysaccharide-based medications. Alginates can be used in wound dressings, drug sustained-release carriers, etc. Ultrapure alginate can be used to develop human implant materials, such as vascular stents.

5.5.2.4 Chemical Products

Seaweed polysaccharides can be used as an active ingredient and thickening agent in cosmetics, enhancing the moisturizing and hydrating effects of the cosmetics. The combination of alginate and diatomite can be used for wastewater treatment.

5.6 Market Prospects of Seaweed

Seaweed is gradually expanding from traditional food to multiple high-value fields, showing huge market potential. The $4.7 billion worldwide algae products market is predicted to increase at a compound yearly growth rate of 6.3% to $6.4 billion by 2026.

Seaweed is rich in a variety of bioactive ingredients, which makes it widely used in food, biopharmaceutical, and agricultural processing industries, and is closely related to human life and health. In addition, from a sustainable development perspective, the development of seaweed applications is in line with the Sustainable Development Goals of the United Nations today, including food security, health and well-being, ecosystem protection, and bioenergy development.

To meet the huge consumer demand, the industry needs to fully utilize its potential, overcome the existing problems, and make a successful transformation. From a technical perspective, the transition from labor-intensive to a combination of genetic breeding and artificial intelligence processes; from a product perspective, from raw material exports to high-value products; from a policy perspective, the integration of standard systems, such as the joint establishment of a global certification system by ISO and FAO, to standardize the industrial structure and regulation. In addition, due to the potential hazards of heavy metals or high iodine in some edible seaweeds, the safety and health of seaweed use should be studied in the future. At the same time, future research should also focus on the fermentation, encapsulation, and processing methods of seaweed to improve the biodigestibility and bioavailability of the nutrients contained.

References

1 Hsu, H.Y. and Hwang, P.A. (2019). Clinical applications of fucoidan in translational medicine for adjuvant cancer therapy. *Clinical and Translational Medicine* 8 (1): e15.

2 Bergstrom, E., Lahnstein, J., Collins, H. et al. (2023). Cell wall organic matrix composition and biomineralization across reef-building coralline algae under global change. *Journal of Phycology* 59 (1): 111–125.

3 El-Manaway I.M. and Rashedy S.H. (2022). Sustainable global resources of seaweeds volume 1: bioresources, cultivation, trade and multifarious applications, In: *The Ecology and Physiology of Seaweeds: An Overview* (ed. A. Ranga Rao, G.A. Ravishankar), 3–16. Cham: Springer.

4 Xing, R., Ma, W., Shao, Y. et al. (2019). Factors that affect the growth and photosynthesis of the Filamentous Green Algae, *Chaetomorpha Valida*, in Static Sea cucumber aquaculture ponds with high salinity and high pH. *PeerJ* 7: e6468.

5 Zhang, J., Yao, J., Hu, Z.M. et al. (2019). Phylogeographic diversification and postglacial range dynamics shed light on the conservation of the Kelp *Saccharina Japonica*. *Evolutionary Applications* 12 (4): 791–803.

6 Dos Santos, G.S., de Souza, T.L., Teixeira, T.R. et al. (2023). Seaweeds and corals from the Brazilian Coast: review on biotechnological potential and environmental aspects. *Molecules* 28 (11): 4285.

7 Krishnan, L., Ravi, N., Mondal, A.K. et al. (2024). Seaweed-based polysaccharides–review of extraction, characterization, and bioplastic application. *Green Chemistry* 26 (10): 5790–5823.

8 Peñalver, R., Lorenzo, J.M., Ros, G. et al. (2020). Seaweeds as a functional ingredient for a healthy diet. *Marine Drugs* 18 (6): 301.

9 Pereira, L., Cotas, J., and Goncalves, A.M. (2024). Seaweed proteins: a step towards sustainability? *Nutrients* 16 (8): 1123.

10 MacArtain, P., Gill, C.I., Brooks, M. et al. (2007). Nutritional value of edible seaweeds. *Nutrition Reviews* 65 (12): 535–543.

11 Premarathna, A.D., Tuvikene, R., Fernando, P.H.P. et al. (2022). Comparative analysis of proximate compositions, mineral and functional chemical groups of 15 different seaweed species. *Scientific Reports* 12 (1): 19610.

12 Bayomy, H.M. (2022). Effects of culinary treatments on the physicochemical properties of *Ulva Lactuca* collected from Tabuk Coast of Red Sea in Saudi Arabia. *Saudi Journal of Biological Sciences* 29 (4): 2355–2362.

13 Andersen, S., Noahsen, P., Rex, K.F. et al. (2019). Iodine in edible seaweed, its absorption, dietary use, and relation to iodine nutrition in arctic people. *Journal of Medicinal Food* 22 (4): 421–426.

14 Santos, S.A., Félix, R., Pais, A.C. et al. (2019). The quest for phenolic compounds from macroalgae: a review of extraction and identification methodologies. *Biomolecules* 9 (12): 847.

15 Hao, J., Zhang, J., and Wu, T. (2024). Fucoxanthin extract ameliorates obesity associated with modulation of bile acid metabolism and gut microbiota in high-fat-diet fed mice. *European Journal of Nutrition* 63 (1): 231–242.

16 Wu, Q., Wu, S., Cheng, Y. et al. (2021). *Sargassum fusiforme* fucoidan modifies gut microbiota and intestinal metabolites during alleviation of hyperglycemia in type 2 diabetic mice. *Food & Function* 12 (8): 3572–3585.

17 Li, S., Liu, W., Li, Y. et al. (2024). Extraction, purification, structural characterization and anti-hyperlipidemia activity of fucoidan from *Laminaria digitata*. *International Journal of Biological Macromolecules* 279: 135223.

18 Oliyaei, N., Zekri, S., Iraji, A. et al. (2025). Health benefits of algae and marine-derived bioactive metabolites for modulating ulcerative colitis symptoms. *Journal of Functional Foods* 125: 106690.

19 Deniaud-Bouët, E., Hardouin, K., Potin, P. et al. (2017). A review about brown algal cell walls and fucose-containing sulfated polysaccharides: cell wall context, biomedical properties and key research challenges. *Carbohydrate Polymers* 175: 395–408.

20 Wang, X., Zhang, L., Qin, L. et al. (2022). Physicochemical properties of the soluble dietary fiber from *Laminaria japonica* and its role in the regulation of type 2 diabetes mice. *Nutrients* 14 (2): 329.

21 Murphy, M.J. and Dow, A.A. (2021). Clinical studies of the safety and efficacy of macroalgae extracts in cosmeceuticals. *The Journal of Clinical and Aesthetic Dermatology* 14 (10): 37–41.

22 Koopmann, I.K., Möller, S., Elle, C. et al. (2022). Optimization of astaxanthin recovery in the downstream process of *Haematococcus pluvialis*. *Foods* 11 (9): 1352.

23 Wei, W., Zhang, X., Hou, Z. et al. (2021). Microbial regulation of deterioration and preservation of salted kelp under different temperature and salinity conditions. *Foods* 10 (8): 1723.

24 Jönsson, M., Allahgholi, L., Sardari, R.R. et al. (2020). Extraction and modification of macroalgal polysaccharides for current and next-generation applications. *Molecules* 25 (4): 930.

25 Hussain, M., Gantumur, M.A., Manzoor, M.F. et al. (2023). Sustainable emerging high-intensity sonication processing to enhance the protein bioactivity and bioavailability: an updated review. *Ultrasonics Sonochemistry* 97: 106464.
26 Roohinejad, S., Koubaa, M., Barba, F.J. et al. (2017). Application of seaweeds to develop new food products with enhanced shelf-life, quality and health-related beneficial properties. *Food Research International* 99: 1066–1083.
27 Ali, M.Q., Azhar, M.A., Munaim, M.S.A. et al. (2025). Recent advances in edible seaweeds: ingredients of functional food products, potential applications, and food safety challenges. *Food and Bioprocess Technology* 18: 4947–4974.
28 Cassani, L., Gomez-Zavaglia, A., Jimenez-Lopez, C. et al. (2020). Seaweed-based natural ingredients: stability of phlorotannins during extraction, storage, passage through the gastrointestinal tract and potential incorporation into functional foods. *Food Research International* 137: 109676.
29 Costa, M.M., Pestana, J.M., Carvalho, P. et al. (2022). Effect on broiler production performance and meat quality of feeding *Ulva lactuca* supplemented with carbohydrases. *Animals* 12 (13): 1720.

6

Microalgae

Carlo Esposito[1,2], Sara Pozzo[1], and Francesco Secundo[1]

[1] *Istituto di Scienze e Tecnologie Chimiche "Giulio Natta", Milan, Italy*
[2] *Department of Pharmaceutical Sciences, University of Milan, Milan, Italy*

6.1 Introduction

Interest in microalgae as food ingredients has accelerated in recent years, driven by the growing demand for sustainable, health-promoting alternatives to conventional food sources. Microalgae offer multiple advantages: they can be cultivated in diverse environments, including nonarable land, use fewer resources than traditional agriculture, and contribute to CO_2 sequestration. Their biochemical profile—rich in proteins, lipids, pigments, polysaccharides, vitamins, and minerals—makes them promising candidates for functional foods, nutraceuticals, and other high-value applications.

The global microalgae market is projected to grow from approximately USD 841 million in 2025 to around USD 1.58 billion by 2034, at a compound annual growth rate (CAGR) of about 7.25%. This growth is primarily driven by increasing demand in sectors such as food, feed, and nutraceuticals. (Source: Fortune Business Insights, https://www.fortunebusinessinsights.com/microalgae-market-110314).

However, achieving widespread, consolidated application in the food sector still faces multiple challenges. Key among them are the high costs and energy demands associated with harvesting, extraction, and drying processes, which significantly affect the economic viability of microalgal foods. Regulatory barriers also remain substantial: despite thousands of microalgal species having been identified, only a limited number have received regulatory approval in major markets. Furthermore, variations in nutritional composition depending on growth conditions, seasons, and cultivation systems add complexity to standardization and product consistency.

Another essential dimension is consumer acceptance, which depends on factors including familiarity, sensory attributes (taste, color, aroma, texture), perceived health benefits, and transparency (e.g., labeling and sustainability claims). Poor sensory perception and strong or unfamiliar flavors/colors can act as major barriers. Strategies to address these include optimizing sensory profiles, employing masking or blending techniques, and clearly communicating nutritional and environmental advantages.

This chapter will examine the following topics in depth:

1) Technological aspects: methods of cultivation, harvesting, extraction, and drying; emerging technologies aimed at improving yield, reducing energy consumption, and preserving bioactive compounds.
2) Bioactive compound classes: proteins, pigments (e.g., carotenoids, chlorophylls, and phycobiliproteins), lipids, and polysaccharides; their nutritional and functional roles; and approaches to extraction and purification.
3) Consumer-related aspects: acceptance, sensory challenges, labeling, and communication strategies, as well as the ethical, environmental, and economic perceptions.
4) Current challenges: regulatory approval, cost bottlenecks, variation in quality, and scalability of emerging processes.
5) Future perspectives: biorefinery approaches, market growth, new applications in food products (including integration into both familiar and novel food matrices), and sustainability considerations (resource use, environmental impact).

By addressing these issues, the objective is to map the current state of microalgae in food, identify key barriers and enablers, and point toward promising pathways for making microalgae-based foods more economically viable, technologically robust, and broadly acceptable to consumers (Fig. 6.1).

6.2 Main Microalgae Used in Food

Microalgae represent a sustainable and versatile source of functional ingredients for the food industry as sources of various active molecules or bioactive compounds, which include polysaccharides, fatty acids (FAs), proteins, peptides, polyphenols, diterpenes, steroids, and alkaloids in large quantities [1].

Over 30,000 identified species, only a few have received approval as safe ingredients for human consumption by regulatory bodies such as the FDA (Food and Drug Administration) and EFSA (European Food Safety Authority) [2]. The most widely used species have a unique nutritional profile and well-established industrial applications.

Among the most used microalgae for food purposes, Spirulina (*Arthrospira platensis, A. maxima*), whose global annual production in 2019 was estimated to be over 56,000 tons (dry mass), with current estimates for China alone around 10,000 tons. Spirulina is a filamentous cyanobacterium known for its high-protein content (55–70% of dry biomass), carotenoids, chlorophylls, phycocyanin, and B vitamins, including B_{12} [3, 4]. It is classified into two categories based on its habitat: freshwater spirulina (*Spirulina plantensis*) grows in freshwater or less saline environments and may have a different nutritional profile and growth rate than saltwater varieties, while saltwater spirulina (*Spirulina maxima* and *S. fusiformis*) thrives in more saline and alkaline conditions, generally with higher mineral and sodium content [5]. Used in powder or tablet form, it is considered a superfood for athletes, vegetarians, and the elderly. They are used in products such as snacks, bars, functional pasta, and plant-based drinks. It is classified as safe (GRAS) by the FDA and approved in the EU.

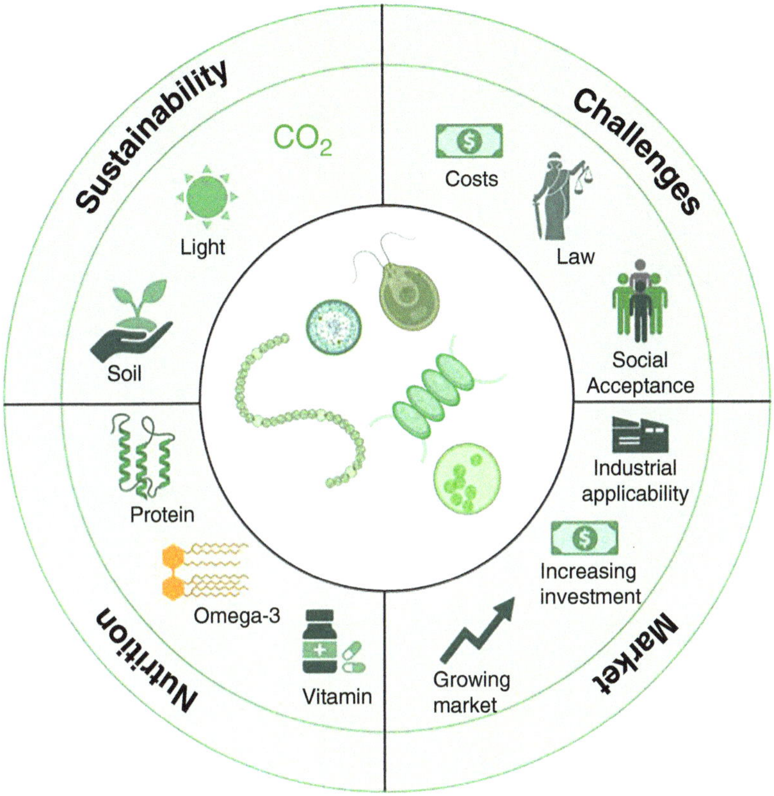

Figure 6.1 Overview of the role of microalgae in sustainable biotechnology. The figure illustrates their contribution to sustainability (use of CO_2, light, and soil), nutritional benefits (proteins, omega-3, vitamins), and market opportunities (growing demand, investments, industrial applicability), while also highlighting current challenges such as production costs, legislation, and social acceptance.

Chlorella (*Chlorella vulgaris*) is a green microalgae rich in chlorophyll, protein (up to 60%), iron, and fiber (Table 6.1) [6, 7]. It is known for its detoxifying effect and immune support. It is sold in powder, tablets, or as an ingredient in drinks and supplements [7]. It is one of the best-selling algae in the European market.

Dunaliella salina is a halophilic microalgae known for its very high β-carotene content (up to 10% of dry biomass) [8]. Finding application in both food and cosmetics, it is used in supplements for skin and vision protection [9, 10]. It is approved by EFSA as a natural source of carotenoids.

Haematococcus pluvialis is the main natural source of astaxanthin, a powerful antioxidant with protective effects on cardiovascular, neurological, and skin health [11]. It is grown in photobioreactors and marketed as a lipid extract or powder for supplements, energy drinks, and nutraceuticals [12]. EFSA has also approved it as a safe source of carotenoids.

Nannochloropsis sp. is a marine microalgae rich in EPA (eicosapentaenoic acid), an essential omega-3 (Table 6.1). It is used as a substitute for fish oil for the fortification

Table 6.1 Protein, carbohydrates, lipids, total dietary fiber, and carotenoid compositions expressed as %w/w on a dry basis ± SD (n = 3) of *A. platensis*, *Scenedesmus almeriensis, H. pluvialis* (green phase), *H. pluvialis* (red phase), *Chlorella vulgaris*, *D. salina*, *Nannochloropsis* sp.

Composition (%w/w on a dry basis ± SD)	*Arthrospira platensis*	*Scenedesmus almeriensis*	*Haematococcus pluvialis* (green phase)	*Haematococcus pluvialis* (red phase)	*Chlorella vulgaris*	*Dunaliella salina*	*Nannochloropsis* sp.
Protein	46.76 ± 0.95	12.93 ± 0.69	32.59 ± 1.20	25.69 ± 1.27	45.64 ± 1.20	10.03 ± 0.57	26.67 ± 1.10
Carbohydrates	3.32 ± 0.05	4.51 ± 0.41	0.13 ± 0.01	6.30 ± 0.24	5.30 ± 0.50	25.31 ± 1.55	32.05 ± 0.70
Lipids	1.40 ± 0.12	2.05 ± 0.12	3.24 ± 0.11	2.60 ± 0.10	3.13 ± 0.21	3.49 ± 0.10	15.30 ± 0.24
Total dietary fiber	42.82 ± 1.20	22.60 ± 1.50	34.56 ± 0.90	58.52 ± 2.56	35.04 ± 1.60	8.97 ± 0.50	17.67 ± 0.80
Carotenoids	<LdL*	0.30 ± 0.05	<LdL*	2.87 ± 0.15	<LdL*	3.46 ± 0.15	<LdL*

<LdL* = lower than the detection limit.
Source: [13]/ MDPI / CC BY 4.0.

of functional foods and baby foods. It is sold as an oil, flour, or microencapsulated ingredient. Protein content ranges from 18% to 35% [14].

The current market for microalgae-based products focuses primarily on the production of whole dried biomass from *Chlorella, Arthrospira,* and *Schizochytrium*, as well as specific compounds such as beta-carotene, astaxanthin, phycocyanin, EPA, and DHA [15, 16]. Although these products are still less competitive than conventional alternatives, they offer unique technological and application advantages in some sectors.

In addition to products already marketed, there is a growing landscape of new applications, which can be classified into three levels of technological maturity: (1) complete products, authorized but not yet on the market (e.g., arachidonic acid, phycoerythrin, and microalgae-enriched foods); (2) products in an advanced stage of development (e.g., fresh or frozen biomass from *Nannochloropsis*, *Isochrysis*, *Pavlova*, and others; fucoxanthin, polysaccharides); (3) early-stage products such as lutein and bulk oils.

All of the above-mentioned species are approved for food use and are among the innovative ingredients for the development of functional and sustainable foods. The use of microalgae also extends to "Phycogastronomy" an emerging field that integrates algae into gourmet dishes, elevating them from a nutritional ingredient to an element of haute cuisine.

6.3 Cultivation and Harvesting for Food Application

Harvesting microalgal biomass can account for a significant portion of total production costs, potentially up to 30%, as a single liter of culture contains only a few grams of biomass, requiring large volumes to achieve economically viable yields. Various harvesting strategies exist, often combined sequentially to improve efficiency. These methods can be divided into conventional techniques, already used at commercial scale, and non-conventional approaches, still largely in the research phase. Estimated costs for conventional methods range from approximately $28 per ton for alum-based coagulation/flocculation to approximately $500 per ton for centrifugation. By comparison, emerging technologies currently cost approximately $1,350 per ton (17), indicating that conventional approaches remain more practical and cost-effective for large-scale operations (Fig. 6.2).

6.3.1 Filtration Techniques

Filtration is one of the main techniques used for microalgae harvesting and is mainly divided into *screening* and *membrane filtration*. The first involves the use of mechanical filters to select microalgae based on their size (typically >100 to >25 μm). However, these filters require frequent cleaning and are not efficient for small cells [17, 18]. Filters may be inclined or vibrating. Membrane filtration, on the other hand, uses micro- or ultrafiltration, which is more energy-efficient compared to centrifugation. Nevertheless, fouling, caused by microalgae producing EPS

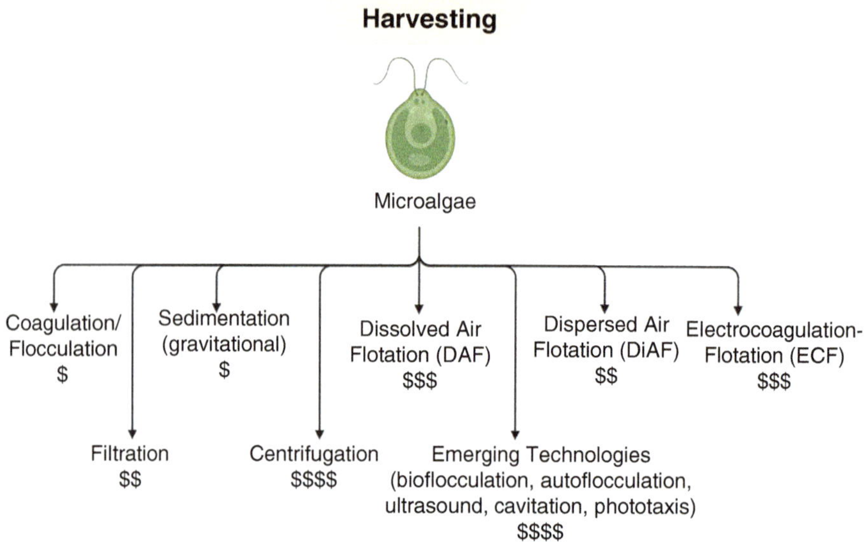

Figure 6.2 Downstream processing of microalgae biomass after harvesting, leading to different applications such as biofuels, feed, nutraceuticals, and pharmaceuticals. The relative cost of each technique is indicated by the number of dollar signs ($–$$$$), with a greater number denoting a higher cost.

(extracellular polymeric substances), limits its performance. To mitigate fouling, cross-flow filtration is preferred over dead-end filtration, as it reduces clogging and yields more concentrated slurries [19, 20].

6.3.2 Sedimentation and Flocculation Techniques

6.3.2.1 Sedimentation

Gravitational sedimentation separates denser microalgal cells from the clarified supernatant [21]. This technique is low-cost but only suitable for large, nonmotile species. The process speed is governed by Stokes' law. It requires long processing times and has a limited final yield (2–3% total solids) [20, 22]. Sedimentation is usually implemented through tanks, clarifiers, or inclined settlers and is often used downstream of flocculation processes.

6.3.2.2 Coagulation and Flocculation

Several flocculation methods exist, including electric, chemical, biobased, or magnetic flocculation, allowing aggregation of small cells (2–20 μm). Electric-based methods like electrophoresis promote aggregation due to the negative surface charge of microalgae, allowing cells to concentrate under an applied electric field [23]. Chemical coagulants (e.g., aluminum or iron salts) or synthetic polymers neutralize surface charges, enabling coagulation [24]. Although effective, they may contaminate biomass and water. Natural biopolymers such as chitosan are promising and have been tested on various algal species (e.g., *C. vulgaris* and *Nannochloropsis* sp.) with up to 90% efficiency. However, they remain too costly [24, 25]. The effectiveness of these methods depends on multiple factors, including algal species, water chemistry, and operating conditions.

6.3.3 Flotation Techniques

Flotation relies on the adhesion of air bubbles to cells to carry them to the surface for removal [20, 25]. Several types of flotation techniques have been developed:

DAF (Dissolved Air Flotation): Very effective (>99% removal with flocculants), producing bubbles with diameters from 50 to 100 μm [26, 27]. Despite its efficiency, it is energy-intensive and less suitable for saline environments.

DiAF (Dispersed Air Flotation): Produces larger bubbles (700–1500 μm) using dispersed air [25]. Lower efficiency requires the addition of surfactants or flocculants. Also known as *foam fractionation*.

ECF (Electrocoagulation-Flotation): Produces microbubbles (15–45 μm) through electrolysis of water. Uses sacrificial electrodes (releasing metallic flocculants) or carbon-based electrodes [28]. Suitable for saline cultures but requires careful tuning to avoid the formation of unwanted compounds.

6.3.4 Centrifugation

Centrifugation is highly efficient (recovering 80–90% of biomass) and does not require chemical additives. However, it demands significant energy input (up to 20 kWh/m^3) and is unsuitable for fragile species such as *D. salina*, due to cell rupture [29]. Therefore, it is mainly used for high-value biomass or as a concentration step following other harvesting methods.

6.3.5 Emerging Technologies

Bioflocculation: Uses EPS produced by microalgae or other microorganisms (e.g., fungi) to form flocs that aggregate via electrostatic interactions. Effective, but often introduces contaminants from auxiliary species.

Autoflocculation: Relies on spontaneous EPS production triggered by abiotic or biotic stress (e.g., pH shifts and predators). Some species flocculate in response to grazers like *Daphnia* or *Tetrahymena*. However, the underlying mechanisms remain poorly understood [30].

Ultrasound and Cavitation: These techniques induce bubble collapse, releasing EPS and enhancing flocculation. Promising but energy-intensive and more commonly applied for cell disruption [31].

Phototaxis: This technique exploits the light-induced movement of motile microalgae (e.g., *Chlamydomonas*), enabling chemical-free harvesting. Efficiency remains low and is limited to phototactic species. It is influenced by light intensity, direction, and wavelength [31].

6.4 Drying

After harvesting, the stabilization of microalgal biomass is a crucial step to preserve its biochemical properties and ensure its suitability for various applications. In this context, drying represents a fundamental phase of the process, as it allows for the preservation of sensitive bioactive compounds such as antioxidants, polyphenols, and anti-inflammatory agents, as well as essential macronutrients like proteins, lipids, and carbohydrates. The removal of water from the biomass also prevents the degradation of moisture-sensitive molecules, improves product shelf life, and promotes lipid concentration, thereby making the biomass more suitable for the production of biofuels, dietary supplements, cosmetics, and biotechnological applications. Drying also contributes to enhancing the physical stability of the biomass, facilitating its storage and transport, particularly in large-scale production settings [32]. The dried biomass can subsequently be processed into powders, extracts, or other formulations for use in various sectors, including food, cosmetics, and pharmaceuticals [33]. However, it is important to note that this stage, along with extraction, can account for up to 60% of the total production costs [34]. Numerous drying techniques are available for microalgae, each with varying suitability depending on the characteristics of the desired final product, but all aim to reduce the moisture content in the biomass (Fig. 6.3).

6.4.1 Oven Drying

This technique relies on the application of heat to remove moisture from microalgae [35] and allows for precise control of both temperature and processing time. It is a simple and widely accessible method, making it particularly suitable for laboratory applications or small-scale industrial production [36]. The operating temperature

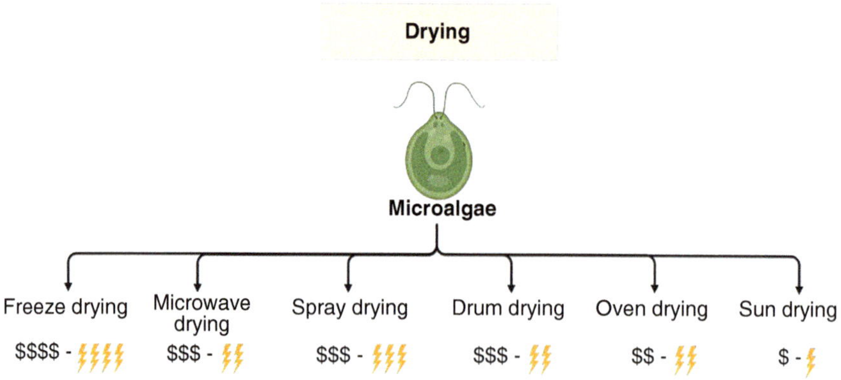

Figure 6.3 Comparison of drying methods for microalgae biomass, illustrating differences in economic costs and energy requirements associated with each technique. The relative cost and energy demand of each technique are indicated by the number of dollar signs ($–$$$$) and lightning symbols, respectively, with a greater number denoting higher values.

typically ranges between 40 and 70 °C, depending on the microalgal species involved, and must be carefully monitored to prevent the degradation or alteration of bioactive compounds in the biomass [37, 38]. Several studies have successfully applied this method to species such as *Tetradesmus obliquus*, *Scenedesmus*, and *Chlorella pyrenoidosa* [39–41], demonstrating its effectiveness in preserving lipids, proteins, and other valuable biochemical compounds.

6.4.2 Sun Drying

Solar drying is a traditional technique that has been used for centuries to stabilize wet algal biomass. In this process, the energy required for water evaporation is provided directly by solar radiation, while moisture removal occurs through natural airflow. As it requires minimal infrastructure and simple equipment, this method is particularly suitable for small-scale applications or in settings with limited resources [42]. It is also a highly cost-effective and environmentally sustainable solution, relying solely on sunlight as an energy source [43]. However, this technique has certain limitations. Its effectiveness is strongly dependent on environmental conditions, and it cannot be reliably applied in the absence of sufficient sunlight. Additionally, the direct exposure of biomass to the open environment increases the risk of microbial or environmental contamination during the drying process [45, 47, 48].

Despite these drawbacks, solar drying has shown a significant reduction in energy consumption and operational costs when compared to more advanced methods such as spray drying [44]. The optimal operating temperature ranges from 34 to 42 °C [45, 46], with an average thermal efficiency of 62%, and a total process duration of approximately 3 to 5 hours. Recent studies have demonstrated that this method can significantly enhance the levels of bioactive compounds, with phycocyanin and flavonoids in microalgae increasing by 43% and 28%, respectively, while also contributing to CO_2 savings.

Microalgae such as *Spirulina* [47] and *Dunaliella* [48] are particularly well suited to this method, due to their natural resistance to high light intensity and prolonged solar exposure, which does not significantly compromise their biochemical properties [49].

In particular, the application of this technique to *C. vulgaris* has yielded excellent results, showing only a slight reduction in vitamin B_6 and some mineral content when compared to freeze-drying [50].

6.4.3 Drum Drying

The process involves passing the microalgal biomass through a rotating drum with heated metal walls to remove moisture [51]. The ability to operate continuously and handle large volumes makes this technique particularly appealing from an industrial perspective [52]. Moreover, drum drying can be easily integrated with other existing processing systems, making it a suitable technology for upgrading already operational facilities [53]. However, parameters such as optimal drum loading must be carefully monitored to achieve high yields [46]. A major drawback is the high operating temperature of the drum (150–180 °C), which can lead to protein denaturation and cause changes in color and taste (e.g., burning of the product). As a result, this method is unsuitable for products that are high in sugar or salt content [54].

6.4.4 Freeze-Drying

Freeze-drying is a widely recognized and commonly used drying method for the dehydration of microalgal biomass [33, 52, 55]. The combination of low temperatures and vacuum conditions facilitates the sublimation of ice directly from the solid to the vapor phase, bypassing the liquid state entirely [56]. One of the main advantages of this technique is its ability to preserve the nutritional content and intrinsic bioactive compounds of microalgae. However, the procedure is time-consuming and energy-intensive, primarily due to the essential freezing and sublimation stages [55]. Optimizing various process parameters, such as freezing rate, drying temperature, and pressure, plays a crucial role in achieving a high-quality product while reducing processing time and energy consumption [35, 55]. Several studies have demonstrated that this technique effectively preserves the biochemical characteristics of the final product [57, 58].

6.4.5 Spray Drying

In this system, the microalgal suspension is atomized into small droplets inside a drying chamber, where it undergoes a rapid dehydration process at high temperatures (up to 220 °C) using hot air or gas [38, 59]. The result is a fine powder that is easily storable and applicable across various industrial sectors, including chemical (e.g., fertilizers and detergents), pharmaceutical (e.g., drug formulations), and food (e.g., milk powder production) industries [60, 61]. One of the main advantages of this technique is the ability to precisely control critical parameters such as processing temperature, droplet size, and airflow. This allows for the effective preservation of bioactive compounds, including lipids, carbohydrates, proteins, and FAs. Several studies have shown that, despite the high temperatures involved, the nutritional and functional quality of the final product can be comparable to that obtained through freeze-drying. However, heat-sensitive compounds such as vitamins and pigments may degrade during the process [33, 62]. For these reasons, spray drying is considered a viable alternative to freeze-drying, offering the added advantages of speed and scalability [63]. This technology has been successfully applied to various microalgae species, including *C. vulgaris*. In one study, the drying process was conducted at 93 °C, and no major alterations in the biochemical profile were observed, aside from a reduction in sugar content and browning of the final product, likely due to Maillard reactions [50]. Despite its effectiveness, spray drying has demonstrated higher economic and environmental costs compared to techniques such as freeze-drying, solar drying, and pulse combustion drying. In other applications, such as with *Tetraselmis chui*, the process preserved 80%–92% of β-carotene content and 46%–81% of phenolic compounds, indicating good retention of bioactive molecules [62].

6.4.6 Microwave Drying

The process leverages the ability of microwaves to interact with water molecules within the biomass, generating heat rapidly and uniformly from the inside out. The main advantages of this technique lie in the speed of the dehydration process, which significantly reduces heat exposure time. This is crucial for preserving the integrity of bioactive compounds, the nutritional composition, and the functional properties of the final product [64, 65]. In addition to these benefits, microwave drying also offers high energy efficiency and generally more uniform and repeatable processing results. Behera and Balasubramanian [37] evaluated the performance of microwave drying using a power range between 360 and 900 W and compared it with conventional oven drying (OD) at temperatures ranging from 40 to 100 °C. The effectiveness of microwave systems is attributed to their ability to heat biomass evenly and volumetrically through dipolar interaction with water molecules, resulting in greater effective diffusivity compared to traditional methods. Other studies have also explored this technique, highlighting that a power range between 20 and 30 W g^{-1} represents an optimal compromise for reducing moisture content while preserving the biochemical stability of the biomass [66, 67].

6.5 Extraction of Bioactive Compounds

Microalgae represent a valuable source of high-value-added bioactive compounds, including natural pigments, antioxidants, proteins, and lipids. Various species are currently cultivated and processed for the targeted extraction of these molecules, which are used in sectors such as functional foods, dietary supplements, and cosmetics. For example, phycocyanin, a blue pigment extracted from *Spirulina*, is isolated using food-grade solvents or filtration and employed as a natural colorant. Astaxanthin, on the other hand, is a powerful antioxidant obtained from *H. pluvialis*, which preserves its molecular integrity. Microalgae are also a promising source of proteins and omega-3 FAs, which can be recovered through mechanical or enzymatic cell disruption. The resulting extracts are used in an expanding range of innovative and sustainable products.

6.5.1 Protein

The protein content of microalgae varies significantly across species and is strongly influenced by growth conditions [14, 68]. Even within the same species, protein composition can change considerably due to various biotic and abiotic factors, including the growth phase, cultivation system and mode, and environmental conditions such as temperature, light intensity, and pH [69, 70]. Additionally, substrate composition, such as the carbon source, nitrogen and phosphorus availability, and salinity, plays a critical role [14, 71].

Protein recovery strategies from microalgae are closely linked to the intended end-use of the product. A key distinction is made between whole-cell protein-rich biomass and extracted protein ingredients, which can be further classified based on their purity and degree of processing into protein concentrates, isolates, hydrolysates, and bioactive peptides [72]. Whole microalgae are commonly marketed as dietary supplements and generally do not require extensive processing beyond cultivation, harvesting, and drying. However, to obtain more refined protein products, such as concentrates or isolates, cell disruption is necessary to enhance the release of intracellular compounds. Several approaches can be employed for this purpose, including mechanical methods (e.g., bead milling, high-pressure homogenization (HPH), ultrasound, or microwave treatment), nonmechanical techniques (such as enzymatic lysis or chemical treatments), and hybrid strategies [73].

Traditional extraction techniques, such as solid–liquid extraction for proteins or Soxhlet extraction for lipids, are widely used but present several limitations. These include long processing times, high solvent and energy consumption, and the need for multiple steps to achieve acceptable yields. To overcome these challenges, innovative and emerging technologies, such as ultrasound-assisted extraction (UAE), microwave-assisted extraction (MAE), and pulsed electric fields (PEFs), are being explored as more efficient, environmentally friendly, and sustainable alternatives. In particular, the effectiveness of various UAE systems, including water baths and ultrasonic probes operating at variable frequencies, has been extensively investigated for the recovery of proteins from microalgae.

Mechanical disruption methods, such as freeze–thaw cycles and HPH, have proven particularly effective, achieving protein yields above 20 g/100 g in species such as *C. vulgaris*, *Nannochloropsis oceanica*, and *T. chui* [74]. UAE, especially under alkaline conditions, further increased yields, surpassing 30%. In contrast, enzyme-assisted extraction generally resulted in lower yields (<10 g/100 g).

Combined approaches, such as UAE with sodium deoxycholate, have shown enhanced efficacy in breaking down rigid cell walls, improving protein solubility, and reaching extraction efficiencies up to 86% in *Chlorella* and *Spirulina* [75, 76]. Lupatini et al. optimized UAE conditions for *A. platensis*, achieving a protein yield of 75.76% and carbohydrate recovery of 41.52% using a pH of 9, temperature of 30 °C, 35 minutes of sonication at full power, and agitation at 80 rpm for 50 minutes. Similarly, Hildebrand et al. [77] reported even higher yields, up to 79.1%, from *C. vulgaris* by applying alkaline UAE with 0.4 M NaOH, significantly outperforming simple maceration. These results confirm the effectiveness of alkaline UAE as a highly efficient method for microalgal protein recovery.

PEF treatment has also proven to be an energy-efficient method for the selective release of proteins. Coustets et al. [78] demonstrated that efficiency is species-dependent, with fields of 3–6 kV/cm required for *C. vulgaris* and *N. salina*, respectively. In *H. pluvialis*, the cumulative number of pulses had a greater impact than field intensity. Similarly, using a cell wall-deficient mutant of *Chlamydomonas reinhardtii*, 'T Lam et al. [79] achieved up to 70% release of water-soluble proteins with minimal energy input, further enhanced by enzymatic pretreatment.

Other studies highlighted the kinetic advantages of combining PEF with thermal incubation. Katsimichas et al. [80] showed that *Spirulina* biomass pretreated with PEF and incubated at 40 °C achieved near-complete protein extraction in just 2.3 hours, compared to over 10 hours in controls. Similarly, Perez et al. [81] found that PEF combined with incubation significantly outperformed both ultrasound and autofermentation, increasing protein yield by 79% and demonstrating energy efficiency 1.7 times greater than HPH, making it a promising strategy for biorefinery applications.

MAE is also emerging as a promising strategy for the efficient recovery of microalgal proteins, due to its ability to induce rapid and selective heating. The dielectric effect accelerates cell wall disruption and enhances protein solubilization while minimizing thermal degradation compared to conventional heat-based methods [82, 83]. In addition to achieving high yields, MAE uses lower solvent volumes and reduces environmental impact, making it a versatile and potentially scalable technique for industrial applications.

Several studies have demonstrated its effectiveness. Mahali and Sibi [84] obtained a protein yield of 78% from *A. platensis*, with maximum solubility of 74.9% at pH 9.0, highlighting the critical role of pH in optimizing extraction efficiency. Chew et al. [64] proposed a microwave-assisted three-phase partitioning method for *C. vulgaris*, achieving a protein recovery of 63.2% and a separation efficiency of 67.2% under optimized conditions. Finally, Motlagh et al. [85] applied MAE to *N. oceanica* using ionic liquids, reaching a 65% yield under moderate conditions (0.5 g biomass, 2% [Ch][Ac], 40 °C, 30 minutes).

Overall, these findings confirm the potential of MAE as an efficient, selective, and sustainable technique for protein extraction from various microalgal species, while also demonstrating how innovative extraction technologies can maximize protein yields and offer faster, more eco-friendly solutions for industrial-scale microalgal biorefineries.

Purification Methods. After the extraction phase, the next step is purification, a process that is often costly and whose feasibility strongly depends on the target market and the commercial value of the proteins extracted from microalgae. Purification techniques include various chromatographic methods such as ion-exchange chromatography, affinity chromatography, gel permeation, and size exclusion chromatography. These methods allow for the separation of molecules based on physicochemical properties such as molecular weight, hydrophobicity, and electric charge, making them particularly useful for isolating specialized products like bioactive peptides [86]. For example, ion-exchange chromatography has been effectively used to remove chlorophyll, which can negatively affect the color, taste, and structure of protein extracts [87].

More generally, protein ingredients extracted from microalgae are initially purified by removing cellular debris through centrifugation. Subsequently, soluble proteins are often concentrated either by precipitation—induced by adjusting the pH of the medium—or by membrane filtration systems such as ultrafiltration and diafiltration.

Ursu et al. [88] studied a combined approach based on isoelectric point precipitation and tangential ultrafiltration. Protein precipitation yielded recovery rates of 98% and 71% (w/w) at pH 12 and pH 7, respectively. Fractionation by tangential ultrafiltration using a 300 kDa membrane resulted in approximately 87% and 95% protein retention at pH 7 and 12, respectively.

Safi et al. [89] also investigated ultrafiltration/diafiltration processes using membranes with different molecular weight cutoffs (1,000, 500, and 300 kDa). They observed that a combination of enzymatic cell lysis (with proteases) followed by ultrafiltration/diafiltration led to soluble protein yields of about 25% in the filtrate obtained with the 300 kDa membrane. Moreover, increasing the membrane cutoff did not significantly improve process performance.

Liu et al. [90] studied the fractionation of lipids, soluble proteins, and other compounds from *Parachlorella kessleri*. After cell disruption by bead milling (pH 7, residence time 5.6 minutes) and removal of debris by centrifugation (3,000 g for 20 minutes), the permeate obtained through dynamic filtration with a flat-sheet polysulfone membrane (0.1 μm cutoff, 0.3 bar pressure) allowed recovery of 18% of the initial dry biomass weight. This permeate contained 63% ash and 10% soluble proteins on a dry weight basis. The authors also highlighted that the purity of all water-soluble compounds could be further improved by applying an additional diafiltration step.

6.5.2 Carotenoids

Carotenoids are lipophilic pigments widely distributed in microalgae and play a key role both in photosynthesis and as high-value bioactive compounds with nutraceutical potential. These compounds are classified into two main categories: carotenes, composed exclusively of carbon and hydrogen atoms ($C_{40}H_{56}$), and xanthophylls, which are derived from carotenes through oxidation and contain oxygenated functional groups (e.g., carbonyl, carboxyl, hydroxyl, and epoxy groups). Huang et al. [91] compiled the presence and distribution of 42 xanthophylls and 5 carotenes in microalgae, showing variability across chlorophytes, dinoflagellates, and prasinophytes (Fig. 6.4).

Among the most relevant carotenoids from both nutritional and technological perspectives are: β-carotene, lycopene, zeaxanthin, lutein, violaxanthin, and astaxanthin. These compounds are found in many microalgal species, although their abundance can vary significantly depending on cultivation conditions.

The growing demand for natural carotenoids has driven the development of more efficient and sustainable extraction techniques.

Shokrkar and Nedaei [92] investigated the extraction kinetics of chlorophylls and carotenoids from mixed microalgae using methanol as a solvent. Optimal conditions (40 °C, pH 7, 300 rpm, 4 g/L biomass) yielded extraction efficiencies of 94.10% for chlorophylls and 95.43% for carotenoids, demonstrating the effectiveness of the protocol under controlled conditions.

A more environmentally friendly approach was proposed by Zazirna et al. [93], who compared UAE methods (probe and bath) for recovering carotenoids from various microalgae, initially focusing on *Microchloropsis gaditana*, and later extending the study to *Tisochrysis lutea*, *Porphyridium cruentum*, *Phaeodactylum tricornutum*, and defatted biomass. Ultrasonic probe extraction for 5 minutes in ethanol yielded results comparable to conventional long-term extraction (24-hour agitation in acetone), with values of 9.4 ± 2.5 and 9.6 ± 3.2 mg/g DW, respectively.

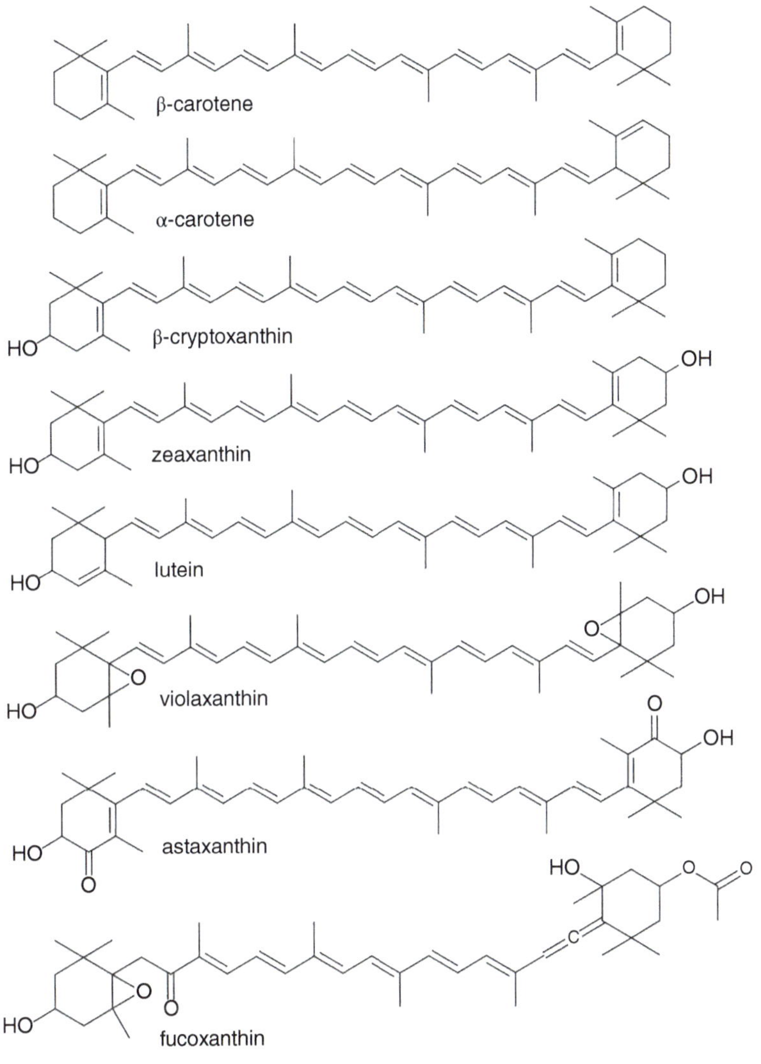

Figure 6.4 Structure of representative carotenoids commonly found in microalgae.

The ultrasonic method also proved significantly more sustainable (AGREEprep™ score: 0.62 vs. 0.22), achieving good yields across microalgae with diverse cell wall structures and in pretreated biomass, with recovery rates ranging from 41% to 100%.

A further significant contribution comes from Fernandes et al. [94], who optimized the extraction of carotenoids and chlorophylls from *Scenedesmus obliquus* and *A. platensis* using ethanol and ionic liquids as green solvents. Optimal conditions included a solid-to-liquid ratio of 1:10 and three 3-minute sonication cycles, resulting in yields of 2,602 and 764 μg/g DW for carotenoids, and 22 and 5.8 mg/g DW for chlorophylls, respectively, for *S. obliquus* and *A. platensis*. The ethanolic extracts demonstrated high antioxidant activity, up to 26 times greater than α-tocopherol, making this methodology particularly promising for the sustainable recovery of bioactive microalgal ingredients.

6.5.3 Chlorophyll

Chlorophyll is a fundamental pigment in photosynthetic processes, located both in the reaction centers and in the light-harvesting complexes of photosystems [95]. These molecules are valued not only for their biological function but also for their chromatic and nutraceutical properties, making them particularly important in the biorefinery sector. Although over 30,000 microalgal species have been identified, only a few are used industrially for pigment production: *Spirulina* and *Chlorella* are mainly employed as dietary supplements and feed additives (in tablet or powder form), whereas *H. pluvialis* and *D. salina* are used for the production of purified pigments.

The efficiency of chlorophyll extraction from algal biomass depends on several factors, including cell wall resistance, the type of physical pretreatment applied, solvent characteristics (such as penetration ability and solvation properties), contact time, and biomass storage conditions [96]. Pretreatments such as grinding, sonication, ultrasonication, or homogenization have been shown to enhance chlorophyll release, improving extraction yields. Moreover, freeze-drying the biomass prior to solvent extraction has proven to be more effective than simply freezing it after filtration [97]. The growth phase also influences pigment content: microalgae harvested in the stationary phase exhibit significantly higher concentrations of chlorophyll *a* than those collected during the logarithmic phase [98].

Traditionally, intracellular chlorophyll is extracted using organic solvents that penetrate the cell membrane and dissolve the lipids and proteins of the chloroplast membranes. Among commonly used solvents, dimethylformamide (DMF) has shown the highest efficiency, followed by methanol, ethanol, and acetone [99, 100]. However, methanol may induce degradation of chlorophyll *a*, while pure acetone, though less efficient in terms of yield, significantly limits the formation of degradation products [96]. In some cases, DMF has proven particularly effective for extracting chlorophyll from cyanobacteria and coccoid green algae, but its high toxicity limits its large-scale application.

In recent years, the need for more sustainable extraction approaches has led to the development of alternative technologies to conventional solvents. Among these, PEF-assisted extraction, MAE, and UAE have gained attention for their efficiency and lower environmental impact. PEF treatment applied to *C. vulgaris* demonstrated promising results: a field strength of 20 kV/cm for 75 μs increased chlorophyll *a* and *b* yields by 0.7 and 0.8 times immediately after treatment, while extraction performed one hour later resulted in higher increases (1.6 and 2.1 times, respectively). This enhancement was strongly correlated with irreversible electroporation of cell membranes, as indicated by a correlation coefficient of $R = 0.93$ [101, 102].

MAE relies on the interaction between microwave radiation and polar molecules, generating internal heat and promoting cell disruption. Orthesin et al. [103] tested a low-power version (LP-MAE) on dried *Spirulina platensis*, optimizing operating parameters. Using 90% methanol, a soaking time of 1 hour, 40 W for 5 minutes, and a solid-to-solvent ratio of 1:20, a chlorophyll yield of 5045 μg/g was achieved, without degrading the pigment. In a related study, Georgiopoulou et al. [104] applied MAE to *C. vulgaris* using aqueous ethanol (90%) at 60 °C, 300 W for 14 minutes. This process yielded 11.1% extract with high chlorophyll content (63.4 mg/g), carotenoids (24.9 mg/g), and phenolic compounds (9.3 mg GA/g), along with significant antioxidant activity. While conventional solvent extraction (SLE) achieved slightly higher overall yields, MAE and SFE provided extracts of comparable or superior quality in significantly shorter processing times.

Paterson et al. [105] applied CO_2-SFE to *Nannochloropsis gaditana*, obtaining up to 60.2 mg/g of chlorophyll and 21.2 mg/g of carotenoids. These yields were further enhanced by pretreatments such as freeze-thaw cycles and ultrasonication, as well as the addition of cosolvents like ethanol and ethyl acetate. According to Tzima et al. [106], SFE represents an effective and sustainable solution, particularly suitable for industrial-scale production of high-quality pigments. Supercritical CO_2 extraction (SFE) offers several advantages, including solvent-free extracts, high product purity, fewer processing steps, and mild thermal conditions that help preserve pigment stability. However, optimal conditions must be tailored to each microalgal species, as SFE efficiency is highly sensitive to pressure and temperature.

Overall, these findings highlight how innovative extraction technologies not only increase chlorophyll yields significantly but also ensure faster, more selective, and eco-friendly processes, paving the way for the industrial implementation of microalgal biorefineries.

6.5.4 Phycobiliproteins

Phycobiliproteins are classified into three main groups: phycoerythrin (red), phycocyanin (blue), and allophycocyanin (blue). Phycocyanin includes both C-phycocyanin (C-PC) and R-phycocyanin (R-PC). Currently, there is growing biotechnological and industrial interest in using natural colorants, especially blue pigments. The main known source is phycocyanin (PC), although its instability limits its application in certain industrial food processes, as it degrades at temperatures above 45 °C and is sensitive to light and pH changes [107, 108].

Previous studies on phycobiliproteins have focused on preventing color loss due to degradation. In this regard, several preservatives have been tested, including citric acid, fructose, beet pectin, crosslinking with methylglyoxal (MGO), and anionic micelles [109]. For example, phycobiliproteins from *S. platensis* showed improved stability in the presence of citric acid (4 mg/mL) at 35 °C for 15 days.

Among the various extraction methods for phycocyanins, both conventional and nonconventional approaches have been explored, with varying degrees of industrial applicability.

Among the different cell disruption methods used for C-PC extraction, freeze-thaw cycles have proven effective, yielding high purity values (up to 0.87) [110]. However, 3–4 cycles appear to offer the best compromise between efficiency and purity, as excessive cycles tend to release contaminating proteins [110]. Despite the high quality of the extracts, this method is time-consuming and energy-intensive, making it suitable mainly for laboratory-scale applications.

Simple mixing, while easy to implement, requires prolonged extraction times (up to 24 hours) and yields are generally modest. Homogenization, such as that performed using Ultra-Turrax, offers faster extraction due to high shear forces, but it is less selective and often leads to the co-extraction of cell debris [111, 112]. Increasing the temperature during the process can improve yield but negatively affects the purity of C-PC [113].

Bead milling stands out for its high disruption efficiency and is often used as a reference method to quantify total C-PC content [114, 115]. The results strongly depend on operational parameters such as bead size and density, agitation speed, and medium viscosity [73, 116, 117]. This method is particularly suitable for industrial applications due to its high biomass throughput and low energy consumption. However, similar to homogenization, its low selectivity requires additional purification steps.

HPH has proven effective for cell disruption but entails high energy consumption, especially at low cell concentrations (0.01–0.85% w/w). This method is also nonselective and necessitates further purification steps to isolate C-PC [118].

High-pressure processing (HPP) has shown that pressures up to 200 MPa enhance C-PC extraction, while higher pressures (e.g., 600 MPa) lead to significant protein degradation and up to 50% color loss [119]. These findings highlight the importance of pressure optimization to avoid damage to C-PC.

UAE offers the advantage of short processing times (2.5–16.2 minutes), with high yields achieved at moderate amplitude intensities (40–60%). Higher amplitudes do not improve yield and may reduce extract purity. Ultrasonic baths require longer durations (20–40 minutes) but can be combined with other methods to improve extraction performance [110, 120–123]. In the case of MAE, some studies have reported higher yields than bead milling or autoclaving. For example, 1400 W for 120 seconds outperformed 2 hours of milling at 60 rpm or 30 minutes of autoclaving at 121 °C [124]. However, microwaves pose a risk of overheating, which can degrade C-PC, as seen with 133 W for 166 seconds, where mixing at 7,000 rpm for 10 minutes resulted in purer extracts than either MW or US alone [112].

PEF stand out for producing selective blue-colored extracts, indicating the release of C-PC with minimal chlorophyll extraction. Posttreatment extraction times range from 15 minutes to 3 hours, suggesting that electroporation causes gradual membrane permeabilization that promotes partial and selective release of intracellular compounds [114, 125].

Lastly, moderate electric field (MEF) technology, which combines electrical and thermal effects, has shown improved extraction yields at 44 °C after 30 minutes compared to traditional methods such as freeze–thaw or untreated controls. However, at higher temperatures (e.g., 51 °C), yields decrease due to C-PC degradation, underscoring the need for precise control of temperature and treatment duration [126].

6.5.5 Lipids

Fatty acids (FAs), in addition to being fundamental components of membranes (as glycerolipids or sphingolipids), play a key role in energy metabolism and cellular signaling, especially when oxidized into oxylipins [127].

FAs are carboxylic acids with aliphatic chains ranging from 8 to 30 carbon atoms, and they can be saturated (SFAs), monounsaturated (MUFAs), or polyunsaturated (PUFAs), depending on the number of double bonds [128]. The application properties of FAs depend on chain length and degree of unsaturation; for instance, short-chain SFAs and MUFAs (such as 16:0 and 16:1) are suitable for biofuel and green chemistry applications [107].

Polyunsaturated FAs like arachidonic acid (ARA, omega-6), EPA, and docosahexaenoic acid (DHA) (both omega-3) are highly valuable for nutritional, medical, and biotechnological applications [129]. Furthermore, EPA, DHA, and ARA are precursors of bioactive signaling molecules such as eicosanoids, resolvins, maresins, protectins, and isoprostanoids, involved in the regulation of inflammation and immune responses [130–132].

6.5.5.1 Folch Extraction Method

A widely known lipid extraction method is the Folch technique [133], which uses a chloroform/methanol solvent system. This method has been extensively applied to various biological samples, such as tissues, plasma, and microalgae, and is considered ideal for lipid class separation [134, 135]. The original approach is based on biphasic liquid extraction, separating lipids into the organic phase and polar molecules into the aqueous phase.

The Bligh and Dyer method, for example, is considered more efficient for extracting specific lipid classes. While the Folch method remains a gold standard for comprehensive lipid extraction, it may not fully meet lipidomics requirements.

Bligh and Dyer Method. Developed in 1959, the Bligh and Dyer method is one of the most widely used protocols for extracting lipids from biological biomass. It uses a mixture of chloroform, methanol, and water to solubilize lipids, based on the "like dissolves like" principle. Compared to the Folch method, long considered the gold standard for total lipid extraction, the Bligh and Dyer protocol employs a different solvent-to-water ratio and has proven more efficient under certain conditions, with lipid yields up to 44% higher [136].

Recent improvements have enhanced its yield and adaptability. For instance, increasing methanol content has shown effectiveness, especially in green microalgae. Acid-phase modifications have improved the extraction of polar lipids like phosphatidic and lysophosphatidic acids [137].

Hybrid strategies combining Bligh and Dyer with innovative technologies have also emerged. The use of ionic liquids and subcritical water has improved lipid recovery from wet biomass without significantly altering the lipid profile [138]. MAE has also shown promising results, increasing efficiency and reducing operational time [139].

Overall, the Bligh and Dyer method remains a solid reference for lipid extraction, especially where high yields, selective efficiency, and compatibility with emerging technologies are required.

6.5.5.2 Bead Beating Extraction

Bead beating is a promising mechanical technique for extracting bioactive compounds from microalgae, due to its efficiency and cost-effectiveness [140, 141]. The method uses microspheres (glass-ceramic or steel) that, during agitation, break the algal cell walls and release intracellular content [142, 143]. It is effective for extracting lipids, proteins, pigments, and polysaccharides.

Optimization focuses on parameters such as bead size and material, agitation intensity, and duration, all of which influence yield and product quality. Combining bead beating with techniques such as flotation or flocculation further enhances biomass recovery. This method shows high potential for industrial-scale applications in algal biorefineries.

6.5.5.3 Green Solvent Extraction

Extraction with green solvents involves safer organic solvents compared to traditional chloroform and methanol, such as ethanol, isopropanol, butanol, and 2-ethoxyethanol. These solvents offer higher lipid recovery and improved safety standards while reducing environmental hazards [144, 145].

6.5.5.4 Solvent-Free Techniques

Solvent-free methods offer an eco-friendly alternative by minimizing the use of toxic compounds. Deep eutectic solvents (DES), when combined with subcritical water extraction (SWE), can disrupt tough algal cell walls and enhance lipid yields even from wet biomass [146, 147].

6.5.5.5 Osmotic Pressure Extraction

This method leverages hypertonic stress to induce cell lysis in microalgae, facilitating lipid release [148, 149]. Osmotic agents such as sodium acetate have shown good performance with species like *Tetraselmis marina* and *D. salina* [150]. Additionally, enzymatic pretreatment can improve extraction efficiency, and the use of wet biomass reduces energy costs associated with drying [151, 152].

6.5.5.6 Ultrasound-Assisted Extraction (UAE)

UAE uses acoustic cavitation to break cell walls and efficiently release lipids. UAE has been shown to improve both lipid yield and quality compared to other methods [153, 154]. When combined with MAE or pressurized liquids, it further enhances the extraction of omega-3 FAs from microalgae such as *Isochrysis galbana* [155]. Studies on wet biomass have shown that UAE significantly increases lipid yield and energy efficiency compared to methods like MAE [156, 157]. For example, UAE achieved a 21% higher yield than MAE from *Chlorococcum novae-angliae*, particularly favoring the extraction of saturated FAs, making it more suitable for biofuel production [157]. Moreover, UAE showed selectivity in lipid composition, while

UAE favors saturated FAs, MAE or other methods are more effective in extracting polyunsaturated FAs, relevant for nutraceutical and health applications [157, 158]. The latter study also introduced a floating microsphere-assisted variant, improving cell disruption and lipid recovery while generating a lipid profile suitable for biodiesel production [158].

6.5.5.7 Isotonic Extraction

Isotonic extraction uses osmotically balanced solutions to gently rupture cells, facilitating lipid release without damaging bioactive compounds [159]. This method is particularly suitable for wet biomass and has demonstrated higher yields compared to conventional techniques. Enzymatic pretreatment has also improved both the quality and quantity of extracted lipids [155, 164].

6.5.5.8 Microwave-Assisted Extraction (MAE)

MAE applies electromagnetic energy to heat the biomass, breaking cell structures and facilitating lipid release. It offers reduced extraction times, low solvent consumption, and is applicable to both wet and dry biomass [148, 165, 166]. Lipid yields are further improved when integrated with ionic liquids, brine extraction, and other hybrid approaches [148, 167]. MAE is a flexible and scalable technique with successful results across different microalgal species [168].

6.5.5.9 Enzyme-Assisted Lipid Extraction (EALE)

EALE employs specific enzymes to degrade algal cell walls, facilitating lipid release. Enzymes such as Celluclast and Viscozyme have proven effective in increasing the lipid yield of *I. galbana* [155]. The method's effectiveness increases when combined with UAE, MAE, or pressurized liquid extraction, due to synergistic effects [148, 169]. Moreover, the use of DES and bio-based solvents in combination with EALE supports green chemistry principles [146]. Enzymatic extraction not only enhances process sustainability but is also effective for subsequent biodiesel production, due to the compatibility of the extracted lipids with transesterification reactions [170].

In summary, emerging technologies for lipid extraction from microalgae are increasingly moving toward integrated, eco-friendly, and high-efficiency approaches, with significant implications for the bioeconomy and the sustainable production of biofuels and high-value products.

6.6 Consumer Acceptance of Food Products Containing Microalgae

Although the inherent biochemical traits and cultivation systems of microalgae align closely with the increasing demand for sustainable and health-oriented food production, consumer acceptance of microalgae-based food products, particularly in Western countries, remains limited. This hesitation is shaped by a complex interplay of personal, social, economic, ethical, environmental, and safety-related factors, all of which influence perception and purchasing behavior [171].

Consumer acceptance of these products depends on prior experiences and familiarity with microalgae. Acceptance and willingness to purchase vary among countries. Findings from consumer acceptance studies—whether based on panelists or consumer surveys—are context-specific to the population studied, although certain common patterns may emerge. To enhance consumer acceptance of microalgae-based foods, producers need to overcome sensory and cost-related barriers, strengthen consumer awareness through education, and integrate microalgae into more familiar and appealing food products in order to reduce barriers such as neophobia [172, 173].

Among the various drivers of consumer acceptance of new food products, knowledge stands out as a particularly influential factor for microalgae-based foods. Well-informed consumers are more likely to overcome initial personal and social barriers and to be open to novel food experiences. Understanding the nutritional profile and functional benefits of foods containing microalgae, along with awareness of the sustainability of their production, can strongly shape consumer perceptions and encourage their integration into the diet especially as concerns about environmental damage and limited resources grow, and the number of environmentally conscious consumers increases accordingly [172, 174].

Consumers are more likely to accept a microalgae-based food product when they are informed about the beneficial properties of the microalgae and their components. Some specific cases are reported below.

1) *S. platensis* and *C. vulgaris* are valuable sources of protein and essential amino acids. The poor protein digestibility caused by thick cell walls can be mitigated through mechanical and nonmechanical processes, and their high protein content and favorable amino acid profiles position microalgae as a realistic alternative in the context of a growing global population in need of new protein sources.
2) *N. oceanica* can provide omega-3 FAs suitable as a food supplement, helping to achieve a proper omega-3 to omega-6 ratio in human diets, which is important for cardiovascular health and inflammatory state control. Other microalgae also contain significant amounts of polyunsaturated FAs, the composition of which can be modulated under specific growth conditions.
3) Polysaccharides from microalgae contribute not only to nutritional properties, but some, such as β-1,3-glucans from *Chlorella*, *I. galbana*, and *Euglena*, also provide health benefits, including immunostimulatory, antioxidant, and cholesterol-lowering effects.
4) Microalgae contain high levels of bioactive molecules such as pigments, lipids and water-soluble vitamins, and carotenoids with health-promoting properties, making microalgae well-suited to meet the increasing consumer demand for functional foods and nutraceutical ingredients.
5) Compared to traditional protein sources such as crops and livestock, microalgae can have a lower impact in terms of liters of water required per kilogram of protein produced (up to about 30 times less for wet *Spirulina* vs. beef) and a lower space requirement per kilogram of protein.

6) Wastewater could be used as a medium for microalgal cultivation, although further processing is required to ensure the safety levels necessary for microalgae to be included in food.

In this context, labeling of new food products containing microalgae helps build consumer trust by providing clear and comprehensive information. This includes details on product characteristics (such as nutritional value, health benefits, organic certification, or vegan suitability) as well as production aspects (such as sustainability and ethical practices). Transparent labeling can motivate consumers to pay more for products that meet their expectations from multiple perspectives [172, 174].

Sensorial attributes act as both drivers and barriers to consumer acceptance, influencing the willingness to try, purchase, and repurchase microalgae-based products, which is essential for establishing their position in the market. Microalgae and microalgal-based food products are characterized by diverse colors, tastes, aromas, and textures resulting from variations in nonvolatile and volatile compounds such as pigments, amino acids, nucleotides, polysaccharides, peptides, and organic volatile compounds.

Sensorial properties and consumer perception depend not only on the microalgal species but also on the geographical conditions of cultivation and of purchase, as composition and perception may vary [173, 175]. Since color, aroma, taste, and texture all contribute to flavor, they can enhance palatability but may also discourage untrained consumers.

Color is one of the most important traits of food appearance, which significantly influences consumer acceptance. Microalgae are rich in pigments, and their use as food ingredients may impart peculiar, sometimes unusual colors that consumers may perceive differently: some may find an unexpected color intriguing and a colorful product attractive, while others may reject it, not recognizing it as familiar or related to a natural product. Knowledge can help, as awareness of food's health properties can encourage acceptance of its unconventional color. In some cases, color is directly associated with healthiness and can therefore support consumer purchases.

In addition, carotenoid pigments are precursors of molecules contributing to the typical aroma profile of microalgal ingredients, such as ionones and cyclocitral. When included in food products, pigment-rich microalgae might therefore positively or negatively affect flavor. This largely depends on the quantity used and the characteristics of the food matrix, which may mitigate the microalgal flavor [176].

A large assortment of odorous volatile organic compounds, along with nonvolatile taste compounds, play a role in microalgal flavors, which have been described as containing sulfurous, meaty, fishy, marine, fermented, and oily as well as herbaceous, floral, fruity, woody, and spicy notes [177].

The presence and concentration of nonvolatile molecules such as free amino acids, nucleotides, and pigments largely determine the taste profile. These molecules (Table 6.2) can act synergistically, contributing to umami, sweet, sour, bitter, salty, and sulfur notes in microalgae and in foods containing them. The most effective molecules for the distinctive umami taste are glutamic acid and guanosine

Table 6.2 Key taste active compounds contributing to microalgal flavor [175, 177, 178].

Class	Molecules	Attribute	Microalgae
Free amino acids	Glutamic acid—glutamate	Acid/Umami	*H. pluvialis, P. tricornutum, S. costatum, T. chui, C. vulgaris, A. platensis B. braunii*
	Aspartic acid—aspartate	Acid/Umami	*H. pluvialis, S. costatum, A. platensis, A. acuminatus, B. braunii, T. lutea*
	Alanine	Sweet	*H. pluvialis, P. tricornutum, T. lutea, S. costatum, A. platensis, A. acuminatus, B. braunii; C. vulgaris*
	Glycine	Sweet	*H. pluvialis, S. costatum, A. platensis, A. acuminatus, B. braunii, C. vulgaris*
	(Serine)	Sweet	*C. vulgaris*
	(Threonine)	Sweet	*C. vulgaris*
	Arginine	Bitter/(sweet)	*H. pluvialis, T. chui, T. lutea, P. tricornutum, S. costatum, A. platensis, B. braunii*
	Valine	(Sweet)/Bitter	*T. lutea, P. tricornutum*
	Methionine	Bitter/ Sulfureous	*C. vulgaris*
	Cysteine	Bitter/ Sulfureous	*platensis*
	Cystine	Sulfureous	*C. vulgaris*
5′-nucleotides	AMP	Umami	*T. lutea, R. salina, S. costatum, T. chui, C. vulgaris, A. platensis*
	GMP	(Sweet)/ Umami	*S. costatum, P. tricornutum, T. lutea, C. vulgaris*
Salts	Na^+	Salt	*T. chui, R. salina, N. oceanica, T. lutea, S. costatum*
	K^+	Salt	*P. tricornutum, S. costatum, T. chui, T. lutea, C. vulgaris*

monophosphate, though other compounds such as the umami amino acid aspartic acid and additional nucleotides, particularly adenosine monophosphate and inosine monophosphate also play a non-negligible role [175, 178]. Environmental factors, including geographical origin and cultivation milieu, have been shown to influence microalgae composition. In certain food applications, freshwater algae are preferred over marine varieties, as the latter often impart a stronger fishy taste. Additionally, compounds from sea-floor deposits, such as trimethylamine and dimethylamide, can cause undesirable off-flavors when they enter the food chain.

The odorous compounds (Table 6.3) include both intrinsic microalgal secondary metabolites (related to species, origin, and growth milieu) and molecules formed

Table 6.3 Key odor active compounds contributing to microalgal flavor [175, 178].

Class	Molecules	Attribute	Microalgae species
Carotenoid derivatives			
Terpenoid aldehydes	β-Cyclocitral	Grape, floral, sweet tobacco	*D. salina*
	Safranal	Sweet, green-floral, tobacco-herbaceous	*D. salina*
Terpenoid ketones	α-Ionone	Floral, violet-like scents, warm, berry	*D. salina*
	β-Ionone	Floral, violet-like scents, woody note for the β-isomer	*D. salina*
	Isophorone	Cedarwood, sweet, green, camphor, fruity, musty, tobacco	*D. salina (C. vulgaris)*
S-compounds			
	Dimethyl sulfide	Sulfurous, cabbage, onion, sweet corn, vegetable, tomato, green radish	*P. tricornutum, T. lutea, S. costatum, T. chui*
	Methional	Cooked potato-like	*T. lutea*
	Methanethiol	Sulfuric, cabbage-like	*T. lutea, R. salina, S. costatum, T. chui*
N-compounds			
Tertiary amine	Trimethyl amine	Fishy, ammonia	*D. salina, R. salina, T. chui, P. tricornutum*
Pyrazines	3-Ethyl-2,5-dimethyl pyrazine	Roasty, nutty, earthy	*C. vulgaris*
	2,5-Dimethylpyrazine	Nutty, burnt, cocoa	*A. platensis*
Pyridines	3-Methyl pyridine	Fish-like, sweetish	*D. salina, S. costatum, C. vulgaris*
Fatty acid derivatives			
Saturated aldehydes	Propanal	Fruity	*S. costatum, N. oceanica, D. salina, A. platensis*
	Hexanal	Green, grassy	*N. oceanica, S. costatum, D. salina, C. vulgaris, T. lutea*
	Heptanal	Grassy, fatty, green, citrus-like	*S. costatum, N. oceanica, D. salina*
	Octanal	Citrus, floral, fruity	*D. salina, S. costatum*
	Nonanal	Grassy, fatty, tallow-like, citrus	*D. salina, R. salina*

(Continued)

Table 6.3 (Continued)

Class	Molecules	Attribute	Microalgae species
Unsaturated aldehydes	(*E*,*E*)-2,4-Heptadienal	Green, fatty, fishy	*S. costatum, D. salina, N. oceanica*
	2,6-Nonadienal (*E*,*E*)	Tallow-like	*S. costatum, N. oceanica, T. lutea, D. salina*
	2,4-Decadienal (*E*,*E*)	Fatty, deep-fried	*N. oceanica, D. salina*
	2,4-Decadienal (*E*,*Z*)	Fatty, deep-frying fat-like, green	*N. oceanica, D. salina, S. costatum*
Saturated ketones	2-Heptanone	Soap, fruity, rancid	*D. salina, C. vulgaris, T. lutea, N. oceanica*
Unsaturated ketones	3,5-Octadien-2-one (*E*,*E*)	Musty, fatty, earthy, mushroom	*S. costatum, N. oceanica, T. lutea, C. vulgaris, R. salina, T. chui, D. salina*
	1-Octen-3-one	Mushroom-like	*N. oceanica, D. salina*
Unsaturated alcohols	1-Hexen-3-ol	pungent	*S. costatum*
	1-Octen-3-ol	Mushroom-like, earthy, fishy, oily	*N. oceanica, T. lutea, A. plastensis, T. chui, C. vulgaris*
Furans	2-Pentylfuran	Earthy, green, beany, vegetable, and fish-like	*D. salina, N. oceanica, C. vulgaris*
	2-Propylfura	Ethereal, green	*S. costatum*
	2-Ethylfuran	Beany, chemical, liquorice	*S. costatum, N. oceanica, D. salina*
Others			
	Benzaldehyde	Bitter almond-like	*S. costatum, T. lutea, C. vulgaris, D. salina, N. oceanica*
	2,3-Butanedione	Butter-like, sweet	*P. tricornutum, T. lutea, C. vulgaris, R. salina*

postharvest (depending on processing and storage methods). Only a limited number of volatiles exceed their odor threshold and directly affect the aroma of microalgae. Other molecules present at lower levels can nonetheless contribute to odor and flavor. This occurs when their total concentration rises above the threshold due to additive effects in a food product.

The fishy aroma trait of microalgae results from the combined effect of many odor-active molecules and their level. Most of the volatile compounds carrying fishy/marine notes originate from polyunsaturated FA degradation (by lipoxygenase enzymes or autoxidation), including linear aldehydes (both saturated and unsaturated), ketones, and alcohols. Nitrogen- or sulfur-containing compounds such as pyrazines, trimethylamine, and dimethyl sulfides affect the flavor with nutty, roasted, fishy, and cabbage notes. Other molecules, such as terpenoids, contribute to the microalgal aromatic profile with more floral and fruity notes; for example, β-cyclocitral, which derives from β-carotene [175, 177, 178].

It should also be noted that products containing microalgae rich in polyunsaturated fatty acids (PUFAs) must be stored under controlled, low-oxygen conditions. Oxygen exposure can trigger the formation of nonvolatile and volatile PUFA oxidation products, resulting in bitter or fishy off-flavors, which would reduce shelf life and compromise consumer acceptance.

Finally, when incorporated as biomass or ingredients (such as proteins and glucans), microalgae modify food structure, rheology, texture, and breakdown, thereby affecting also sensory attributes such as mouthfeel and flavor, as texture modulates the release of aromatic and taste compounds, ultimately affecting consumer acceptance [179–181].

Incorporating microalgae into products such as cookies, snacks, and dairy is hence generally acceptable as long as they don't bring a strange (umami, earthy, or fishy) taste, which could negatively affect consumer perception. Therefore, controlled small-scale inclusion is essential. Selecting the appropriate species and quantity can help achieve a more palatable, sweet taste suitable for different categories of products. Most consumer studies indicate, nevertheless, that lower inclusion levels generally yield higher acceptance scores [172].

Consumer acceptance is facilitated for specific microalgal-based food products: it has been suggested that microalgae that show a fishy taste—especially marine variants—are better suited for supplementation in fish, fish-derived products, or seafood alternatives as they would have a minor impact on taste perception as their flavor blends naturally with the product [178, 182]. In addition, some nonvolatile (free amino acids) and volatile molecules (pyrazines) can improve a meaty flavor, making microalgae an eligible flavor enhancer for plant-based meat analogues.

Different approaches have been developed to improve consumer acceptance. These include technological processes and formulation strategies (Table 6.4) designed to reduce the strong color, taste, and odor associated with the use of microalgae for food product applications [172, 176]. By addressing these issues, larger quantities could be incorporated, making it possible to achieve meaningful nutritional, health, and sustainability benefits [177].

Considering their characteristic technological, bioactive and nutritional features, microalgae and cyanobacteria can be added to food products (as biomass or selected compounds) as food thickener, colorants, emulsifier, flavoring agent, and source of nutritional and bioactive molecules. Certain food products containing microalgae are no longer commercially available and are currently listed as either "out of stock" or absent from major commercial and sector-specific platforms, as well as from company websites. Today's market includes several product categories such as bakery, beverages, confectionery, and others (Table 6.5). Microalgae as food ingredients align well with current consumer demands for functional food (e.g., Detox Smoothie mix), gluten-free (e.g., Gluten-free Spirulina Protein Cookies), and vegan products (e.g., Vegan Sea Crackers with *Chlorella* and *T. chui*).

Table 6.4 Technological and formulation strategies (adapted from [172, 177, 183].

Technological strategies	
Process	**Outturn**
Random mutagenesis	*Example*: chlorophyll-deficient mutants of *Chlorella vulgaris* • Improved color and sensorial properties (pale microalgae) *At the same time* • Reduced health benefits (reduced content of antioxidant molecules) • Reduced production sustainability (mutants are obligate heterotrophs)
Controlled DNA techniques	• Altered genetic information to produce target molecules to improve sensorial profile *At the same time* • Strict regulation for food application
Growth condition control Photo-, hetero-, mixotrophic	*Example*: heterotrophically grown *Chlorella* • Improved color and sensorial properties (lighter green color and smooth flavor due to reduced chlorophyll content)
Medium composition	*Example*: controlled medium N content • Reduced VOCs production
Drying Solar-, freeze-, spray drying	*Example*: solar and freeze-dried *T. chui* and *M. oceanica* • Improved color (reduced pigment content due to heat degradation, oxygen and light exposure)
Cell disruption Drying, mechanical, enzymatic, and chemical methods	• Modification of sensorial and functional properties (due to release of intracellular compounds and increased oxidation)
Extraction Mechanical, chemical, ultrasound	*Example*: ethanol-extraction of chlorophyll from microalgal biomass • Extraction of high-value compounds • Improved sensorial properties of bread containing ethanol-extracted biomass of *T. chui*, *C. vulgaris*, and *M. gaditana* (due to total extraction of chlorophyll and a more appreciated bread color) *At the same time* • Reduction of health properties and product shelf life (due to total extraction of carotenoids and reduction of phenolic/antioxidant compounds)
Encapsulation	*Example*: coating microalgae cells through spray drying • Improved sensorial properties (due to removal/masking of off-odors/flavors) • Improved rheological properties • Improved stability over storage and processing (delay in degradation of bioactive compounds)

Table 6.4 (Continued)

	Technological strategies
Process	**Outturn**
Fermentation	• Improved sensorial properties (due to off-flavor volatiles reduction and new molecules production by LAB or yeasts fermentation and enzymatic activities)
Enzymatic methods	• Deodorization and decoloration through bioprocessing of microalgal biomass
Storage O_2-controlled storage conditions	• Improved sensorial properties (due to PUFA-oxidation products reduction)
	Formulation strategies
Food matrix and microalgae selection	• Marine species for fish, fish-derived products, or seafood alternatives • Pyrazine-containing microalgae for plant-based meat analogues • Food matrix and microalgae species with matching colors (example: *Spirulina, Chlorella,* and *Tetraselmis* included in broccoli soup)
Microalgae added quantity	• According to food matrix, to obtain a product with sensorial properties accepted by consumer
Other ingredient selection	• Masking approach: mixing microalgae with other ingredients (salts, sugars, spices, and flavor compounds) in order to change the sensory profile and hide the distinctive microalgal taste. *Example*: masking fishy taste with acidic taste through citrus juice • Pairing approach: combining foods or ingredients that taste good together and ascribe to microalgal distinctive taste a key role in the final food product. *Example*: enhancing microalgal aroma, adding sweetness

The most commonly represented algae in commercial food products are *Spirulina* sp. and *Chlorella* sp., while other algae such as *Schizochytrium* sp. and *Euglena* sp. are included in a more limited number of products; *Schizochytrium sp.* is used as a source of healthy polyunsaturated FAs (e.g., infant formula powder and plant-based salmon fillet), *Euglena* sp. is included mainly in Japanese food (e.g., dried noodles with *Euglena* & *Chlorella*). Other algae such as *Dunaliella*, *Haematococcus,* and *Nannochloropsis* are mainly commercialized as ingredients and dietary supplements or are utilized in other non-food applications such as cosmetics.

Spirulina is used in various forms, including extract (aqueous), concentrate, powder, paste, crumbs, and flakes in concentrations ranging from 0.05% (e.g., Fresh Smoothie Spinach, Peppermint, and Spirulina) to 17% (e.g., Crunchy topping with

Table 6.5 Examples of commercial food products containing microalgae available in 2025.

Food category: BEVERAGES

Product	Microalga	Food product	Producer	Content; other information
Fermented beverage	*Spirulina*	Spirulina Kombucha	Malaysia	ND; spirulina extract
Fermented beverage	*Spirulina*	Kombucha with Spirulina and peppermint drink	France	0.1%
Drink	*Spirulina*	Plant-based drinks with spirulina	France	0.4%
Drink	*Spirulina*	Organic drink with spirulina	Italy	7%; aqueous extract of spirulina
Smoothie	*Spirulina*	Fresh smoothie banana, yuzu and blue spirulina	Germany	0.1%
Smoothie	*Spirulina*	Fresh smoothie spinach, peppermint, and spirulina	Germany	0.05%; blue spirulina powder
Premix for drinks	*Spirulina*	Milk make melon	Japan	ND; spirulina blue declared as coloring agent
Premix for drinks	*Spirulina*	Detox smoothie mix	Estonia	ND; product declared as dietary supplement
Smoothie	*Spirulina*, *Chlorella*	Green smoothie	Canada	ND
Premix for shakes	*Spirulina*, *Chlorella*, and *Schizochytrium*	Green elixir shakes powder (with spirulina and chlorella)	UK	Sulfite-1% *A. plantensis* and 0.8% *C. vulgaris*; algae oil powder (Schizochytrium—6% DHA and 3% EPA oil)

Food category: BAKED GOODS

Product	Microalga	Food product	Producer	Content; other information
Crackers	*Spirulina*	Spirulina-filled crackers	Malaysia	1.8%
Crackers	*Spirulina*	Spirulina and spring onion soda cracker	Taiwan region of China	ND
Crackers	*Spirulina*	Spirulina cracker	Italy	ND
Crackers	*Spirulina*	Organic raw crackers spirulina	Bulgaria	1.1%

Crackers	*Chlorella*	Matcha *Senbei* rice crackers	Japan	ND
Crackers	*Chlorella*	Organic Algae Crackers Rosemary and Sea Salt	Austria	5%
Crackers	*Chlorella, T. chui*	Vegan Sea Crackers with Chlorella and *T. chui*	Portugal	ND
Breadsticks	*Chlorella*	Chlorella breadsticks	Italy	ND
Bakery	*Spirulina*	Veggie wrap: spirulina	USA	ND
Bakery	*Spirulina*	Pizza Crust: spirulina	USA	ND
Snacks— bar	*Spirulina*	Spirulina and lemon protein bars	Bulgaria	3%
Snacks—bar	*Spirulina*	Spirulina/lemon bar	Germany	5%
Snacks—bar	*Spirulina*	Raw sprouted millet bar with lemon and spirulina	Italy	0.15%
Snacks—bar	*Spirulina*	Spirulina organic quinoa cakes	Italy	3%
Snacks—bar	*Spirulina*	Chia protein lemon spirulina Buddha bar	Portugal	2%; spirulina is declared as a source of nutrients
Biscuits/cookies	*Spirulina*	Gluten-free spirulina pregnancy choco cookies	India	ND; Kodo millet and quinoa (VitB9 sources)
Biscuits/cookies	*Spirulina*	Gluten-free spirulina protein cookies	India	ND
Biscuits/cookies	*Spirulina*	Spirulina biscuits	France	3%
Biscuits/cookies	*Chlorella*	Sweets sandwich rich matcha	Japan	ND; chlorella declared as coloring agent
Biscuits/cookies	*Chlorella*	Organic chlorella orange cookies	UK	0.6%
Pastries, snacks	*Chlorella*	Marukin thick cut matcha baum	Japan	ND; chlorella not included in coloring agents
Pastries, snacks	*Chlorella*	Cocoa Valencianas	Spain	0.4%
Puffed snacks	*Chlorella*	Choco corn matcha latte flavor Shimi	Japan	ND; chlorella declared as coloring agent
Puffed snacks	*Chlorella*	Matcha and lime organic puffs	Romania	ND
Bean-based snacks	*Chlorella*	Grilled edamame truffle-flavored salt	Japan	ND

(*Continued*)

Table 6.5 (Continued)

Food category: NOODLES and PASTA				
Product	**Microalga**	**Food product**	**Producer**	**Content; other information**
Noodles	*Chlorella*	Noodles (Hiyamugi)	Japan	ND; chlorella declared as coloring agent
Noodles	*Chlorella*	Rice noodle chlorella flavor	Taiwan region of China	ND; health-oriented product series
Noodles	*Chlorella*	Chlorella noodles	Japan	2 grams of chlorella/serving (declared healthy ingredient)
Noodles	*Euglena, Chlorella*	Dried noodles with euglena and chlorella	Japan	ND
Instant noodles	*Chlorella*	Negira chicken salt ramen	Japan	ND; chlorella extract (in noodles)
Instant noodles	*Euglena, Spirulina*	Instant noodles with euglena topping	Japan	ND; euglena as topping, spirulina as coloring agent
Gluten-free pasta	*Spirulina*	Spirulina spaghetti with brown rice flour	Italy	2%
Gluten-free pasta	*Spirulina*	Rice and spirulina fusilli	Italy	1.5%
Pasta	*Spirulina*	White corn and spirulina shell pasta	Italy	5%
Pasta	*Spirulina*	Organic hulled wheat pasta with spirulina	Italy	1%
Pasta	*Spirulina*	Blue spaghetti with spirulina and bergamot essential oil	Italy	1% Spirulina concentrate (natural extract), blue color
Pasta	*Spirulina*	Fusilli with spirulina	Italy	1.9%
Pasta	*Chlorella*	Chlorella pasta	Italy	ND

Food category: CONFECTIONERY

Product	Microalga	Food product	Producer	Content; other information
Gums	*Spirulina*	True gum mint	Philippines	ND; spirulina extract
Chewy Candies	*Spirulina*	DUO chardonnay Soda	Japan	ND; spirulina blue declared as coloring agent
Chewy Candies	*Spirulina*	Gummy bears with spirulina	Italy	5%
Candies	*Spirulina*	Green tea and spirulina algae candy	Italy	0.05%
Chocolate	*Spirulina*	Spirulina chocolate bar	France	5%; spirulina flakes
Chocolates	*Spirulina*	Marble chocolate confectionery	Japan	ND; spirulina blue declared as coloring agent
Chocolates	*Spirulina*	Choco Spaceships dragee	Bulgaria	ND; spirulina concentrate declared as coloring agent
Chocolates	*Chlorella*	Meito Japanese Tea Chocolate Assortment	Japan	ND; chlorella declared as coloring agent

Food category: VARIOUS

Product	Microalga	Food product	Producer	Content; other information
Instant soup	*Spirulina*	Instant spirulina and spinach soup	Spain	2%
Instant soup	*Spirulina*	Instant spirulina and nettle soup	France	2%
Cereals	*Spirulina*	Spirulina cereal	India	ND
Cereals	*Spirulina*	Apple cinnamon crispy muesli with spirulina	Austria	6%; spirulina-paste
Cereals	*Chlorella*	Chlorella cereal	Malaysia	1%
Topping/Seasoning	*Spirulina*	Crunchy topping with spirulina algae and strawberry	Germany	12%; spirulina algae
Topping/Seasoning	*Spirulina*	Crunchy topping with spirulina algae	Germany	17%; spirulina algae crumbs

(Continued)

Table 6.5 (Continued)

Food category: VARIOUS				
Product	**Microalga**	**Food product**	**Producer**	**Content: other information**
Topping/Seasoning	*Spirulina*	Flaked food yeast with spirulina	Italy	5%; seasoning for pasta, soups, and sauces
Topping/Seasoning	*Spirulina*	Organic caramelized sesame with spirulina	Spain	ND
Compote	*Spirulina*	Spirulina apple compotes	France	1%
Yogurt	*Spirulina*	Spirulina coconut flavor yogurt	Singapore	ND; spirulina gently balanced by sweetness of coconut
Ice-cream base	*Spirulina*	Fruity and veggy base for sorbets	SM	ND
Infant formula	*Schizochytrium*	Infant formula powder	USA	ND; *Schizochytrium* Sp. oil (source of DHA)
Dairy-free cheese	*Spirulina*	Blue dairy-free cheese	UK	ND; blue veins of spirulina in the product
Veggie burger	*Spirulina*	Veggie burger with olives and spirulina algae	Italy	1.5%
Vegburger premix	*Spirulina*	Vegburger mix with ginger, Nori Alga, and spirulina	Italy	1.5%
Gnocchi premix	*Chlorella*	Chlorella gnocchi premix	Italy	ND
Plant-based fish	*Schizochytrium*	Plant-based salmon fillet	USA	ND; DHA and EPA algal oil from *Schizochytrium*

spirulina algae). Less information is available regarding the specific forms of *Chlorella* used in products. Labels indicate whether *Chlorella* is included as coloring agent or as healthy ingredient, with reported concentrations ranging from 0.4% (e.g., Cocoa Valencianas pastries) to 5% (e.g., Organic Algae Crackers Rosemary & Sea Salt).

Microalgae are incorporated into beverages as a source of nutrients, bioactive molecules, and natural colorants, contributing to those visual and health-related attributes in high demand among consumers today. Their inclusion in baked goods has shown a relative success due to the effective masking of microalgal flavor by other ingredients. Consequently, sensorial properties (texture, flavor, and color), dough quality, as well as nutrients (protein) and bioactive composition of pasta, cookies, snacks, pastries, and crackers can be improved [184].

The selection of additional ingredients and the type of food product reflect the application of formulation strategies (Table 6.4) aimed at meeting consumer demands and increasing consumer acceptance. Ingredients such as lemon, bergamot, matcha, chocolate, and coconut are used as part of masking and pairing strategies (e.g., Spirulina/Lemon bar and Spirulina Coconut Yogurt). Familiar products—such as pasta for Italians, and soups and chocolate for the French—are chosen as means for microalgae inclusion.

6.7 Current Problems and Challenges

Widespread and consolidated application of microalgae in the food sector faces numerous challenges. Although approximately 30,000 microalgae species have been identified, few have received approval from international agencies such as GRAS [2, 185, 186]. This is due to several factors, the main one being the lengthy and costly regulatory process, which limits the introduction of new organisms into the food sector.

Technical issues, such as harvesting and product stabilization, which can account for up to 60% of production costs, are also challenges to be addressed to make microalgae a competitive food or source of food ingredients. Current techniques, such as extraction and drying, are particularly energy-intensive and unsustainable. Specific limitations of individual technologies must also be considered, such as reduced sedimentation efficiency or the difficulty of flocculation of smaller microalgae. Although emerging techniques such as cavitation, phototaxis, and ultrasound are under development, they require further study and are often difficult to scale up industrially.

In addition to technical issues, there are critical issues related to the biochemistry of microalgae, particularly the variation in their nutritional profile depending on the growth medium, the seasons, the species, and environmental stress [187]. Finally, considering the individual molecules of industrial interest that can be extracted from microalgae, prices are still uncompetitive compared to existing alternatives, such as fish oil, soybean oil, or synthetic dyes.

6.8 Future and Perspectives

Future prospects for the use of microalgae as food ingredients focus primarily on technological innovation, market expansion, and environmental sustainability. Technologically, attention is focused on developing more efficient and environmentally friendly extraction techniques, such as extraction assisted by ultrasound, microwaves, or PEF, as well as the use of supercritical fluids. At the same time, the biorefinery approach, which aims to fully valorize biomass by exploiting all its components—from proteins to lipids, pigments, and polysaccharides—represents a key strategy for reducing costs and increasing the competitiveness of microalgal products.

Applications in the pharmaceutical, food, and industrial sectors are emerging as application areas with numerous opportunities. These include the nutraceutical sector, thanks to the production of high-value bioactive compounds such as omega-3s and carotenoids, and the field of innovative materials, with the development of biopolymers and biodegradable packaging. The integration of microalgae into fine dining, known as phycogastronomy, also shows growing potential to transform them from a simple functional ingredient to a distinctive element of gourmet products.

The application sector will work closely with research, which will continue to focus on the selection of high-yielding strains and metabolic engineering, with the aim of optimizing the production of metabolites of interest and ensuring an increasingly reliable and sustainable supply.

From a regulatory and commercial perspective, the list of species approved as safe is expected to gradually expand, thus facilitating the introduction of new products to the market. Finally, sustainability is a key element for this sector; cultivation in brackish or wastewater will further reduce resource consumption, while the high CO_2 fixation capacity will contribute to mitigating the effects of climate change.

References

1 Bachheti, A., Bachheti, R.K., and Husen, A. (2023). *Aquatic Medicinal Plants*, 1e. Boca Raton: CRC Press.

2 Mobin, S. and Alam, F. (2017). Some promising microalgal species for commercial applications: a review. *Energy Procedia* 110: 510–517. https://doi.org/10.1016/j.egypro.2017.03.177.

3 Gunes, S., Tamburaci, S., Dalay, M.C., and Deliloglu Gurhan, I. (2017). *In vitro* evaluation of *Spirulina platensis* extract incorporated skin cream with its wound healing and antioxidant activities. *Pharmaceutical Biology* 55: 1824–1832. https://doi.org/10.1080/13880209.2017.1331249.

4 Suwono, L.V. and Suryoprabowo, S. (2025). Nutrition and health functions of *Spirulina* and its application in food industry: a review. *IOP Conference Series: Earth and Environmental Science* 1488: 012101. https://doi.org/10.1088/1755-1315/1488/1/012101.

5 Olguín, E.J., Galicia, S., Camacho, R. et al. (1997). Production of *Spirulina* sp. in sea water supplemented with anaerobic effluents in outdoor raceways under temperate climatic conditions. *Applied Microbiology and Biotechnology* 48: 242–247. https://doi.org/10.1007/s002530051045.

6 Gouveia, L. (1996). Evolution of pigment composition in *Chlorella vulgaris*. *Bioresource Technology* 57: 157–159. https://doi.org/10.1016/0960-8524(96)00058-2.

7 Safi, C., Zebib, B., Merah, O. et al. (2014). Morphology, composition, production, processing and applications of *Chlorella vulgaris*: a review. *Renewable and Sustainable Energy Reviews* 35: 265–278. https://doi.org/10.1016/j.rser.2014.04.007.

8 Sousa, V., Maciel, F., Vicente, A.A. et al. (2024). Development of highly effective growth strategies aiming at improving the content of carotenoids in *Dunaliella salina* IFDSAL-JY215. *Sustainable Food Technology* 2: 1735–1746. https://doi.org/10.1039/d4fb00229f.

9 Félix, F.C.C.D.S., Hidalgo, V.B., De Carvalho, A.K.F. et al. (2024). Assessing the application of marine microalgae *Dunaliella salina* in a biorefinery context: production of value-added biobased products. *Biofuels, Bioproducts and Biorefining* 18: 439–452. https://doi.org/10.1002/bbb.2577.

10 Araj-Shirvani, M., Honarvar, M., Jahadi, M., and Mizani, M. (2024). Biochemical profile of *Dunaliella* isolates from different regions of Iran with a focus on pharmaceutical and nutraceutical potential applications. *Food Science & Nutrition* 12: 4914–4926. https://doi.org/10.1002/fsn3.4137.

11 Mularczyk, M., Michalak, I., and Marycz, K. (2020). Astaxanthin and other nutrients from *Haematococcus pluvialis*—multifunctional applications. *Marine Drugs* 18: 459. https://doi.org/10.3390/md18090459.

12 Koller, M., Muhr, A., and Braunegg, G. (2014). Microalgae as versatile cellular factories for valued products. *Algal Research* 6: 52–63. https://doi.org/10.1016/j.algal.2014.09.002.

13 Molino, A., Iovine, A., Casella, P. et al. (2018). Microalgae characterization for consolidated and new application in human food, animal feed and nutraceuticals. *The International Journal of Environmental Research and Public Health* 15: 2436. https://doi.org/10.3390/ijerph15112436.

14 Wang, Y., Tibbetts, S., and McGinn, P. (2021). Microalgae as sources of high-quality protein for human food and protein supplements. *Foods* 10: 3002. https://doi.org/10.3390/foods10123002.

15 Chang, M., Zhang, T., Guo, X. et al. (2020). Optimization of cultivation conditions for efficient production of carotenoid-rich DHA oil by *Schizochytrium* sp. S31. *Process Biochemistry* 94: 190–197. https://doi.org/10.1016/j.procbio.2020.04.007.

16 Wang, C., Onyeaka, H., Miri, T., and Soltani, F. (2024). *Chlorella vulgaris* as a food substitute: applications and benefits in the food industry. *Journal of Food Science* 89: 8231–8247. https://doi.org/10.1111/1750-3841.17529.

17 Habib, M.A.B., Parvin, M., Huntington, T.C., and Hasan, M.R. (2008). *A Review on Culture, Production and Use of Spirulina as Food for Humans and Feeds for Domestic Animals and Fish*. Rome: Food and Agriculture Organization of the United Nations.

18 Kurpan, D., Idà, A., Körner, F.G. et al. (2024). Long-term evaluation of productivity and harvesting efficiency of an industrial Spirulina (*Arthrospira platensis*) production facility. *Bioresource Technology Reports* 25: 101741. https://doi.org/10.1016/j.biteb.2023.101741.

19 Depraetere, O., Pierre, G., Noppe, W. et al. (2015). Influence of culture medium recycling on the performance of *Arthrospira platensis* cultures. *Algal Research* 10: 48–54. https://doi.org/10.1016/j.algal.2015.04.014.

20 Roselet, F., Vandamme, D., Muylaert, K., and Abreu, P.C. (2019). Harvesting of microalgae for biomass production. In: *Microalgae Biotechnology for Development of Biofuel and Wastewater Treatment* (ed. M.A. Alam and Z. Wang), 211–243. Singapore: Springer.

21 Nie, C., Pei, H., Jiang, L. et al. (2018). Growth of large-cell and easily-sedimentation microalgae Golenkinia SDEC-16 for biofuel production and campus sewage treatment. *Renewable Energy* 122: 517–525. https://doi.org/10.1016/j.renene.2018.02.005.

22 Ahmad, S., Iqbal, K., Kothari, R. et al. (2022). A critical overview of upstream cultivation and downstream processing of algae-based biofuels: opportunity, technological barriers and future perspective. *Journal of Biotechnology* 351: 74–98. https://doi.org/10.1016/j.jbiotec.2022.03.015.

23 Kumar, N., Banerjee, C., Negi, S., and Shukla, P. (2023). Microalgae harvesting techniques: updates and recent technological interventions. *Critical Reviews in Biotechnology* 43: 342–368. https://doi.org/10.1080/07388551.2022.2031089.

24 Mathimani, T. and Mallick, N. (2018). A comprehensive review on harvesting of microalgae for biodiesel – key challenges and future directions. *Renewable and Sustainable Energy Reviews* 91: 1103–1120. https://doi.org/10.1016/j.rser.2018.04.083.

25 Laamanen, C.A., Ross, G.M., and Scott, J.A. (2016). Flotation harvesting of microalgae. *Renewable and Sustainable Energy Reviews* 58: 75–86. https://doi.org/10.1016/j.rser.2015.12.293.

26 Hanotu, J., Bandulasena, H.C.H., and Zimmerman, W.B. (2012). Microflotation performance for algal separation. *Biotechnology and Bioengineering* 109: 1663–1673. https://doi.org/10.1002/bit.24449.

27 Xu, K., Zou, X., Chang, W. et al. (2021). Microalgae harvesting technique using ballasted flotation: a review. *Separation and Purification Technology* 276: 119439. https://doi.org/10.1016/j.seppur.2021.119439.

28 Barros, A.I., Gonçalves, A.L., Simões, M., and Pires, J.C.M. (2015). Harvesting techniques applied to microalgae: A review. *Renewable and Sustainable Energy Reviews* 41: 1489–1500. https://doi.org/10.1016/j.rser.2014.09.037.

29 Ghazvini, M., Kavosi, M., Sharma, R., and Kim, M. (2022). A review on mechanical-based microalgae harvesting methods for biofuel production. *Biomass and Bioenergy* 158: 106348. https://doi.org/10.1016/j.biombioe.2022.106348.

30 Muir, E., Grossman, A.R., Chisti, Y. et al. (2024). Self-aggregation for sustainable harvesting of microalgae. *Algal Research* 83: 103685. https://doi.org/10.1016/j.algal.2024.103685.

31 Zhu, J., Wakisaka, M., Omura, T. et al. (2024). Advances in industrial harvesting techniques for edible microalgae: recent insights into sustainable, efficient methods and future directions. *Journal of Cleaner Production* 436: 140626. https://doi.org/10.1016/j.jclepro.2024.140626.

32 Alibas, I., Yilmaz, A., Alibas, K. et al. (2023). The effect of different drying methods and microalgae species on the quality parameters of biodiesel obtained by transesterification technique. *Biomass and Bioenergy* 168: 106688. https://doi.org/10.1016/j.biombioe.2022.106688.

33 Fábio, d.F.N. and Demarco, M. (2020). Drying and Quality of Microalgal. In: *Microalgae: From Physiology to Application*, 71. London: https://doi.org/10.5772/intechopen.83737.

34 Deprá, M.C., Severo, I.A., Dos Santos, A.M. et al. (2020). Environmental impacts on commercial microalgae-based products: sustainability metrics and indicators. *Algal Research* 51: 102056. https://doi.org/10.1016/j.algal.2020.102056.

35 Aljabri, H., Cherif, M., Siddiqui, S.A. et al. (2023). Evidence of the drying technique's impact on the biomass quality of *Tetraselmis subcordiformis* (Chlorophyceae). *Biotechnology for Biofuels and Bioproducts* 16: 85. https://doi.org/10.1186/s13068-023-02335-x.

36 Oliveira, C.Y.B.D., Viegas, T.L., Lopes, R.G. et al. (2020). A comparison of harvesting and drying methodologies on fatty acids composition of the green microalga *Scenedesmus obliquus*. *Biomass and Bioenergy* 132: 105437. https://doi.org/10.1016/j.biombioe.2019.105437.

37 Behera, B. and Balasubramanian, P. (2021). Experimental and modelling studies of convective and microwave drying kinetics for microalgae. *Bioresource Technology* 340: 125721. https://doi.org/10.1016/j.biortech.2021.125721.

38 Zhang, H., Gong, T., Li, J. et al. (2022). Study on the effect of spray drying process on the quality of microalgal biomass: a comprehensive biocomposition analysis of Spray-Dried *S. acuminatus*. *Biomass and Bioenergy Research* 15: 320–333. https://doi.org/10.1007/s12155-021-10343-8.

39 Bagchi, S.K., Rao, P.S., and Mallick, N. (2015). Development of an oven drying protocol to improve biodiesel production for an indigenous chlorophycean microalga *Scenedesmus* sp. *Bioresource Technology* 180: 207–213. https://doi.org/10.1016/j.biortech.2014.12.092.

40 Bagchi, S.K., Patnaik, R., Rao, P.S. et al. (2022). Establishment of an efficient tray-drying process for qualitative biodiesel production from a locally isolated microalga *Tetradesmus obliquus* cultivated in polyhouse raceway ponds. *Algal Research* 64: 102674. https://doi.org/10.1016/j.algal.2022.102674.

41 Biz, A.P., Cardozo-Filho, L., and Zanoelo, E.F. (2019). Drying dynamics of microalgae (*Chlorella pyrenoidosa*) dispersion droplets. *Chemical Engineering and Processing - Process Intensification* 138: 41–48. https://doi.org/10.1016/j.cep.2019.03.007.

42 Kalaiselvan, N. and Mathimani, T. (2022). Design and fabrication of box-type passive solar dryer (BTPSD) with thermal insulation material for valorizing biomass and neutral lipids of marine *Chlorella vulgaris* for biodiesel application. *Scientific Reports* 12: 6046. https://doi.org/10.1038/s41598-022-09665-0.

43 Sung, Y.J., Yu, B.S., Yang, H.E. et al. (2022). Microalgae-derived hydrogen production towards low carbon emissions via large-scale outdoor systems. *Bioresource Technology* 364: 128134. https://doi.org/10.1016/j.biortech.2022.128134.

44 López Pastor, R., Pinna-Hernández, M.G., and Acién Fernández, F.G. (2023). Technical and economic viability of using solar thermal energy for microalgae drying. *Energy Reports* 10: 989–1003. https://doi.org/10.1016/j.egyr.2023.07.040.

45 Hosseinizand, H., Sokhansanj, S., and Lim, C.J. (2018). Studying the drying mechanism of microalgae *Chlorella vulgaris* and the optimum drying temperature to preserve quality characteristics. *Dry Technology* 36: 1049–1060. https://doi.org/10.1080/07373937.2017.1369986.

46 Silva, J.P.S., Veloso, C.R.R., De Souza Barrozo, M.A., and Vieira, L.G.M. (2021). Indirect solar drying of *Spirulina platensis* and the effect of operating conditions on product quality. *Algal Research* 60: 102521. https://doi.org/10.1016/j.algal.2021.102521.

47 Benedetti, M., Vecchi, V., Barera, S., and Dall'Osto, L. (2018). Biomass from microalgae: the potential of domestication towards sustainable biofactories. *Microbial Cell Factories* 17: 173. https://doi.org/10.1186/s12934-018-1019-3.

48 Xi, Y., Wang, J., Chu, Y. et al. (2020). Effects of different light regimes on *Dunaliella salina* growth and β-carotene accumulation. *Algal Research* 52: 102111. https://doi.org/10.1016/j.algal.2020.102111.

49 Maltsev, Y., Maltseva, K., Kulikovskiy, M., and Maltseva, S. (2021). Influence of light conditions on microalgae growth and content of lipids, carotenoids, and fatty acid composition. *Biology* 10: 1060. https://doi.org/10.3390/biology10101060.

50 Baune, M.-C., Januschewski, E., Bussa, M. et al. (2025). Innovative vs. classical methods for drying heterotrophic *Chlorella vulgaris*: impact on the nutritional properties, safety, sustainability and costs. *Algal Research* 86: 103913. https://doi.org/10.1016/j.algal.2025.103913.

51 Loke Show, P. (2022). Global market and economic analysis of microalgae technology: status and perspectives. *Bioresource Technology* 357: 127329. https://doi.org/10.1016/j.biortech.2022.127329.

52 Barbosa-Cánovas, G., Ortega-Rivas, E., Juliano, P., and Yan, H. (2005). *Food Powders: Physical Properties, Processing, and Functionality*. New York, NY (US): Springer.

53 Chen, C.-L., Chang, J.-S., and Lee, D.-J. (2015a). Dewatering and drying methods for microalgae. *Dry Technology* 33: 443–454. https://doi.org/10.1080/07373937.2014.997881.

54 Karthik, P., Chhanwal, N., and Anandharamakrishnan, C. (2017). Drum drying. In: *Handbook of Drying for Dairy Products*, 1e (ed. C. Anandharamakrishnan), 43–56. Wiley.

55 Schmid, B., Navalho, S., Schulze, P.S.C. et al. (2022). Drying microalgae using an industrial solar dryer: a biomass quality assessment. *Foods* 11: 1873. https://doi.org/10.3390/foods11131873.

56 Foo, S.C., Mok, C.Y., Ho, S.Y., and Khong, N.M.H. (2023). Microalgal culture preservation: progress, trends and future developments. *Algal Research* 71: 103007. https://doi.org/10.1016/j.algal.2023.103007.

57 Su, Y. (2024). The drying technologies, physical properties and storage stability of edible microalgae powders. In: *Handbook of Food Powders*, 557–574. Sawston, Cambridge: Woodhead Publishing https://doi.org/10.1016/B978-0-323-98820-9.00011-9.

58 Amin, M., Chetpattananondh, P., Cheng, C.-K. et al. (2021). Drying characteristics and impacts on quality of marine *Chlorella* sp. biomass and extracts for fuel applications. *Journal of Environmental Chemical Engineering* 9: 106386. https://doi.org/10.1016/j.jece.2021.106386.

59 Raposo, M.F.J., Morais, A.M.M.B., and Morais, R.M.S.C. (2012). Effects of spray-drying and storage on astaxanthin content of *Haematococcus pluvialis* biomass. *World Journal of Microbiology and Biotechnology* 28: 1253–1257. https://doi.org/10.1007/s11274-011-0929-6.

60 Troyano, J., Çamur, C., Garzón-Tovar, L. et al. (2020). Spray-drying synthesis of MOFs, COFs, and related composites. *Accounts of Chemical Research* 53: 1206–1217. https://doi.org/10.1021/acs.accounts.0c00133.

61 Moradi Maryamnegari, S., Ashrafizadeh, A., and Jaskulski, M. (2023). A novel approach for the reduction of powder deposition in spray drying chambers. *Dry Technology* 41: 1017–1030. https://doi.org/10.1080/07373937.2022.2140353.

62 Bonilla-Ahumada, F.D.J., Khandual, S., and Lugo-Cervantes, E.D.C. (2018). Microencapsulation of algal biomass (*Tetraselmis chui*) by spray-drying using different encapsulation materials for better preservation of beta-carotene and antioxidant compounds. *Algal Research* 36: 229–238. https://doi.org/10.1016/j.algal.2018.10.006.

63 Vilatte, A., Spencer-Milnes, X., Jackson, H.O. et al. (2023). Spray drying is a viable technology for the preservation of recombinant proteins in microalgae. *Microorganisms* 11: 512. https://doi.org/10.3390/microorganisms11020512.

64 Chew, K.W., Chia, S.R., Lee, S.Y. et al. (2019). Enhanced microalgal protein extraction and purification using sustainable microwave-assisted multiphase partitioning technique. *Chemical Engineering Journal* 367: 1–8. https://doi.org/10.1016/j.cej.2019.02.131.

65 Polat, D.G., Durmaz, Y., Konar, N. et al. (2022). Using spray-dried and encapsulated *Nannochloropsis oculata* biomasses in white spread. *Journal of Applied Phycology* 34: 375–383. https://doi.org/10.1007/s10811-021-02669-9.

66 Mayol, A.P., Culaba, A.B., Villagracia, A.R. et al. (2015). Investigation of the drying characteristics of microalgae using microwave irradiation. In: *2015 International Conference on Humanoid, Nanotechnology, Information Technology, Communication and Control, Environment and Management (HNICEM)*, 1–6. Cebu City: IEEE.

67 Villagracia, A.R.C., Mayol, A.P., Ubando, A.T. et al. (2016). Microwave drying characteristics of microalgae (*Chlorella vulgaris*) for biofuel production. *Clean Technologies and Environmental Policy* 18: 2441–2451. https://doi.org/10.1007/s10098-016-1169-0.

68 Becker, E.W. (2007). Micro-algae as a source of protein. *Biotechnology Advances* 25: 207–210. https://doi.org/10.1016/j.biotechadv.2006.11.002.

69 de Vree, J.H., Bosma, R., Janssen, M. et al. (2015). Comparison of four outdoor pilot-scale photobioreactors. *Biotechnol Biofuels* 8: 215. https://doi.org/10.1186/s13068-015-0400-2.

70 Sun, H., Kasdin, N.J., and Vanderbei, R. (2018). Identification and adaptive control of a high-contrast focal plane wavefront correction system. *J Astron Telesc Instrum Syst* 4: 049006. https://doi.org/10.1117/1.JATIS.4.4.049006.

71 Chen, H., Zheng, Y., Zhan, J. et al. (2017). Comparative metabolic profiling of the lipid-producing green microalga Chlorella reveals that nitrogen and carbon metabolic pathways contribute to lipid metabolism. *Biotechnol Biofuels* 10: 153. https://doi.org/10.1186/s13068-017-0839-4.

72 Soto-Sierra, L., Stoykova, P., and Nikolov, Z.L. (2018). Extraction and fractionation of microalgae-based protein products. *Algal Research* 36: 175–192. https://doi.org/10.1016/j.algal.2018.10.023.

73 Günerken, E., D'Hondt, E., Eppink, M.H.M. et al. (2015). Cell disruption for microalgae biorefineries. *Biotechnology Advances* 33: 243–260. https://doi.org/10.1016/j.biotechadv.2015.01.008.

74 Moreira, C., Ferreira-Santos, P., Nunes, R. et al. (2025). Influence of different processing techniques on microalgal protein extraction. *Algal Research* 86: 103958. https://doi.org/10.1016/j.algal.2025.103958.

75 Guadalupi, L.S., Bianco, M., Cataldi, T.R.I. et al. (2025). Ultrasound-assisted protein extraction for deep proteome analysis of *Spirulina* and *Chlorella microalgae*. *LWT* 222: 117647. https://doi.org/10.1016/j.lwt.2025.117647.

76 Lupatini, A.L., De Oliveira, B.L., Colla, L.M. et al. (2017). Protein and carbohydrate extraction from *S. platensis* biomass by ultrasound and mechanical agitation. *Food Research International* 99: 1028–1035. https://doi.org/10.1016/j.foodres.2016.11.036.

77 Hildebrand, G., Poojary, M.M., O'Donnell, C. et al. (2020). Ultrasound-assisted processing of *Chlorella vulgaris* for enhanced protein extraction. *Journal of Applied Phycology* 32: 1709–1718. https://doi.org/10.1007/s10811-020-02105-4.

78 Coustets, M., Joubert-Durigneux, V., Hérault, J. et al. (2015). Optimization of protein electroextraction from microalgae by a flow process. *Bioelectrochemistry* 103: 74–81. https://doi.org/10.1016/j.bioelechem.2014.08.022.

79 T Lam, G.P., Van Der Kolk, J.A., Chordia, A. et al. (2017). Mild and selective protein release of cell wall deficient microalgae with pulsed electric field. *ACS Sustainable Chemistry & Engineering* 5: 6046–6053. https://doi.org/10.1021/acssuschemeng.7b00892.

80 Katsimichas, A., Limnaios, A., Dimitrakopoulos, K. et al. (2024). Pulsed electric fields assisted extraction of proteins and phycocyanin from *Arthrospira platensis* biomass: a kinetic study. *Food and Bioproducts Processing* 147: 304–314. https://doi.org/10.1016/j.fbp.2024.07.012.

81 Perez, B., Zermatten, C., Haberkorn, I., and Mathys, A. (2024). Enhancing protein extraction from heterotrophic *Auxenochlorella protothecoides* microalgae through emerging cell disruption technologies combined with incubation. *Bioresource Technology* 407: 131099. https://doi.org/10.1016/j.biortech.2024.131099.

82 Kapoore, R., Butler, T., Pandhal, J., and Vaidyanathan, S. (2018). Microwave-assisted extraction for microalgae: from biofuels to biorefinery. *Biology* 7: 18. https://doi.org/10.3390/biology7010018.

83 Ali, S.S., Al-Tohamy, R., Al-Zahrani, M. et al. (2025). Advancements and challenges in microalgal protein production: a sustainable alternative to conventional protein sources. *Microbial Cell Factories* 24. https://doi.org/10.1186/s12934-025-02685-1.

84 Mahali, M.G.S. (2019). Extraction methods and functional properties of protein from *Arthospira platensis* for bioavailability of algal proteins. *International Journal of Pharmaceutical and Chemical Sciences* 5: 20. https://doi.org/10.11648/j.ijpc.20190502.12.

85 Motlagh, S.R., Elgharbawy, A.A., Khezri, R. et al. (2023). Ionic liquid-based microwave-assisted extraction of protein from *Nannochloropsis* sp. biomass. *Biomass Convers Biorefinery* 13: 8327–8338. https://doi.org/10.1007/s13399-021-01778-2.

86 Das, R.S., Tiwari, B.K., and Garcia-Vaquero, M. (2022). Bioactive peptides from algae. In: *Bioactive Peptides from Food* (ed. L.M.L. Nollet and S. Ötleş), 31–54. CRC Press.

87 Schwenzfeier, A., Wierenga, P.A., and Gruppen, H. (2011). Isolation and characterization of soluble protein from the green microalgae *Tetraselmis* sp. *Bioresource Technology* 102: 9121–9127. https://doi.org/10.1016/j.biortech.2011.07.046.

88 Ursu, A.-V., Marcati, A., Sayd, T. et al. (2014). Extraction, fractionation and functional properties of proteins from the microalgae *Chlorella vulgaris*. *Bioresource Technology* 157: 134–139. https://doi.org/10.1016/j.biortech.2014.01.071.

89 Safi, C., Olivieri, G., Campos, R.P. et al. (2017). Biorefinery of microalgal soluble proteins by sequential processing and membrane filtration. *Bioresource Technology* 225: 151–158. https://doi.org/10.1016/j.biortech.2016.11.068.

90 Liu, S., Rouquié, C., Lavenant, L. et al. (2022). Coupling bead-milling and microfiltration for the recovery of lipids and proteins from *Parachlorella kessleri*: impact of the cell disruption conditions on the separation performances. *Separation and Purification Technology* 287: 120570. https://doi.org/10.1016/j.seppur.2022.120570.

91 Huang, J.J., Lin, S., Xu, W., and Cheung, P.C.K. (2017). Occurrence and biosynthesis of carotenoids in phytoplankton. *Biotechnology Advances* 35: 597–618. https://doi.org/10.1016/j.biotechadv.2017.05.001.

92 Shokrkar, H. and Nedaei, L. (2024). Chlorophyll and carotenoid extraction from mixed microalgae; experimental and kinetic study. *Biomass Convers Biorefinery* 14: 17891–17903. https://doi.org/10.1007/s13399-023-04068-1.

93 Zazirna, M., Tischler, S., Marko, D. et al. (2024). Ultrasound-based strategies for the recovery of microalgal carotenoids: Insights from green extraction methods to UV/MS-based identification. *Food Research International* 187: 114354. https://doi.org/10.1016/j.foodres.2024.114354.

94 Fernandes, A.S., Caetano, P.A., Jacob-Lopes, E. et al. (2024). Alternative green solvents associated with ultrasound-assisted extraction: a green chemistry approach for the extraction of carotenoids and chlorophylls from microalgae. *Food Chemistry* 455: 139939. https://doi.org/10.1016/j.foodchem.2024.139939.

95 Saini, D.K., Pabbi, S., and Shukla, P. (2018). Cyanobacterial pigments: perspectives and biotechnological approaches. *Food and Chemical Toxicology* 120: 616–624. https://doi.org/10.1016/j.fct.2018.08.002.

96 Viera, I. and Roca, M. (2020). Analytical protocols in chlorophyll analysis. In: *Pigments from Microalgae Handbook* (ed. E. Jacob-Lopes, M.I. Queiroz, and L.Q. Zepka), 127–149. Cham: Springer International Publishing.

97 Hagerthey, S.E., William Louda, J., and Mongkronsri, P. (2006). Evaluation of pigment extraction methods and a recommended protocol for periphyton chlorophyll a determination and chemotaxonomic assessment. *Journal of Phycology* 42: 1125–1136. https://doi.org/10.1111/j.1529-8817.2006.00257.x.

98 Schumann, R., Häubner, N., Klausch, S., and Karsten, U. (2005). Chlorophyll extraction methods for the quantification of green microalgae colonizing building facades. *International Biodeterioration & Biodegradation* 55: 213–222. https://doi.org/10.1016/j.ibiod.2004.12.002.

99 Hosikian, A., Lim, S., Halim, R., and Danquah, M.K. (2010). Chlorophyll extraction from microalgae: a review on the process engineering aspects. *International Journal of Chemical Engineering* 2010: 1–11. https://doi.org/10.1155/2010/391632.

100 Ahmadi, K., Omidi, H., Mehrnia, M., and Ur Rehman, H. (2019). Comparative extraction efficacy of different organic solvents for leaf chlorophylls and carotenoids in (<em>*Portulaca oleracea*</em> L.) varying with growth behavior and stress type. Preprints, 2019090152. https://doi.org/10.20944/preprints201909.0152.v1.

101 Luengo, E., Condón-Abanto, S., Álvarez, I., and Raso, J. (2014). Effect of pulsed electric field treatments on permeabilization and extraction of pigments from *Chlorella vulgaris. The Journal of Membrane Biology* 247: 1269–1277. https://doi.org/10.1007/s00232-014-9688-2.

102 Lafarga, T. and Aguiló-Aguayo, I. (2022). Chapter 9 - Pulsed electric fields for the extraction of lipids, pigments, and polyphenols from cultured microalgae. In: *Innovative and Emerging Technologies in the Bio-Marine Food Sector* (ed. M. Garcia-Vaquero and G. Rajauria), 197–221. Cambridge, MA: Academic Press (US). https://doi.org/10.1016/B978-0-12-820096-4.00010-9.

103 Orthesin, N., Hidayat, I.T., Wahyuni, W.T. et al. (2024). Optimization of chlorophyll extraction from dried *Spirulina platensis* using low power microwave assisted extraction method. *IOP Conference Series: Earth and Environmental Science* 1359: 012021. https://doi.org/10.1088/1755-1315/1359/1/012021.

104 Georgiopoulou, I., Tzima, S., Louli, V., and Magoulas, K. (2023). Process optimization of microwave-assisted extraction of chlorophyll, carotenoid and phenolic compounds from chlorella vulgaris and comparison with conventional and supercritical fluid extraction. *Applied Sciences* 13: 2740. https://doi.org/10.3390/app13042740.

105 Paterson, S., Gómez-Cortés, P., Villanueva-Bermejo, D. et al. (2025). Effect of supercritical CO2 extraction on the recovery of bioactive lipids and pigments from pretreated *Nannochloropsis gaditana* biomass. *LWT* 223: 117748. https://doi.org/10.1016/j.lwt.2025.117748.

106 Tzima, S., Georgiopoulou, I., Louli, V., and Magoulas, K. (2023). Recent advances in supercritical CO2 extraction of pigments, lipids and bioactive compounds from microalgae. *Molecules* 28: 1410. https://doi.org/10.3390/molecules28031410.

107 Saini, D.K., Chakdar, H., Pabbi, S., and Shukla, P. (2020). Enhancing production of microalgal biopigments through metabolic and genetic engineering. *Critical Reviews in Food Science and Nutrition* 60: 391–405. https://doi.org/10.1080/10408398.2018.1533518.

108 Chaiklahan, R., Chirasuwan, N., Srinorasing, T. et al. (2022). Enhanced biomass and phycocyanin production of *Arthrospira* (*Spirulina*) *platensis* by a cultivation management strategy: light intensity and cell concentration. *Bioresource Technology* 343: 126077. https://doi.org/10.1016/j.biortech.2021.126077.

109 Pez Jaeschke, D., Rocha Teixeira, I., Damasceno Ferreira Marczak, L., and Domeneghini Mercali, G. (2021). Phycocyanin from *Spirulina*: a review of extraction methods and stability. *Food Research International* 143: 110314. https://doi.org/10.1016/j.foodres.2021.110314.

110 Tavanandi, H.A., Mittal, R., Chandrasekhar, J., and Raghavarao, K.S.M.S. (2018). Simple and efficient method for extraction of C-phycocyanin from dry biomass of *Arthospira platensis*. *Algal Research* 31: 239–251. https://doi.org/10.1016/j.algal.2018.02.008.

111 Chaiklahan, R., Chirasuwan, N., Loha, V. et al. (2011). Separation and purification of phycocyanin from *Spirulina* sp. using a membrane process. *Bioresource Technology* 102: 7159–7164. https://doi.org/10.1016/j.biortech.2011.04.067.

112 İlter, I., Akyıl, S., Demirel, Z. et al. (2018). Optimization of phycocyanin extraction from *Spirulina platensis* using different techniques. *Journal of Food Composition and Analysis* 70: 78–88. https://doi.org/10.1016/j.jfca.2018.04.007.

113 Silveira, S.T., Burkert, J.F.M., Costa, J.A.V. et al. (2007). Optimization of phycocyanin extraction from *Spirulina platensis* using factorial design. *Bioresource Technology* 98: 1629–1634. https://doi.org/10.1016/j.biortech.2006.05.050.

114 Jaeschke, D.P., Mercali, G.D., Marczak, L.D.F. et al. (2019). Extraction of valuable compounds from *Arthrospira platensis* using pulsed electric field treatment. *Bioresource Technology* 283: 207–212. https://doi.org/10.1016/j.biortech.2019.03.035.

115 Suarez Garcia, E., Lo, C., Eppink, M.H.M. et al. (2019). Understanding mild cell disintegration of microalgae in bead mills for the release of biomolecules. *Chemical Engineering Science* 203: 380–390. https://doi.org/10.1016/j.ces.2019.04.008.

116 Postma, P.R., Miron, T.L., Olivieri, G. et al. (2015). Mild disintegration of the green microalgae *Chlorella vulgaris* using bead milling. *Bioresource Technology* 184: 297–304. https://doi.org/10.1016/j.biortech.2014.09.033.

117 Montalescot, V., Rinaldi, T., Touchard, R. et al. (2015). Optimization of bead milling parameters for the cell disruption of microalgae: process modeling and application to *Porphyridium cruentum* and *Nannochloropsis oculata*. *Bioresource Technology* 196: 339–346. https://doi.org/10.1016/j.biortech.2015.07.075.

118 de Carvalho, J.C., Magalhães, A.I., de Melo Pereira, G.V. et al. (2020). Microalgal biomass pretreatment for integrated processing into biofuels, food, and feed. *Bioresource Technology* 300: 122719. https://doi.org/10.1016/j.biortech.2019.122719.

119 Li, Y., Zhang, Z., Paciulli, M., and Abbaspourrad, A. (2020). Extraction of phycocyanin—A natural blue colorant from dried spirulina biomass: influence of processing parameters and extraction techniques. *Journal of Food Science* 85: 727–735. https://doi.org/10.1111/1750-3841.14842.

120 Moraes, C.C., Sala, L., Cerveira, G.P., and Kalil, S.J. (2011). C-phycocyanin extraction from *Spirulina platensis* wet biomass. *Brazilian Journal of Chemical Engineering* 28: 45–49. https://doi.org/10.1590/S0104-66322011000100006.

121 Ores, J.d.C., Amarante, M.C.A.d., and Kalil, S.J. (2016). Co-production of carbonic anhydrase and phycobiliproteins by *Spirulina* sp. and *Synechococcus nidulans*. *Bioresource Technology* 219: 219–227. https://doi.org/10.1016/j.biortech.2016.07.133.

122 Vernès, L., Abert-Vian, M., El Maâtaoui, M. et al. (2019). Application of ultrasound for green extraction of proteins from spirulina. Mechanism, optimization, modeling, and industrial prospects. *Ultrasonics Sonochemistry* 54: 48–60. https://doi.org/10.1016/j.ultsonch.2019.02.016.

123 Pan-utai, W. and Iamtham, S. (2019). Extraction, purification and antioxidant activity of phycobiliprotein from *Arthrospira platensis*. *Process Biochemistry* 82: 189–198. https://doi.org/10.1016/j.procbio.2019.04.014.

124 Larrosa, A.P.Q., Camara, Á.S., Moura, J.M., and Pinto, L.A.A. (2018). *Spirulina* sp. biomass dried/disrupted by different methods and their application in biofilms production. *Food Science and Biotechnology* 27: 1659–1665. https://doi.org/10.1007/s10068-018-0397-y.

125 Martínez, J.M., Luengo, E., Saldaña, G. et al. (2017). C-phycocyanin extraction assisted by pulsed electric field from *Artrosphira platensis*. *Food Research International* 99: 1042–1047. https://doi.org/10.1016/j.foodres.2016.09.029.

126 Ferreira-Santos, P., Nunes, R., De Biasio, F. et al. (2020). Influence of thermal and electrical effects of ohmic heating on C-phycocyanin properties and biocompounds recovery from *Spirulina platensis*. *LWT* 128: 109491. https://doi.org/10.1016/j.lwt.2020.109491.

127 Cook, R., Lupette, J., and Benning, C. (2021). The Role of Chloroplast Membrane Lipid Metabolism in Plant Environmental Responses. *Cells* 10: 706. https://doi.org/10.3390/cells10030706.

128 Lupette, J. and Maréchal, E. (2018). Phytoplankton glycerolipids: challenging but promising prospects from biomedicine to green chemistry and biofuels. In: *Blue Biotechnology: Production and Use of Marine Molecules* (ed. L.B.S.-S.B. Stephen), 191–215. Wiley-VCH Verlag GmbH & Co. KGaA.

129 Lupette, J. and Benning, C. (2020). Human health benefits of very-long-chain polyunsaturated fatty acids from microalgae. *Biochimie* 178: 15–25. https://doi.org/10.1016/j.biochi.2020.04.022.

130 Di Dato, V., Orefice, I., Amato, A. et al. (2017). Animal-like prostaglandins in marine microalgae. *ISME Journal* 11: 1722–1726. https://doi.org/10.1038/ismej.2017.27.

131 Meyer, N., Rettner, J., Werner, M. et al. (2018). Algal oxylipins mediate the resistance of diatoms against algicidal bacteria. *Marine Drugs* 16: 486. https://doi.org/10.3390/md16120486.

132 Vigor, C., Oger, C., Reversat, G. et al. (2020). Isoprostanoid profiling of marine microalgae. *Biomolecules* 10: 1073. https://doi.org/10.3390/biom10071073.

133 Folch, J., Lees, M., and Stanley, G.H.S. (1957). A simple method for the isolation and purification of total lipides from animal tissues. *Journal of Biological Chemistry* 226: 497–509. https://doi.org/10.1016/s0021-9258(18)64849-5.

134 Ulmer, C.Z., Jones, C.M., Yost, R.A. et al. (2018). Optimization of Folch, Bligh-Dyer, and Matyash sample-to-extraction solvent ratios for human plasma-based lipidomics studies. *Analytica Chimica Acta* 1037: 351–357. https://doi.org/10.1016/j.aca.2018.08.004.

135 Wong, M.W.K., Braidy, N., Pickford, R. et al. (2019). Comparison of single phase and biphasic extraction protocols for lipidomic studies using human plasma. *Frontiers in Neurology* 10. https://doi.org/10.3389/fneur.2019.00879.

136 Fiset, C., Liefer, J., Irwin, A.J., and Finkel, Z.V. (2017). Methodological biases in estimates of macroalgal macromolecular composition. *Limnology and Oceanography: Methods* 15: 618–630. https://doi.org/10.1002/lom3.10186.

137 Triebl, A., Trötzmüller, M., Eberl, A. et al. (2014). Quantitation of phosphatidic acid and lysophosphatidic acid molecular species using hydrophilic interaction liquid chromatography coupled to electrospray ionization high resolution mass spectrometry. *Journal of Chromatography A* 1347: 104–110. https://doi.org/10.1016/j.chroma.2014.04.070.

138 Chen, X., Hu, L., Xing, R. et al. (2015b). Ionic liquid-assisted subcritical water promotes the extraction of lipids from wet microalgae *Scenedesmus* sp. *European Journal of Lipid Science and Technology* 117: 1192–1198. https://doi.org/10.1002/ejlt.201400189.

139 Yang, F., Xiang, W., Sun, X. et al. (2014). A novel lipid extraction method from wet microalga *Picochlorum* sp. at room temperature. *Marine Drugs* 12: 1258–1270. https://doi.org/10.3390/md12031258.

140 Sosa-Hernández, J.E., Escobedo-Avellaneda, Z., Iqbal, H.M.N., and Welti-Chanes, J. (2018). State-of-the-art extraction methodologies for bioactive compounds from algal biome to meet bio-economy challenges and opportunities. *Molecules* 23: 2953. https://doi.org/10.3390/molecules23112953.

141 Ng, J.W., Teh, T.M., Melanie, W., and Talukder, M.M.R. (2024). Integrating bead milling and alkaline solubilization for enhanced protein recovery from microalgae: a comprehensive approach. *Future Foods* 9: 100385. https://doi.org/10.1016/j.fufo.2024.100385.

142 Setyo Wibowo, C., Sugiarto, B., Zikra, A. et al. (2019). The effect of bioethanol-varying gasoline blends on performance and emission of SI engine 150 CC. *AIP Conference Proceedings* 2062: 020020.

143 Alhattab, M., Moorthy, L.S., Patel, D. et al. (2024). Oleaginous microbial lipids' potential in the prevention and treatment of neurological disorders. *Marine Drugs* 22: 80. https://doi.org/10.3390/md22020080.

144 Mallick, N., Bagchi, S.K., Koley, S., and Singh, A.K. (2016). Progress and challenges in microalgal biodiesel production. *Front Microbiol* 7. https://doi.org/10.3389/fmicb.2016.01019.

145 Mustapha, S.I. and Isa, Y.M. (2020). Utilization of quaternary solvent mixtures for extraction of lipids from Scenedesmus obliquus microalgae. *Cogent Eng* 7: 1788877. https://doi.org/10.1080/23311916.2020.1788877.

146 Danilović, B.R., Đorđević, N.G., Karabegović, I.T. et al. (2021). Enhancing lipid extraction from green microalgae *Chlorella* sp. using a deep eutectic solvent pretreatment. *Chemical Industry & Chemical Engineering Quarterly* 27: 313–317.

147 Moreno Martínez, P., Ortiz-Martínez, V.M., Sánchez Segado, S. et al. (2022). Deep eutectic solvents for the extraction of fatty acids from microalgae biomass: recovery of omega-3 eicosapentaenoic acid. *Separation and Purification Technology* 300: 121842. https://doi.org/10.1016/j.seppur.2022.121842.

148 Zghaibi, N., Omar, R., Mustapa Kamal, S.M. et al. (2019). Microwave-assisted brine extraction for enhancement of the quantity and quality of lipid production from microalgae *Nannochloropsis* sp. *Molecules* 24: 3581. https://doi.org/10.3390/molecules24193581.

149 Nasution, R.W., Hambali, E., and Suprihatin (2019). Utilization of palm oil mill effluent (POME) from the residual biogas power plant for microalgae production as raw material of biodiesel. *International Journal of Engineering and Management Research* 9: 141–145. https://doi.org/10.31033/ijemr.9.5.20.

150 Moussa, I.D.B., Bouaziz, M., and Dhouib, A. (2018). Production of lipids from tunisian wild microalgae: influence of cell disruption, extraction methods and UV-B radiation exposure. *Biomedical Journal of Scientific & Technical Research* 9: 001–011. https://doi.org/10.26717/BJSTR.2018.09.001814.

151 Moon, M., Park, W.-K., Suh, W.I. et al. (2019). Biological carbon recovery from sugar refinery washing water into microalgal DHA: medium optimization and stress induction. *Scientific Reports* 9: 19959. https://doi.org/10.1038/s41598-019-56406-x.

152 Wang, Q., Oshita, K., and Takaoka, M. (2020). A novel method for extraction of lipids from liquid microalgae without dewatering. *Preprint* (Version 1) available at Research Square. https://doi.org/10.21203/rs.3.rs-90843/v1.

153 Fattah, I.M.R., Noraini, M.Y., Mofijur, M. et al. (2020). Lipid extraction maximization and enzymatic synthesis of biodiesel from microalgae. *Applied Sciences* 10: 6103. https://doi.org/10.3390/app10176103.

154 Rokicka, M., Zieliński, M., Dudek, M., and Dębowski, M. (2021). Effects of ultrasonic and microwave pretreatment on lipid extraction of microalgae and methane production from the residual extracted biomass. *BioEnergy Research* 14: 752–760. https://doi.org/10.1007/s12155-020-10202-y.

155 Señoráns, M., Castejón, N., and Señoráns, F.J. (2020). Advanced Extraction of Lipids with DHA from *Isochrysis galbana* with Enzymatic Pre-Treatment Combined with Pressurized Liquids and Ultrasound Assisted Extractions. *Molecules* 25: 3310. https://doi.org/10.3390/molecules25143310.

156 Patel, A., Arora, N., Pruthi, V., and Pruthi, P.A. (2019). A novel rapid ultrasonication-microwave treatment for total lipid extraction from wet oleaginous yeast biomass for sustainable biodiesel production. *Ultrasonics Sonochemistry* 51: 504–516. https://doi.org/10.1016/j.ultsonch.2018.05.002.

157 Alpural, A., Dincoglu, B., and Imamoglu, E. (2024). Ultrasound-assisted extraction vs. microwave-assisted extraction for sustainable development goals: selecting the ideal lipid extraction and fatty acid profile. *Chemical Engineering and Processing: Process Intensification* 205: 110035. https://doi.org/10.1016/j.cep.2024.110035.

158 Zou, X., Xu, K., Chang, W. et al. (2021). Rapid extraction of lipid from wet microalgae biomass by a novel buoyant beads and ultrasound assisted solvent extraction method. *Algal Research* 58: 102431. https://doi.org/10.1016/j.algal.2021.102431.

159 Ramola, B., Kumar, V., Nanda, M. et al. (2019). Evaluation, comparison of different solvent extraction, cell disruption methods and hydrothermal liquefaction of *Oedogonium* macroalgae for biofuel production. *Biotechnology Reports* 22: e00340. https://doi.org/10.1016/j.btre.2019.e00340.

160 Muir, E., Guieysse, B., and Plouviez, M. (2025). Large-scale harvesting of microalgae. In: *Microalgae Horizons* (ed. I.A. Severo, J.C. Ordóñez, A.B. Mariano, and J.V.C. Vargas), 377–399. Cham: Springer Nature Switzerland.

161 Badmus, U.O., Taggart, M.A., and Boyd, K.G. (2019). The effect of different drying methods on certain nutritionally important chemical constituents in edible brown seaweeds. *Journal of Applied Phycology* 31: 3883–3897. https://doi.org/10.1007/s10811-019-01846-1.

162 Agbede, O.O., Oke, E.O., Akinfenwa, S.I. et al. (2020). Thin layer drying of green microalgae (*Chlorella* sp.) paste biomass: drying characteristics, energy requirement and mathematical modeling. *Bioresource Technology Reports* 11: 100467. https://doi.org/10.1016/j.biteb.2020.100467.

163 Vítová, M. (ed.) (2020). *Microalgae - From Physiology to Application*. IntechOpen.

164 Hammann, W., Ross, A., and Seames, W. (2024). Sequential extraction of carbohydrates and lipids from chlorella vulgaris using combined physical and chemical pre-treatments. *ChemEngineering* 8: 11. https://doi.org/10.3390/chemengineering8010011.

165 Zghaibi, N., Omar, R., Mustapa Kamal, S.M. et al. (2020). Kinetics study of microwave-assisted brine extraction of lipid from the microalgae *Nannochloropsis* sp. *Molecules* 25: 784. https://doi.org/10.3390/molecules25040784.

166 Quesada-Salas, M.C., Delfau-Bonnet, G., Willig, G. et al. (2021). Optimization and comparison of three cell disruption processes on lipid extraction from microalgae. *Processes* 9: 369. https://doi.org/10.3390/pr9020369.

167 Rezaei Motlagh, S., Harun, R., Awang Biak, D.R. et al. (2019). Screening of suitable ionic liquids as green solvents for extraction of eicosapentaenoic acid (EPA) from microalgae biomass using COSMO-RS model. *Molecules* 24: 713. https://doi.org/10.3390/molecules24040713.

168 de Moura, R.R., Etges, B.J., dos Santos, E.O. et al. (2018). Microwave-assisted extraction of lipids from wet microalgae paste: a quick and efficient method. *European Journal of Lipid Science and Technology* 120: 1700419. https://doi.org/10.1002/ejlt.201700419.

169 Blanco-Llamero, C. and Señoráns, F.J. (2021). Biobased solvents for pressurized liquid extraction of *Nannochloropsis gaditana* omega-3 lipids. *Marine Drugs* 19: 107. https://doi.org/10.3390/md19020107.

170 Blanco-Llamero, C., García-García, P., and Señoráns, F.J. (2021). Combination of synergic enzymes and ultrasounds as an effective pretreatment process to break microalgal cell wall and enhance algal oil extraction. *Foods* 10: 1928. https://doi.org/10.3390/foods10081928.

171 Villaró, S., Viñas, I., and Lafarga, T. (2021). Consumer acceptance and attitudes toward microalgae and microalgal-derived products as food. In: *Cultured Microalgae for the Food Industry*, 367–385. In: Elsevier.

172 Olsen, M.L., Olsen, K., and Jensen, P.E. (2024). Consumer acceptance of microalgae as a novel food - where are we now? And how to get further. *Physiology Plant* 176: e14337. https://doi.org/10.1111/ppl.14337.

173 Laureati, M., De Boni, A., Saba, A. et al. (2024). Determinants of consumers' acceptance and adoption of novel food in view of more resilient and sustainable food systems in the EU: a systematic literature review. *Foods* 13: 1534. https://doi.org/10.3390/foods13101534.

174 Bürck, M., Lemes, A.C., Egea, M.B., and Braga, A.R.C. (2024). Exploring the potential and challenges of fermentation in creating foods: a spotlight on microalgae. *Fermentation* 10: 649. https://doi.org/10.3390/fermentation10120649.

175 Chua, W.C.L., Yeo, A.Y.Y., Yuan, W. et al. (2024). Flavour characterization of twelve species of edible algae. *Algal Research* 80: 103540. https://doi.org/10.1016/j.algal.2024.103540.

176 Queiroz, M.I. and Mitterer-Daltoé, M.L. (2023). Sensorial characters of microalgae biomass and its individual components. In: *Handbook of Food and Feed from Microalgae*, 537–546. In: Elsevier.

177 Nunes, M.C., Ferreira, J., and Raymundo, A. (2023). Volatile fingerprint impact on the sensory properties of microalgae and development of mitigation strategies. *Current Opinion in Food Science* 51: 101040. https://doi.org/10.1016/j.cofs.2023.101040.

178 Coleman, B., Van Poucke, C., Dewitte, B. et al. (2022). Potential of microalgae as flavoring agents for plant-based seafood alternatives. *Future Foods* 5: 100139. https://doi.org/10.1016/j.fufo.2022.100139.

179 Day, L. and Golding, M. (2016). Food structure, rheology, and texture. In: *Encyclopedia of Food Chemistry*, 125–129. Cambridge, MA: Elsevier.

180 Bernaerts, T.M.M., Gheysen, L., Foubert, I. et al. (2019). The potential of microalgae and their biopolymers as structuring ingredients in food: a review. *Biotechnology Advances* 37: 107419. https://doi.org/10.1016/j.biotechadv.2019.107419.

181 Bernaerts, T.M.M. and Van Loey, A.M. (2021). Chapter 10 - Microalgae as structuring ingredients in food. In: *Cultured Microalgae for the Food Industry* (ed. T. Lafarga and G. Acién), 265–286. Cambridge, MA: Academic Press https://doi.org/10.1016/B978-0-12-821080-2.00012-5.

182 Siddiqui, S.A., Ucak, İ., Afreen, M. et al. (2024). Microalgae as a potential raw material for plant-based seafood alternatives: a comprehensive review. *Food Science & Nutrition* 12: 8559–8593. https://doi.org/10.1002/fsn3.4313.

183 Girotto, F. and Scapini, A. (2025). Microalgal biomass in the European food industry: navigating regulation, technological innovation, and consumer acceptance. *Algal Research* 91: 104288. https://doi.org/10.1016/j.algal.2025.104288.

184 Senapati, P.K., Singh, M., and Karmi, M. (2024). Algae-derived food and feed products. In: *Cyanobacteria Biotechnology* (ed. M.A. Mehmood, S. Malik, S.G. Musharraf, and R. Boopathy), 115–137. Cham: Springer Nature Switzerland.

185 Khan, M.I., Shin, J.H., and Kim, J.D. (2018). The promising future of microalgae: current status, challenges, and optimization of a sustainable and renewable industry for biofuels, feed, and other products. *Microbial Cell Factories* 17: 36. https://doi.org/10.1186/s12934-018-0879-x.

186 Torres-Tiji, Y., Fields, F.J., and Mayfield, S.P. (2020). Microalgae as a future food source. *Biotechnology Advances* 41: 107536. https://doi.org/10.1016/j.biotechadv.2020.107536.

187 Adams, C., Godfrey, V., Wahlen, B. et al. (2013). Understanding precision nitrogen stress to optimize the growth and lipid content tradeoff in oleaginous green microalgae. *Bioresource Technology* 131: 188–194. https://doi.org/10.1016/j.biortech.2012.12.143.

7

Sea Cucumber

Rong Cao and Yong Sun

Yellow Sea Fisheries Research Institute, Academy of Fishery Sciences, Shandong, China

Sea cucumber is the common name for animals within the class Holothuroidea, belonging to the phylum Echinodermata [1]. The name "Echinodermata" derives from the Greek words "echinos" (meaning "spine") and "derma" (meaning "skin"), thus signifying "spiny-skinned animals". Characteristically, echinoderms possess an endoskeleton embedded within their body wall; skeletal elements often protrude through the body surface, forming spines, which gives the phylum its name. Echinoderms represent an ancient lineage, with origins traceable to the Cambrian period of the Paleozoic era over 500 million years ago [2]. They are widely distributed in the ocean, and almost all of them lead a benthic lifestyle (Fig. 7.1).

The extant phylum Echinodermata comprises five classes: Crinoidea (sea lilies), Asteroidea (sea stars), Ophiuroidea (brittle stars), Echinoidea (sea urchins and sand dollars), and Holothuroidea (sea cucumbers) (Fig. 7.2). Among these, Holothuroidea is the class of greatest economic importance [3].

7.1 Species of Sea Cucumbers and Their Distribution Areas

Sea cucumbers can be found in oceans across the globe. They are mainly distributed in tropical and temperate regions, with around 1,400 species [4]. The sea cucumber resources in tropical waters are more diverse, mainly occurring in the tropical waters of the Pacific and Indian Oceans. Among these areas, the Indo-West Pacific region has the largest number of sea cucumber species and the greatest resource volume currently known. Genera such as *Stichopus*, *Holothuria*, *Actinopyga*, and *Bohadschia* of the order Aspidochirotida are common in this region [5].

Species of the genus *Stichopus* include *Stichopus chloronotus* and *Stichopus hermanni*; species of the genus *Holothuria* include *Holothuria scabra*, *Holothuria arenicola*, *Holothuria impatiens*, *Holothuria edulis*, *Holothuria hilla*, *Holothuria leucospilota*, and *Holothuria nobilis*; species of the genus *Actinopyga* include *Actinopyga*

Figure 7.1 Sea cucumber (*Apostichopus japonicus*).

Figure 7.2 Other typical echinoderms.

lecanora, *Actinopyga echinites*, *Actinopyga mauritiana*, and *Actinopyga miliaris*; species of the genus *Bohadschia* include *Bohadschia marmorata* and *Bohadschia argus*. The species of sea cucumbers in the order Dendrochirotida are not very abundant. Commonly seen ones are *Cladolabes aciculus*, *Ohshimella ehrenbergi*, and *Afrocucumis africana*.

The sea cucumber resources in temperate seas are relatively single in species, mainly distributed on the east and west coasts of the Pacific Ocean. As listed in Table 7.1, high-quality and high-economic value varieties are mainly distributed along the coasts of the Northern Hemisphere's Pacific Ocean, the coasts of Latin America, and the coasts of the Arctic Ocean, and mostly belong to the family Stichopodidae [5]. For instance, *Stichopus japonicus* and *Apostichopus japonicus* are found in the Yellow Sea and Bohai Sea of China, the Japanese Archipelago, and the Korean Peninsula [6]; *Parastichopus californicus* is distributed along the North American coast [7]; *Isostichopus badionotus* is distributed in Latin America, the Caribbean Sea, and the Mexican coast [8]; *Isostichopus fuscus* is distributed from the California Peninsula to the Ecuadorian mainland [9]; and *Holothuria tubulosa* is distributed in the Mediterranean Sea and the eastern Atlantic Ocean [10]. There are relatively more sea cucumber species in the Southern Hemisphere, yet most of the varieties have low edible value [5].

Table 7.1 Typical sea cucumber varieties and their global distribution.

Species	Classification	Morphology	Geographical distribution
Apostichopus japonicus	Class—Holothuroidea Order—Aspidochirotida Family—Stichopodidae		Mainly distributed in the western Pacific Ocean, such as the Yellow Sea, the Sea of Japan, and the Sea of Okhotsk. The northern limit of its geographical distribution is the coastal areas of Sakhalin Island, Russia, and Alaska, and the southern boundary is Tanega-shima Island in Japan.
Stichopus chloronotus	Class—Holothuroidea Order—Aspidochirotida Family—Stichopodidae		Mainly distributed in the West Indies, Mascarene Islands, East Africa, Madagascar, Maldives, Sri Lanka, Bay of Bengal, East Indies, northern Australia, the Philippines, southern Japan, and most of the islands in the central and western Pacific Ocean.
Stichopus herrmanni	Class—Holothuroidea Order—Aspidochirotida Family—Stichopodidae		Mainly distributed in the Mascarene Islands, East Africa, Madagascar, the Red Sea, the southeastern part of the Arabian Peninsula, the Gulf of Aqaba, the Persian Gulf, the Maldives, Sri Lanka, the Bay of Bengal, the East Indies, northern Australia, the Philippines, and southern Japan.
Stichopus horrens	Class—Holothuroidea Order—Aspidochirotida Family—Stichopodidae		Mainly distributed in the Red Sea, East Africa, Maldives, East Indies, Philippines, southern Japan, and northern Australia.

(Continued)

Table 7.1 (Continued)

Species	Classification	Morphology	Geographical distribution
Isostichopus badionotus	Class—Holothuroidea Order—Aspidochirotida Family—Stichopodidae	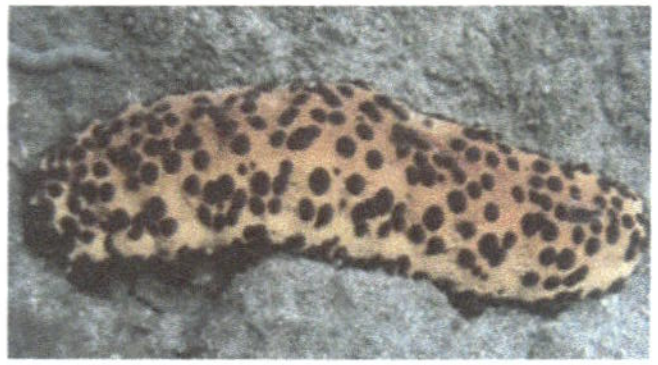	Mainly distributed in Brazil, Venezuela, Colombia, Panama, Yucatan (Mexico), southern Florida, the Bahamas, and South Carolina (USA), the Ascension Islands in the central Atlantic Ocean, and the Gulf of Guinea off the western coast of Africa.
Parastichopus californicus	Class—Holothuroidea Order—Aspidochirotida Family—Stichopodidae		Mainly distributed along the Pacific coast of North America, from the Aleutian Islands, Alaska, to the Gulf of California.
Thelenota ananas	Class—Holothuroidea Order—Aspidochirotida Family—Stichopodidae		Mainly distributed in East Africa, Madagascar, New Caledonia, the Mascarene Islands, the Maldives Islands, Indonesia, the Ryukyu Islands of Japan, and northern Australia.
Isostichopus fuscus	Class—Holothuroidea Order—Aspidochirotida Family—Stichopodidae		Mainly distributed from the California Peninsula to the Ecuadorian mainland.

Actinopyga mauritiana	Class—Holothuroidea Order—Aspidochirotida Family—Holothuriidae		Mainly distributed in the West Indies, Red Sea, East Africa, Bay of Bengal, northern Australia, the Philippines, and southern Japan.
Acinopyga leanora	Class—Holothuroidea Order—Aspidochirotida Family—Holothuriidae		Mainly distributed in the Mascarene Islands, East Africa, Red Sea, Madagascar, Sri Lanka, Bay of Bengal, East Indies, northern Australia, Philippines, southern China, southern Japan, and islands in the South Pacific.
Actinopyga miliaris	Class—Holothuroidea Order—Aspidochirotida Family—Holothuriidae		Mainly distributed in the islands of western Indian Ocean, Mascarene Islands, East Africa, Madagascar, Red Sea, Sri Lanka, Bay of Bengal, East Indies, north Australia, the Philippines, China, and southern Japan, South Pacific Islands east to French Polynesia.
Bohadschia argus	Class—Holothuroidea Order—Aspidochirotida Family—Holothuriidae		Mainly distributed in Sri Lanka, Bay of Bengal, East Indies, northern Australia, the Philippines, China, southern Japan, and south Pacific Islands.

(Continued)

Table 7.1 (Continued)

Species	Classification	Morphology	Geographical distribution
Bohadschia marmorata	Class—Holothuroidea Order—Aspidochirotida Family—Holothuriidae		Widely distributed throughout the Indian Ocean to East Africa, including the Mascarene Islands, Madagascar, the Red Sea, Sri Lanka, the Bay of Bengal, the East Indies, northern Australia, the Philippines, China, southern Japan, and the South Pacific Islands.
Holothuria atra	Class—Holothuroidea Order—Aspidochirotida Family—Holothuriidae		Widely distributed in the Indo-Pacific region. From Mozambique and Madagascar in the west, to Tahiti in the east, to southern Japan in the north, and to Lord Howe Island in Australia in the south.
Holothuria leucospilota	Class—Holothuroidea Order—Aspidochirotida Family—Holothuriidae		Widely distributed in the Indo-West Pacific region, from East Africa to the Hawaiian Islands and Society Islands, northward to southern Japan, and southward to Lord Howe Island and Shark Bay in Australia.
Holothuria mexicana	Class—Holothuroidea Order—Aspidochirotida Family—Holothuriidae		Mainly distributed in the Florida Keys, the Bahamas, Cuba, Puerto Rico, Jamaica, Barbados, Tobago, Aruba, the Yucatan Peninsula, Belize, Bonaire, Venezuela, and the Colombian Islands.

Holothuria scabra	Class—Holothuroidea Order—Aspidochirotida Family—Holothuriidae		Widely distributed between 30°N and 30°S, mainly concentrated in the tropical Indian Ocean-Pacific region.
Holothuria fuscopunctata Jaeger	Class—Holothuroidea Order—Aspidochirotida Family—Holothuriidae		Mainly distributed in the central and western Pacific Ocean and the Indian Ocean and has also been found in Tonga.
Cucumaria frondosa	Class—Holothuroidea Order—Dendrochirotida Family—Cucumariidea		Mainly distributed in the North Atlantic Ocean, from the Arctic to Cape Cod, as well as in Scotland and the Orkney Islands. It is also distributed in Iceland, along the coast of Russia, and off the coast of Greenland.
Cucumaria japonica	Class—Holothuroidea Order—Dendrochirotida Family—Cucumariidea		Mainly distributed in the northeastern part of the Yellow Sea, the northeastern coast of Honshu Island, the Russian coast of the Sea of Japan, the Sea of Okhotsk, the Kuril Islands, the Kamchatka Peninsula, and the Bering Sea.

7.2 The Nutrition and Function of Sea Cucumbers

7.2.1 Basic Chemical Composition of Sea Cucumbers

Sea cucumbers are typical aquatic products with high protein and low fat. Sea cucumbers contain various vitamins such as vitamins A, B, D, and E. Although the lipid content of sea cucumbers is not high, the types of essential fatty acids are complete, such as linoleic acid, linolenic acid, EPA, and DHA. These fatty acids have a variety of physiological functions. Sea cucumbers contain a variety of essential macro and trace elements for the human body, especially with relatively high iron and zinc content. In addition, sea cucumbers contain very unique functional components such as polysaccharides, saponins, and active lipids, which are beneficial to promoting human health [11–13].

A. japonicus is the main variety in the Chinese sea cucumber market [14]. Numerous factors can influence the basic nutritional composition of *A. japonicus*, including region, season, the age of the sea cucumber, and the cultivation mode [12, 15, 16]. By summarizing the research reports on the basic nutritional composition of the *A. japonicus* over the past 10 years and the test data of our team, and converting these data to a dry basis for statistical analysis, the following conclusions have been drawn: the protein content of the body wall of *A. japonicus* is $(52.92 \pm 7.41)\%$, the ash content is $(28.48 \pm 7.72)\%$, the lipid content is $(4.30 \pm 2.12)\%$, the polysaccharide content is $(7.12 \pm 3.07)\%$, and the saponin content is $(0.40 \pm 0.17)\%$ (Fig. 7.3).

Whether there are differences in nutrition, quality, and safety between artificially cultured and wild *A. japonicus* has always been a concern for the sea cucumber industry and consumers [17, 18]. It should be noted that "wild sea cucumbers" in China do not refer to wild-caught sea cucumbers naturally bred, but rather to "bottom-seeded sea cucumbers." "Bottom-seeded" refers to the aquaculture method where sea cucumber juveniles are sown directly onto the seabed for natural growth, distinct from pond or raft farming.

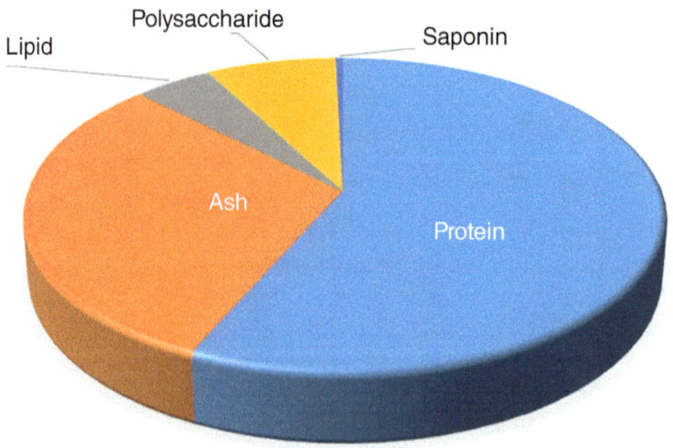

Figure 7.3 Basic nutritional composition of the body wall of *Apostichopus japonicus*.

Table 7.2 Comparative analysis results of the chemical composition of wild and cultured *Apostichopus japonicus* (on a dry basis, %).

Nutritional components	Wild *A. japonicus*	Cultured *A. japonicus*
Ash	25.73 ± 0.20**	20.74 ± 0.17
Salt	18.47 ± 0.41*	15.40 ± 0.69
Crude protein	54.82 ± 0.39**	59.13 ± 0.71
Collagen	39.43 ± 0.38	40.37 ± 1.41
Polysaccharide	10.19 ± 0.17*	9.27 ± 0.29
Saponin	0.81 ± 0.11*	0.62 ± 0.10

Note: * indicates a significant difference ($P < 0.05$), and ** indicates an extremely significant difference ($P < 0.01$).

After the cultivation of aquatic economic animals, due to the changes in the growth environment and feed, their nutritional components will have certain differences compared with the wild population. Through a comparative analysis of the chemical composition of bottom—seeded (i.e., "wild") and cultured *A. japonicus* (as shown in Table 7.2), it was found that compared with wild *A. japonicus*, the cultured *A. japonicus* has a higher crude protein content, while the contents of polysaccharides and saponins are slightly lower. The contents of elements such as potassium, sodium, magnesium, and calcium are all lower than those of wild *A. japonicus*, while the difference in the composition of essential amino acids is relatively small [19].

7.2.2 Active Ingredients in Sea Cucumbers

7.2.2.1 Collagen

Protein is a substance with the highest content among various nutrients in sea cucumbers. The protein in the body wall of sea cucumbers is mainly collagen, accounting for up to 70%. After being highly cross-linked, collagen forms collagen fibers, which play important physiological roles in the life activities of sea cucumbers. The biochemical properties of collagen also have a significant impact on the processing and product storage of sea cucumbers after harvest [20, 21].

Collagen is a type of biopolymer and a structural protein within the extracellular matrix. It exhibits strong extensibility, along with excellent water retention and emulsifying properties. Collagen is insoluble in cold water, dilute acids, and dilute alkali solutions and is not readily hydrolyzed by common proteases. Nevertheless, it can be degraded by collagenase, and the resulting fragments are easily hydrolyzed by ordinary proteases. Regular proteins typically possess a double-helix structure, whereas collagen has a triple-helix structure composed of three polypeptide chains. Specifically, three left-handed polypeptide chains form a left-handed helical structure, which then tightly interlock through hydrogen bonds to form a right-handed superhelical structure. This triple-helix structure is a fundamental characteristic of collagen (Fig. 7.4). The most notable feature of the helical region is the periodic

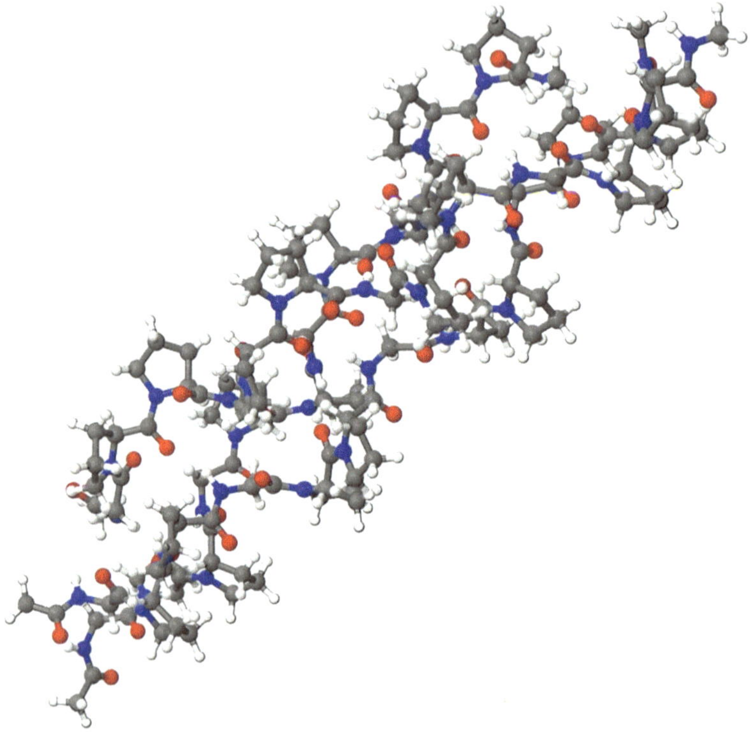

Figure 7.4 Schematic diagram of the triple-helix structure of sea cucumber collagen.

arrangement of amino acids in the pattern of Gly-X-Y. In the X and Y positions, proline (Pro) and hydroxyproline (Hyp) are frequently found, and these are amino acids unique to collagen. When compared to terrestrial animals, the total content of proline and hydroxyproline in the collagen of aquatic animals is lower [22].

To date, as many as 21 types of collagen have been identified within the collagen family. Among them, the common types are I, II, III, V, and XI [23]. Type I collagen is primarily distributed in tissues like the skin and tendons. It is also the most abundant protein in the by-products (skin, bones, and scales) of aquatic product processing and has the most extensive applications in the medical field. In fish, one of the most notable characteristics of type I collagen is its low thermal stability, and it displays species-specific features. Type II collagen is synthesized by chondrocytes. Type V collagen generally refers to extracellular collagen and is abundant in connective tissues. Type XI collagen is a minor constituent of cartilage. It plays a crucial role in the formation of cartilage collagen fibers and the composition of the cartilage matrix, and it often coexists with type II collagen.

Regarding the structural types of sea cucumber collagen, they are still not very clear so far. Researchers discovered that the collagen type of *S. japonicus* produced in Japan is similar to type I collagen of vertebrates, with a composition of $(\alpha_1)2\alpha_2$; the collagen of *A. japonicus* produced in China is composed of three α_1 chains [20, 21].

In terms of amino acid composition, the collagen of *A. japonicus* aligns with that of typical collagen. Among these amino acids, the content of Gly is the highest, making up approximately 1/3 of the total amino acid content. The contents of Ala, Glu, Pro, Hyp, and Asp are relatively high, while those of His, Tyr, and Met are relatively low. Proline and hydroxyproline are characteristic amino acids of collagen. Hydroxyproline does not participate in the biosynthesis of collagen in a direct form but is converted from proline through the action of hydroxylase [24].

7.2.2.2 Polysaccharide

Polysaccharide is an important active ingredients in sea cucumbers. Sea cucumber polysaccharides can be divided into two categories. These polysaccharides can be classified into two primary categories: (1) holothurian glycosaminoglycan (HG), a branched heteropolysaccharide comprising D-*N*-acetylgalactosamine, D-glucuronic acid, and L-fucose, which exhibits structural similarity to chondroitin sulfate [25]; and (2) holothurian fucoidan (HF), a linear homopolysaccharide consisting exclusively of L-fucose [26]. Although the monosaccharide compositions of the two are different, some of the hydroxyl groups on the sugar chains are sulfated, and the degree of sulfation is around 30% for both.

Monosaccharides, including fucose, galactose, glucose, and mannose, constitute critical components of sea cucumber polysaccharides. Among these monosaccharides in sea cucumber polysaccharides, fucose exhibits the highest content, ranging from 10.97% to 20.25%, followed by galactose with a content of 1.35%–3.80%, and mannose, which accounts for 0.38%–0.95% [27, 28].

Sulfate groups represent another pivotal category of constituents in sea cucumber polysaccharides. The sulfation of hydroxyl groups in polysaccharides not only enhances the solubility of polysaccharides but also alters their conformational structures, thereby emerging as a key determinant of their biological activities. The sulfate content in several common sea cucumber polysaccharides varies from 19.54% to 29.28%. Specifically, the polysaccharides from the Mexican sea cucumber (*Holothuria mexicana*) possess a relatively high sulfate group content, whereas those from *A. japonicus* (cultured in China and sourced from Japan) exhibit relatively lower levels, at 19.54% and 20.13%, respectively. Nevertheless, a lower sulfate group content does not equate to weaker biological activity, as the activity of sulfate groups is not solely dependent on their content but is also governed by their substitution positions within the polysaccharide structure [28, 29] (Fig. 7.5).

7.2.2.3 Saponins

Sea cucumber saponins are secondary metabolites produced by sea cucumbers and serve as the chemical basis for their chemical defense mechanisms. When sea cucumbers are attacked by predators, they extend the Cuvierian tubules from their coelomic cavity and subsequently release a white or wine-colored fluid containing saponins as a defensive agent [30]. To date, the structures of over 100 sea cucumber saponins have been elucidated, and these compounds have been demonstrated to possess diverse physiological activities, including antitumor,

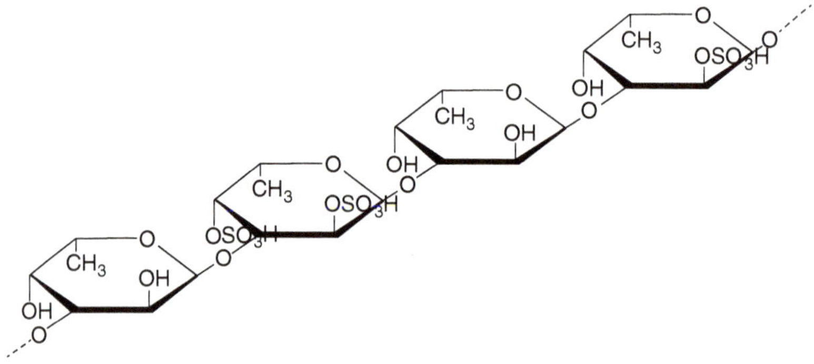

Figure 7.5 Chemical structure of the tetrasaccharide repeating unit in fucoidan from *Acaudina molpadioidea*.

hemolytic, and antibacterial properties, thereby qualifying as a class of important bioactive constituents in sea cucumbers [31].

The saponin content varies significantly among different sea cucumber species. For instance, *Pearsonothuria graeffei* (Semper, 1868) exhibits a relatively high saponin content, reaching 3.514%, whereas *H. scabra* (Jaeger, 1833) contains the lowest level, at only 0.205%, representing a 17-fold difference between the two. Even within the same species, saponin content may fluctuate depending on factors such as geographical origin and the age of the sea cucumber [32, 33].

Sea cucumber saponins are composed of two moieties: an aglycone and an oligosaccharide chain, with molecular weights ranging from 600 to 1500 Da. The chemical structures of sea cucumber saponins are characterized by complexity and diversity. Factors contributing to structural variations in the aglycone include the configuration of the side chain, modifications of the lactone ring, the position of double bonds, and substituents at the 12th, 16th, and 17th carbon positions. For the oligosaccharide chain, structural variability arises from the linkage positions and sequence of monosaccharides, the composition and number of monosaccharide units, as well as the positions and quantities of sulfate ester groups [34, 35].

To date, the majority of identified sea cucumber saponin aglycones possess lanostane-type triterpenoid structures, which are classified into alkane-type and non-alkane-type [36]. Most sea cucumber saponins fall into the alkane-type category (Fig. 7.6). Over 70 alkane-type triterpenoid saponins from sea cucumbers have been characterized. Based on the substitution pattern at the 16th carbon atom, these saponins are further divided into two subgroups: those without substitution at this position (Fig. 7.7) and those bearing a ketone or acetyl group (Fig. 7.8) [37].

Non-alkane-type sea cucumber saponins have been less frequently identified. Their structural characteristics are defined by the presence or absence of an 18(16) lactone bond in the aglycone. For instance, a saponin isolated from *Pentacta australis* lacks a lactone bond in its aglycone (Fig. 7.9), whereas saponins derived from *Pentamera calcigera* possess an 18(16) lactone structure in their aglycones (Fig. 7.10) [37, 38].

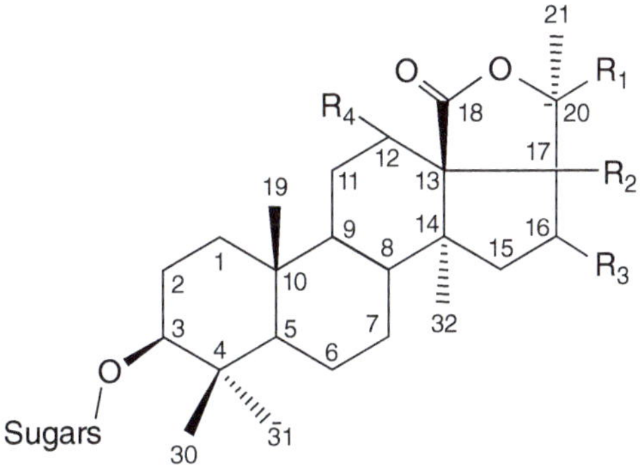

Figure 7.6 Alkane-type glycoside structure of sea cucumber saponins.

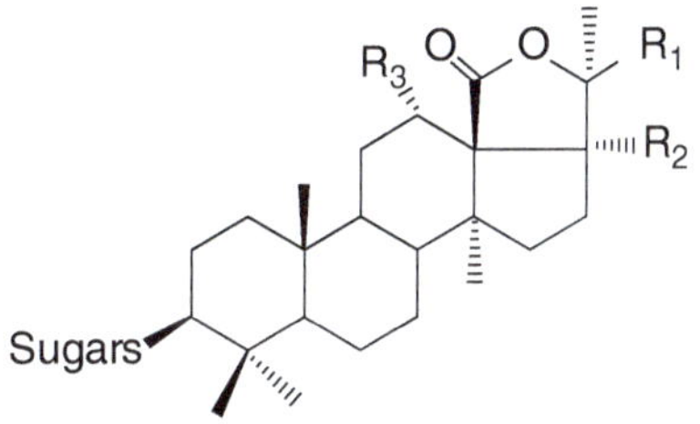

Figure 7.7 Structure of an unsubstituted sea cucumber alkane-type glycoside at C16 position.

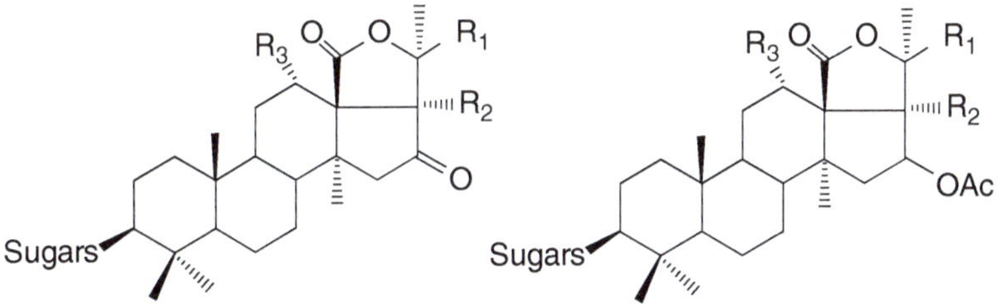

Figure 7.8 Structures of sea cucumber alkane-type glycosides with C16 keto (left) or acetyl (right) substitutions.

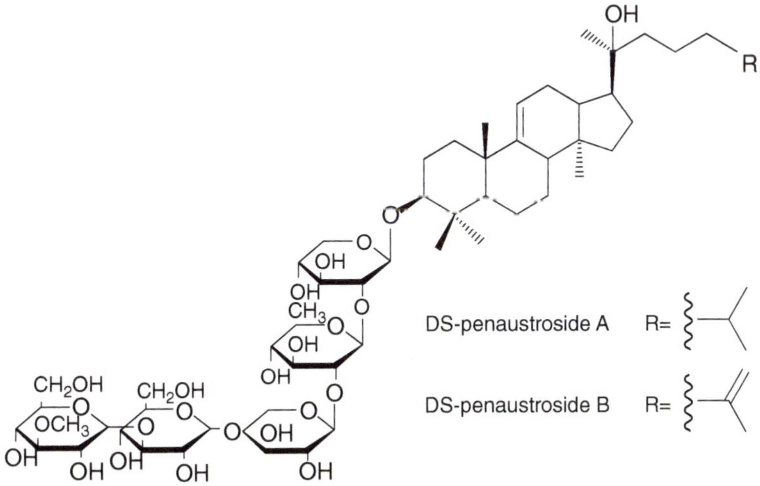

Figure 7.9 Non-alkane triterpenoid saponins from *Pentacta australis*.

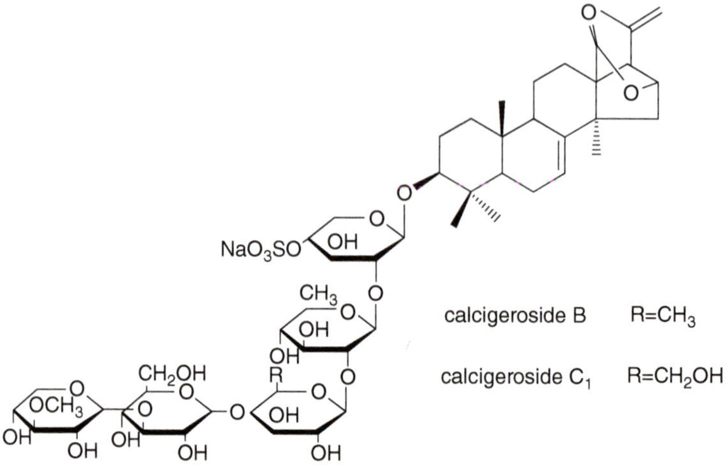

Figure 7.10 Non-alkane triterpenoid saponins from *Pentacta calcigera.*

The sugar chain of sea cucumber saponins is linked to the third carbon position of the aglycone via a β-*O*-glycosidic bond, with xylose serving as the initial connecting monosaccharide. Typically, the sugar chain comprises 2–6 monosaccharide units, including xylose (Xyl), 3-*O*-methylxylose (3-*O*-MeXyl), quinovose (Qui), glucose (Glc), and 3-*O*-methylglucose (3-*O*-MeGlc), among others. Hydroxyl groups on certain monosaccharides are frequently sulfated. For example, Frondoside B, a saponin isolated from *Cucumaria frondosa*, contains two sulfate substitutions (Fig. 7.11) [39]. In contrast, Cucumarioside A_7-1, A_7-2, and A_7-3, saponins derived from *Cucumaria japonica*, each possess three sulfate substitutions (Fig. 7.12) [40]. Saponins lacking sulfate group substitutions on their sugar chains are primarily

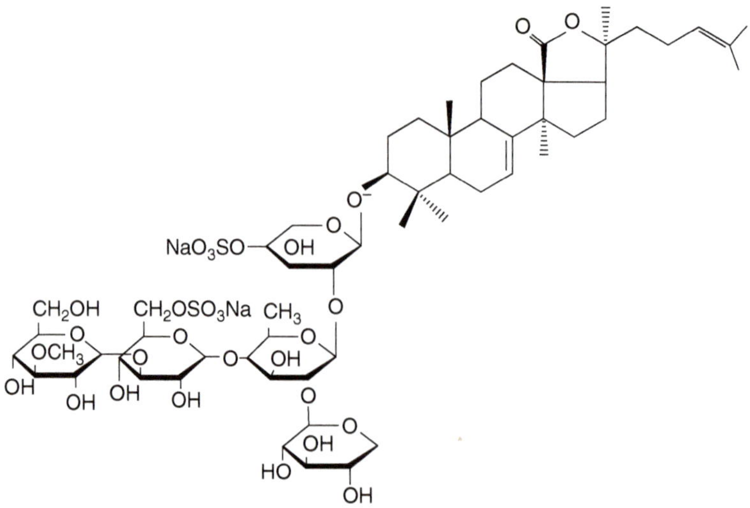

Figure 7.11 Structure of a saponin Frondooside B from *Cucumaria frondose* [39] / with permission of American Chemical Society.

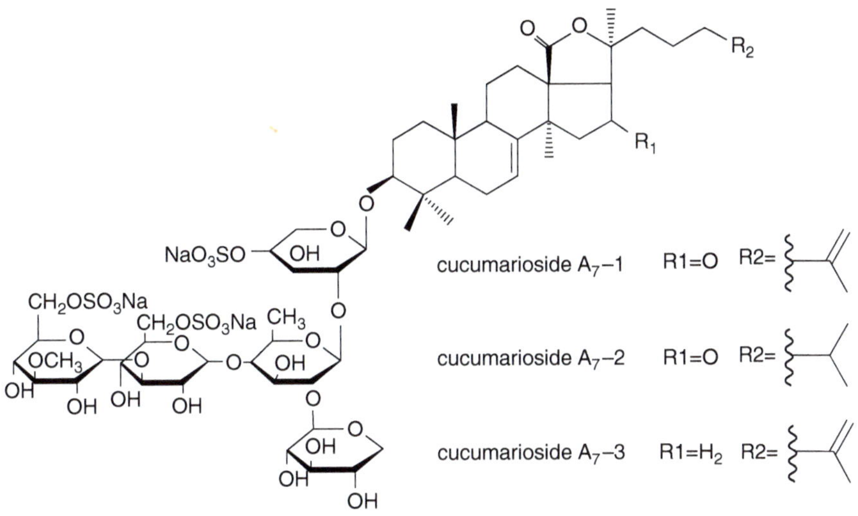

Figure 7.12 Structure of saponins Cucumarioside A_7-1, A_7-2, and A_7-3 from *Cucumaria japonica* [40] Springer Nature.

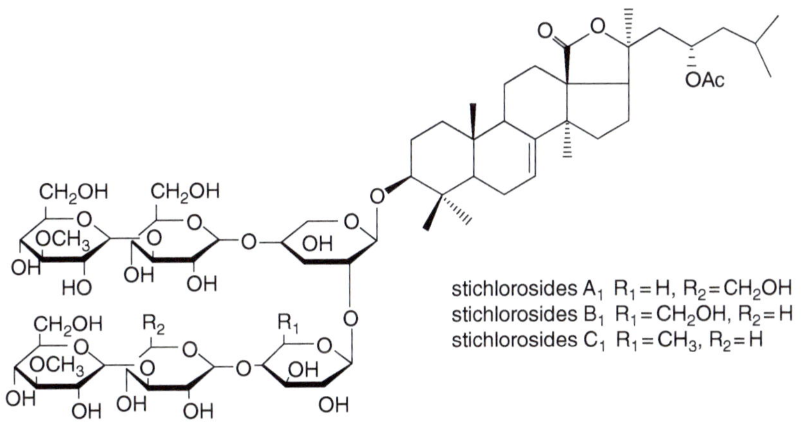

Figure 7.13 Structure of saponins Stichlorosides A_I, B_1, and C_1 from *Stichopus variegatus* [41] / Springer Nature.

found in the genus *Apostichopus*; for instance, Stichloroside A1, B1, and C1, isolated from *Apostichopus variegatus*, belong to this category (Fig. 7.13) [41].

7.2.3 Health Benefits of Sea Cucumbers

In traditional Chinese medicine (TCM) theory, sea cucumber is attributed with the properties of tonifying the kidney and replenishing essence, augmenting vital *qi*, and nourishing blood. Its judicious consumption is believed to facilitate cellular regeneration and the repair of damaged tissues, while also enhancing the body's immune function, thereby contributing to health preservation and physical fortification [42].

Over the past few decades, anecdotal reports and clinical observations have documented favorable outcomes in the management of aplastic anemia and diabetes mellitus when sea cucumber serves as a primary component in medicinal diets or therapeutic formulations [43, 44]. Moreover, its efficacy in invigorating the spleen, moistening dryness, fortifying bones, enhancing intelligence, improving visual acuity, enriching the blood, nourishing the complexion, blackening the hair, and promoting longevity has gained widespread acknowledgment in TCM practice.

Additionally, sea cucumber has been extensively incorporated into various medicinal diets and therapeutic regimens specifically formulated for postsurgical patients and for women during the postpartum recovery period, underscoring its role in supporting convalescence and restorative processes.

With the advancement of scientific research and the continuous refinement of analytical technologies, scholars have carried out extensive investigations into the biological activities of sea cucumber employing multidisciplinary technical approaches, including chemistry, pharmacology, molecular biology, cytology, and proteomics [44–48]. These studies have not only validated the healthcare functionalities of sea cucumber, as summarized from the clinical experience of TCM, but also identified a multitude of novel biological activities.

7.2.3.1 Immunity Enhancement

Numerous bioactive components in sea cucumber exhibit immunomodulatory effects, particularly in enhancing immune function. Animal studies have demonstrated that sea cucumber saponins can augment cellular immunity and nonspecific immune functions in normal mice. For mice with immunosuppression, sea cucumber saponins significantly increase serum hemolysin levels, promote humoral immune responses and delayed-type hypersensitivity, and enhance the proliferative capacity of splenic lymphocytes, thereby improving their cellular immune function [49]. Additionally, sea cucumber saponins markedly elevate the phagocytic rate and phagocytic index of peritoneal macrophages in mice, further reinforcing nonspecific immunity [50].

Furthermore, sea cucumber polysaccharides have been shown to significantly enhance both cellular and nonspecific immune functions of the organism. They are capable of ameliorating and reversing the immunosuppressive states in animal models induced by hormonal agents or chemical drugs, thereby restoring immune homeostasis [51].

7.2.3.2 Blood Glucose Regulation

Diabetes mellitus is a clinical syndrome arising from the interplay of genetic and environmental factors. In recent years, its global prevalence has exhibited a continuous upward trend, emerging as the third major disease posing a severe threat to human health, following cardiovascular and cerebrovascular diseases as well as malignancies. As estimated by the World Health Organization, the global population of diabetic patients will exceed 300 million by 2025, rendering the management of diabetes and its complications a critical challenge confronting nations worldwide [52].

A multitude of bioactive components in sea cucumber possess blood glucose-regulating properties. For instance, sulfated fucoidan derived from sea cucumber can significantly reduce fasting blood glucose levels and insulin concentrations in diabetic mice, thereby exerting hypoglycemic effects and ameliorating insulin resistance [53]. Sea cucumber saponins regulate glucose metabolism by increasing lipocalin content and inhibiting α-glucosidase activity, which notably alleviates glucose intolerance and insulin resistance [54]. Additionally, sea cucumber cerebrosides and their primary structural unit, ceramide, have also been demonstrated to improve insulin resistance [55].

7.2.3.3 Lowering of Blood Lipids

Blood lipids are a general term for neutral fats (triglycerides) and lipids (phospholipids, glycolipids, sterols, steroids) in plasma. High blood lipid levels can directly cause some diseases, such as atherosclerosis, coronary heart disease, and pancreatitis, which seriously jeopardize human health.

Sea cucumber can significantly reduce the content of triglyceride in the liver of normal rats and has the same lipid metabolism regulating ability for experimental hypercholesterolemic rats [56]; sea cucumber polysaccharide has obvious inhibitory effect on the elevation of blood lipids in mice caused by high-fat diet, and it can effectively prevent the danger of hyperlipidemia and atherosclerosis [57]; sea cucumber fucoidan sulfate can effectively inhibit the increase of body weight and fat accumulation, and significantly reduce the accumulation of serum and hepatic lipid and has a lowering effect of blood lipids [58].

7.2.3.4 Anticoagulation and Anti-thrombosis

Heparin, the most commonly used clinical anticoagulant, is widely employed in the prevention and treatment of thromboembolic diseases; however, it is associated with significant side effects. Blood coagulation is characterized as a cascade process involving the sequential activation of a series of serine proteases. Sea cucumber polysaccharides exhibit multifaceted actions in the coagulation process; in addition to exerting anticoagulant effects and reducing blood viscosity, they also promote lipid metabolism without obvious side effects [59]. Furthermore, sea cucumber chondroitin sulfate can act on the intrinsic coagulation pathway to inhibit thrombosis, achieved through suppressing the production of thrombospondin and regulating the content of hemophilic factors [60].

7.2.3.5 Improvement of Dementia

The global prevalence of dementia is on a year-on-year rise, with approximately 36 million individuals currently affected worldwide. Among the various subtypes of dementia, Alzheimer's disease (AD) stands as the most prevalent form.

Notably, cerebrosides and phospholipids derived from sea cucumbers exhibit remarkable neuroprotective properties. These effects are mediated through multiple mechanisms: reducing the leakage of intracellular lactate dehydrogenase, preserving the integrity of cellular membranes, enhancing the antioxidant capacity of cells, downregulating the expression of proapoptotic genes, and inhibiting neuronal apoptosis. Furthermore, sea cucumber phospholipids have been demonstrated to ameliorate learning and memory impairments in murine models of AD [61].

7.2.3.6 Antitumor

Tumor is the local manifestation of a class of systemic diseases, and it will continue to proliferate even after the cancer-causing factors stop working. At present, the treatment of tumors generally adopts radiotherapy, chemotherapy, and surgery. In addition, medicinal diet or food therapy is an important way to assist in the treatment or prevention of tumors.

Sea cucumber contains a variety of antitumor active ingredients, such as sea cucumber saponins, sea cucumber polysaccharides, and sea cucumber active lipids. These active ingredients can play an inhibitory role in different stages of tumor occurrence, development, and metastasis and are expected to be developed into new anticancer drugs [50, 62, 63].

With the development of science and the continuous progress of analytical technology, researchers have conducted extensive studies on the bioactivity of sea cucumber by using the technical means of chemistry, pharmacology, molecular biology, cytology, histology, etc., which not only confirmed the healthcare functions of sea cucumber summarized by the clinical experience of Chinese medicine, but also discovered many new activities of sea cucumber. As a result, long-term consumption of sea cucumber is beneficial to human health care and disease prevention.

7.3 Sea Cucumber Processing and Products

7.3.1 Sea Cucumber Processing

Live sea cucumbers undergo "autolysis" promptly after harvest, necessitating immediate processing [64]. The following exemplifies the processing procedure for dried sea cucumbers (Fig. 7.14).

7.3.1.1 Degutting

Live sea cucumbers with excessive surface impurities should be rinsed to remove contaminants such as sediment and grit. The abdomen of the sea cucumber is incised starting from the region near the tail end, with the length of the incision approximately one-third of the body length. Subsequently, internal organs, including intestinal glands and reproductive glands, as well as other impurities within the coelomic cavity, are excised and removed.

7.3.1.2 Cleaning and Preliminary Selection

The sea cucumbers, post-evisceration, are subjected to washing to eliminate residual dirt. Specimens that are excessively large or small are sorted out to ensure uniformity in size specifications.

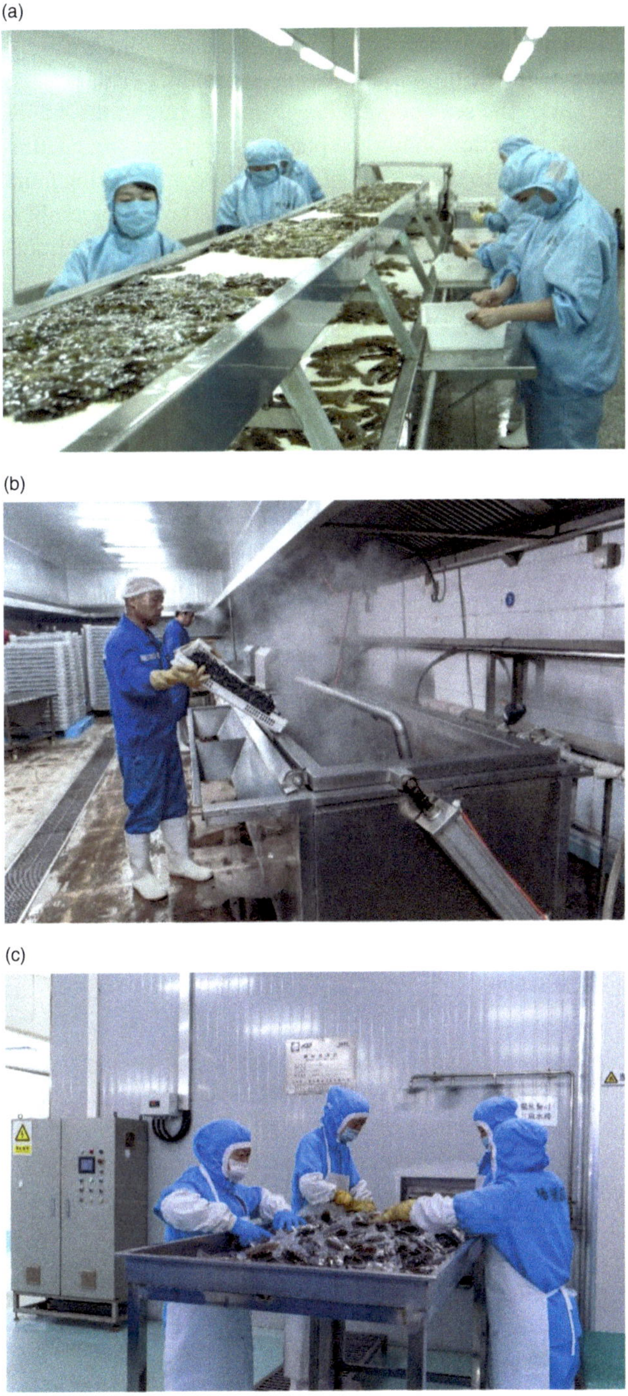

Figure 7.14 Sea cucumber processing scenes (a: the sea cucumber pre-treatment, b: boiling, c: packaging workshop).

7.3.1.3 Precooking

The primarily selected sea cucumbers are immersed in water at a temperature range of 70–100 °C for precooking, with the duration lasting 8–30 minutes. This process continues until the outer skin of the sea cucumbers becomes tight and the spines turn hard. During precooking, attention should be paid to stirring the sea cucumbers to prevent them from adhering to the bottom of the pot, and any floating foam should be removed promptly. Upon completion of precooking, the sea cucumbers can proceed directly to the drying process. If salting is required, they can be transferred to the salting process.

7.3.1.4 Salting

The precooked sea cucumbers are taken out and placed in a food-grade container. An appropriate amount of salt is added, followed by thorough mixing. The mixture is then allowed to stand at room temperature for 8–12 hours. Alternatively, saturated brine may be added. After the soaking period, the salted sea cucumbers can be removed and directly transferred to the drying process. Salted sea cucumbers that cannot be processed immediately should be stored in a cold storage facility at −18 °C or lower for subsequent use.

7.3.1.5 Desalination

Salted sea cucumbers intended for desalination are immersed in water at a temperature below 40°C, which is then heated to approximately 80 °C for soaking. The desalination process is maintained for 1.5–4 hours.

7.3.1.6 Shaping

Sea cucumbers are immersed in boiling water, with the duration of boiling adjusted according to the thickness of their body wall and the firmness of their spines. They are then retrieved after approximately 1–5 minutes.

7.3.1.7 Drying

Drying is typically performed via mechanical or natural methods. For mechanical drying, the temperature must be kept below 40 °C, with intense airflow to facilitate the process. Sea cucumbers are laid flat on drying curtains and placed in drying yards or rooms, with continuous ventilation maintained. To optimize drying efficiency, dried sea cucumbers may be sealed in containers (or bags) for 12–48 hours if needed, allowing moisture to redistribute evenly from the interior to the surface. Regular inspections of sea cucumbers' color and moisture content are required. If necessary, shaping may be conducted multiple times during drying, following the specified shaping protocol.

7.3.1.8 Sorting and Inspection

Dried products undergo size sorting, followed by grade sorting after inspection.

7.3.1.9 Packaging and Storage

Qualified products shall be hermetically packaged. Inner packaging materials must exhibit pressure resistance and toughness, with cushioning material added to preserve the visual integrity of dried sea cucumbers. All packaging materials must comply with relevant food safety standards. Each batch shall include an inspection certificate.

Products must be stored in cool, dry conditions protected from moisture, sunlight, pests, harmful contaminants, and other damage.

7.3.2 Processed Sea Cucumber Products

As illustrated in Table 7.3, sea cucumber products encompass morphologically preserved varieties, such as salted sea cucumbers, frozen-boiled sea cucumbers, dried sea cucumbers (including salt-dried, low-salt dried, freeze-dried, and instant rehydration [cook-free] dried types), ready-to-eat sea cucumbers, and preprepared sea cucumber dishes, as well as processed derivatives including sea cucumber oral liquids, sea cucumber powder (with capsules included), and sea cucumber extracts [65]. Salted and frozen-boiled sea cucumbers function as intermediate products in sea cucumber processing, which can be further processed into various dried sea cucumbers, ready-to-eat sea cucumbers, and other products according to production requirements. Dried sea cucumbers, characterized by convenience in transportation and circulation, as well as a long shelf life, stand as the mainstream products in the market. Ready-to-eat sea cucumbers, favored by young consumers, have gained popularity due to their ease of consumption.

7.4 Sea Cucumber-Related Standards

Outside of China, the only international standard related to sea cucumbers is Japan's Agricultural Standard for Dried Sea Cucumbers, which was implemented in 1956 and revised in 1969. In this standard, quality evaluation primarily relies on sensory indicators, with only one physicochemical index specified: moisture content must be less than 22%. No reports on quality standards for sea cucumbers (specifically *A. japonicus*) have been found in other countries or regions.

At present, China has formulated relevant standards covering such fields as sea cucumber germplasm, seedlings, breeding, feed, processing technical specifications, detection methods, and products. In 2011, the state initiated the formulation of the *National Food Safety Standard for Dried Sea Cucumbers*, which was drafted by the Yellow Sea Fisheries Research Institute of the Chinese Academy of Fishery Sciences. After five years of efforts, this standard has been released (standard number: GB 31602-2015) and was formally implemented on November 13, 2016 [66]. In accordance with the provisions of Article 25 of the *Food Safety Law* that "food safety standards are mandatory standards," the *National Food Safety Standard for Dried*

Table 7.3 Types of sea cucumber products and their processing characteristics (taking the Chinese sea cucumber market as an example).

Types	Product pictures	Processing techniques and characteristics
Salted sea cucumber		Salted sea cucumbers are processed from live specimens through a series of procedures, including degutting, cleaning, precooking, and salting. This product addresses the issue of the poor storability of fresh sea cucumbers. It primarily serves as an intermediate raw material for the processing of various dried and ready-to-eat sea cucumbers, while also being a frequently consumed product among consumers in coastal regions. Prior to consumption, desalination and cleaning are required, followed by boiling and soaking until reaching an edible state. Salted sea cucumbers are relatively low in cost and necessitate low-temperature storage.
Frozen boiled sea cucumbers		Frozen boiled sea cucumbers are prepared from live specimens via degutting, cleaning, and boiling, followed by cooling and rapid freezing. They also function as intermediate raw materials for the processing of various sea cucumber products and can be further processed into dried sea cucumbers, ready-to-eat sea cucumbers, and other derived products. In comparison to salted sea cucumbers, frozen boiled sea cucumbers omit the step of salt incorporation, thereby simplifying the processing procedure. Nevertheless, they require rapid cooling and have more stringent storage conditions than salted sea cucumbers.
Salted dried sea cucumbers		Salted dried sea cucumbers are typically produced from salted sea cucumbers through drying processes. They are characterized by simple processing equipment, low production costs, and the ability to be stored for an extended period at room temperature. During processing, salted dried sea cucumbers utilize a large quantity of salt, and their size is generally larger than that of low-salt dried sea cucumbers.

Low-salt dried sea cucumbers		Low-salt dried sea cucumbers are generally produced from salted sea cucumbers via processes, including desalination, shaping, and drying; alternatively, they can be manufactured from fresh sea cucumbers through procedures such as degutting, cleaning, boiling, shaping, and drying. In comparison to salted dried sea cucumbers, low-salt dried sea cucumbers are typically smaller in size, yet they possess lower salt content, higher protein content, and a superior rehydration effect during rehydration.
Freeze-dried sea cucumbers		Freeze-dried sea cucumbers are prepared by rehydrating and processing salted sea cucumbers or dried sea cucumbers, followed by vacuum freeze-drying. They are characterized by low water content and convenient consumption, as they only require a short soaking in water to become ready for eating. However, their processing cost is relatively high, and after rehydration, they exhibit a soft texture with poor elasticity.
Instant rehydration (cook-free) dried sea cucumbers		Instant rehydration (cook-free) dried sea cucumbers are processed using low-temperature curing and low-temperature drying techniques. Their most distinctive feature is that they do not require tedious rehydration prior to consumption. Simply place the product in an insulated container, add hot water, and it will be ready for consumption after 8–12 hours. The quality indices of instant rehydration (cook-free) dried sea cucumbers are superior to those of traditional salted dried and low-salt dried sea cucumbers, though their production costs and market prices are correspondingly slightly higher.

(*Continued*)

Table 7.3 (Continued)

Types	Product pictures	Processing techniques and characteristics
Ready-to-eat sea cucumbers		Ready-to-eat sea cucumbers are products manufactured from raw materials such as fresh sea cucumbers, salted sea cucumbers, frozen-boiled sea cucumbers, or dried sea cucumbers via processes including cleaning, degutting, boiling, rehydration, seasoning (or without seasoning), and sterilization (or without sterilization). Ready-to-eat sea cucumbers are categorized into two types: those stored at room temperature and those stored frozen. Room-temperature ready-to-eat sea cucumbers are packaged in vacuum or nitrogen-filled packaging, require refrigeration, can be consumed immediately upon opening, and generally have a shelf life of no more than 3 months. Frozen, ready-to-eat sea cucumbers need to be stored under frozen conditions, typically with a shelf life of over 12 months, and are consumed after thawing.
Sea cucumber preprepared dishes		Sea cucumber preprepared dishes refer to finished or semifinished dishes that take sea cucumbers as the main raw material. They are produced through preprocessing (such as cutting, stirring, marinating, tumbling, shaping, and seasoning) and/or precooking (such as stir-frying, frying, roasting, boiling, and steaming), and are prepackaged. Sea cucumber preprepared dishes not only possess nutritional and healthcare properties but also cater to consumers' demands for convenience and efficiency and are expected to become an important economic growth driver in the sea cucumber industry in the future.

Sea cucumber oral liquid		Sea cucumber oral liquid is prepared by enzymatically hydrolyzing raw sea cucumbers, followed by processes such as filtration, flavoring, filling, and sterilization. Some products also incorporate other ingredients with healthcare functions. Sea cucumber oral liquid is characterized by convenient consumption, rich nutrients, and easy absorption. However, such products have deviated from the original form of sea cucumbers, leading consumers to often adopt a highly cautious attitude when making purchases.
Sea cucumber functional ingredient extracts	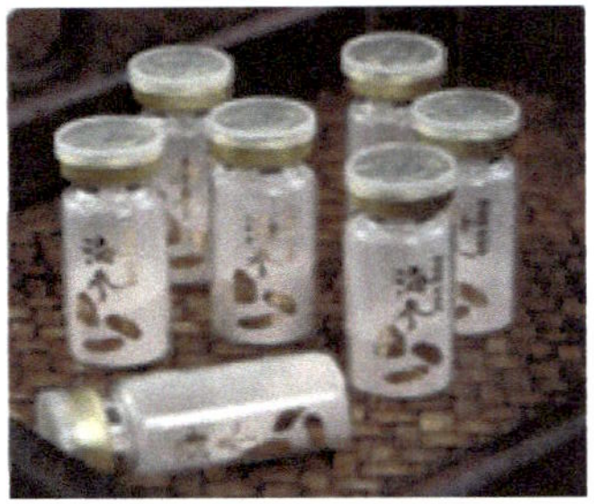	Sea cucumber functional ingredient extracts, such as sea cucumber peptide and sea cucumber polysaccharide preparations, focus on the healthcare functions of sea cucumbers. The processing process mainly involves enzymatic hydrolysis, separation, and purification of sea cucumbers or by-products from sea cucumber processing. After extracting the required ingredients from sea cucumbers, they are formulated into tablets, capsules, and other forms, either alone or in combination with other functional ingredients. Such products are relatively high-priced in the market and primarily target specific groups with healthcare demands.

Table 7.4 Key quality indicators in the Chinese national standard for dried sea cucumber (GB31602-2015).

Subjects	Requirements
Color	Natural colors such as blackish brown, blackish gray, gray, or yellowish brown, with a white frost on the surface and relatively uniform color.
Odor	Distinctive fresh and fishy smell of sea cucumber, with no unpleasant odor.
Appearance	Natural appearance of a sea cucumber, with a small amount of calcareous ossicles exposed, and the spines are straight and basically intact.
Protein/(g/100 g)	≥40
Moisture/(g/100 g)	≤15
Salt/(g/100 g)	≤40
Total water-soluble sugars/(g/100 g)	≤3
Dry weight ratio after rehydration/(g/100 g)	≥40
Sand content/(g/100 g)	≤3

Sea Cucumbers is a mandatory standard that all producers and operators must implement, and all dried sea cucumbers on the market must meet the requirements of this standard. As presented in Table 7.4, the main quality indicators of the GB 31602-2015 standard for dried sea cucumbers are as follows:

The dry weight ratio after rehydration refers to the percentage of dry matter obtained by rehydrating dried sea cucumbers through processes such as soaking and boiling, during which substances added to the dried sea cucumbers, such as salt, sugar, and gelatin, dissolve in water. The sea cucumbers are then dried, and the percentage of dry matter obtained is calculated as a percentage of the original dry weight of the sea cucumbers. As an important indicator for evaluating the quality of sea cucumbers, a higher dry weight ratio after rehydration indicates fewer exogenous additives (such as sugar and salt) in the sea cucumbers, which can well characterize the quality of sea cucumbers.

7.5 Rehydration Technique for Dried Sea Cucumber

Live sea cucumbers are predominantly processed into dried sea cucumbers, salted sea cucumbers, freeze-dried sea cucumbers, ready-to-eat sea cucumbers, sea cucumber capsules, and sea cucumber oral liquids. Among these, dried sea cucumbers account for approximately 70% of production, representing the dominant product category. As dried sea cucumbers require rehydration before consumption, scientific rehydration techniques are crucial for their edibility [67]. Currently, diverse methods for rehydrating dried sea cucumbers exist domestically; however, the lack of scientific and systematic research has left consumers without reliable guidance.

7.5.1 Home Rehydration Method for Dried Sea Cucumber

7.5.1.1 Soaking

Place dried sea cucumbers in a clean container, submerge in water (ensuring enough coverage above specimens), and refrigerate at 4 ± 1 °C for 2–4 days. Replace water two to three times daily until specimens achieve pliable texture.

Notes

1) Purpose: Hydration and desalination.
2) Water specifications: Potable tap water or purified water is acceptable.
3) Refrigeration rationale: Prevents epidermal sloughing and microbial proliferation.

7.5.1.2 Teeth Removal and Cleaning

The so-called sea cucumber teeth are small white calcareous ring-shaped ossicles around the mouth of the sea cucumber, located below the head of the sea cucumber (Fig. 7.15).

Make a longitudinal incision along the ventral aperture. Excise the teeth and rinse thoroughly with potable water.

7.5.1.3 Thermal Processing (Boiling)

Submerge cleaned sea cucumbers in boiling water within a clean vessel. Maintain a gentle simmer for 40–70 minutes. Subsequently, cease heating and steep covered for 60–120 minutes.

Notes

1) Doneness criterion: Sea cucumber suspended centrally by chopsticks exhibits bidirectional hang down with flexible tremor (Fig. 7.16). Rigidity indicates insufficient processing.
2) Processing time: High-quality dried sea cucumbers (e.g., low-salt dried sea cucumbers) usually require 60–70 minutes, while larger-sized sea cucumbers may require a longer processing time.
3) High-altitude adjustment: Boiling point reduction necessitates time extension or pressure cooking.
4) Physicochemical objective: Achieve complete textural softening for subsequent rehydration. Post-thermal specimens are edible but suboptimal in size.

Figure 7.15 Morphology and location of sea cucumber teeth.

Figure 7.16 Doneness criterion for thermal processing of sea cucumber.

5) Color: Green/brownish-red pigmentation is normal for water after thermal processing of sea cucumbers.

7.5.1.4 Hydration Expansion

Transfer thermally processed sea cucumbers to distilled/purified water. Refrigerate (4 ± 1 °C) for 24–72 hours with 2–3 daily water changes until optimal hydration is achieved (Fig. 7.17).

Notes

1) Two to three times of daily water replacement is essential, otherwise the sea cucumbers will not expand properly.

Figure 7.17 Comparison of dried sea cucumbers (low-salt) before and after rehydration.

2) Pure/distilled water can also be replaced with tap water, but the rehydrated sea cucumbers will be smaller in size.
3) Commercial overhydration compromises texture. For home preparation, it is recommended to rehydrate sea cucumbers to about twice their original length to achieve the optimal soft, sticky, and elastic taste.
4) One to three days range of rehydration accommodates textural preference.

7.5.1.5 Storage

Once rehydrated, sea cucumbers should be consumed within 2–3 days under refrigeration (≤4 °C), or frozen (−18 °C) for extended storage.

7.5.2 Simplified Rehydration Method for Dried Sea Cucumber

1) Soaking
   Place dried sea cucumbers in a clean container. Submerge in water and refrigerate at 4±1 °C for about 2 days. Replace water two to three times daily until specimens attain pliable texture.
2) Teeth Removal and Cleaning
   Perform longitudinal incision along the ventral aperture. Excise the calcareous ring and rinse thoroughly with potable water.
3) Thermal Retention Processing
   Transfer cleaned specimens to a vacuum-insulated container (e.g., thermos flask). Add boiling water (100 °C) and maintain thermal retention for 12–20 hours.
   Note: Resultant specimens are edible but exhibit suboptimal hydration expansion.
4) Hydration Expansion
   Immerse thermally retained specimens in distilled/purified water. Refrigerate (4 ± 1 °C) for 24–48 hours with two to three daily water changes until consumption readiness.

7.5.3 Accelerated Rehydration Method for Dried Sea Cucumber

7.5.3.1 Primary Thermal Processing

Place dried sea cucumbers directly into a vacuum-insulated container (e.g., thermos flask). Add boiling water (100 °C), ensuring complete submersion. Maintain thermal retention for 16–24 h.

7.5.3.2 Post-Processing Preparation

Extract specimens and perform longitudinal incision along the ventral axis. Excise the calcareous ring. Rinse thoroughly under running potable water.

Note: Specimens are now edible but may exhibit residual salinity.

7.5.3.3 Optional Desalination/Expansion

For elevated salinity: Immerse in distilled water (4 °C) for 2–4 hours.

For enhanced hydration: Transfer to distilled/purified water and refrigerate (4 ±1 °C) for 1–2 days with two to three daily water changes.

7.6 Sea Cucumber Market and Consumption

From a global perspective, sea cucumber farming has only developed on a large scale in China. Japan, South Korea, Australia, and other countries have attempted to cultivate sea cucumbers, but production has been low, mainly due to the lack of a local consumer market to support it. Except for China, other countries mainly rely on wild harvesting for sea cucumbers, and their products are mainly exported to China and some Southeast Asian countries [3, 5].

China is the world's largest producer and consumer of sea cucumbers. With the continuous increase in sea cucumber farming production and the expansion of the industry, the supply of sea cucumbers can basically meet market demand. According to the 2024 China Fisheries Statistical Yearbook, China's total sea cucumber farming production reached 292,045 tons in 2023 (Fig. 7.18), with an estimated total industry value of 150 billion RMB [68].

The consumption of sea cucumber products is primarily concentrated in the catering, gift-giving, and personal use sectors. According to data from iiMedia Research Group, among the motivations for Chinese consumers to purchase sea cucumber products, 66.5% are for gifting to elders as a nutritional supplement, 53.5% are for personal consumption, 52.8% are for gifting to friends and family, and 34.7% are for business gifting. The primary demographic for sea cucumber purchasers is aged 26–45, with a trend toward younger consumers. Young consumers are showing a rapid increase in demand for ready-to-eat sea cucumber products. Additionally, the health benefits of sea cucumber products are gradually being recognized and accepted by the public, leading to a gradual increase in consumer acceptance of nontraditional sea cucumber products (such as sea cucumber oral liquids, sea cucumber extract capsules, and sea cucumber tablets). As health awareness and purchasing power continue to grow, sea cucumber products are expected to occupy a larger share of the future health food market.

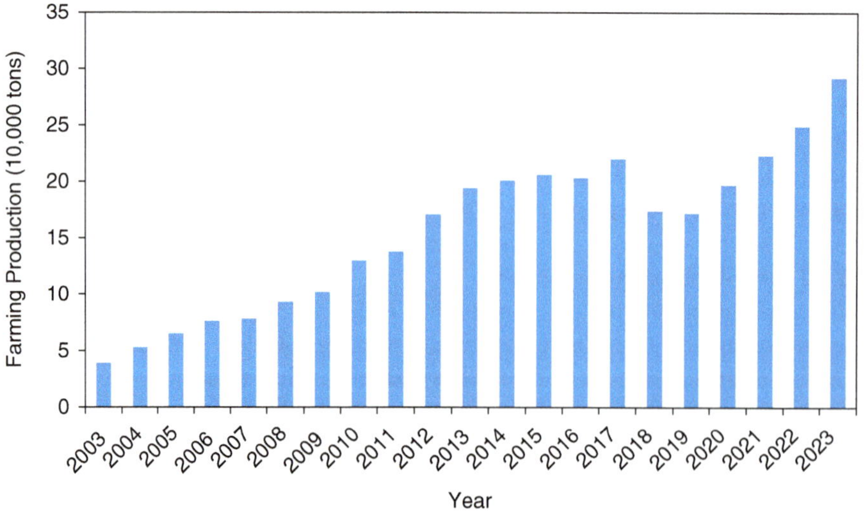

Figure 7.18 Sea cucumber farming production in China (2003–2023).

References

1 Miller, A. K., Kerr, A. M., Paulay, G., Reich, M., Wilson, N. G., Carvajal, J. I., & Rouse, G. W. (2017). Molecular phylogeny of extant Holothuroidea (Echinodermata). *Molecular Phylogenetics and Evolution*, 111, 110–131. https://doi.org/10.1016/j.ympev.2017.02.014

2 Rahman, I.A. and Zamora, S. (2024). Origin and early evolution of echinoderms. *In Annual Review of Earth and Planetary Sciences* 52: 295–320. https://doi.org/10.1146/annurev-earth-031621-113343.

3 Toral-Granda, V., Lovatelli, A., and Vasconcellos, M. (2008). *Sea Cucumbers: A Global Review of Fisheries and Trade*. FAO.

4 Kerr, AM, Kim, J. (2001). Phylogeny of Holothuroidea (Echinodermata) inferred from morphology. *Zoological Journal of the Linnean Society*, 133(1), 63–81. https://doi.org/10.1006/zjls.2000.0280

5 Purcell, S. W., Samyn, Y., & Conand, C. (2012). Commercially important Sea Cucumbers of the World . *FAO*.

6 Choe, S., Ohshima, Y (1961). On the morphological and ecological differences between two commercial forms, "green"; and "red" of the Japanese common sea cucumber, *Stichopus japonicus selenka Nippon Suisan Gakkaishi*, 27(2), 97–106. https://doi.org/10.2331/suisan.27.97

7 Montgomery, E. M., Suhrbier, A. D., & Pearce, C. M. (2024). Chapter 42—Biology, ecology, aquaculture, and commercial products of *Apostichopus californicus*. In A. Mercier, J.-F. Hamel, A. D. Suhrbier, & C. M. Pearce (Eds.), *The World of Sea Cucumbers* (pp. 653–675). Kidlington, England: Academic Press. https://doi.org/10.1016/B978-0-323-95377-1.00042-4.

8 Olvera-Novoa, M. A., Arjona-Cambranes, K., Felaco, L., Martinez-Milian, G., & Sanchez- Tapia, I. (2024). Chapter 46—Aquaculture of the chocolate chip sea cucumber, *Isostichopus badionotus*, in Yucatan, Mexico: progress and challenges. In A. Mercier, J.-F. Hamel, A. D. Suhrbier, & C. M. Pearce (Eds.), *The World of Sea Cucumbers* (pp. 717–736). Kidlington, England:Academic Press. https://doi.org/10.1016/B978-0-323-95377-1.00048-5

9 Reyes-Bonilla, H., & Herrero-Pérezrul, M. D. (2003). Population parameters of an exploited population of *Isostichopus fuscus* (Holothuroidea) in the southern Gulf of California, Mexico, *Fisheries Research*, 59(3), 423–430. https://doi.org/10.1016/S0165-7836(02)00023-1

10 Kyritsi, M., Tsiolas, G., Michailidou, S., Koukaras, K., & Argiriou, A. (2023). Genomic and 16S metabarcoding data of *Holothuria tubulosa* Gmelin, 1791. *Data in Brief*, 48, 109171. https://doi.org/10.1016/j.dib.2023.109171

11 Xu, C., Zhang, R., & Wen, Z. (2018). Bioactive compounds and biological functions of sea cucumbers as potential functional foods. *Journal of Functional Foods*, 49, 73–84. https://doi.org/10.1016/j.jff.2018.08.009

12 Bordbar, S., Anwar, F., & Saari, N. (2011). High-value components and bioactives from sea cucumbers for functional foods—a review. *Marine Drugs*, 9(10), 1761–1805. https://doi.org/10.3390/md9101761

13 Mazlan, N. B., Abd Rahman, N. N. B., Shukhairi, S. S. B., & Nazahuddin, M. N. A. B. (2023). Sea cucumbers: source of nutritional, medicinal, and cosmeceutical

products. In M. D. Shah, J. Ransangan, & B. A. Venmathi Maran (Eds.), Marine Biotechnology: Applications in Food, Drugs and Energy (pp. 171–188). Springer Nature, Singapore. https://doi.org/10.1007/978-981-99-0624-6_8

14 Xu, D., Su, L., & Zhao, P. (2015). Chapter 21—*Apostichopus japonicus* in the worldwide production and trade of sea cucumbers. In H. Yang, J.-F. Hamel, & A. Mercier (Eds.), *Developments in Aquaculture and Fisheries Science* (pp. 383–398). New York, USA: Elsevier. https://doi.org/10.1016/B978-0-12-799953-1.00021-0

15 Song, Z., Li, P., Hu, S., Liu, C., Hao, T., & Han, X. (2023). Influence of dietary phosphorus on the growth, feed utilization, proximate composition, intestinal enzymes, and oxidation resistance of sea cucumber *Apostichopus japonicus*. *Aquaculture Nutrition*, 2023(1), 2266191. https://doi.org/10.1155/2023/2266191

16 Zhao, G., Zhao, W., Han, L., Ding, J., & Chang, Y. (2020). Metabolomics analysis of sea cucumber (*Apostichopus japonicus*) in different geographical origins using UPLC–Q-TOF/MS. *Food Chemistry*, 333, 127453. https://doi.org/10.1016/j.foodchem.2020.127453

17 Imbs, A. B., Svetashev, V. I., & Rodkina, S. A. (2022). Differences in lipid class and fatty acid composition between wild and cultured sea cucumbers, *Apostichopus japonicus*, explain modification and deposition of lipids. *Aquaculture Research*, 53(3), 810–819. https://doi.org/10.1111/are.15617

18 Zhao, X., Liu, Y., Li, Y., Zhang, X., & Qi, H. (2018). Authentication of the sea cucumber (*Apostichopus japonicus*) using amino acids carbon stable isotope fingerprinting. *Food Control*, 91, 128–137. https://doi.org/10.1016/j.foodcont.2018.03.041

19 Yang, H., Hamel, J.-F., & Mercier, A. (2015). The sea cucumber *Apostichopus japonicus*: History, biology and aquaculture. https://api.semanticscholar.org/CorpusID:83208774

20 Saito, M., Kunisaki, N., Urano, N., & Kimura, S. (2002). Collagen as the major edible component of sea cucumber (*Stichopus japonicus*). *Journal of Food Science*, 67(4), 1319–1322. https://doi.org/10.1111/j.1365-2621.2002.tb10281.x

21 Cui, F., Xue, C., Li, Z., Zhang, Y., Dong, P., Fu, X., & Gao, X. (2007). Characterization and subunit composition of collagen from the body wall of sea cucumber *Stichopus japonicus*. *Food Chemistry*, 100(3), 1120–1125. https://doi.org/10.1016/j.foodchem.2005.11.019

22 Lim, Y.-S., Ok, Y.-J., Hwang, S.-Y., Kwak, J.-Y., & Yoon, S. (2019). Marine collagen as a promising biomaterial for biomedical applications. *Marine Drugs*, 17(8). https://doi.org/10.3390/md17080467

23 Gelse, K., Pöschl, E., & Aigner, T. (2003). Collagens—structure, function, and biosynthesis. *Advanced Drug Delivery Reviews*, 55(12), 1531–1546. https://doi.org/https://doi.org/10.1016/j.addr.2003.08.002

24 Oh, G.-W., Ko, S.-C., Lee, D. H., Heo, S.-J., & Jung, W.-K. (2017). Biological activities and biomedical potential of sea cucumber (*Stichopus japonicus*): A review. *Fisheries and Aquatic Sciences*, 20(1), 28. https://doi.org/10.1186/s41240-017-0071-y

25 Yin, R., Zhou, L., Gao, N., Lin, L., Sun, H., Chen, D., Cai, Y., Zuo, Z., Hu, K., Huang, S., Liu, J., & Zhao, J. (2021). Unveiling the disaccharide-branched glycosaminoglycan and anticoagulant potential of its derivatives. *Biomacromolecules*, 22(3), 1244–1255. https://doi.org/10.1021/acs.biomac.0c01739

26 Zayed, A., El-Aasr, M., Ibrahim, A.-R. S., & Ulber, R. (2020). Fucoidan characterization: determination of purity and physicochemical and chemical properties. *Marine Drugs*, 18(11). https://doi.org/10.3390/md18110571

27 Ustyuzhanina, N. E., Bilan, M. I., Dmitrenok, A. S., Shashkov, A. S., Nifantiev, N. E., & Usov, A. I. (2017). The structure of a fucosylated chondroitin sulfate from the sea cucumber *Cucumaria frondosa*. *Carbohydrate Polymers*, 165, 7–12. https://doi.org/10.1016/j.carbpol.2017.02.003

28 Hossain, A., Dave, D., & Shahidi, F. (2023). Sulfated polysaccharides in sea cucumbers and their biological properties: a review. *International Journal of Biological Macromolecules*, 253, 127329. https://doi.org/10.1016/j.ijbiomac.2023.127329

29 Felix, A. L., Penno, S. M., Bezerra, F. F., & Mourão, P. A. S. (2024). Fucosylated chondroitin sulfate, an intriguing polysaccharide from sea cucumber: past, present, and future. *Glycobiology*, 35(2), cwae098. https://doi.org/10.1093/glycob/cwae098

30 Thimmappa, R., Wang, S., Zheng, M., Misra, R. C., Huang, A. C., Saalbach, G., Chang, Y., Zhou, Z., Hinman, V., Bao, Z., & Osbourn, A. (2022). Biosynthesis of saponin defensive compounds in sea cucumbers. *Nature Chemical Biology*, 18(7), 774–781. https://doi.org/10.1038/s41589-022-01054-y

31 Zhao, Y.-C., Xue, C.-H., Zhang, T.-T., & Wang, Y.-M. (2018). Saponins from sea cucumber and their biological activities. *Journal of Agricultural and Food Chemistry*, 66(28), 7222–7237. https://doi.org/10.1021/acs.jafc.8b01770

32 Khattab, R. A., Elbandy, M., Lawrence, A., Paget, T., Rae-Rho, J., Binnaser, Y. S., & Ali, I. (2018). Extraction, identification and biological activities of saponins in sea cucumber *Pearsonothuria graeffei*. *Combinatorial Chemistry & High Throughput Screening*, 21(3), 222–231. https://doi.org/10.2174/1386207321666180212165448

33 Van Dyck, S., Gerbaux, P., & Flammang, P. (2010). Qualitative and quantitative saponin contents in five sea cucumbers from the Indian ocean. *Marine Drugs*, 8(1), 173–189. https://doi.org/10.3390/md8010173

34 Kamyab, E., Kellermann, M. Y., Kunzmann, A., & Schupp, P. J. (2020). Chemical biodiversity and bioactivities of saponins in echinodermata with an emphasis on sea cucumbers (holothuroidea). In S. Jungblut, V. Liebich, & M. Bode-Dalby (Eds.), *YOUMARES 9—The Oceans: Our Research, Our Future: Proceedings of the 2018 conference for YOUng MArine RESearcher in Oldenburg, Germany* (pp. 121–157). Cham, Switzerland: Springer International Publishing. https://doi.org/10.1007/978-3-030-20389-4_7

35 Caulier, G., Van Dyck, S., Gerbaux, P. et al. (2011). Review of saponin diversity in sea cucumbers belonging to the family holothuriidae. *Pacific Community (SPC) Beche-de-Mer Information Bulletin* 31: 48–54.

36 Kim, S.-K., & Himaya, S. W. A. (2012). Chapter 20—Triterpene glycosides from sea cucumbers and their biological activities. In S.-K. Kim (Ed.), *Marine Medicinal Foods* (Vol. 65, pp. 297–319). California, USA: Academic Press. https://doi.org/10.1016/B978-0-12-416003-3.00020-2

37 Miyamoto, T., Togawa, K., Higuchi, R., Komori, T., & Sasaki, T. (1992). Structures of four new triterpenoid oligoglycosides: DS-penaustrosides a, B, C, and D from the sea cucumber *Pentacta australis*. *Journal of Natural Products*, 55(7), 940–946. https://doi.org/10.1021/np50085a014

38 Avilov, S. A., Antonov, A. S., Drozdova, O. A., Kalinin, V. I., Kalinovsky, A. I., Stonik, V. A., Riguera, R., Lenis, L. A., & Jiménez, C. (2000). Triterpene glycosides from the far-eastern sea cucumber *Pentamera calcigera*. 1. Monosulfated glycosides and cytotoxicity of their unsulfated derivatives. *Journal of Natural Products*, 63(1), 65–71. https://doi.org/10.1021/np9903447

39 Findlay, J. A., Yayli, N., & Radics, L. (1992). Novel sulfated oligosaccharides from the sea cucumber *Cucumaria frondosa*. *Journal of Natural Products*, 55(1), 93–101. https://doi.org/10.1021/np50079a014

40 Drozdova, O. A., Avilov, S. A., Kalinovskii, A. I., Stonik, V. A., Mil'grom, Yu. M., & Rashkes, Ya. V. (1993). Trisulfated glycosides from the holothurian *Cucumaria japonica*. *Chemistry of Natural Compounds*, 29(3), 309–313. https://doi.org/10.1007/BF00630529

41 Stonik, V. A., Mal'tsev, I. I., Kalinovskii, A. I., & Elyakov, G. B. (1982). Glycosides of marine invertebrates. XII. Structure of a new triterpene oligoglycoside from holothurians of family Stichopodidae. *Chemistry of Natural Compounds*, 18(2), 182–186. https://doi.org/10.1007/BF00577189

42 Xia, S., Li, M., Shen, T., Gao, Y., Ma, X., & Yuan, X. (2024). Chapter 5—Sea cucumbers in Chinese daily life. In A. Mercier, J.-F. Hamel, A. D. Suhrbier, & C. M. Pearce (Eds.), *The World of Sea Cucumbers* (pp. 65–75). London, England: Academic Press. https://doi.org/10.1016/B978-0-323-95377-1.00044-8

43 Wang, T., Zheng, L., Zhao, T., Zhang, Q., Liu, Z., Liu, X., & Zhao, M. (2020). Anti-diabetic effects of sea cucumber (*Holothuria nobilis*) hydrolysates in streptozotocin and high-fat-diet induced diabetic rats via activating the PI3K/Akt pathway. *Journal of Functional Foods*, 75, 104224. https://doi.org/10.1016/j.jff.2020.104224

44 Tang, L., Cai, S., Xiao, M., Li, D., & Mou, H. (2025). A novel exploration of the extraction of sea cucumber glycoproteins and their potential mechanism in prediabetes intervention. *Food Research International*, 218, 116881. https://doi.org/10.1016/j.foodres.2025.116881

45 Hossain, A., Dave, D., & Shahidi, F. (2022). Antioxidant potential of sea cucumbers and their beneficial effects on human health. *Marine Drugs*, 20(8). https://doi.org/10.3390/md20080521

46 Sang, X., Xing, Z., Zhou, B., Wang, Y., Guan, X., Wang, F., Li, Y., Zhao, Q., & Li, Z. (2025). Sea cucumber polysaccharides promote gut–liver axis health by modulating microbiota, *Metabolism, and Gene Expression in Mice. Foods*, 14(17). https://doi.org/10.3390/foods14172962

47 Mondol, M. A. M., Shin, H. J., Rahman, M. A., & Islam, M. T. (2017). Sea cucumber glycosides: chemical structures, producing species and important biological properties. *Marine Drugs*, 15(10). https://doi.org/10.3390/md15100317

48 Shou, Y., Feng, C., Lu, Q., Mao, X., Huang, H., Su, Z., Guo, H., & Huang, Z. (2023). Research progress on the chemical components and biological activities of sea cucumber polypeptides. *Frontiers in Pharmacology*, Volume 14-2023. https://doi.org/10.3389/fphar.2023.1290175

49 Wang, J.-F., Fu, J., Wang, Y.-M. et al. (2010). Effects of Saponins of *Pearsonothuria graeffei* on immune regulation in mice (in Chinese). *Periodical of Ocean University of China* 40 (2): 28–32.

50 Janakiram, N. B., Mohammed, A., Bryant, T., Lightfoot, S., Collin, P. D., Steele, V. E., & Rao, C. V. (2015). Improved innate immune responses by Frondanol A5, a sea cucumber extract, prevent intestinal tumorigenesis, *Cancer Prevention Research*, 8(4), 327–337. https://doi.org/10.1158/1940-6207.CAPR-14-0380

51 Monmai, C., Park, S. H., You, S., & Park, W. J. (2018). Immuno-enhancement effect of polysaccharide extracted from *Stichopus japonicus* on cyclophosphamide-induced immunosuppression mice. *Food Science and Biotechnology*, 27(2), 565–573. https://doi.org/10.1007/s10068-017-0248-2

52 Zhou, B., Rayner, A. W., Gregg, E. W., Sheffer, K. E., Carrillo-Larco, R. M., Bennett, J. E., Shaw, J. E., Paciorek, C. J., Singleton, R. K., Pires, A. B., Stevens, G. A., Danaei, G., Lhoste, V. P., Phelps, N. H., Heap, R. A., Jain, L., Brisis, Y. D. D., Galeazzi, A., Kengne, A. P., . . . Ezzati, M. (2024). Worldwide trends in diabetes prevalence and treatment from 1990 to 2022: a pooled analysis of 1108 population-representative studies with 141 million participants. *The Lancet*, 404(10467), 2077–2093. https://doi.org/10.1016/s0140-6736(24)02317-1

53 Wang, Y., Wang, J., Zhao, Y., Hu, S., Shi, D., & Xue, C. (2016). Fucoidan from sea cucumber *Cucumaria frondosa* exhibits anti-hyperglycemic effects in insulin resistant mice via activating the PI3K/PKB pathway and GLUT4. *Journal of Bioscience and Bioengineering*, 121(1), 36–42. https://doi.org/10.1016/j.jbiosc.2015.05.012

54 Fu, X., Wen, M., Han, X., Yanagita, T., Xue, Y., Wang, J., Xue, C., & Wang, Y. (2016). Effect and potential mechanism of action of sea cucumber saponins on postprandial blood glucose in mice. *Bioscience, Biotechnology, and Biochemistry*, 80(6), 1081–1087. https://doi.org/10.1080/09168451.2016.1153950

55 Yang, J.-Y., Zhang, T.-T., Dong, Z., Shi, H.-H., Xu, J., Mao, X.-Z., Wang, Y.-M., & Xue, C.-H. (2021). Dietary supplementation with exogenous sea-cucumber-derived ceramides and glucosylceramides alleviates insulin resistance in high-fructose-diet-fed rats by upregulating the IRS/PI3K/Akt signaling pathway. *Journal of Agricultural and Food Chemistry*, 69(32), 9178–9187. https://doi.org/10.1021/acs.jafc.0c06831

56 Xu, Z.-Q., Xu, J., Xue, Y., Li, Z.-J., Wang, J.-F., Wang, J.-H., Xue, C.-H., & Wang, Y.-M. (2012). Effects of bioactive components of sea cucumber on the serum, liver lipid profile and lipid absorption. *Bioscience, Biotechnology, and Biochemistry*, 76(12), 2214–2218. https://doi.org/10.1271/bbb.120495

57 Liu, X., Sun, Z., Zhang, M., Meng, X., Xia, X., Yuan, W., Xue, F., & Liu, C. (2012). Antioxidant and antihyperlipidemic activities of polysaccharides from sea cucumber *Apostichopus japonicus*. *Carbohydrate Polymers*, 90(4), 1664–1670. https://doi.org/10.1016/j.carbpol.2012.07.047

58 Li, S., Li, J., Zhi, Z., Wei, C., Wang, W., Ding, T., Ye, X., Hu, Y., Linhardt, R. J., & Chen, S. (2017). Macromolecular properties and hypolipidemic effects of four sulfated polysaccharides from sea cucumbers. *Carbohydrate Polymers*, 173, 330–337. https://doi.org/10.1016/j.carbpol.2017.05.063

59 Luo, L., Wu, M., Xu, L., Lian, W., Xiang, J., Lu, F., Gao, N., Xiao, C., Wang, S., & Zhao, J. (2013). Comparison of physicochemical characteristics and anticoagulant activities of polysaccharides from three sea cucumbers. *Marine Drugs*, 11(2), 399–417. https://doi.org/10.3390/md11020399

60 Li, Q., Cai, C., Chang, Y., Zhang, F., Linhardt, R. J., Xue, C., Li, G., & Yu, G. (2018). A novel structural fucosylated chondroitin sulfate from *Holothuria mexicana* and its effects on growth factors binding and anticoagulation. *Carbohydrate Polymers*, 181, 1160–1168. https://doi.org/10.1016/j.carbpol.2017.10.100

61 Li, Q., Che, H.-X., Wang, C.-C., Zhang, L.-Y., Ding, L., Xue, C.-H., Zhang, T.-T., & Wang, Y.-M. (2019). Cerebrosides from sea cucumber improved Aβ1–42-induced cognitive deficiency in a rat model of Alzheimer's disease. *Molecular Nutrition & Food Research*, 63(5), 1800707. https://doi.org/10.1002/mnfr.201800707

62 Janakiram, N. B., Mohammed, A., & Rao, C. V. (2015). Sea cucumbers metabolites as potent anti-cancer agents. *Marine Drugs*, 13(5), 2909–2923. https://doi.org/10.3390/md13052909

63 Lei, D., Zhao-Jie, L., Jie, X., Jing-Feng, W., Yong, X., Chang-Hu, X., Koretaro, T., & Yu-Ming, W. (2012). The anti-tumor activities of cerebrosides derived from sea cucumber *Acaudina molpadioides* and starfish *Asterias amurensis* in vitro and in vivo. *Journal of Oleo Science*, 61(6), 321–330. https://doi.org/10.5650/jos.61.321

64 Yan, T., Sun, J., Zheng, J., & Yang, J. (2024). An analysis combining proteomics and transcriptomics revealed a regulation target of sea cucumber autolysis. *Comparative Biochemistry and Physiology Part D: Genomics and Proteomics*, 52, 101274. https://doi.org/10.1016/j.cbd.2024.101274

65 Mercier, A., Hamel, J.-F., Suhrbier, A. D., & Pearce, C. M. (Eds.). (2024). *The World of Sea Cucumbers: Challenges, Advances, and Innovations* (1st ed.). Kidlington, England: Nikki P. Levy.

66 National Health Commission of the People's Republic of China. (2015). National food safety standard—Dried sea cucumber GB 31602-2015.

67 Wang, X., Su, Y., Wang, Y., Chen, X., Chen, X., & Liu, Z. (2023). The effect of ultrasound on the rehydration characteristics of semi-dried salted *Apostichopus japonicus. Foods*, 12(24). https://doi.org/10.3390/foods12244382

68 Ministry of Agriculture and Rural Affairs of People's Republic of China. (2024). *China Fisheries Statistical Yearbook* 2024. Beijing, China: China Agriculture Press.

8

Other Aquatic Products

Hu Hou[1], *Lulu Zhu*[2], *and Yan Fan*[3]

[1] *State Key Laboratory of Marine Food Processing & Safety Control, College of Food Science and Engineering, Ocean University of China, Qingdao, China*
[2] *College of Food Science and Engineering, Hainan University, Haikou, China*
[3] *College of Marine Life Sciences, Ocean University of China, Qingdao, China*

Global demand for diverse and nutrient-rich aquatic products continues to rise, driving increased attention to underutilized marine resources beyond conventional fish, shellfish, and crustaceans. Representative species such as sea urchins, starfish, sea anemones, and jellyfish possess unique biological traits, rich nutritional components (including high-quality proteins, bioactive compounds, and special polysaccharides), and distinct sensory attributes, holding significant potential for food development, functional nutrition, and industrial application. However, current research on these species remains highly fragmented, with most studies focusing narrowly on nutritional composition, while systematic investigations into their classification, global distribution, flavor formation mechanisms, physicochemical stability, and market potential are notably lacking. These gaps limit their industrial promotion and hinder the diversification of the aquatic product supply chain. To address these challenges, this chapter provides a comprehensive overview of the species characteristics, distribution, nutritional profiles, flavor attributes, and physicochemical properties of these four aquatic organisms, alongside an analysis of their current market status and future prospects. It aims to provide foundational insights for researchers, industry practitioners, and stakeholders seeking to unlock the value of these underutilized aquatic resources.

8.1 Species and Classification

8.1.1 Sea Urchins

Sea urchins (*Echinoidea*) are ancient marine invertebrates of significant ecological, economic, and edible value, distributed in oceans worldwide [1]. These bottom-dwelling organisms exhibit remarkable biodiversity, with over 800 known species, displaying diverse morphologies and vibrant colorations ranging from purple and black to vivid reds and greens [2]. Their globular bodies are organized into five-part symmetry (pentaradial symmetry), a hallmark of echinoderms.

Due to the different classification bases of scholars, the classification methods of sea urchins also vary. Most of the currently common international classification systems are based on the work of T. Mortensen (1928–1955), which classified the sea urchins into 11 orders. As of 2016, the International Integrated Taxonomic Information System (ITIS) classifies sea urchins into 2 suborders, 4 general orders, 13 orders, 10 suborders, 46 families, 71 genera, and 137 species. As of 2024, the Catalogue of Life classifies sea urchins into 4 subclasses, 4 general orders, 19 orders, 8 suborders, 160 families, 1,047 genera, and 4,271 species, with new additions such as Camarodonta, Echinolampadacea, and Echinoneoida. In addition, there are two common classification methods in China: (1) Based on the external morphology of sea urchins, they are classified into three orders: Centrochinoida, Tatataenoidea, and Spatangoida. (2) Based on the combination of shape and other morphological features, the sea urchins are classified into eight orders: Cidaroida, Lepidocentroida, Stirodonta, Aulodonta, Camarodonta, Holectypoida, Clypeastroida, and Spatangoida.

8.1.2 Starfish

Starfish, or sea stars (Phylum: Echinodermata; Class: Asteroidea), are marine invertebrates. They are planktonic in the larval stage and become benthic in adulthood. They typically exhibit pentaradial symmetry with a central disc and usually five arms, although some species can have up to forty arms. The arms are not distinctly separated from the central disc, and the coelomic cavity extends into them, housing various internal organs. The mouth is located on the underside (oral surface), while the anus is situated on the upper side (aboral surface). The underside of each arm features ambulacral grooves lined with rows of tube feet equipped with suction cups, facilitating locomotion and prey capture [3].

The designation Asteroidea was established in 1830 by the French zoologist de Blainville. This highly diverse class encompasses more than 1,900 extant species, taxonomically organized into seven living orders: Brisingida, Forcipulatida, Notomyotida, Paxillosida, Spinulosida, Valvatida, and Velatida. The classification also includes the extinct lineages of Calliasterellidae and Trichasteropsida.

8.1.3 Sea Anemone

Sea anemones are a general term for the order Actiniaria of the class Anthozoa. Although sea anemones look like flowers, they are actually predators. Some sea anemones are transparent, and their yellow-brown or even red and green colors are produced by symbiotic algae and their own pigments, so they mainly rely on photosynthesis to provide energy. This invertebrate has no skeleton and anchors to fixed objects on the seabed, such as rocks and corals. While they can move slowly, certain species are capable of swimming by swinging their bodies in response to emergencies. Notably, sea anemones exhibit remarkable longevity, often living for several centuries, with some individuals documented to be over 2,000 years old. Hermit crabs sometimes carry sea anemones on their shells as camouflage, and numerous marine organisms engage in symbiotic relationships with them, even utilizing them for defense purposes.

The order Actiniaria, comprising sea anemones, falls under the subclass Hexacorallia within the class Anthozoa and phylum Cnidaria [4]. Based on extensive DNA analyses, Rodriguez et al. introduced a restructured classification of Actiniaria that divides the order into two suborders: Anenthemonae, which includes the superfamilies Edwardsioidea and Actinernoidea, and Enthemonae, comprising the superfamilies Actinostoloidea, Actinioidea, and Metridioidea.

8.1.4 Jellyfish

Jellyfish, belonging to the phylum Cnidaria, are gelatinous marine organisms classified into four main classes: Scyphozoa, Cubozoa, Hydrozoa, and Staurozoa. Among these, the edible species *Rhopilema esculentum*, commonly known as "sea blister," is a commercially important scyphozoan predominantly found in Asian coastal waters such as the Yellow Sea and Bohai Sea.

Scyphozoa, or "true jellyfish," are characterized by a dominant medusa stage with a bell-shaped umbrella, often with tentacles and lappets along the margin. Species such as *Aurelia aurita* and *Nemopilema nomurai* are known for large blooms. Although many possess stinging cells, most are not highly dangerous to humans.

Cubozoa, or box jellyfish, have a cube-shaped medusa and are among the most venomous marine animals. Species like *Chironex fleckeri* possess toxins that can be fatal to humans [5]. They are strong swimmers and exhibit a complex life cycle with both polyp and medusa stages.

Hydrozoa are highly diverse and include both solitary and colonial forms, such as the Portuguese man o' war (*Physalia physalis*). Many hydrozoans have a life cycle alternating between polyp and medusa stages. Their medusae are generally smaller than those of scyphozoans [6].

Staurozoa, or stalked jellyfish, are benthic and attached via a stalk. They lack a free-swimming medusa stage and capture prey using tentacles arranged around an inverted bell-shaped body [7]. They represent a smaller and less diverse class within Cnidaria.

Globally, over 2,000 jellyfish species have been identified. They are among the oldest extant metazoans, with fossil evidence dating back to the Cambrian [8, 9]. All jellyfish possess cnidocytes—stinging cells used for prey capture and defense. Although only about 3% of species pose a serious threat to humans, some cubozoans and hydrozoans are highly venomous.

8.2 Distribution of Other Aquatic Products

8.2.1 Distribution of Sea Urchins

Sea urchins are widely distributed throughout the world's oceans, with the greatest number of species especially in the Indian and Western Pacific Oceans. They extend from the intertidal zone to depths of up to 5,000 meters. Among the economically important sea urchin species globally recognized, several stand out for their distinctive characteristics and culinary value (red sea urchin, purple sea urchin, and short-spined sea urchin) [10].

The purple sea urchin (*Heliocidaris crassispina*), predominantly found in the coastal waters of Japan, Korea, and China, features a deep purple or purplish-black test measuring 5–8 cm in diameter with elongated, sharp spines in matching hues. Its orange-yellow gonads are prized for their rich, sweet flavor and creamy texture, making them a premium choice for Japanese delicacies like sushi and uni donburi.

Known as the "King of Sea Urchins" in Japan, the short-spined sea urchin (*Hemicentrotus pulcherrimus*) inhabits coastal regions from Hokkaido to Kyushu as well as the Yellow Sea and East China Sea. This species is easily identified by its 3–5 cm green or brownish-green test resembling horse manure and short, densely packed light-colored spines. The vividly orange-red gonads offer a delicate texture and exceptional sweetness that have earned them this regal title.

The northern purple sea urchin (*Strongylocentrotus nudus*), distributed across Hokkaido, the Russian Far East, and the Korean Peninsula, boasts one of the largest commercial sizes with its deep purple to blackish-purple test reaching up to 10 cm in diameter and robust, medium-length spines. Its pale yellow to orange gonads have a distinctive savory quality with slight metallic bitterness, often preferred in processed forms like salted or fermented uni.

Mainly found from Honshu to Kyushu in Japan and along southeastern China's coast, the red sea urchin (*Pseudocentrotus depressus*) displays a characteristic flattened, red or reddish-brown test (4–6 cm diameter) with short, blunt spines of similar coloration. The deep yellow gonads are renowned for their intense umami flavor, frequently used in uni paste or served as sashimi due to their robust taste profile.

Widely distributed across the tropical Indo-Pacific, including Australia, Hawaii, and Southeast Asia, the collector sea urchin (*Tripneustes gratilla*) stands out with its large (8–12 cm) white or light gray test and short spines featuring distinctive white-ringed tips. The pale-yellow gonads offer a mild, subtly sweet flavor that is less intense than temperate species, commonly consumed fresh in tropical coastal cuisines or prepared in local specialties like Hawaiian uni shooters.

8.2.2 Distribution of Starfish

Starfish, a major class within the phylum Echinodermata, are key members of marine ecosystems and play a fundamental role in maintaining the balance of benthic ecology. Like other echinoderms, they cannot survive in freshwater environments due to their need to maintain a precise internal electrolyte balance that mirrors that of seawater. The distribution and habitats of starfish are significantly impacted by marine environments, as they can be found in various oceanic regions, ranging from tropical coral reefs to frigid polar waters, and from intertidal areas to deep abyssal plains up to 6,000 meters below the surface [11]. Globally, the North Pacific Ocean exhibits the highest species diversity and biomass of starfish. Their habitats are diverse, including rocky shores, tidal pools, mudflats, sandy bottoms, kelp forests, and deep-sea floors, with the greatest species diversity occurring in coastal areas.

Different species of starfish orders display strikingly distinct geographic distributions due to their unique physiological structures and environmental adaptations.

The Brisingida are classic deep-sea dwellers, characterized by a small, rigid disc and multiple long, slender arms. They feed by suspension feeding and live almost exclusively in deep-sea environments, although a few species have been found in shallow waters around Antarctica [12].

The Forcipulatida are common representatives of temperate seas, noted for their distinctive pincer-like pedicellariae and robust bodies. They are widely distributed throughout temperate regions worldwide, with the common starfish (*Asterias vulgaris*) along North Atlantic coasts and in tidal pools serving as a typical example. This order also includes species adapted to cold and deep-sea environments.

Notomyotida species are flexible deep-sea explorers. Inhabiting deep waters, these starfish possess longitudinal muscle bands along the inner surfaces of their arms, granting them exceptional flexibility to navigate complex deep-sea habitats.

The Paxillosida are primitive, soft-bottom dwellers found mainly on sandy or muddy substrates. Lacking tube feet, suckers, and external stomach eversion for feeding, they represent a relatively ancestral group. A common member is the comb star (*Astropecten polyacanthus*), which is widely distributed across the tropical Indo-Pacific region.

Spinulosida species are finely-skeletonized nearshore inhabitants. They feature delicate skeletal structures covered with clusters of short spines. Typically found in coastal environments, the red starfish (*Echinaster sepositus*) is often seen on rocks and seagrass beds in the eastern Atlantic and Mediterranean.

Valvatida is the most species-rich order of starfish. As the largest starfish order, its members exhibit a wide range of morphologies, from pentagonal to cushion-shaped forms. They are widely distributed across global seas, including cushion stars in tropical coral reefs and leather stars on temperate rocky shores.

Velatida are common in cold-water and deep-sea habitats, with a global distribution, mainly comprising species adapted to these environments. They typically appear in pentagonal or star-shaped forms with five to fifteen arms. For example, some species of the genus Solaster belong to this order and are commonly found in the cold waters of the North Pacific.

8.2.3 Distribution of Sea Anemone

Different sea anemone species exhibit marked geographic differences in distribution, reflecting their unique physiological traits and environmental adaptations. Most sea anemones live in shallow seas and puddles or cracks on rocky shores, while some are adapted to the deep sea, with depths ranging from 100 m to over 10,000 m. Notably, giant forms measuring up to one meter across are confined to tropical coral reefs. Over 1,000 known species inhabit substrates such as mud or rock crevices, each exhibiting distinct regional patterns.

Cribrinopsis crassa is restricted to the Mediterranean Sea, adorning rocky crevices at depths of 30 to 50 m. With its bright orange-yellow disc and distinctive purple-tipped tentacles, it favors soft substrates along deep coastal edges.

Condylactis passiflora is found in the Caribbean Sea and along the Atlantic coast, typically inhabiting waters at depths of less than 30 meters, including the Caribbean Sea, Florida, Bahamas, and Bermuda. It prefers coral reef fissures and strong light zones, and is noted for its thick tentacles and potent nematocysts.

Entacmaea quadricolor inhabits eastern Pacific reefs around Hawaii at depths of 2–20 m. It features short tentacles with fleshy protrusions comprising symbiotic algae, and often coexists with shrimps and clownfish.

Found across the Indian and Pacific Oceans, including the Red Sea and Samoa, *Heteractis magnifica* inhabits shallow waters down to 10 m, displaying vibrant colors on its smooth or bubble-textured column, while its tufted yellow or green tentacles provide a home for symbionts such as clownfish and cleaner shrimp.

Stichodactyla mertensii, one of the world's largest anemones, is widespread in tropical Indo-Pacific seas from the Comoros and Madagascar to southern Japan, Australia, and Fiji. It anchors on rocky or coral substrates at depths down to 20 m and supports a diverse community of host fish.

Urticina felina inhabits the colder waters of the northeastern Atlantic, North Sea, and Baltic coasts, ranging from shallow subtidal zones down to 30 m. Short, banded tentacles encircle the oral disc, and individuals typically occur alone on hard substrates, where they may form loose associations with shrimps or sea urchins. This species favors brightly lit habitats, thriving in areas that receive up to 50,000 lux of illumination.

8.2.4 Distribution of Jellyfish

Jellyfish (*Medusozoa*) are widely distributed across global marine environments, from coastal waters to the open ocean, with a limited number of species inhabiting freshwater systems. Their distribution is primarily influenced by environmental factors such as temperature, salinity, ocean currents, and prey availability [13, 14].

Globally, jellyfish inhabit all latitudinal zones. Warm tropical and subtropical waters support high species diversity, including cubozoans and hydrozoans such as *Physalia physalis*, along with scyphozoans like Rhopilema spp. in the Indo-Pacific region [5, 15]. Temperate regions host seasonally abundant populations including *Aurelia aurita* and Cyanea capillata [16, 17].

The marginal seas of China demonstrate clear environmental gradients from temperate to tropical regimes, hosting diverse jellyfish communities. In the temperate Bohai and Yellow Seas, jellyfish abundance shows strong seasonality. The region experiences periodic mass occurrences of *Nemopilema nomurai*, which intrude from the East China Sea in summer and autumn, impacting local ecosystems [18]. Vertical stratification is evident with Aurelia coerulea in surface waters and *Nemopilema nomurai* in deeper layers [19]. Horizontal distribution correlates with temperature, salinity, and nutrient gradients. The subtropical East China Sea serves as an ecotone supporting both temperate and warm-water species, functioning as a source region for *Nemopilema nomurai* blooms [19].

Beyond marine environments, the freshwater hydrozoan *Craspedacusta sowerbyi* occurs sporadically in Chinese lakes and reservoirs. Long-term observations indicate changes in jellyfish populations within Chinese waters, with increasing species records in the South China Sea and East China Sea since the late 1990s, potentially linked to environmental changes such as climate warming [20]. Modern monitoring techniques, particularly environmental DNA metabarcoding, have become valuable tools for mapping jellyfish distribution and population dynamics.

8.3 Nutrients of Other Aquatic Products

8.3.1 Nutrients of Sea Urchins

The gonads (commonly called "roe" or "uni") of sea urchins represent the primary commercial product, constituting about 10% of total body weight [21]. These prized delicacies feature a distinctive half-moon shape and yellow-orange hue, with biochemical composition dominated by moisture, protein, carbohydrates, and lipids [22]. Sea urchin gonads represent a nutritionally dense marine food source, offering an exceptional combination of high-quality protein, beneficial lipids, and essential micronutrients [23]. With approximately 20 g of protein/100 g, nearly double the content found in conventional protein sources like chicken, pork, fish, and eggs. These gonads contain functionally important proteins, including actin, myosin, dyneins, and various filamentous proteins [24, 25].

The lipid profile of sea urchin gonads is particularly noteworthy, containing 7.2 g/100 g with a predominance of health-promoting unsaturated fatty acids. These include significant amounts of polyunsaturated fatty acids (PUFAs) such as arachidonic acid (AA; 20:4n-6) and eicosapentaenoic acid (EPA; 20:5n-3) [26], along with diverse phospholipids like phosphatidylcholine (PC), phosphatidylethanolamine (PE), phosphatidylserine (PS), phosphatidylinositol, and cardiolipin [27, 28]. The approximately 40 identified fatty acids, combined with cholesterol and various sterols, contribute to demonstrated health benefits, including cardiovascular protection, cholesterol modulation, and nervous system support, largely mediated by compounds like lecithin (PC) with its serum lipid-regulating properties.

Micronutrient content is equally impressive, with sea urchin gonads providing essential B vitamins (thiamine/B_1, riboflavin/B_2, niacin) crucial for metabolism and neurological function, along with vitamin A and an array of minerals including iron, zinc, calcium, magnesium, selenium, germanium, strontium, copper, manganese, and molybdenum [10]. Unique bioactive components further enhance their nutritional value, particularly sialoglycolipids (ceramides and glucopyranosyl substances) that support neural tissue nutrition and cognitive function, alongside health-promoting polysaccharides such as sulfated fucose and galactose, plus carotenoids and various sterols including sitosterol [2, 25, 28].

With approximately 14.9 g of carbohydrates/100 g serving as efficient energy sources, sea urchin gonads present a well-rounded nutritional profile. However, this composition exhibits natural variability influenced by seasonal fluctuations (peaking in autumn and declining through winter), environmental conditions including food availability and water temperature, and reproductive cycles [22, 29].

This unique combination of nutrients positions sea urchin gonads not merely as a culinary delicacy but as a functional food with particular benefits for cardiovascular and neurological health. While seasonal variations occur, they consistently provide bioavailable essential nutrients in concentrated forms, making them a nutritionally valuable marine resource.

8.3.2 Nutrients of Starfish

Starfish are marine invertebrates that, as predators or scavengers in many marine ecosystems, contain a variety of bioactive compounds such as proteins, polysaccharides, lipids, vitamins, minerals, and unique saponins [30]. Given the biological activities and nutritional values associated with these substances, starfish demonstrate significant potential as functional food ingredients and sources for novel drugs, despite not being a conventional food source.

8.3.2.1 Protein

The body wall of starfish contains approximately 7% protein, of which collagen is the main component. Studies on the body wall collagen of the starfish *Acanthaster planci* show that starfish collagen is a trimer containing $(\alpha_1)_2\alpha_2$. The body wall of starfish can be used as a raw material for extracting gelatin. The resulting starfish gelatin products have good color and can be eaten directly. They have the effects of delaying aging, improving immune function, lowering blood lipids, and being anti-inflammatory.

8.3.2.2 Polysaccharides

The polysaccharides in starfish are primarily structurally complex sulfated polysaccharides (SPSs), which are not only structural components but are also of great interest for their significant biological activities. The polysaccharide content varies by species, typically accounting for 5–20% of the dry weight.

Starfish polysaccharides possess a wide range of biological activities. For instance, polysaccharides extracted from the starfish *Asterias rollestoni* have been confirmed to have potent immunomodulatory activity, capable of activating macrophages and promoting cytokine production [31]. Other studies have also shown that various starfish polysaccharides exhibit antitumor, antiviral, anticoagulant, anti-inflammatory, and antioxidant properties [32–34]. These unique effects make them highly promising natural polymers for biomedical and functional food development.

8.3.2.3 Lipids

Starfish exhibit a relatively high phospholipid content, typically accounting for 35%–43% of their total lipids. The predominant phospholipid species include 1-*O*-alkyl-1,2-acyl-PE and 1,2-diacyl-PC. Additionally, the gonadal tissues of starfish have been found to contain significant levels of EPA and DHA, comprising approximately 11.34% and 6.25% of the total fatty acids, respectively. These essential fatty acids play a crucial role in cardiovascular health, brain development promotion, and inflammatory response regulation.

8.3.2.4 Sterols

Although research on marine steroid compounds has been ongoing for many years, recent studies have continued to reveal a multitude of novel steroid compounds with unique structures isolated from marine organisms. Among these organisms, starfish rank second only to sponges in terms of the abundance of steroid compounds, making them one of the most intensively studied groups in this field. These steroid compounds are typically present as mixtures in starfish tissues, which significantly complicates their isolation and purification. Nevertheless, over 400 steroid compounds have now been isolated from more than 60 starfish species.

Starfish are also known for their high sterol content. They utilize polyhydroxy sterols, sulfated polyhydroxy sterols, and disulfated sterols in their bodies to synthesize saponins [35]. In 1997, Marino et al. isolated 13 novel steroid compounds from the starfish (*Acodontaster conspicuous*), including 9 new steroid oligoglycosides and 4 new polyhydroxysteroid compounds [36].

8.3.2.5 Saponins

The most unique chemical constituents in starfish are asterosaponins, which are structurally complex steroidal or triterpenoid glycosides. Saponins are the main secondary metabolites of starfish and serve as an important weapon in their chemical defense system, protecting them from predators and pathogens.

Asterosaponins exhibit extremely broad and potent biological activities, including hemolytic, cytotoxic, antifungal, antiviral, anti-inflammatory, and antitumor effects. Due to their strong cytotoxicity, asterosaponins have received significant attention in the development of anticancer drugs, with several saponin molecules having been shown to effectively induce apoptosis in cancer cells [37, 38]. However, it is precisely this biological activity that can pose a risk to humans if starfish are consumed raw and unprocessed, which is one of the reasons they are not widely eaten.

8.3.3 Nutrients of Sea Anemone

Sea anemones are not only an important component of marine ecosystems but are also gaining attention as a promising, albeit underexploited, source of nutrition. Their body tissues are rich in high-quality protein, structurally unique bioactive polysaccharides, special fatty acids, and various micronutrients, showcasing their potential application in the fields of functional foods and biomedicine.

8.3.3.1 Polysaccharides and Glycosides

Polysaccharides are closely associated with various biological activities, including immune regulation, intercellular substance transport, cell–cell recognition, and the diagnosis and treatment of cancer. Numerous studies have confirmed that polysaccharides derived from marine animals possess antitumor, antiviral, cardiovascular protective, antioxidant, and immunomodulatory properties. Two cardiovascular-active polysaccharides, designated as Polysaccharides I and II, were isolated from *Zoanthus stephensoni*, a species inhabiting the Xisha Islands in the South China Sea. Polysaccharide I is composed of ribose and galactose in a molar ratio of approximately 2.2:1.0, while Polysaccharide II consists of ribose, galactose, and *N*-acetylgalactosamine in a molar ratio of approximately 3.2:2.0:1.0. In isolated rabbit hearts, Polysaccharide I significantly reduced coronary flow while enhancing myocardial contractility, whereas Polysaccharide II markedly increased coronary flow and also enhanced contractile force. Both polysaccharides demonstrated significant antihypertensive activity in rats.

Polysaccharides extracted from *Edwardsia sipinculoides* using flash extraction techniques exhibited notable in vitro immunological activity. Specifically, these polysaccharide components significantly upregulated TNF-α expression in THP-1 cells, indicating a potential role in immune regulation. In another study, acidic mucopolysaccharides were isolated from *Anthopleura midori* via alkaline extraction. Additionally, a purified sea anemone polysaccharide was obtained from *Anthopleura pacifica*, with a molecular weight of 22 kDa. This polysaccharide was composed of seven types of pyranose-form monosaccharides, including D-(+)-galactosamine hydrochloride, *N*-acetylglucosamine, glucosamine, D-glucose, D-mannose, D-galactose, and D-fucose, in a molar ratio of 1.1:1.5:7.2:7.3:1.0:2.7:2.1.

8.3.3.2 Fatty Acids and Glycerides

Fatty acids and glycerides extracted from sea anemones have garnered attention for their potential bioactivities. From the ethyl acetate extract of *Anthopleura xanthogrammica*, five compounds were isolated. These comprised two fatty acids: linoleic acid and 9,14-docosadienoic acid, and three glycerides: 1-σ-hexadecanoyl-2-σ-(9,12-octadecadie noyl)glyceride, 1-σ-hexadecanoyl-3-σ- (14-eicosenoyl)glyceride, and 1-σ(9,12-octadecadienoyl)-2-σ-(9,12-octadecadienoyl)glyceride. Similarly, from *A. pacifica*, methyl esters of palmitic acid and arachidic acid, along with 1-o-Cerylglycerol, were identified. In addition, the lipid-lowering effects of sea anemone oil were evaluated in male Sprague–Dawley rats. At a low dose (8 mg/kg), no significant difference in lipid levels was observed compared to the control group, indicating minimal hypolipidemic effect. However, at a higher dose (33 mg/kg), a significant reduction in lipid levels was noted, demonstrating a pronounced lipid-lowering activity.

8.3.3.3 Proteins and Bioactive Peptides

Sea anemones are rich in high-quality, digestible protein and bioactive polypeptides, including cytolysins (actinoporins) and ion-channel toxins. Notable examples include anthopleurins from *A. xanthogrammica*, which modulate sodium channels and may support cardiac function [39], and neurotoxins from *Heteractis crispa*, which exhibit anti-inflammatory and neuroactive properties [40]. These molecules are being explored for pharmaceutical applications and functional food development.

8.3.3.4 Edible Potential and Challenges

Despite their promising nutritional profile, the widespread consumption of sea anemones is limited by the presence of natural toxins (such as cytolytic or neurotoxic polypeptides), which can be harmful to humans. However, traditional cooking methods in some coastal cultures, such as the manual detoxification of specific species to prepare local dishes, provide valuable experience for developing safe processing protocols. Combined with modern food science technologies such as membrane separation, high-pressure processing, and enzymatic detoxification, it is possible to effectively reduce or eliminate harmful components while preserving their nutritional value. Therefore, with proper species selection, rigorous safety assessment, and standardized processing procedures, sea anemones hold considerable potential for future integration into functional foods, nutraceuticals, or gourmet cuisine, contributing to both dietary diversity and the valorization of marine resources.

8.3.4 Nutrients of Jellyfish

Jellyfish, particularly edible species such as Rhopilema, are low-calorie marine organisms characterized by high-protein content and a notable mineral profile. They consist of 95%–98% water, with the dry matter composed mainly of collagen, minerals, and trace bioactive compounds [41]. Their nutritional value is often described as "three lows and one high"—low fat, low carbohydrates, low calories, and high minerals—making them suitable for both traditional consumption and modern functional food applications [42].

8.3.4.1 Proteins and Collagen

Protein constitutes a significant portion of the dry weight, with collagen accounting for 60–90% of the total protein content [43]. Jellyfish collagen exhibits a typical triple-helix structure, rich in glycine, proline, and hydroxyproline, which contributes to its structural stability [44]. Compared to terrestrial collagen, it demonstrates higher bioavailability and promotes skin fibroblast adhesion and proliferation [43]. Enzymatically hydrolyzed jellyfish collagen peptides show diverse biological activities, including antioxidant, anti-inflammatory, and wound-healing effects [45].

8.3.4.2 Lipids and Fatty Acids

Lipid content is generally low (<1% of fresh weight), but the fatty acid profile contains beneficial PUFAs, including DHA and EPA [43, 46]. These PUFAs help balance omega-3/omega-6 ratios and support cardiovascular health and anti-inflammatory responses [47].

8.3.4.3 Minerals

Jellyfish contain substantial amounts of minerals, including calcium, magnesium, selenium, and iodine [43, 48]. The calcium content is particularly notable, with 182 mg/100 g of fresh product, making it a valuable dietary source for bone health. These minerals occur in bioavailable forms and contribute to various physiological functions, including antioxidant defense and thyroid support.

8.3.4.4 Carbohydrates

Carbohydrate content is relatively low, primarily existing as SPSs including glycosaminoglycans and hyaluronic acid analogues [49]. These compounds demonstrate bioactive properties, including macrophage activation and antioxidant effects [49, 50].

8.4 Flavor of Other Aquatic Products

8.4.1 Flavor of Sea Urchins

The flavor of sea urchin gonads, often called "uni" or "sea urchin roe," is defined by their cool, silky texture reminiscent of cream or egg yolk and a layered aromatic profile. As the gonads dissolve on the palate, they release a rich, buttery fragrance akin to high-quality cheese or butter, balanced by a fleeting sweetness similar to mild lactose. This unique taste stems from their biochemical composition, particularly free amino acids (FAAs) (glutamine, asparagine, glycine, alanine, and arginine), nucleotide-related compounds, and organic acids, which collectively shape umami, sweetness, and bitterness [51–53]. Glutamine enhances umami, glycine amplifies sweetness, arginine introduces bitterness, and higher ratios of alanine to arginine generally improve flavor balance [54].

Species-specific variations further diversify sea urchin flavors. Short-spined sea urchins (*Heliocidaris crassispina*), prized in Hokkaido, Japan, deliver bold umami and pronounced sweetness, while purple sea urchins (*Strongylocentrotus purpuratus*) offer subtler oceanic notes with a refreshing finish. Northern purple sea urchins (*Strongylocentrotus nudus*), known for their refined texture, are favored in raw preparations or haute cuisine. Preparation methods critically influence culinary outcomes—raw preparations such as sashimi maintain the gonads' signature creamy texture and natural sweetness, while cooked applications like tempura or steaming develop richer umami profiles and textural complexity. These contrasts become particularly pronounced when paired with aromatic accompaniments like perilla leaves.

Geographic origin influences flavor through factors such as water temperature, diet, and habitat. Cold-water Hokkaido short-spined urchins develop robust, concentrated umami, while Dalian northern purple urchins from China exhibit delicate sweetness and refined texture. Volatile compounds further define the sensory experience: butyl acetate contributes green, sour, and fruity notes, while ethylbenzene adds marine freshness [55, 56]. Cultivated sea urchins often emit sweeter, fruitier aromas compared to wild counterparts, though high sulfide levels in lower-quality gonads can produce unpleasant sulfurous odors [56].

To fully appreciate sea urchin, begin by savoring its velvety texture, then focus on unfolding layers of aroma and sweetness. Pairing with sake enhances umami, soy sauce adds saltiness, and lemon juice brightens the richness. Fresh accompaniments like perilla leaves or cucumber balance its intensity.

In summary, sea urchins' allure lies in their complex interplay of species, origin, preparation, and biochemistry. Creamy texture, buttery richness, and nuanced sweetness make it a luxury delicacy, while cooking techniques and pairings expand its culinary versatility. Quality hinges on balanced amino acids, aromatic volatiles, and minimal sulfides, with farmed varieties offering sweeter profiles. Whether enjoyed raw or cooked, sea urchin delivers a multisensory experience cherished by gourmets globally.

8.4.2 Flavor of Starfish

The flavor of starfish, primarily derived from its edible gonads or coelomic fluid, is defined by its firm, slightly gelatinous texture and a layered, complex aromatic profile. Unlike the melt-in-your-mouth creaminess of sea urchin, the texture of starfish is more resilient. As it is chewed, it releases a rich umami akin to a blend of seafood and chicken broth, accompanied by a unique, faint bitterness and a mellow, herbal aftertaste. This distinct taste stems from its biochemical composition, particularly its high content of FAAs (glutamic acid, alanine, glycine), nucleotide-related compounds (like AMP), and its hallmark asterosaponins, which collectively shape its complex flavor interplay of umami, sweetness, and bitterness. Glutamic acid and AMP form the foundation of its intense umami, glycine and alanine contribute a subtle sweetness, while asterosaponins are the source of its signature bitterness and medicinal notes, a key characteristic that distinguishes starfish from other seafood.

Species-specific variations further diversify the flavors of starfish. For instance, the Amur starfish (*Asterias amurensis*), consumed in parts of Asia, delivers intense umami with pronounced minerality and a noticeable bitterness, making it suitable for stews or dried as a seasoning to leverage its strong flavor. In contrast, the comb sea star (*Astropecten polyacanthus*) has a relatively looser texture with milder umami and weaker bitterness, and is typically deep-fried or boiled to achieve a more accessible palate.

Cooking methods critically influence the final taste experience. Due to the presence of saponins, starfish is rarely eaten raw. Traditional cooking methods, such as prolonged stewing or steaming, can partially break down its saponins, reducing bitterness while releasing a deeper, more mellow umami as its proteins and polysaccharides hydrolyze. The texture also becomes softer and more tender. Deep-frying, on the other hand, creates a crisp exterior that contrasts with the firm interior, while the high heat helps to transform its flavor compounds.

Geographic origin influences flavor through factors like water temperature, diet, and habitat. For example, starfish from cold waters (like Hokkaido) accumulate more glycogen and amino acids to withstand low temperatures, resulting in more concentrated and intense umami and sweetness. In contrast, starfish from warmer waters (like the South China Sea) may have a softer texture with a more pronounced briny or "fishy" flavor. Its flavor is also linked to volatile compounds: bromophenols contribute typical oceanic and iodine-like aromas, while dimethyl sulfide (DMS) provides a cooked seafood flavor. Some terpenoids related to saponin degradation may be the source of its herbal or medicinal scents.

To fully appreciate starfish, one must begin by tempering its bitterness through proper cooking. When tasting, first appreciate its unique, resilient texture, then focus on identifying the umami foundation, and finally, recognize the signature, bittersweet aftertaste that unfolds at the back of the palate. In terms of pairings, rich dipping sauces, such as garlic soy sauce, black bean paste, or heavily spiced broths, are typically used to balance its strong intrinsic flavor, rather than delicate seasonings.

In summary, the allure of starfish lies in the complex and challenging interplay between its umami and bitterness, shaped by its unique biochemistry, species, and preparation. Its core characteristics are rich umami, a signature saponin bitterness, and a firm, chewy texture, making it an "acquired taste." Cooking is the key to unlocking its deliciousness and tempering its bitterness, with different methods greatly expanding its culinary dimensions. Quality hinges on the balance between the abundance of umami-rich amino acids and the intensity of its saponin bitterness. Whether used in a soup or as a main course, starfish offers a memorable gustatory adventure for the daring food explorer.

8.4.3 Flavor of Sea Anemone

The cooking method is the key factor that determines the flavor and safety of anemones. Due to the presence of stinging cell toxins, anemones must not be eaten raw. High-temperature frying is the most classic cooking method that best reflects its charm. The snakelocks anemone (*Anemonia viridis*) is considered a delicacy in southwestern Spain and Sardinia, where it is marinated in vinegar, coated in a batter resembling calamari, and fried in olive oil. Fisherman communities in Sabah, Borneo, and the Thousand Islands (known as *rambu-rambu*) in Southeast Asia, as well as in Taizhou, Zhejiang (known as Shasuan), also rely on anemones as a food source.

8.4.4 Flavor of Jellyfish

The flavor profile of edible jellyfish (primarily *Rhopilema* spp. and *Aurelia aurita*) is characterized by a delicate balance of marine freshness, umami, saltiness, and a uniquely crisp yet elastic texture. This distinctive sensory experience arises from their biochemical composition and processing techniques.

Jellyfish are rich in FAAs and nucleotides that contribute to umami and sweetness. Glutamic acid and aspartic acid are major umami contributors, while alanine and glycine impart subtle sweet notes. The synergy between glutamic acid and nucleotides, like inosine monophosphate (IMP), significantly enhances umami perception. Mineral ions (sodium, potassium, magnesium) derived from processing salts moderate saltiness and mask undesirable odors [57, 58].

The aroma is dominated by fresh marine notes, complemented by cucumber-like freshness. Key volatile compounds include alcohols (e.g., 1-octen-3-ol), esters, and sulfur-containing molecules such as DMS. Aldehydes like hexanal and nonanal contribute green, grassy, or citrus hints. The near-absence of fat in jellyfish helps maintain a pure, clean flavor profile [59, 60].

The signature crisp and elastic texture resulting from collagen cross-linking serves as an important carrier for flavor compounds. Modern processing techniques employing calcium-ion cross-linking or acid immersion (e.g., vinegar) not only improve safety but also enhance flavor release and juiciness [61, 62].

Geographic origin influences flavor profiles: jellyfish from cold, high-salinity waters (e.g., the Yellow Sea) develop a crispier texture, while those from warmer regions (e.g., the South China Sea) accumulate more FAAs, yielding a stronger umami taste [20].

Marinating in acidic solutions (e.g., vinegar) releases FAAs and promotes ester formation, adding complexity while softening texture. Salt suppresses bitterness and amplifies umami perception. These processing methods significantly transform the raw material into a desirable culinary ingredient [41].

8.5 Physical and Chemical Characteristics of Other Aquatic Products

8.5.1 Physical and Chemical Characteristics of Sea Urchin

Sea urchin gonads are highly valued in the market for their desirable qualities, such as substantial size, a vivid yellowish-orange hue, a flavor profile that ranges from sweet to mildly briny, and a texture that is both tender and firm [63]. These gonads, nestled within the sea urchin's hemispherical carapace alongside spines and other bodily structures, make up a mere 2%–6% of the total body weight, with the bulk consisting of the shell and viscera. It is important to note that the size and weight of the gonads are significantly dictated by the phase of gametogenesis [64].

The quality of sea urchin gonads is closely linked to their texture and color. Premium quality gonads are known for their compact, dense texture. However, as they mature, they tend to become softer and more slippery, likely due to an increase in water content [65]. The color spectrum of gonads, which can range from green, red, brown, white, to purple, is influenced by pigments that may possess medicinal properties [66–68]. For instance, the gonads of *Paracentrotus lividus* exhibit vibrant orange/yellow shades during the growing phase, which are preferred for marketing, but they transform into dark red/brown during spawning, making them less appealing [69].

The properties of sea urchin gonads are substantially influenced by environmental and physiological factors. The water and ash content in gonads is determined by the conditions of their habitat, the availability of minerals, and their food sources [64]. Gonads that are larger and have a yellow hue typically have higher protein content [23]. The color of the gonads is also associated with their fat content and the presence of pigments such as carotenoids (like β-carotene and echinenone), naphthoquinones, and melanin [70, 71]. For instance, carotenoids such as β-echinenone are responsible for the yellow-red colors in gonads [72].

The market grading of sea urchin roe is often a subjective process, primarily based on the assessment of color and texture. In California, for instance, the grading standards can vary with the supply levels: colors that are less favored might be classified as grade B when the supply is abundant, or as grade A during times of scarcity [73]. The color of the roe is also indicative of its flavor, with bright yellow *Strongylocentrotus franciscanus* gonads typically having a sour taste, whereas greenish hues suggest bitterness, related to certain amino acids [74]. Texture, a crucial quality indicator, is influenced by the reproductive stage of the sea urchin. Gonads in the growing phase are preferred for their moderate firmness, which is firmer than that of mature gonads but softer than that of those in the recovering stage [53, 75].

The nutritional and functional attributes of sea urchin gonads significantly contribute to their value. These gonads are known for their high antioxidant content, as observed in species like *Strongylocentrotus droebachiensis* and *Strongylocentrotus nudus* [76]. However, there is a lack of universal quantitative or qualitative standards for assessing gonad quality. Typical evaluations focus on parameters such as size, color, texture, and the content of FAAs [77, 78]. Consumer preferences tend to lean toward gonads that are large, vividly colored (orange/yellow), have a medium firmness, high alanine content (for sweetness), and low arginine content (to avoid bitterness) [79]. On the other hand, gonads originating from barren habitats are deemed commercially worthless due to their small size, inferior color, and unpalatable texture and taste [80].

8.5.2 Physical and Chemical Characteristics of Starfish

The market value of starfish is predicated on a unique combination of attributes, primarily its intense umami flavor, a signature bitter aftertaste, and a firm, resilient texture achieved after cooking. The edible portions, chiefly the gonads (roe) and sometimes parts of the body wall, constitute a small fraction of the organism's total mass (typically 5–10%), with the majority comprised of the endoskeleton of calcareous ossicles and viscera. The size, weight, and chemical composition of the gonads are critically dictated by the gametogenesis cycle, reaching their peak in mass and quality during the reproductive season.

A defining physicochemical characteristic of starfish tissue is its high susceptibility to autolysis. Unlike the relatively stable gonads of sea urchins, starfish tissues contain highly active endogenous proteolytic enzymes that, upon the organism's death, initiate a rapid process of self-digestion. This enzymatic activity leads to a swift and irreversible transformation of the tissue from a firm, structured state to a liquefied, viscous slurry, rendering it entirely devoid of commercial and culinary value. Consequently, the desirable firm and chewy texture, which is a key quality indicator, is not an intrinsic property of the raw tissue but rather a state that must be preserved through immediate and effective processing (e.g., rapid heating or salting) to inactivate these enzymes. The color of the gonads, typically ranging from pale yellow to orange-red, also serves as an indicator of developmental stage and quality, with more vibrant hues generally corresponding to a firmer texture and optimal maturity.

The properties of starfish are largely shaped by their unique chemical defense mechanisms, most notably the presence of asterosaponins. These complex steroidal glycosides are the principal source of the starfish's characteristic bitterness and its potential toxicity when consumed raw, making thermal processing an absolute prerequisite for safe consumption. In parallel, the intense umami flavor is derived from a high concentration of FAAs, with glutamic acid being a primary contributor. The interplay between the desirable umami from amino acids and the challenging bitterness from saponins constitutes the core flavor profile. The precise concentrations of these compounds, along with water and ash content, are influenced by environmental factors such as habitat conditions and diet.

The quality evaluation of starfish is an entirely processing-centric paradigm. Its commercial viability is not inherent in the raw product but is created through successful postharvest handling. The primary objectives of processing are twofold: first, to rapidly inactivate proteolytic enzymes to prevent autolysis and thereby fix the firm texture; and second, to degrade or reduce saponin levels to a palatable and safe threshold, managing the bitterness. Therefore, any starfish that has undergone autolysis or has been improperly processed to retain excessive bitterness is considered commercially worthless.

The nutritional and functional attributes of starfish present a unique duality of value. For culinary purposes, consumer preference is for a product that has been expertly processed to maximize umami while balancing the bitterness, resulting in a complex and intriguing flavor experience. From a biomedical and biotechnological perspective, however, the value lies in the high concentration of the very compounds that pose a culinary challenge: the bioactive asterosaponins, which have demonstrated potent antitumor, anti-inflammatory, and antifungal properties. Consequently, starfish deemed inferior for consumption due to their small size, poor processing outcomes, or excessively high bitterness may still be valuable raw materials for the extraction of these high-value functional compounds.

8.5.3 Physical and Chemical Characteristics of Sea Anemone

Sea anemones, as members of the phylum Cnidaria, exhibit a range of distinct physical characteristics that adapt them to their marine environments. Typically, they possess a cylindrical polyp body plan, with a basal disc at one end that allows attachment to substrates such as rocks, coral reefs, or even the shells of hermit crabs, facilitating firm anchorage in dynamic aquatic settings. The opposite end features an oral disc, surrounded by numerous tentacles that radiate outward in radial symmetry, a trait common to cnidarians. The number, length, and arrangement of these tentacles vary among species, with some having only a few short tentacles while others possess dozens of long, slender ones, often adorned with specialized structures known as nematocysts. The size of sea anemones can differ significantly; smaller species may measure merely a few millimeters in diameter, whereas larger ones can reach diameters of over a meter, depending on factors such as species and environmental conditions. Their body coloration is equally diverse, ranging from vibrant hues of red, orange, green, and purple to more muted shades of brown or gray, which is often attributed to the presence of pigments in their ectodermal cells and, in many cases, the symbiotic relationship with photosynthetic algae like zooxanthellae, whose chlorophyll can impart a greenish tint [81]. Internally, the body is composed of two germ layers—an outer ectoderm and an inner endoderm—with a gelatinous mesoglea layer in between, providing structural support. The body wall contains both longitudinal and circular muscle fibers, enabling the polyp to contract or expand its body and tentacles, a mechanism crucial for feeding, defense, and responding to environmental stimuli.

In terms of chemical characteristics, sea anemones are notable for their production of various bioactive compounds, primarily associated with their cnidocytes—specialized cells that house nematocysts, the organelles responsible for toxin delivery [82, 83]. These nematocysts release a complex mixture of toxins when triggered, which serve multiple functions: immobilizing prey, deterring predators, and competing with other organisms for space. The toxins typically consist of peptides, proteins, and small molecules, with specific compositions varying among species; some may target the nervous systems of prey by blocking ion channels, while others induce cellular damage through enzymatic activity. Additionally, many sea anemones engage in symbiotic relationships with photosynthetic dinoflagellates, such as zooxanthellae, which reside within their endodermal cells. Through photosynthesis, these algae produce organic carbon compounds, such as glucose, which are transferred to the host anemone, supplementing its nutritional intake. This symbiosis also influences the chemical environment within the anemone's tissues, as the exchange of nutrients and waste products (e.g., carbon dioxide and ammonia) modulates metabolic processes. Furthermore, sea anemones secrete mucopolysaccharide-rich mucus from their epidermal cells, which aids in adhesion to substrates, protects against desiccation in intertidal zones, and may contain antimicrobial agents that prevent colonization by pathogenic microorganisms.

8.5.4 Physical and Chemical Characteristics of Jellyfish

Jellyfish are marine invertebrates characterized by a soft, gelatinous body with radial symmetry, typically in the form of an umbrella or bell. Their most distinctive physical property is an exceptionally high water content, comprising 90%–98% of their total mass. This results in near-neutral buoyancy, translucency, and a soft, elastic yet fragile texture. The body consists mainly of a collagen-rich mesoglea layer, which forms a porous, three-dimensional network of collagen fibers and SPSs. This structure provides mechanical elasticity, with an elastic modulus of 1–5 kPa, and serves to absorb energy from water movements [84, 85].

Lacking any skeletal structure, jellyfish are highly susceptible to ocean currents. Their tentacles contain specialized stinging cells called nematocysts, which can deliver venom composed of proteins, peptides, serotonin, histamine, and various enzymes. These toxins serve for prey capture and defense and can cause severe reactions in humans [86].

The primary organic component of jellyfish is collagen, predominantly type I, which represents a major portion of the dry matter. Its structure is stabilized by a characteristic triple-helix conformation formed by repeating Gly-X-Y amino acid sequences, rich in glycine, proline, and hydroxyproline. Compared to mammalian collagen, jellyfish collagen has lower thermal stability, denaturing at around 25–30°C, due to fewer intermolecular cross-links [46, 87].

Other significant biochemical constituents include various amino acids, unsaturated fatty acids, glycoproteins, and polysaccharides. The elemental composition of jellyfish reflects that of seawater, featuring minerals such as sodium, potassium, magnesium, calcium, and trace metals, which contribute to various biological functions [43].

Jellyfish tissue undergoes rapid autolysis after death due to the action of endogenous proteases. To prevent spoilage, traditional processing methods using alum and salt are employed to reduce water content, stabilize the structure, and detoxify the product. This process also imparts a characteristic crisp texture. Control of aluminum residue is necessary, often achieved through vinegar immersion [88, 89].

Jellyfish also contain bioactive compounds with demonstrated antioxidant, anti-inflammatory, and anticoagulant properties. Collagen-derived peptides and polysaccharides exhibit free radical scavenging activities, while certain venom components have shown potential for pharmaceutical development [90–92].

8.6 Market Prospects of Other Aquatic Products

8.6.1 Market Prospects of Sea Urchins

As a high-value marine resource, sea urchins have demonstrated significant growth potential in global food and pharmaceutical markets in recent years. Historically, they have held an esteemed position in culinary practices, frequently featured in sushi bars, upscale restaurants, and wedding banquets, though household consumption remains limited to small quantities [93]. Emerging ready-to-eat products and shelf-stable pastes are now bridging this gap, making sea urchins more accessible to home cooks through online grocery platforms. Their distinctive taste even propelled them into the top ten food trends in 2016. Driven by rising demand for premium seafood and functional foods, the industry is transitioning from traditional harvesting to sustainable aquaculture and deep processing. However, ecological and technological challenges persist.

The worldwide demand for sea urchins is substantial, estimated at 60,000–70,000 tonnes annually. Japan dominates the market, accounting for approximately 90% of global consumption and importing 80% of its supply [94]. Notably, Southeast Asian markets like Thailand and Vietnam are emerging as new consumption hubs, driven by tourism-driven sushi demand and local fusion cuisine. Here, traditional dishes like sushi and sashimi drive massive demand for fresh sea urchins, with annual imports exceeding $300 million. Beyond Japan, markets in China, South Korea, Europe, and the United States are growing rapidly. China's emerging middle class, for instance, has fueled a 10% annual increase in imports. Europe has also seen rising interest, particularly in sea urchin roe, with France leading the continent's market [95].

Processed products—such as frozen gonads and seasoned sea urchin paste—are gaining traction through e-commerce and retail channels, expanding accessibility. Innovations like freeze-dried gonad powders for dietary supplements and shelf-stable urchin-infused oils are creating niche markets in the health food sector. Additionally, the growing influence of minority populations and the global spread of sushi restaurants have further boosted demand. The high-value, low-volume catering segment has also shown increased interest, albeit with notable market volatility. Sea urchin's culinary versatility is evident in innovative applications, from pasta sauces and bread spreads in Mediterranean cuisine to experimental uses in custards and ice creams.

Wild sea urchin harvesting, which peaked during the 1960s sushi boom, has led to severe population declines. Over six decades of intensive harvesting have pushed more than 40 species to the brink of overexploitation, necessitating a shift toward aquaculture and underutilized species [96]. For example, European species like *Psammechinus miliaris* and *Sphaerechinus granularis* are being explored, though many others remain untapped [97].

To address these challenges, countries are advancing aquaculture technologies. Norway and Canada are refining artificial breeding techniques, optimizing water quality, and feed formulations to improve survival rates. China has piloted raft-based aquaculture in Shandong and Liaoning, leveraging ecological symbiosis between sea urchins and algae to reduce costs and environmental impact. However, obstacles like long cultivation cycles (2–3 years) and inadequate disease control hinder large-scale production.

Beyond culinary uses, deep processing is unlocking new opportunities. Bioactive compounds such as EPA, DHA, and polysaccharides extracted from sea urchin gonads are used in health supplements and pharmaceuticals. Japanese companies produce anti-fatigue capsules from sea urchin extracts, while European and American firms explore applications in treating neurodegenerative diseases. Additionally, sea urchin shells—rich in calcium carbonate—are repurposed into bone repair scaffolds or eco-friendly coatings, enabling full resource utilization.

Despite these prospects, the market confronts challenges. Overharvesting leads to ecological imbalances, for instance, the formation of "urchin barrens" in California, where excessive grazing disrupts marine ecosystems. Market volatility also poses risks; for instance, the prices of Hokkaido's red sea urchins surged by 30% in 2022 due to production declines resulting from high temperatures.

To ensure sustainable growth, the industry must prioritize aquaculture efficiency, diversify product lines, and establish global certification systems. Balancing resource exploitation with conservation will be critical as demand continues to rise.

8.6.2 Market Prospects of Starfish

Decades of global research have established the foundational utility of starfish across evolutionary biology, food science, and industrial manufacturing. By shifting the perspective on starfish from ecological pests to harvestable resources, it is possible to simultaneously mitigate their disruption of marine ecosystems and foster sustainable economic applications.

8.6.2.1 Biomedical Applications

Echinoderms are a prolific source of marine natural products, contributing nearly one-third of all known marine bioactive compounds. Starfish, in particular, are rich in lipids, fatty acids, peptides, and saponins, marking them as significant candidates for pharmaceutical development. The therapeutic value of these organisms is primarily derived from polyhydroxysterols and water-soluble saponins concentrated in the stomach and pyloric caecum. Notably, saponin-based formulations have demonstrated clinical efficacy in oncology. Moreover, numerous echinoderms, such as starfish, demonstrate impressive regenerative capacities. Research on the regrowth of arm tips in *A. rubens* unveiled the phases of regeneration, along with cell proliferation and neuropeptide expression patterns. Consequently, researchers are investigating the regenerative capabilities of echinoderm limbs to enhance comprehension and treatment of neurodegenerative disorders in humans.

8.6.2.2 Food Applications

Starfish gonads (testes and ovaries) represent an underutilized but nutritious food source. These nontoxic organs are rich in essential omega-3 PUFAs, specifically EPA and DHA, which offer crucial nutritional advantages.

The flavor-active amino acids (FAAs) serve as freshness markers for seafood. In the gonads of *Solaster dawsoni*, FAAs make up 51.16% of the total amino acid content, a proportion comparable to that of abalone, indicating that it has high edible value. Promoting starfish consumption converts a fishery bycatch into a marketable asset, benefiting local economies while controlling wild populations.

In coastal cities such as Qingdao City in China, the proliferation of starfish *A. amurensis* has driven its integration into local cuisine. The gonads are cost-effective and typically prepared by steaming with eggs to mimic the texture of crab roe or processed into canned products. During breeding seasons, gonadal yield is substantial, reaching 20–25% of the total body weight, which enhances their economic viability. However, global adoption is hindered by safety concerns regarding heavy metals. As filter feeders consuming shellfish, starfish can bioaccumulate toxins. Research on *A. rubens* near the Scheldt River identified elevated levels of Zinc and Cadmium in the gonads and Lead in the exoskeleton. Although *A. amurensis* in China has not shown similar contamination, the potential for heavy metal enrichment necessitates rigorous screening of origin and species prior to consumption.

8.6.2.3 Industry and Other Applications

Beyond their nutritional profile, starfish offer significant opportunities for valorization through diverse applications, beginning at the biochemical level. Their protein-rich body walls serve as a viable source for collagen, which can be further processed into gelatin or used in various biotechnological applications, such as tissue regeneration, showing beneficial effects on corneal and skin cells. Advanced extraction methods, such as combining high-shear mechanical homogenization with ultrasound, can expedite this process while preserving collagen quality. Furthermore, this collagen can be hydrolyzed into low-molecular-weight peptides, which, due to their excellent water solubility and skin permeability, are highly valued in the cosmetics industry for functions like wound healing and antiaging. Moving from the extraction of specific biomolecules to the utilization of the entire organism, starfish can be processed into a nutrient-dense powder for use as a feed additive in aquaculture. Research has shown that incorporating dried starfish powder into the diets of species such as the sea cucumber *Apostichopus japonicus* and abalone can significantly enhance growth rates, offering a feeding efficacy comparable to conventional baits and presenting a promising strategy for improving feed utilization. On a broader ecological scale, starfish participate in global biogeochemical cycles. Their exoskeletons, formed from calcium carbonate, effectively act as a carbon sink. Upon the organism's death, this embodied carbon is deposited on the seabed, contributing to the long-term reduction of oceanic carbon emissions.

8.6.3 Market Prospects of Sea Anemone

The comprehensive utilization of sea anemones presents many opportunities, with applications far beyond their traditional culinary uses. Sea anemones, as a marine resource, contain bioactive compounds with demonstrated antitumor, antibacterial, anti-inflammatory, immunomodulatory, and antioxidant properties. This makes them a valuable candidate for the formulation of high-end functional foods, dietary supplements, and cosmeceuticals.

In the food sector, sea anemones have long been consumed in coastal regions, prepared in various traditional dishes such as hot and sour soup, dry pot, or stir-fried with sauces. Their distinctive umami flavor is attributed to naturally occurring taste-active compounds such as glutamic acid. In some areas, sea anemones are even combined with milk to produce nutrient-dense formulations aimed at supporting children's growth and development. Beyond their culinary appeal, extracts from sea anemones are being incorporated into functional food products and nutraceuticals designed to support immune function and general health.

From an ecological and industrial standpoint, sea anemones are effective in maintaining the health and resilience of marine ecosystems. As symbiotic partners of coral reefs, they contribute to the stability of reef habitats by limiting algal overgrowth, thereby supporting biodiversity and ecological balance. Furthermore, their biochemical components have promising industrial applications. Polysaccharides and proteins extracted from sea anemone mucus are under investigation as raw materials for developing biocompatible wound dressings and biodegradable materials [39]. Additionally, sea anemone toxins have become indispensable tools in biomedical research, particularly in neurobiology and pharmacology, where they serve as highly specific probes for ion channel function [98].

To fully realize the potential of sea anemones as both a bioresource and a commercial commodity, coordinated efforts across disciplines are essential. This includes systematic species identification and classification, in-depth studies of their metabolomic and proteomic profiles, and the development of scalable, cost-effective extraction and purification technologies. Moreover, robust resource management frameworks and ecological monitoring are needed to prevent habitat degradation and ensure the sustainable harvesting of wild populations, where applicable. As research continues to advance our understanding of their nutritional and pharmacological properties, sea anemones are poised to play an increasingly important role in the sustainable development of the aquatic food sector and related industries.

8.6.4 Market Prospects of Jellyfish

Jellyfish represent a valuable marine resource with significant economic importance, particularly in Asia, where they have been utilized in traditional cuisine for centuries. Growing interest in global health and functional foods has expanded their applications into diverse sectors, including food, biomedicine, cosmetics, and industrial materials. The jellyfish body wall is rich in bioactive compounds such as collagen, polysaccharides, and mucoproteins, underpinning their potential in dietary supplements, functional foods, cosmetics, and biotechnology products [99, 100]. The seasonal abundance of jellyfish provides a reliable raw material base supporting industrial development.

Jellyfish are a low-fat, low-calorie food resource with a long history of consumption in Asian countries such as China, Japan, and Korea [101]. They are traditionally consumed as cold dishes but are increasingly being incorporated into modern food products, including protein beverages, collagen gummies, frozen foods, and sauces [102]. Their high collagen content and favorable amino acid profile make them a valuable nutrient source [103]. Additionally, jellyfish-derived peptides and polysaccharides exhibit antioxidant and anti-inflammatory properties, supporting their use in functional foods [104].

Jellyfish collagen is valued for its biocompatibility, biodegradability, and bioactive functions [45, 105]. It promotes tissue regeneration and mitigates UV-induced oxidative stress [106]. These properties enable its use in wound dressings, tissue engineering scaffolds, and oral supplements [107, 108]. Hydrolyzed jellyfish collagen peptides show promise in inhibiting angiotensin-converting enzyme (ACE) activity, suggesting potential applications in managing hypertension [109, 110].

In cosmetics, jellyfish collagen is used for its moisturizing and antiaging benefits. Its properties improve skin elasticity and barrier function, making it suitable for skincare products [111, 112]. Jellyfish have applications in industrial sectors, including the production of biodegradable packaging materials and hydrogels [113, 114]. Collagen-based films exhibit functional properties that can improve food preservation [115, 116]. Jellyfish also serve as bio-adsorbents for water treatment and can be repurposed into organic fertilizers and feed additives, supporting resource recycling [117, 118].

References

1 Kober, K. M., & Bernardi, G. (2013). Phylogenomics of strongylocentrotid sea urchins. *BMC Evolutionary Biology*, 13(1), 88. https://doi.org/10.1186/1471-2148-13-88

2 Rocha, F., Rocha, A. C., Baião, L. F., Gadelha, J., Camacho, C., Carvalho, M. L., . . . Valente, L. M. P. (2019). Seasonal effect in nutritional quality and safety of the wild sea urchin *Paracentrotus lividus* harvested in the European Atlantic shores. *Food Chemistry*, 282, 84–94. https://doi.org/10.1016/j.foodchem.2018.12.097

3 Brusca, R.C. and Brusca, G.J. (1990). *Invertebrates*. USA: Sinauer Associates, Inc.

4 Daly, M., Brugler, M.R., Cartwright, P. et al. (2007). The phylum Cnidaria: a review of phylogenetic patterns and diversity 300 years after Linnaeus. *Zootaxa* 1668 (1): 127–182. https://doi.org/10.11646/zootaxa.1668.1.11.

5 Lawley, J. W., Ames, C. L., Bentlage, B., Yanagihara, A. A., Goodwill, R., Kayal, E., Hurwitz, K., & Collins, A. G. (2016). Box jellyfish *Alatina alata* has a circumtropical distribution. *The Biological Bulletin*, 231(2), 152–169. https://doi.org/10.1086/690095

6 Sutherland, K. R., Gemmell, B. J., Colin, S. P., & Costello, J. H. (2016). Prey capture by the cosmopolitan hydromedusae, *Obelia* spp., in the viscous regime. *Limnology and Oceanography*, 61(6), 2309–2317. https://doi.org/10.1002/lno.10390

7 Miranda, L. S., Mills, C. E., Hirano, Y. M., Collins, A. G., & Marques, A. C. (2018). A review of the global diversity and natural history of stalked jellyfishes (*Cnidaria, Staurozoa*). *Marine Biodiversity*, 48(4), 1695–1714. https://doi.org/10.1007/s12526-017-0721-4

8 Dunn, F. S., Kenchington, C. G., Parry, L. A., Clark, J. W., Kendall, R. S., & Wilby, P. R. (2022). A crown-group cnidarian from the Ediacaran of Charnwood Forest, UK, *Nature Ecology & Evolution*, 6(8), 1095–1104. https://doi.org/10.1038/s41559-022-01807-x

9 Moon, J., Caron, J. B., & Moysiuk, J. (2023). A macroscopic free-swimming medusa from the middle Cambrian Burgess Shale. *Proceedings of the Royal Society B: Biological Sciences*, 290(1994), 20222490. https://doi.org/10.1098/rspb.2022.2490

10 Archana, A., & Babu, K. R. (2016). Nutrient composition and antioxidant activity of gonads of sea urchin *Stomopneustes variolaris*. *Food Chemistry*, 197, 597–602. https://doi.org/10.1016/j.foodchem.2015.11.003

11 Mah, C., Nizinski, M., & Lundsten, L. (2010). Phylogenetic revision of the Hippasterinae (Goniasteridae; Asteroidea): systematics of deep sea corallivores, including one new genus and three new species. *Zoological Journal of the Linnean Society*, 160(2), 266–301. https://doi.org/10.1111/j.1096-3642.2010.00638.x

12 Downey, M.E. (1986). *Revision of the Atlantic Brisingida (Echinodermata: Asteroidea), with Description of a New Genus and Family*. Washington DC: Smithsonian Institution Press.

13 Oppong-Danquah, E., Miranda, M., Blümel, M., & Tasdemir, D. (2023). Bioactivity profiling and untargeted metabolomics of microbiota associated with mesopelagic jellyfish *Periphylla periphylla*. *Marine Drugs*, 21(2), 129. https://doi.org/10.3390/md21020129

14 Romano, G., Almeida, M., Coelho, A. V., Cutignano, A., Gonçalves, L. G., Hansen, E., Khnykin, D., Mass, T., Ramšak, A., Rocha, M. S., Silva, T. H., Sugni, M., Ballarin, L., & Genevière, A.-M. (2022). Biomaterials and bioactive natural products from marine invertebrates: from basic research to innovative applications. *Marine Drugs*, 20(4), 219. https://doi.org/10.3390/md20040219

15 Gershwin, L. A., & Ekins, M. (2015). A new pygmy species of box jellyfish (Cubozoa: Chirodropida) from sub-tropical Australia. *Marine Biodiversity Records*, 8, e108. https://doi.org/10.1017/S175526721500086X

16 Lucas, C. H., Graham, W. M., & Widmer, C. (2012). Jellyfish life histories: Role of polyps in forming and maintaining scyphomedusa populations. In M. Lesser (Ed.), Advances in Marine Biology (Vol. 63, pp. 133–196). Academic Press. Cambridge, https://doi.org/10.1016/B978-0-12-394282-1.00003-X

17 Gibbons, M. J., & Richardson, A. J. (2008). Patterns of jellyfish abundance in the North Atlantic. *Hydrobiologia*, 616(1), 51–65. https://doi.org/10.1007/s10750-008-9593-8

18 Uye, S. I. (2008). Blooms of the giant jellyfish *Nemopilema nomurai*: a threat to the fisheries sustainability of the East Asian Marginal Seas. *Plankton and Benthos Research*, 3(Suppl), 125–131. https://doi.org/10.3800/pbr.3.125

19 Sun, S., Zhang, F., Li, C., Wang, S., Wang, M., Tao, Z., Wang, Y., Zhang, G., & Sun, X. (2015). Breeding places, population dynamics, and distribution of the giant jellyfish *Nemopilema nomurai* (Scyphozoa: Rhizostomeae) in the Yellow Sea and the East China Sea. *Hydrobiologia*, 754(1), 59–74. https://doi.org/10.1007/s10750-015-2266-5

20 Dong, X., Fan, X., Wang, Y., Qi, L., Liang, S., Yu, C., & Zhu, B. (2018). The effect of different pretreatments on the quality of ready-to-eat jellyfish *Rhopilema esculentum* Kishinouye products. *Fisheries Science*, 84(1), 1–10. https://doi.org/10.1007/s12562-017-1165-2

21 Santos, P. M., Albano, P., Raposo, A., Ferreira, S. M. F., Costa, J. L., & Pombo, A. (2020). The effect of temperature on somatic and gonadal development of the sea urchin *Paracentrotus lividus* (Lamarck, 1816). *Aquaculture*, 528, 735487. https://doi.org/10.1016/j.aquaculture.2020.735487

22 Dincer, T., & and Cakli, S. (2007). Chemical composition and biometrical measurements of the Turkish Sea Urchin (Paracentrotus Lividus, Lamarck, 1816). *Critical Reviews in Food Science and Nutrition*, 47(1), 21–26. https://doi.org/10.1080/10408390600550265

23 Akaerina, F., Tati, N., and Suwandi, R. (2015). Isolasi dan Karakterisasi Senyawa Antibakteri dari Bulu Babi. *Jurnal Pengolahan Hasil Perikanan Indonesia* 18: 61–73. https://doi.org/10.17844/jphpi.2015.18.1.61.

24 Chen, L. Y., Xie, X. N., Li, Y., Xiong, H., & Li, L. (2022). Activation mechanism of whey protein isolate mediated by free radicals generated in the ascorbic acid/hydrogen peroxide system. *Food Chemistry*, 384, 132533. https://doi.org/10.1016/j.foodchem.2022.132533

25 Sibiya, A., Jeyavani, J., Sivakamavalli, J., Ravi, C., Divya, M., & Vaseeharan, B. (2021). Bioactive compounds from various types of sea urchin and their therapeutic effects — a review. *Regional Studies in Marine Science*, 44, 101760. https://doi.org/10.1016/j.rsma.2021.101760

26 Mita, M., & Ueta, N. (1989). Fatty chain composition of phospholipids in sea urchin spermatozoa. *Comparative Biochemistry and Physiology Part B: Comparative Biochemistry*, 92(2), 319-322. https://doi.org/10.1016/0305-0491(89)90285-X

27 Kostetsky, E. Y., Sanina, N. M., & Velansky, P. V. (2014). The thermotropic behavior and major molecular species composition of the phospholipids of echinoderms. *Russian Journal of Marine Biology*, 40(2), 131–139. https://doi.org/10.1134/S1063074014020059

28 Shang, X.-H., Liu, X.-Y., Zhang, J.-P., Gao, Y., Jiao, B.-H., Zheng, H., & Lu, X.-L. (2014). Traditional Chinese medicine——sea urchin. *Mini Reviews in Medicinal Chemistry*, 14. 6. https://doi.org/10.2174/1389557514666140529224147

29 Walker, C.W., Lesser, M.P., and Unuma, T. (2013). Chapter 3 - sea urchin gametogenesis – structural, functional and molecular/genomic biology. In: *Developments in Aquaculture and Fisheries Science* (ed. J.M. Lawrence), 25–43. Amsterdam: Elsevier.

30 van der Heide, M. E., Møller, L. F., Petersen, J. K., & Nørgaard, J. V. (2018). Annual variation in the composition of major nutrients of the common starfish (*Asterias rubens*). *Animal Feed Science and Technology*, 238, 91–97. https://doi.org/10.1016/j.anifeedsci.2018.02.007

31 Liu, Y., Wu, X., Wang, Y., Jin, W., & Guo, Y. (2020). The immunoenhancement effects of starfish *Asterias rollestoni* polysaccharides in macrophages and cyclophosphamide-induced immunosuppression mouse models. *Food & Function*, 11(12), 10700–10708. https://doi.org/10.1039/D0FO01488E

32 Bilan, M. I., Anisimova, N. Y., Tokatly, A. I., Nikogosova, S. P., Vinnitskiy, D. Z., Ustyuzhanina, N. E., Dmitrenok, A. S., Tsvetkova, E. A., Kiselevskiy, M. V., Nifantiev, N. E., & Usov, A. I. (2023). Glycosaminoglycans from the Starfish *Lethasterias fusca*: structures and influence on Hematopoiesis. *Marine Drugs*, 21(4), 205. https://doi.org/10.3390/md21040205

33 Saeid, A., Dave, D., & Shahidi, F. (2025). Polysaccharides from echinoderms: unlocking health benefits and food applications – a review. *Food & Function*, 16(14), 5679–5704. https://doi.org/10.1039/D5FO02177D

34 Zhang, W., Wang, J., Jin, W., & Zhang, Q. (2013). The antioxidant activities and neuroprotective effect of polysaccharides from the starfish *Asterias rollestoni*. *Carbohydrate Polymers*, 95(1), 9–15.https://doi.org/10.1016/j.carbpol.2013.02.035

35 Shiao, M., Lee, K. R., Lin, L., & Wang, C. (1994). Natural Products and Biological Activities of the Chinese Medicinal Fungus *Ganoderma lucidum* American Chemical Society , Washington. 547, 342–354. https://doi.org/10.1021/bk-1994-0547.ch035

36 De Marino, S., Iorizzi, M., Zollo, F., Minale, L., Amsler, C. D., Baker, B. J., & Mcclintock, J. B. (1997). Isolation, structure elucidation, and biological activity of the steroid oligoglycosides and polyhydroxysteroids from the Antarctic starfish *Acodontaster conspicuus*. *Journal of Natural Products*, 60(10), 959–966. https://doi.org/10.1021/np9700578

37 Ha, D. T., Kicha, A. A., Kalinovsky, A. I., Malyarenko, T. V., Popov, R. S., Malyarenko, O. S., Ermakova, S. P., Thuy, T. T. T., Long, P. Q., & Ivanchina, N. V. (2021). Asterosaponins from the tropical starfish *Acanthaster planci* and their cytotoxic and anticancer activities in vitro. *Natural Product Research*, 35(4), 548-555. https://doi.org/10.1080/14786419.2019.1585845

38 Malyarenko, O. S., Malyarenko, T. V., Usoltseva, R. V., Silchenko, A. S., Kicha, A. A., Ivanchina, N. V., & Ermakova, S. P. (2021). Fucoidan from brown algae *Fucus evanescens* potentiates the anti-proliferative efficacy of asterosaponins from starfish

Asteropsis carinifera in 2D and 3D models of melanoma cells. *International Journal of Biological Macromolecules*, 185, 31–39. https://doi.org/10.1016/j.ij.iomac.2021.06.080

39 Menezes, C., & Thakur, N. L. (2022). Sea anemone venom: ecological interactions and bioactive potential. *Toxicon*, 208, 31–46. https://doi.org/10.1016/j.toxicon.2022.01.004

40 Tabakmakher, V. M., Monastyrnaya, M. M., Leichenko, E. V., Gladkikh, I. N., Zelepuga, E. A., Chausova, V. E., Tkacheva, E. S., Agafonova, I. G., & Kozlovskaya, E. P. (2013). Biologically active polypeptides of sea anemones: structure, function, and prospects for application. *Russian Journal of Marine Biology*, 39(5), 311–320. https://doi.org/10.1134/S1063074013050088

41 Bonaccorsi, G., Garamella, G., Cavallo, G., & Lorini, C. (2020). A systematic review of risk assessment associated with jellyfish consumption as a potential novel food. *Foods*, 9(7), 935. https://doi.org/10.3390/foods9070935

42 Raposo, A., Coimbra, A., Amaral, L., Gonçalves, A., & Morais, Z. (2022). Jellyfish as food: a narrative review. *Foods*, 11(18), 2773. https://doi.org/10.3390/foods11182773

43 Khong, N. M. H., Yusoff, F. M., Jamilah, B., Basri, M., Maznah, I., Chan, K. W., & Nishikawa, J. (2016). Nutritional composition and total collagen content of three commercially important edible jellyfish. *Food Chemistry*, 196, 953–960. https://doi.org/10.1016/j.foodchem.2015.09.094

44 Lv, Z., Zhang, C., Song, W., Chen, Q., & Wang, Y. (2022). Jellyfish collagen hydrolysate alleviates inflammation and oxidative stress and improves gut microbe composition in high-fat diet-fed mice. *Mediators of Inflammation*, 2022, 5628702. https://doi.org/10.1155/2022/5628702

45 Zhang, Q., Song, C., Zhao, J., Shi, X., Sun, M., Liu, J., Fu, Y., Jin, W., & Zhu, B. (2018). Separation and characterization of antioxidative and angiotensin converting enzyme inhibitory peptide from jellyfish gonad hydrolysate. *Molecules*, 23(1), 94. https://doi.org/10.3390/molecules23010094

46 Leone, A., Lecci, R. M., Milisenda, G., & Piraino, S. (2015). The bright side of gelatinous blooms: nutraceutical value and antioxidant properties of three Mediterranean jellyfish (Scyphozoa). *Marine Drugs*, 13(8), 4654–4681. https://doi.org/10.3390/md13084654

47 Costa, R., Capillo, G., Albergamo, A., Li Volsi, R., Bartolomeo, G., Bua, G., Ferracane, A., Savoca, S., Gervasi, T., Rando, R., Dugo, G., & Spanò, N. (2019). A multi-screening evaluation of the nutritional and nutraceutical potential of the Mediterranean jellyfish *Pelagia noctiluca*. *Marine Drugs*, 17(3), 172. https://doi.org/10.3390/md17030172

48 Cheng, X., Shao, Z., Li, C., Yu, L., Raja, M. A., & Liu, C. (2017). Isolation, characterization and evaluation of collagen from jellyfish *Rhopilema esculentum* Kishinouye for use in hemostatic applications. *PLOS ONE*, 12(1), e0169731. https://doi.org/10.1371/journal.pone.0169731

49 Liu, X. (2004). Determination of trace mercury in jellyfish by microwave digestion-atomic fluorescence spectrometry. *Journal of Instrumental Analysis* 23 (4): 101–103.

50 Summat, T., Wangtueai, S., You, S., Rod-in, W., Park, W. J., Karnjanapratum, S., Seesuriyachan, P., & Surayot, U. (2023). In vitro anti-inflammatory activity and structural characteristics of polysaccharides extracted from *Lobonema smithii* jellyfish. *Marine Drugs*, 21(11), 559. https://doi.org/https://doi.org/10.3390/md21110559

51 Matveeva, V., Shulgina, L., Prikhodko, Y., Shulgin, Y., Madej, K., & Piekoszewski, W. (2021). Nutritional value of sea urchin roe (Strongylocentrotidae)—study of composition and storage conditions. *Separations*, 8, 174. https://doi.org/10.3390/separations8100174

52 Takagi, S., Murata, Y., Inomata, E., Aoki, M. N., & Agatsuma, Y. (2019). Production of high quality gonads in the sea urchin *Mesocentrotus nudus* (A. Agassiz, 1864) from a barren by feeding on the kelp *Saccharina japonica* at the late sporophyte stage. *Journal of Applied Phycology*, 31(6), 4037–4048. https://doi.org/10.1007/s10.11-019-01895-6

53 Takagi, S., Murata, Y., Inomata, E., Endo, H., Aoki, M. N., & Agatsuma, Y. (2017). Improvement of gonad quality of the sea urchin *Mesocentrotus nudus* fed the kelp *Saccharina japonica* during offshore cage culture. *Aquaculture*, 477, 50–61. https://doi.org/10.1016/j.aquaculture.2017.04.033

54 Phillips, K., Niimi, J., Hamid, N., Silcock, P., Delahunty, C., Barker, M., . . . Bremer, P. (2010). Sensory and volatile analysis of sea urchin roe from different geographical regions in New Zealand. *LWT - Food Science and Technology*, 43(2), 202-213. https://doi.org/10.1016/j.lwt.2009.08.008

55 Phillips, K., Hamid, N., Silcock, P., Delahunty, C., Barker, M., & Bremer, P. (2010). Effect of season on the sensory quality of sea urchin (*Evechinus chloroticus*) roe. *Journal of Food Science*, 75(1), S20–30. https://doi.org/10.1111/j.1750-3841.2009.01388.x

56 Takagi, S., Sato, Y., Kokubun, A., Inomata, E., & Agatsuma, Y. (2020). Odor-active compounds from the gonads of *Mesocentrotus nudus* sea urchins fed *Saccharina japonica* kelp. *PLoS One*, 15(4), e0231673. https://doi.org/10.1371/journal.pone.0231673

57 Carvalho, A. M., Marques, A. P., Silva, T. H., & Reis, R. L. (2018). Evaluation of the potential of collagen from codfish skin as a biomaterial for biomedical applications. *Marine Drugs*, 16(12), 495. https://doi.org/10.3390/md16120495

58 Bonagurio, L. P., Murakami, A. E., Moreira, C. A., Comar, J. F., & Pozza, P. C. (2022). Dietary supplementation with inosine-5′-monophosphate improves the functional, energetic, and antioxidant status of liver and muscle growth in pigs. *Scientific Reports*, 12(1), 350. https://doi.org/10.1038/s41598-021-04023-y

59 Ling, Z. and Li, J. (2012). Flavor identification and analysis of Maillard reaction products from enzymatic hydrolysis of jellyfish skin. *Food Science* 33 (20): 257–261.

60 Youssef, J., Keller, S., & Spence, C. (2019). Making sustainable foods (such as jellyfish) delicious. *International Journal of Gastronomy and Food Science*, 16, 100141. https://doi.org/10.1016/j.ijgfs.2019.10.141

61 Hong, I., Khalid, A. W., Pae, H. C., Cha, J. K., Lee, J. S., Paik, J. W., Jung, U. W., & Choi, S. H. (2020). Distinctive bone regeneration of calvarial defects using biphasic

calcium phosphate supplemented ultraviolet-crosslinked collagen membrane. *Journal of Periodontal & Implant Science*, 50(1), 14–27. https://doi.org/10.50.1/jpis.2020.50.1.14

62 Łepecka, A., Szymański, P., Okoń, A., Siekierko, U., Zielińska, D., Trząskowska, M., Neffe-Skocińska, K., Sionek, B., Kajak-Siemaszko, K., Karbowiak, M., Kołożyn-Krajewska, D., & Do Dong, Z. J. (2023). The influence of the apple vinegar marination process on the technological, microbiological and sensory quality of organic smoked pork hams. *Foods*, 12(8), 1565. https://doi.org/10.3390/foods12081565

63 Lourenço, S., Valente, L. M. P., & Andrade, C. (2019). Meta-analysis on nutrition studies modulating sea urchin roe growth, colour and taste. *Reviews in Aquaculture*, 11(3), 766–781. https://doi.org/10.1111/raq.12256

64 Karnila, R., Iriani, D., Shaarani, S., Yunus, A., & Salma, R. (2022). Nutritional characteristics of sea urchin (*Diadema setosum*) in Bungus, West Sumatera Province. *IOP Conference Series: Earth and Environmental Science*, 1118, 012037. https://doi.org/10.1088/1755-1315/1118/1/012037

65 Padang, A., Nurlina, N., Tuasikal, T., & Subiyanto, R. (2019). Kandungan Gizi Bulu Babi (Echinoidea). *Agrikan: Jurnal Agribisnis Perikanan*, 12, 220. https://doi.org/10.29239/j.agrikan.12.2.220–227

66 Amarowicz, R., Synowiecki, J., & Shahidi, F. (1994). Sephadex LH-20 separation of pigments from shells of red sea urchin (*Strongylocentrotus franciscanus*). *Food Chemistry*, 51(2), 227–229. https://doi.org/10.1016/0308-8146(94)90262-3

67 Kuwahara, R., Hatate, H., Chikami, A., Murata, H., & Kijidani, Y. (2010). Quantitative separation of antioxidant pigments in purple sea urchin shells using a reversed-phase high performance liquid chromatography. *LWT - Food Science and Technology*, 43(8), 1185–1190. https://doi.org/10.1016/j.lwt.2010.03.005

68 Kuwahara, R., Hatate, H., Yuki, T., Murata, H., Tanaka, R., & Hama, Y. (2009). Antioxidant property of polyhydroxylated naphthoquinone pigments from shells of purple sea urchin *Anthocidaris crassispina*. *LWT - Food Science and Technology*, 42(7), 1296–1300. https://doi.org/10.1016/j.lwt.2009.02.020

69 Symonds, R. C., Kelly, M. S., Caris-Veyrat, C., & Young, A. J. (2007). Carotenoids in the sea urchin *Paracentrotus lividus*: occurrence of 9′-cis-echinenone as the dominant carotenoid in gonad colour determination. *Comparative Biochemistry and Physiology Part B: Biochemistry and Molecular Biology*, 148(4), 432–444. https://doi.org/10.1016/j.cbpb.2007.07.012

70 Tsushima, M., & Matsuno, T. (1990). Comparative biochemical studies of carotenoids in sea-urchins—I. *Comparative Biochemistry and Physiology Part B: Comparative Biochemistry*, 96(4), 801–810. https://doi.org/10.1016/0305-0491(90)90235-L

71 Vinnikova, V. V., & Drozdov, A. L. (2011). The ultrastructure of spines in sea urchins of the family Strongylocentrotidae. *Biology Bulletin*, 38(9), 861–867. https://doi.org/10.1134/S1062359011090093

72 Griffiths, M., & Perrott, P. (1976). Seasonal changes in the carotenoids of the sea urchin *Strongylocentrotus dröbachiensis*. *Comparative Biochemistry and Physiology Part B: Comparative Biochemistry*, 55(3), 435–441. https://doi.org/10.1016/0305-0491(76)90317-5

73 McBride, S. C., Price, R. J., Tom, P. D., Lawrence, A. L., & Lawrence, J. M. (2004). Comparison of gonad quality factors: color, hardness and resilience, of *Strongylocentrotus franciscanus* between sea urchins fed prepared feed or algal diets and sea urchins harvested from the Northern California fishery. *Aquaculture*, 233(1),405–422. https://doi.org/10.1016/j.aquaculture.2003.10.014

74 Murata, Y., Sata, N. U., Yokoyama, M., Kuwahara, R., Kaneniwa, M., & Oohara, I. (2001). Determination of a novel bitter amino acid, pulcherrimine, in the gonad of the green sea urchin *Hemicentrotus pulcherrimus*. *Fisheries Science*, 67(2), 341–345. https://doi.org/10.1046/j.1444-2906.2001.00234.x

75 Takagi, S., Murata, Y., Inomata, E. et al. (2018). Dietary effect of kelp (*Saccharina japonica*) on gonad quantity and quality in sea urchins (*Mesocentrotus nudus*) collected from a barren before the fishing season. *Journal of Shellfish Research* 37: 659–669.

76 Zhou, D., Qin, L., Zhu, B., Li, D., Yang, J., Dong, X., & Murata, Y. (2012). Optimisation of hydrolysis of purple sea urchin (Strongylocentrotus nudus) gonad by response surface methodology and evaluation of in vitro antioxidant activity of the hydrolysate. *Journal of the Science of Food and Agriculture*, 92(8), 1694–1701. https://doi.org/10.1002/jsfa.5534

77 Shpigel, M., Schlosser, S. C., Ben-Amotz, A., Lawrence, A. L., & Lawrence, J. M. (2006). Effects of dietary carotenoid on the gut and the gonad of the sea urchin *Paracentrotus lividus*. *Aquaculture*, 261(4), 1269–1280. https://doi.org/10.1016/j.aquaculture.2006.08.029 [78]

78 Woods, C. M. C., James, P. J., Moss, G. A., Wright, J., & Siikavuopio, S. (2008). A comparison of the effect of urchin size and diet on gonad yield and quality in the sea urchin *Evechinus chloroticus* Valenciennes. *Aquaculture International*, 16(1), 49–68. https://doi.org/10.1007/s10499-007-9124-z

79 Walker, C.W., Böttger, S.A., Unuma, T. et al. (2015). Enhancing the commercial quality of edible sea urchin gonads — technologies emphasizing nutritive phagocytes. In: *Echinoderm Aquaculture*, 263–286. Hoboken: Wiley Blackwell.

80 Agatsuma, Y., Sato, M., & Taniguchi, K. (2005). Factors causing brown-colored gonads of the sea urchin *Strongylocentrotus nudus* in northern Honshu, *Japan. Aquaculture*, 249(1), 449–458. https://doi.org/10.1016/j.aquaculture.2005.04.054

81 Davy, S. K., Lucas, I. A. N., & Turner, J. R. (1997). Uptake and persistence of homologous and heterologous zooxanthellae in the Temperate Sea Anemone *Cereus pedunculatus* (Pennant). *Biological Bulletin*, 192(2), 208–216. https://doi.org/10.2307/1542715

82 Mcfarlane, I. D. (1974). Excitatory and inhibitory control of inherent contractions in the sea anemone *Calliactis parasitica*. *Journal of Experimental Biology*, 60(2), 397–422. https://doi.org/10.1242/jeb.60.2.397

83 Shostak, S. (2021). *Cnidaria (Coelenterates)*, 853–872, Wiley. https://doi.org/10.1002/9780470015902.a0029250

84 Miles, J. G., & Battista, N. A. (2019). Naut your everyday jellyfish model: Exploring how tentacles and oral arms impact locomotion. *Fluids*, 4(3), 169. https://doi.org/10.3390/fluids4030169

85 Li, H., Cui, S., Wang, X., Li, Y., & Zhang, S. (2025). Development of a bionic jellyfish robot for collecting polymetallic nodules. *Ocean Engineering*, 324, 120655. https://doi.org/10.1016/j.oceaneng.2024.120655

86 Wang, X., Wang, H., & Brown, H. R. (2011). Jellyfish gel and its hybrid hydrogels with high mechanical strength. *Soft Matter*, 7(1), 211–219. https://doi.org/10.1039/C0SM00632G

87 Smith, I., Sotelo, C. G., Pérez-Martín, R. I., & Silva, T. H. (2023). Characterization of the biophysical properties and cell adhesion interactions of marine invertebrate collagen from *Rhizostoma pulmo*. *Marine Drugs*, 21(2), 59. https://doi.org/10.3390/md21020059

88 Liu, H., Wang, C., Li, C., Qin, Y., Wang, Z., Yang, F., Li, Z., & Wang, J. (2018). A functional chitosan-based hydrogel as a wound dressing and drug delivery system in the treatment of wound healing. *RSC Advances*, 8(14), 7533–7549. https://doi.org/10.1039/C7RA13510F

89 Bleve, G., Ramires, F. A., De Domenico, S., & Leone, A. (2021). An alum-free jellyfish treatment for food applications. *Frontiers in Nutrition*, 8, 718798. https://doi.org/10.3389/fnut.2021.718798

90 Upata, M., Siriwoharn, T., Makkhun, S., & Yarnpakdee, S. (2022). Tyrosinase inhibitory and antioxidant activity of enzymatic protein hydrolysate from jellyfish (*Lobonema smithii*). *Foods*, 11(4), 615. https://doi.org/10.3390/foods11040615

91 Rastogi, A., Bhowmick, R., & Pandey, D. K. (2012). Anticoagulant activity of moon jellyfish (*Aurelia aurita*) tentacle extract. *Toxicon*, 60(4), 719–723. https://doi.org/10.1016/j.toxicon.2012.05.008

92 Zare, A., Ferdowsi, S., Ghavam, M., & Moghaddam, F. D. (2023). Chemical compositions and experimental and computational modeling of the anticancer effects of cnidocyte venoms of jellyfish *Cassiopea andromeda* and *Catostylus mosaicus* on human adenocarcinoma A549 cells. *Marine Drugs*, 21(3), 168. https://doi.org/10.3390/md21030168

93 Sun, J. and Chiang, F.S. (2015). Use and exploitation of sea urchins. In: *Echinoderm Aquaculture* (ed. N.P. Brown and S.D. Eddy), 25–45. Hoboken: John Wiley & Sons.

94 James, P., Evensen, T., Samuelsen, A. (2017). Commercial scale sea urchin roe enhancement in Norway: Enhancement, transport and market assessment.

95 Stefansson, G., Kristinsson, H., Ziemer, N., et al. (2017). Markets for sea urchins: a review of global supply and markets.

96 Andrew, N. L., Agatsuma, Y., Ballesteros, E., Bazhin, A. G., Creaser, E. P., Barnes, D., . . . Xiaoqi, Z. (2002). Status and Management of World Sea Urchin Fisheries. *Oceanography and Marine Biology - An Annual Review*, 40, 343–425. https://doi.org/10.1201/9780203180594.ch7

97 Suckling, C. C. (2021). Responses to environmentally relevant microplastics are species-specific with dietary habit as a potential sensitivity indicator. *Science of the Total Environment*, 751, 142341. https://doi.org/10.1016/j.scitotenv.2020.142341

98 Thangaraj, S., Bragadeeswaran, S., & Gokula, V. (2019). Bioactive compounds of sea anemones: a review. *International Journal of Peptide Research and Therapeutics*, 25(4), 1405–1416. https://doi.org/10.1007/s10989-018-9786-6

99 Cadar, E., Pesterau, A. M., Sirbu, R., Negreanu-Pirjol, B. S., & Tomescu, C. L. (2023). Jellyfishes—significant marine resources with potential in the wound-healing process: a review. *Marine Drugs*, 21(4), 201. https://doi.org/10.3390/md21040201

100 Hu, B., Zong, Z., Han, L., Cao, J., Yang, J., Zheng, Q., Zhang, X., Liu, Y., & Yao, Z. (2025). Jellyfish collagen: a promising and sustainable marine biomaterial with emerging applications in food, cosmetics, and biomedical—a review. *Applied Food Research*, 5(2), 101165. https://doi.org/10.1016/j.afres.2025.10.165

101 Hsieh, Y. H. P., & Rudloe, J. (1994). Potential of utilizing jellyfish as food in Western countries. *Trends in Food Science & Technology*, 5(7), 225–229. https://doi.org/10.1016/0924-2244(94)90253-4

102 Ramires, F. A., Bleve, G., De Domenico, S., & Leone, A. (2022). Combination of solid state and submerged fermentation strategies to produce a new jellyfish-based food. *Foods*, 11(24), 3974. https://doi.org/10.3390/foods11243974

103 Guttuso, P., Nogueira, N., Gueroun, S. K. M., & Akyol, O. (2025). Is jellyfish a suitable ingredient for aquafeed? A comprehensive review of nutritional potential and limitation. *Frontiers in Marine Science*, 12, 1539725. https://doi.org/10.3389/fmars.2025.1539725

104 Ma, Y., Yu, H., Xing, R., Liu, S., & Li, P. (2023). Lipid-lowering activity and mechanism of peptides from jellyfish *Nemopilema nomurai*. *Journal of Functional Foods*, 101, 105421. https://doi.org/10.1016/j.jff.2023.105421

105 Selvamuthukumaran, M., Khanum, F., & Bawa, A. S. (2007). Development of sea buckthorn mixed fruit jelly. *International Journal of Food Science & Technology*, 42(4), 403–410. https://doi.org/10.1111/j.1365-2621.2006.01251.x

106 Ranasinghe, R. A. S. N., Wijesekara, W. L. I., Senanayake, S. A., & Marapana, R. A. U. J. (2025). The impact of in vitro gastrointestinal hydrolysis on the antioxidant activity of gelatin extracted from jellyfish *Acromitus flagellatus*. *Journal of Food Science and Technology. Advance online publication.* https://doi.org/10.1007/s13197-024-06035-6

107 Flaig, I., Radenković, M., Najman, S., Pröhl, A., Jung, O., & Barbeck, M. (2020). In vivo analysis of the biocompatibility and immune response of jellyfish collagen scaffolds and its suitability for bone regeneration. *International Journal of Molecular Sciences*, 21(12), 4518. https://doi.org/10.3390/ijms21124518

108 Pesterau, A. M., Popescu, A., Sirbu, R., Cadar, E., Busuricu, F., Dragan, A. M. L., Pascale, C., Ionescu, A. M., Bogdan-Andreescu, C. F., Radu, M.-D., & Tomescu, C. L. (2025). Marine jellyfish collagen and other bioactive natural compounds from the sea, with significant potential for wound healing and repair materials. *Marine Drugs*, 23(6), 252. https://doi.org/10.3390/md23060252

109 Zhuang, Y., Sun, L., & Li, B. (2012). Production of the angiotensin-I-converting enzyme (ACE)-inhibitory peptide from hydrolysates of jellyfish (*Rhopilema esculentum*) collagen. *Food and Bioprocess Technology*, 5(5), 1622–1629. https://doi.org/10.1007/s11947-010-0439-9

110 Barzideh, Z., Latiff, A.A., Gan, C.-Y. et al. (2014). ACE inhibitory and antioxidant activities of collagen hydrolysates from the ribbon jellyfish (*Chrysaora* sp.). *Food Technology and Biotechnology* 52 (4): 495–504. https://doi.org/10.17113/ftb.52.04.14.3641.

111 Geng, H., Li, R., Teng, L., Yu, C., Wang, W., Gao, K., Li, A., Liu, S., Xing, R., Yu, H., & Li, P. (2024). Exploring the efficacy of hydroxybenzoic acid derivatives in mitigating jellyfish toxin-induced skin damage: Insights into protective and reparative mechanisms. *Marine Drugs*, 22(5), 205. https://doi.org/10.3390/md22050205

112 Zhuang, Y., Hou, H., Zhao, X., Zhang, Z., & Li, B. (2009). Effects of collagen and collagen hydrolysate from jellyfish (*Rhopilema esculentum*) on mice skin photoaging induced by UV irradiation. *Journal of Food Science*, 74(6), H183–H188. https://doi.org/10.1111/j.1750-3841.2009.01236.x

113 Carvalho, D. N., Williams, D. S., Sotelo, C. G., Pérez-Martín, R. I., Mearns-Spragg, A., Reis, R. L., & Silva, T. H. (2022). Marine origin biomaterials using a compressive and absorption methodology as cell-laden hydrogel envisaging cartilage tissue engineering. *Biomaterials Advances*, 137, 212843. https://doi.org/10.1016/j.bioadv.2022.212843

114 Xu, X. F., Dai, M. Q., Yan, J. N., Du, Y. N., Wang, C., Lai, B., & Wu, H. T. (2025). Comparison of the properties of composite films constructed from chitosan and gelatin from two jellyfish species. *Food Hydrocolloids*, 167, 111412. https://doi.org/10.1016/j.foodhyd.2025.111412

115 Tang, P., Ji, B., & Sun, G. (2022). Enhanced physicochemical and functional properties of collagen films cross-linked with laccase oxidized phenolic acids for active edible food packaging. *Food Chemistry*, 393, 133353. https://doi.org/10.1016/j.foodchem.2022.133353

116 Azaza, Y. B., Ben Slimane, S., Abdelhedi, O., Nasri, M., & Jridi, M. (2023). Development and characterization of composite films based on chitosan and collagenous proteins from bluefin tuna: application for peeled shrimp preservation. *Cellulose*, 30(1), 373–395. https://doi.org/10.1007/s10570-022-04893-z

117 D'Ambra, I., & Merquiol, L. (2022). Jellyfish from fisheries by-catches as a sustainable source of high-value compounds with biotechnological applications. *Marine Drugs*, 20(4), 266. https://doi.org/10.3390/md20040266

118 Ranasinghe, R. A. S. N., Wijesekara, W. L. I., Perera, P. R. D., Senanayake, S. A., Pathmalal, M. M., & Marapana, R. A. U. J. (2022). Nutritional value and potential applications of jellyfish. *Journal of Aquatic Food Product Technology*, 31(5), 445–482. https://doi.org/10.1080/10498850.2022.2060717

9

Health Benefits of Aquatic Products

Zwe-Ling Kong, Jerrell Felim, and Delvina

Taiwan Ocean University, Keelung, China

Aquatic foods, a diverse category encompassing fish, shellfish, and seaweed, have long been recognized across various cultures as indispensable components of the human diet. They provide a wealth of high-quality protein, essential lipids, and a wide array of fundamental micronutrients for maintaining and promoting human health and overall well-being [1]. This recognition is deeply rooted in the nutritional profiles of these foods, which offer a unique combination of elements vital for supporting bodily functions and preventing disease.

Aquatic foods possess the inherent potential to be an indispensable part of a balanced and health-conscious dietary regimen, benefiting individual health and aligning with principles of environmental stewardship, social equity, and long-term economic viability [2–4]. However, their critical role within the larger framework of sustainable food systems is frequently underestimated or inadequately recognized by policymakers, researchers, and consumers alike. This underappreciation may hinder the development of comprehensive strategies that fully leverage the potential of aquatic resources for improving global nutrition and health [5].

While aquatic foods are widely appreciated as excellent sources of dietary protein, their nutritional contributions extend far beyond this macronutrient. They are particularly valuable for their rich content of other essential nutrients that are often challenging to obtain in sufficient quantities from different food groups [6]. This includes long-chain omega-3 fatty acids, crucial for various physiological functions, and a diverse array of micronutrients, such as iodine, selenium, calcium, iron, and zinc, all of which play vital roles in maintaining optimal human health [7].

Compared to the production of most terrestrial animal-source foods, the farming and harvesting of aquatic foods generally have a lower environmental impact and a smaller carbon footprint. This makes aquatic food systems a more sustainable option for meeting the growing global demand for nutritious animal-source foods while minimizing ecological consequences and contributing to a more environmentally responsible food supply [6].

Aquatic animals, particularly fish, are highly nutritious and often economically accessible sources of protein. They serve as valuable dietary supplements by providing essential vitamins, proteins, micronutrients, and minerals, particularly for individuals from lower socioeconomic backgrounds. Their affordability and nutritional density make them a crucial food source for vulnerable populations [8].

Aquaculture has evolved into one of the key sectors within global fisheries, making a significant contribution to food security worldwide by increasing the sustainable supply of animal protein and supporting nutritional needs amid a growing global population. Its role in meeting the increasing demand for aquatic foods is becoming ever more important [3, 9].

The Food and Agriculture Organization (FAO), through its various reports, including "The State of World Fisheries and Aquaculture" (SOFIA), highlights the increasing global consumption of aquatic foods and underscores their significant potential in effectively addressing the pressing global issues of food insecurity and malnutrition that affect numerous populations worldwide. These reports provide valuable data and insights into the role of aquatic foods in global food systems [2, 10].

Numerous studies have consistently demonstrated that incorporating seafood into the diet can significantly reduce the risk of developing primary health conditions such as heart attack, stroke, obesity, and hypertension. These findings highlight the profound impact of aquatic foods on overall health and disease prevention. These findings underscore the importance of promoting seafood consumption for its wide-ranging health benefits [11–16].

9.1 Nutrient Density

9.1.1 Definition of Nutrient Density and Its Importance

Nutrient-dense foods are defined as those that provide a high amount of nutrients, such as vitamins, minerals, complex carbohydrates, lean protein, and healthy fats, relative to their calorie content. This concept is crucial in nutrition as it emphasizes the quality of calories consumed rather than just the quantity [17]. Foods high in nutrient density allow individuals to meet their nutritional needs without exceeding their energy requirements, which is essential for maintaining a healthy weight and preventing nutrient deficiencies [18].

The significance of nutrient density lies in its ability to help individuals avoid nutrient deficiencies, which can lead to a range of serious health problems, including anemia and impaired vision. By prioritizing nutrient-dense foods, individuals can ensure they are obtaining the necessary vitamins and minerals for proper physiological function without consuming excessive calories that could contribute to weight gain and other health issues [18–20].

9.1.2 Key Nutrients in Aquatic Products

Aquatic products, including fish, shellfish, and seaweed, are renowned for their rich and diverse array of key nutrients that contribute significantly to human health. Among these, omega-3 fatty acids (eicosapentaenoic acid [EPA] and docosahexaenoic acid [DHA]) stand out for their essential role in multiple physiological functions. These long-chain polyunsaturated fatty acids (PUFAs) are vital for cardiovascular health, brain development, and reducing inflammation throughout the body. Fatty fish, such as salmon, mackerel, and sardines, are particularly rich in EPA and DHA [21, 22].

Aquatic products are also excellent sources of high-quality protein, peptides, and various other bioactive compounds [23–25]. The protein found in fish and shellfish is easily digestible and contains all the essential amino acids the human body requires [26]. Furthermore, the digestion of these proteins yields peptides, some of which possess bioactive properties such as antihypertensive, antioxidant, and anti-inflammatory effects [27–29] (Fig. 9.1). Compared to terrestrial protein sources, aquatic proteins often exhibit higher digestibility and a more favorable amino acid profile, particularly in terms of essential amino acids such as lysine and methionine [30].

Polysaccharides are another important group of bioactive compounds found in aquatic foods, especially in seaweed [31]. These complex carbohydrates exhibit various biological activities, including antioxidant, anticoagulant, antiviral, anti-inflammatory, and antitumor properties [32]. Different types of seaweed, such as brown, green, and red algae, contain various polysaccharides, including fucoidan, alginate, carrageenan, and ulvan, each with unique potential health benefits [33–36] (Fig. 9.2).

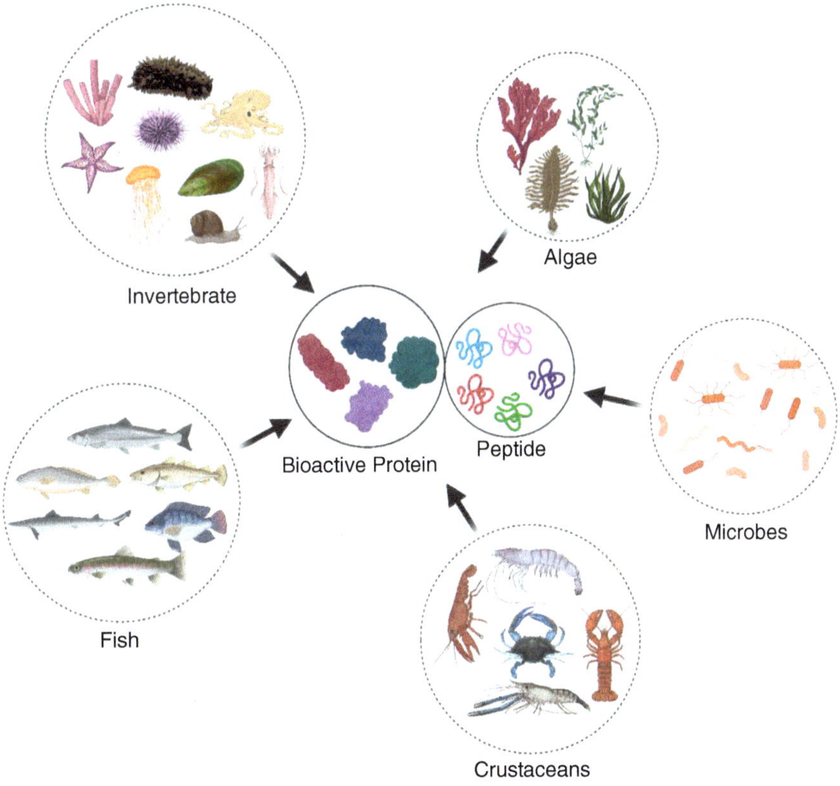

Figure 9.1 Major groups of marine organisms as sources of bioactive proteins and peptides. *Source:* [25]/ MDPI / CC BY 4.0.

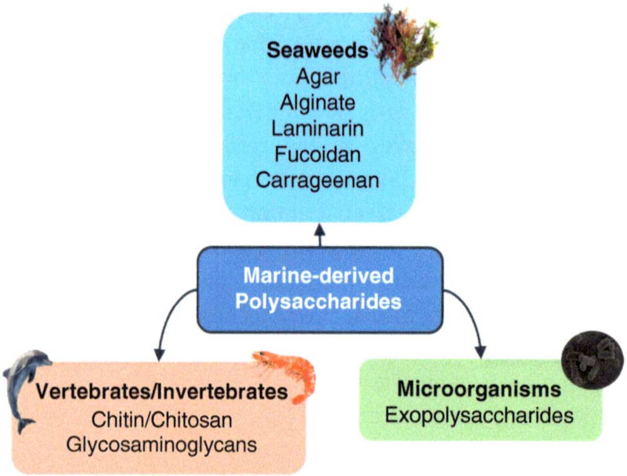

Figure 9.2 Sources of marine polysaccharides. *Source:* [37]/ MDPI / CC BY 4.0.

In addition to these major nutrient categories, aquatic products are rich in other key bioactive compounds that contribute to their overall health-promoting effects. These include astaxanthin, a powerful antioxidant found in salmon and shrimp [38]; phospholipids, essential components of cell membranes present in various seafoods [39]; choline, important for brain health and found in fish and shellfish [40]; glucosamine, known for its joint health benefits and often derived from shellfish [41]; squalene, an antioxidant found in shark liver oil [145]; coenzyme Q10, vital for energy production and present in fatty fish [42]; chitosan, derived from shellfish and studied for its potential effects on weight loss and cholesterol levels [43]; and taurine, an amino acid abundant in seafood with potential cardiovascular benefits [44].

Aquatic products are also significant sources of essential vitamins and minerals. Vitamin D, crucial for bone health, is found in fatty fish. B-complex vitamins, particularly vitamin B_{12}, which is essential for nerve function, are abundant in fish and shellfish. Iodine, vital for thyroid function, is highly concentrated in seafood, especially seaweed. Selenium, an antioxidant that supports thyroid health, is present in various types of seafood. Zinc, which is important for immune function, is found in high levels in shellfish such as oysters [45–49].

9.2 Brain Health

9.2.1 Key Bioactive Components in Aquatic Products for Brain Health

Nutrition and the human brain are closely linked, with increasing awareness of the impact of diet on cognitive function and neurological health [50]. As the global population ages, there is a growing interest in natural compounds that can improve brain function, cognitive performance, and protect against neurological disorders [51].

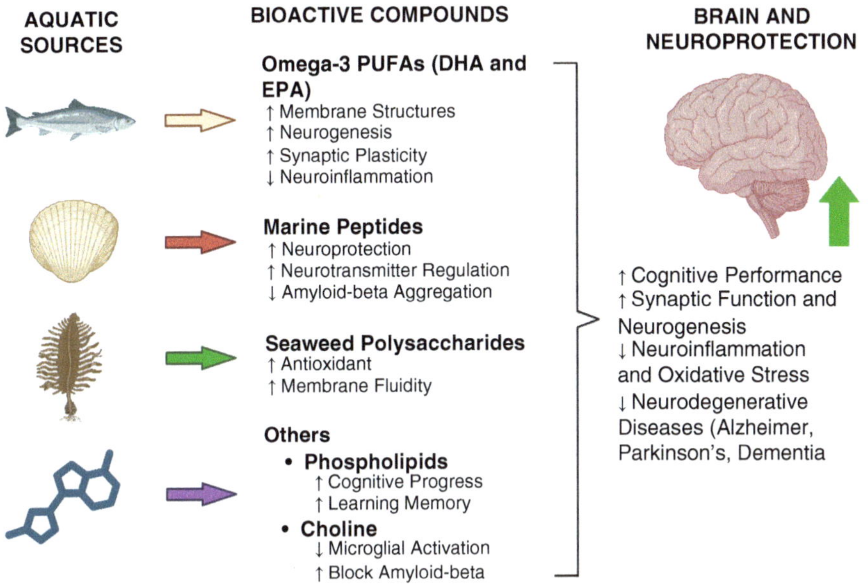

Figure 9.3 Aquatic bioactives—including omega-3 PUFAs, marine peptides, seaweed polysaccharides, and phospholipid derivatives—exert multifaceted neuroprotective effects through modulation of membrane dynamics, neurogenesis, synaptic plasticity, and inflammation. These mechanisms converge to support cognitive performance and mitigate neurodegenerative processes such as Alzheimer's and Parkinson's disease.

Aquatic products, including fish, shellfish, algae, and seaweed, offer a rich reservoir of bioactive components. This report provides an overview of key bioactive compounds found in aquatic products and their roles in promoting brain health (Fig. 9.3).

9.2.2 Omega-3 Polyunsaturated Fatty Acids (DHA and EPA)

9.2.2.1 Structural Component of Brain Cell Membranes and Fluidity

DHA and EPA, both long-chain omega-3 PUFAs (Fig. 9.4), are essential for brain function in adults [53]. These fatty acids are fundamental building blocks for brain cells and are integral to the structure of cell membranes [54]. DHA is the most abundant omega-3 PUFA in the brain, constituting over 90% of the n−3 PUFAs present and approximately 10%–20% of the brain's total lipid content [55]. Research has shown a correlation between dietary imbalances in omega-3 fatty acids and impaired brain function, as well as increased susceptibility to cognitive disorders [56].

The human body's ability to convert plant-derived alpha-linolenic acid (ALA) into DHA is limited, and the formation of DHA from EPA is preferentially utilized for the structural development of the brain, particularly during pregnancy and extending into the third decade of life [57, 58]. Studies have demonstrated that lower levels of DHA in the blood of older adults are associated with smaller brain size, a recognized sign of accelerated brain aging [55, 56].

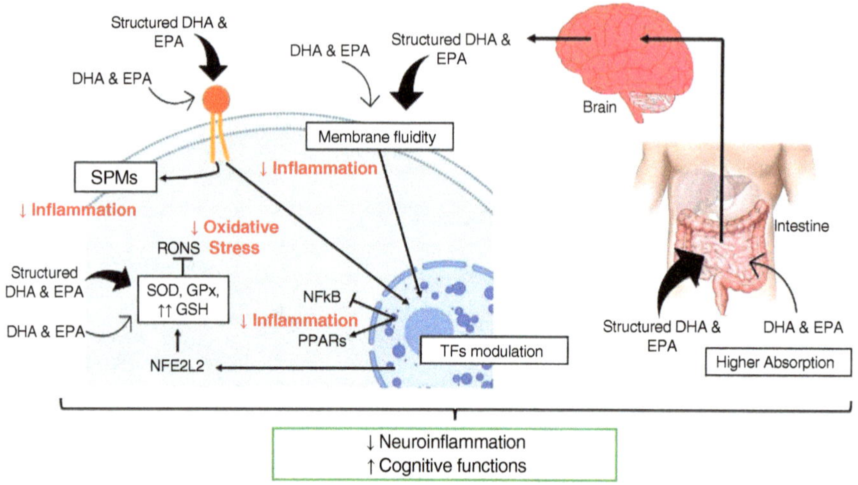

Figure 9.4 Schematic showing how ω-3 LCPUFAs (DHA and EPA) may reduce neuroinflammation and improve cognition by regulating oxidative stress and inflammation in brain cells; thick arrows highlight the stronger effects of their structured forms via transcription factor modulation. *Source:* [52]/ MDPI / CC BY 4.0.

9.2.2.2 Role in Neurogenesis and Synaptic Plasticity

EPA and DHA are essential nutrients that enhance brain function by supporting neurogenesis, synaptic plasticity, and neurotransmission within the central nervous system [52, 59]. DHA promotes neuronal differentiation, neurite outgrowth, and the formation of pre- and postsynaptic proteins, which are crucial for enabling synaptic transmission [60–62].

Omega-3 fatty acids enhance long-term potentiation and regulate the expression of synaptic proteins, promoting dendritic arborization and the formation of new spines. DHA also increases drebrin and PSD95 expression, promoting the structural and functional integrity of synapses [63–65]. Moreover, Omega-3 PUFAs, including DHA, may exert neurological benefits through activating specific receptors such as GPR120 and GPR40, as well as PPARs [58, 59].

9.2.2.3 Impact on Cognitive Performance, Memory, and Learning

DHA is a crucial component of human intelligence, essential for maintaining quality of life, enhancing productivity, and promoting societal progress. However, modern dietary patterns often lead to insufficient intake [66]. Studies have shown a correlation between DHA levels and cognitive function in healthy adults, with higher levels associated with improved nonverbal reasoning, mental flexibility, working memory capacity, and vocabulary skills. EPA and omega-3 fatty acids are associated with delayed recall, processing speed, and brain entorhinal cortex thickness [67]. Ingestion of these fatty acids can enhance learning capacity, improve memory function, promote cognitive well-being, and increase blood flow to the brain.

9.2.2.4 Reduction of Neuroinflammation

EPA and DHA, omega-3 fatty acids, have been found to reduce neuroinflammation in the brain, a common factor in neurological disorders and aging [55]. EPA may reverse the pro-inflammatory profile of cells from Alzheimer's disease (AD) patients and impact mood disorders. DHA inhibits NFκB activation in microglial cells, reducing pro-inflammatory cytokines and oxidative stress [68–70]. It also serves as a precursor for specialized pro-resolving mediators, orchestrating the inflammatory response without suppressing the immune system [68].

9.2.3 Marine-Derived Peptides

9.2.3.1 Neuroprotective Effects

Marine bioactive compounds, particularly peptides, have shown significant neuroprotective effects, including antioxidant, anti-neuroinflammatory, and cholinesterase-inhibitory activities [71]. These peptides have been shown to exhibit higher bioactivity than their terrestrial counterparts and have been found to regulate the balance of free radicals, contributing to antiaging effects in brain health models [72, 73]. Fish peptides may counteract memory impairment by reducing apoptosis, suppressing the production of reactive oxygen species (ROS), and lowering calcium influx [74]. These marine-derived biocompounds have been explored for their neuroprotective capabilities.

9.2.3.2 Potential Role in Reducing Amyloid-Beta Aggregation in Alzheimer's Disease, Dementia, Parkinson's

The self-assembly of peptides and proteins into amyloid fibrils is a hallmark of several diseases, including AD [76]. Marine natural products, including peptides and peptide-like molecules, have shown potential in multi-target interventions in AD, including modulating Aβ aggregation [75] (Fig. 9.5). Preliminary studies

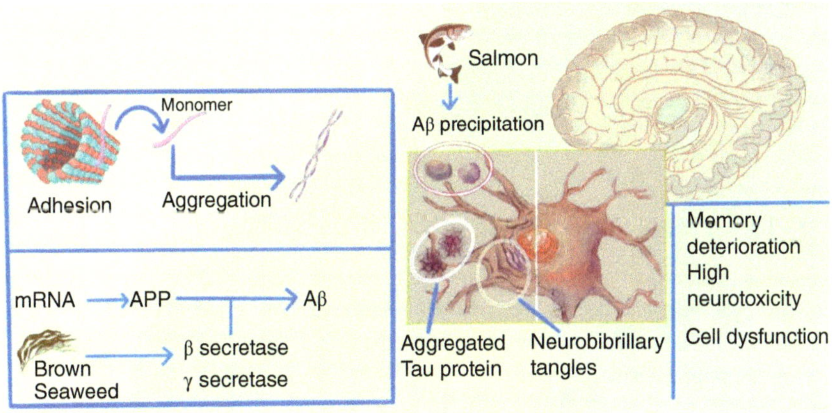

Figure 9.5 Mechanistic overview of Aβ aggregation modulation by bioactive compounds from marine organisms. *Source:* [75]/ MDPI / CC BY 4.0.

suggest that plasma exchange, combined with therapeutic albumin replacement, can induce the mobilization of amyloid β protein, thereby improving memory and language functions [77]. Collagen type VI, a protein found in the extracellular matrix, has been shown to have protective effects against Aβ proteins in both mouse and human brain tissue [78].

9.2.3.3 Improvement in Neurotransmitter Regulation

Marine bioactive peptides have been found to modulate neurotransmitter systems in the central nervous system, interacting with receptors and enzymes involved in neurotransmission [79]. Taurine, a neuromodulator found in scallops, mussels, and clams, plays a role in signal transduction and calcium movement [80–82]. Marine collagen peptides (MCPs) have been shown to facilitate long-term learning and memory in rats with perinatal asphyxia by reducing acetylcholinesterase activity in the brain. Some marine peptides exhibit opioid-like effects, potentially influencing motivation, emotion, behavior, stress, appetite, and pain management. Examples include ω-conotoxin MVIIA, which blocks N-type voltage-sensitive calcium channels, and α-conotoxin TxIB, which blocks α6/α3β2β3 nicotinic acetylcholine receptors, which are potential therapeutic targets for conditions like addiction and Parkinson's disease [25]. Oyster peptides have been shown to promote sleep by increasing inhibitory neurotransmitters, such as 5-hydroxytryptamine and γ-aminobutyric acid, while decreasing excitatory neurotransmitters, including dopamine and norepinephrine [83].

9.2.4 Seaweed-Derived Polysaccharides

9.2.4.1 Fucoidan, Laminarin, and Alginate

Seaweed compounds, including polysaccharides such as fucoidan, laminarin, and alginate, have been found to have potential neuroprotective effects against neurodegenerative diseases, including Alzheimer's and Parkinson's [85]. These compounds have antioxidant and anti-inflammatory properties, which can reduce oxidative stress and inflammation in the brain (Fig. 9.6). Fucoidan, a sulfated polysaccharide found in brown seaweed, has been shown to inhibit tau protein formation and reduce Aβ neurotoxicity [84]. Ulvan, an acidic polysaccharide from green algae, has been shown to inhibit Aβ fibrogenesis and Aβ oligomers, potentially mitigating Aβ-related pathology in AD [75].

9.2.5 Other Key Bioactive Compounds for Brain Health

9.2.5.1 Astaxanthin

9.2.5.1.1 Anti-inflammatory and Antioxidant Properties for Neuroprotection Astaxanthin, a carotenoid pigment found in marine organisms, is a potent antioxidant with anti-inflammatory properties [86]. It can cross the blood-brain barrier, modulating neuroinflammation and influencing adult neurogenesis. Astaxanthin's potency enables it to neutralize free radicals and promote autophagy, a crucial pathway in combating neurodegeneration [87].

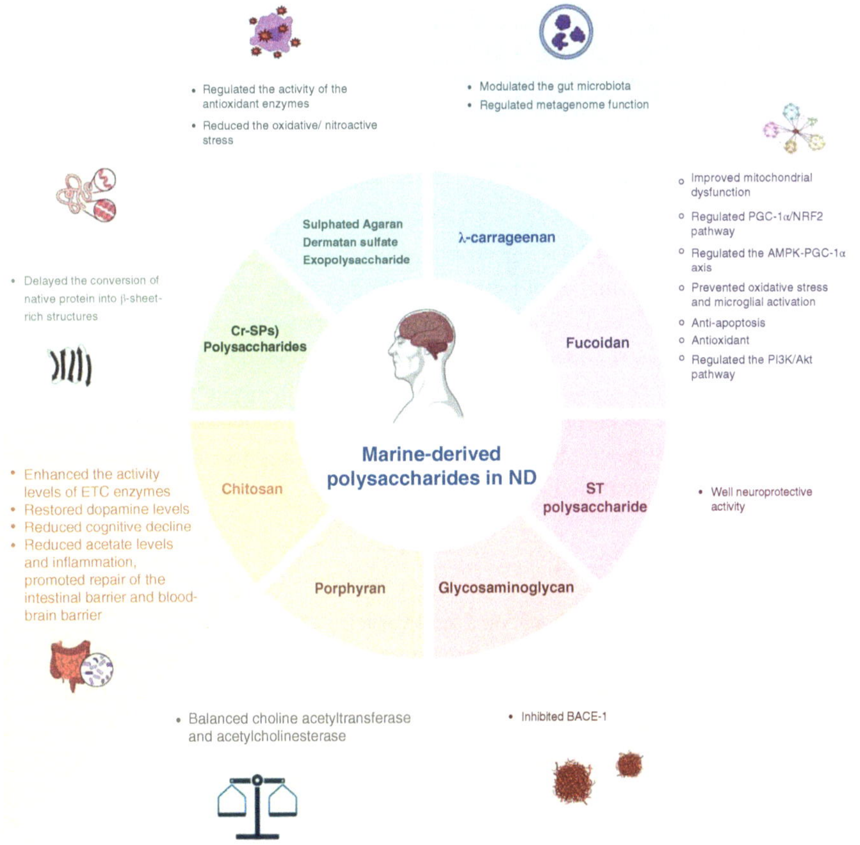

Figure 9.6 Mechanistic snapshot of marine polysaccharides in neurodegeneration, showing gut–brain axis effects, barrier restoration, oxidative stress reduction, and anti-amyloidogenic activity. *Source:* [84]/ Frontiers Media SA / CC BY 4.0.

9.2.5.2 Phospholipids

Phospholipids, a class of lipids that form the foundation of cell membranes, are crucial for maintaining the structural integrity and fluidity of brain cells [88]. Key phospholipids, including phosphatidylcholine, phosphatidylserine, and phosphatidylethanolamine, play specific roles in brain function, such as stimulating neurotransmitter production, maintaining membrane fluidity, and supporting energy production [89]. Dietary phospholipids have been shown to enhance various cognitive processes in both animals and humans, including learning and memory, by undergoing continuous turnover to repair and maintain cell membranes [90].

9.2.5.3 Choline

Choline is a crucial nutrient precursor to acetylcholine, a neurotransmitter involved in brain functions like memory, learning, and attention. It is essential for optimal health and cognitive performance [40, 91]. Choline influences stem cell proliferation, apoptosis, and memory function during fetal development. It is also crucial for the development of the hippocampus and frontal cortex. Lifelong choline supplementation may help prevent AD by blocking the production of amyloid-beta plaques and reducing microglial activation [92].

9.3 Heart Health

Cardiovascular diseases (CVDs) are the leading cause of mortality worldwide, and dietary interventions utilizing bioactive compounds from aquatic environments are increasingly being explored [93]. These bioactive compounds, including omega-3 fatty acids, marine peptides, seaweed polysaccharides, and antioxidants, have demonstrated beneficial effects on cardiovascular health, such as reducing inflammation, improving lipid profiles, regulating blood pressure, and inhibiting thrombosis [94–97]. Understanding these mechanisms is crucial for developing sustainable strategies for CVD prevention and management.

9.3.1 Omega-3 Fatty Acids (EPA, DHA)

9.3.1.1 Anti-inflammatory Effects on Blood Vessels

Omega-3 PUFAs, specifically EPA and DHA, are known for their anti-inflammatory properties [93]. Fatty acids like DHA and EPA can help lower inflammation, prevent blood vessel damage, and reduce the risk of heart disease and stroke by replacing pro-inflammatory omega-6 fatty acids, altering enzyme availability and eicosanoid production, and influencing transcription factors involved in the inflammatory response (Fig. 9.7). These interactions significantly impact the inflammatory environment contributing to CVD. ω-3 LCPUFAs reduce neuroinflammation, inhibit NLRP3 inflammasome activation, and support vascular health through anti-inflammatory and endothelial-protective effects [33, 98, 99] (Fig. 9.8).

9.3.1.2 Reduction of Triglycerides and LDL Cholesterol

Consuming omega-3 fatty acids, particularly EPA and DHA, significantly reduces blood triglyceride concentrations, a risk factor for atherosclerosis and CVD [101]. The impact on low-density lipoprotein (LDL) cholesterol is complex and depends on the type and dosage of the medication. Higher doses, typically 2–4 g/day, show greater efficacy. The primary mechanisms involve reducing liver production of VLDL triglycerides and enhancing their removal from the bloodstream [100, 102]. The composition of omega-3 supplements should be carefully considered with an individual's lipid profile and overall CVD risk factors [103].

9.3.1.3 Improvement in Endothelial Function and Blood Pressure Regulation

EPA and DHA improve endothelial function, a crucial aspect of maintaining healthy blood vessels [13]. Endothelial dysfunction, characterized by reduced vasodilation and a pro-inflammatory state, is an early pathological event in atherosclerosis. Omega-3 fatty acids help regulate blood pressure, particularly in individuals with

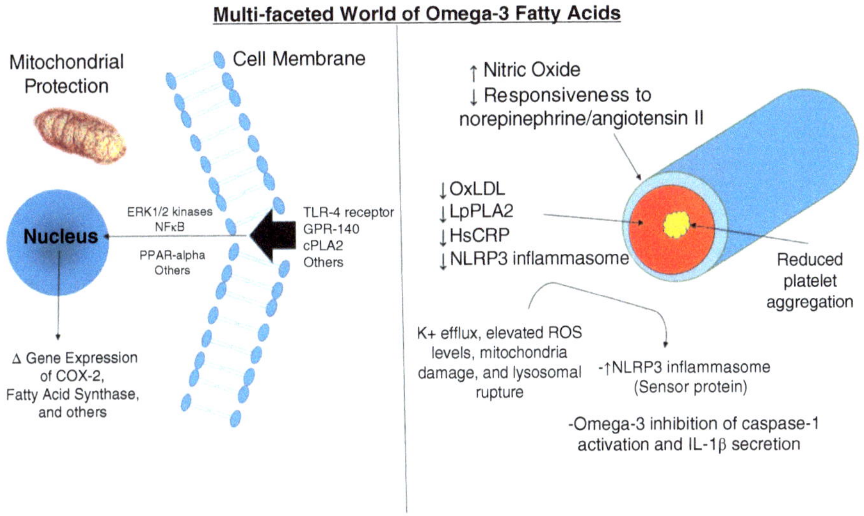

Figure 9.7 Omega-3 modulation of vascular homeostasis: enhanced endothelial function and inhibition of key inflammatory markers and enzymes. *Source:* [98]/ Springer Nature / CC BY 4.0.

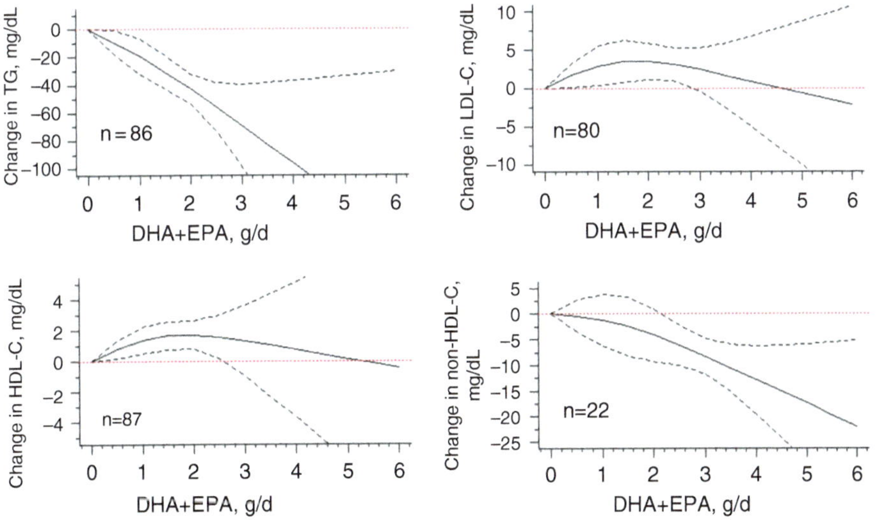

Figure 9.8 Dose–response relationship between changes in lipids and combined intake of DHA+EPA. Marginal average dose–response curve (solid line) with 95% point-wise CIs (dashed lines) estimated by a 1-stage random-effects restricted cubic spline model, using 0 g/day as the reference. Studies included n = 86 for TG, n = 80 for LDL-C, n = 87 for HDL-C, and n = 22 for non-HDL-C. DHA indicates docosahexaenoic acid; EPA, eicosapentaenoic acid; HDL-C, high-density lipoprotein cholesterol; LDL-C, low-density lipoprotein cholesterol; and TG, triglyceride. *Source:* [100]/ American Heart Association / CC BY-NC-ND 4.0.

hypertension. They increase nitric oxide bioavailability, reduce oxidative stress, and modulate calcium signaling, leading to vasodilation. EPA influences sodium excretion in the kidneys [104].

9.3.2 Marine Peptides (Derived from Fish Protein Hydrolysates)

Marine-derived bioactive peptides, including protein hydrolysates from fish species such as anchovy and mackerel, exhibit antihypertensive, antioxidant, and antithrombotic properties [95, 105]. These peptides are promising for developing functional foods and pharmaceutical products that promote cardiovascular health, including antihypertensive properties, protection against cellular damage, and blood clot prevention [106].

9.3.2.1 Antihypertensive Properties (ACE-Inhibitory Activity)

Marine peptides have been found to inhibit the angiotensin-I-converting enzyme (ACE), a key enzyme in the renin–angiotensin system (RAS), leading to vasodilation and reduced blood pressure [95, 107] (Fig. 9.9). In vitro studies have reported IC50 values for peptides from various marine sources, including abalone, anchovy, mackerel, hard clam, blue mussel, and jellyfish [107–116]. In vivo studies on spontaneously hypertensive rats have shown significant reductions in systolic blood pressure following oral administration of hydrolysates and purified peptides 107, 109, 114, 115, 117]. This suggests that the marine environment could be a rich source of natural compounds for managing hypertension, offering safer and more tolerable alternatives to synthetic ACE inhibitor drugs.

9.3.2.2 Improvement in Arterial Flexibility and Vascular Function

Marine peptides, particularly collagen peptides derived from abalone and jellyfish, have been shown to enhance arterial flexibility and improve vascular function [24, 118]. Some marine peptides can inhibit the proliferation and migration of aortic smooth muscle cells, thereby contributing to the development of atherosclerosis. Additionally, some marine peptides can enhance endothelial function, which is crucial for maintaining healthy blood vessels [119]. These peptides also exert broader vasculoprotective actions, potentially benefiting long-term cardiovascular health.

9.3.3 Seaweed Polysaccharides (Fucoidan, Alginate, Carrageenan, Rhamnan Sulfate)

Seaweeds, also known as marine algae, contain complex polysaccharides such as fucoidan, alginate, carrageenan, and ulvan. These polysaccharides have biological activities beneficial for cardiovascular health, including antioxidant, anti-inflammatory, antihypertensive, and antithrombotic effects [97, 120–123].

9.3.3.1 Blood Pressure-Lowering Effects

Seaweed polysaccharides have been found to lower blood pressure, making them potential dietary interventions for hypertension [120]. They inhibit ACE, preventing the formation of vasoconstrictor angiotensin II and leading to vasodilation [121].

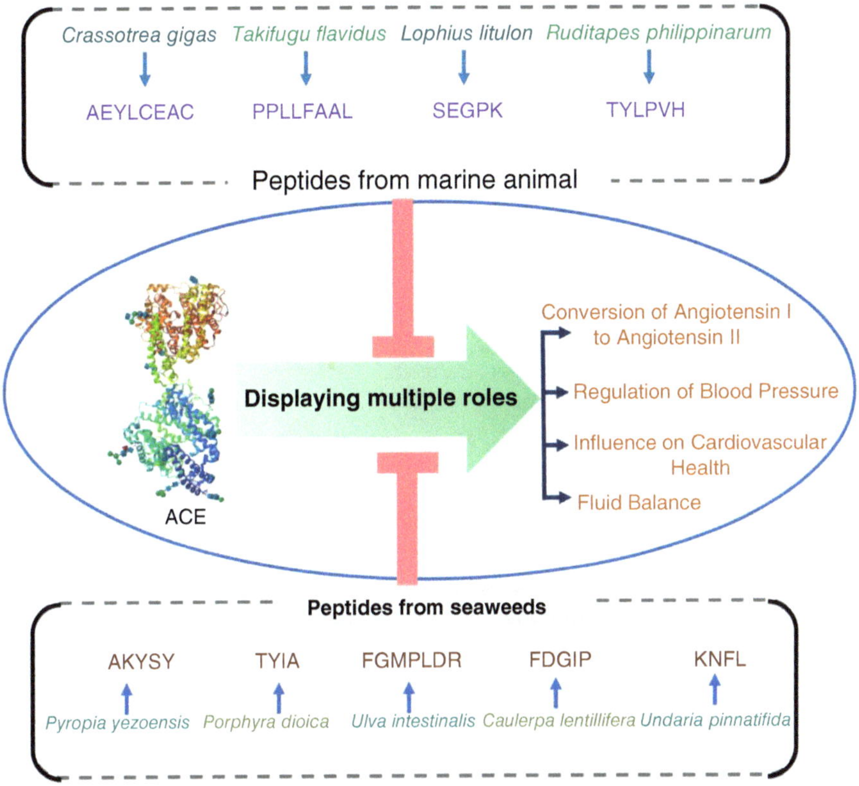

Figure 9.9 Sources and sequences of ACE-inhibitory peptides from oysters, pufferfish, monkfish, clams, and five seaweeds, illustrating their collective roles in cardiovascular homeostasis. *Source:* [95]/ MDPI / CC BY 4.0.

Regular consumption of specific types of seaweed, such as wakame and nori, has been shown to significantly reduce systolic and diastolic blood pressure in patients with hypertension [124, 125].

9.3.3.2 Improvement in Lipid Metabolism and Cholesterol Regulation

Seaweed polysaccharides regulate lipid metabolism and cholesterol levels, thereby contributing to a healthier cardiovascular profile [120, 121, 126]. Soluble fibers in seaweed can bind to intestinal bile acids, thereby promoting the excretion of cholesterol [123]. Fucoidan, a sulfated polysaccharide derived from brown seaweed, reduces total cholesterol and high-density lipoprotein levels while increasing high-density lipoprotein levels [97]. Other seaweed polysaccharides, like alginates and carrageenans, also positively affect serum lipid profiles. Seaweed polysaccharides offer a dietary approach to managing dyslipidemia by influencing liver metabolic processes [122].

9.3.3.3 Antithrombotic Effects for Heart Disease Prevention

Seaweed polysaccharides, such as fucoidan and chitosan, have been shown to exhibit antithrombotic effects, potentially preventing blood clot formation—a major cause of CVD [121]. These polysaccharides prolong blood clotting time and inhibit platelet

Table 9.1 Antithrombotic potency of brown algae polysaccharide.

Group	Dose	APTT (s)	PT (s)
Control		17.63 ± 1.79	8.35 ± 0.68
Low-dose PLG	2.5 mg/kg	18.97 ± 3.78	7.98 ± 0.73
High-dose PLG	5.0 mg/kg	24.96 ± 3.09^{a}	8.43 ± 0.64
Enoxaparin	1.0 U/kg	21.86 ± 3.67^{b}	8.10 ± 0.70

[a] $p < 0.01$ vs the control group.
[b] $p < 0.05$.
PLG (polysaccharide extract from Laminaria japonica) at 5.0 mg/kg produces a significant APTT increase ($p < 0.01$) with PT remaining at control levels.
Source: [133]/ John Wiley & Sons.

aggregation, which are crucial steps in blood clot formation [127–130]. As shown in Table 9.1, polysaccharides extracted from Laminaria japonica significantly extend the APTT value without altering PT levels, demonstrating strong antithrombotic potency. They work by interfering with the complex cascade of reactions leading to blood clot formation and inhibiting the activation and aggregation of platelets. Thus, seaweed polysaccharides hold promise as natural antithrombotic agents [131, 132].

9.3.4 Other Key Bioactive Compounds for Cardiovascular Health

9.3.4.1 Astaxanthin and Other Marine Antioxidants

Astaxanthin, a red carotenoid pigment found in marine organisms, is a potent antioxidant with anti-inflammatory properties [96]. It helps protect against heart diseases caused by oxidative stress and contributes to overall cardiovascular protection [94]. Astaxanthin inhibits the oxidation of LDL cholesterol, a key process in atherosclerosis development, and may increase high-density lipoprotein (HDL) cholesterol levels [134]. It also reduces atherosclerotic plaques within arteries and improves the structural integrity of the arterial wall. Other marine antioxidants, like phycocyanin in blue-green algae, also exhibit cardioprotective effects [135, 136].

9.3.4.2 Taurine (High in Fish and Shellfish)

Taurine, a non-protein amino acid found in fish and shellfish, is essential for maintaining cardiovascular health and preventing CVD [44]. It regulates blood pressure, protects heart tissues from oxidative stress, and influences calcium ion concentration. Taurine can reduce systolic and diastolic blood pressure, particularly in patients with hypertension, and improve heart function [137, 138]. Its diverse physiological roles underscore its importance in preventing CVD (Fig. 9.10).

9.3.4.3 Coenzyme Q10 (Found in Fatty Fish and Shellfish)

Coenzyme Q10 (CoQ10), found in fatty fish and shellfish, is essential for heart cell function, electron transport chain production, and antioxidant protection [139]. It has been suggested to reduce the risk of heart failure and hypertension, two major CVDs (Fif. 9.10). Clinical research shows that CoQ10 can improve symptoms of

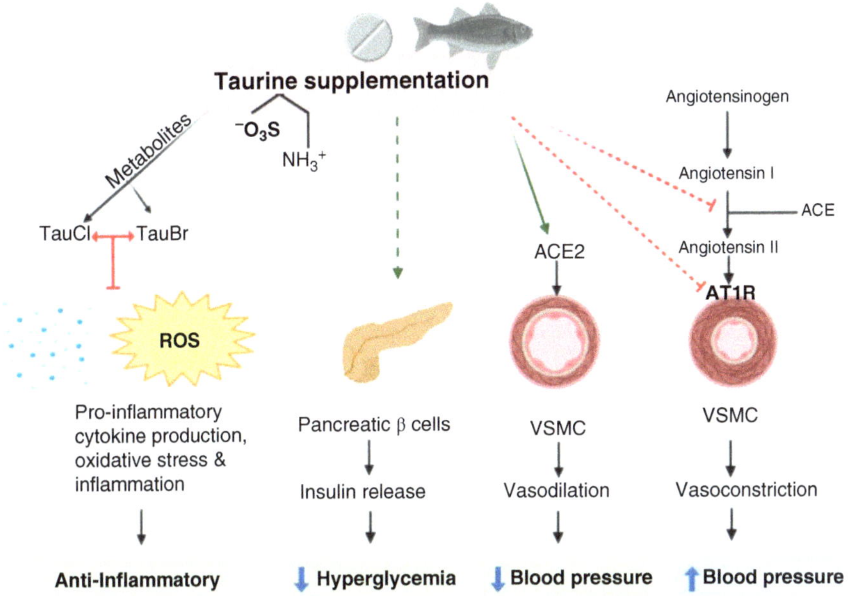

Figure 9.10 Possible sites of action of taurine as a beneficial agent in treating cardiovascular disease (CVD) and its anti-inflammatory properties. The taurine metabolites, TauCl and TauBr, exhibit anti-inflammatory properties by inhibiting the production of pro-inflammatory cytokines, mitigating oxidative stress (ROS), and reducing inflammation. Taurine itself may directly reduce hyperglycemia, activate the protective arm of the renin–angiotensin system (RAS) by increasing ACE2 expression, and inhibit the harmful axis of the RAS by reducing Ang II production, which is currently one of the primary targets in treating CVD and coronary heart disease (CAD). It must be emphasized that taurine supplementation eliciting insulin release may be detrimental/have no effect in patients who are obese and have already elevated insulin levels. TauCl, taurine–chloramine; TauBr, taurine–bromamine; ROS, reactive oxygen species; VSMC, vascular smooth muscle cell; ACE, angio-converting enzyme; AT1R, AngII type 1 receptor. *Source:* [44]/ MDPI / CC BY 4.0.

congestive heart failure and lower blood pressure in patients with hypertension. It also enhances endothelial function and reduces oxidative stress, which are crucial for the prevention and management of CVD [140–144].

9.3.4.4 Squalene (Found in Shark Liver Oil and Some Fish Oils)

Squalene, a naturally occurring triterpene found in shark liver oil and fish oils, has been found to have cholesterol-lowering properties and cardioprotective effects [145]. It inhibits the enzyme 3-hydroxy-3-methylglutaryl coenzyme A (HMG-CoA) reductase, a key enzyme in cholesterol synthesis [146]. Human trials have demonstrated that squalene supplementation can lower total cholesterol and LDL cholesterol levels, while exhibiting antioxidant properties, making it a potential natural agent for cholesterol management and cardiovascular protection [145] (Fig. 9.11).

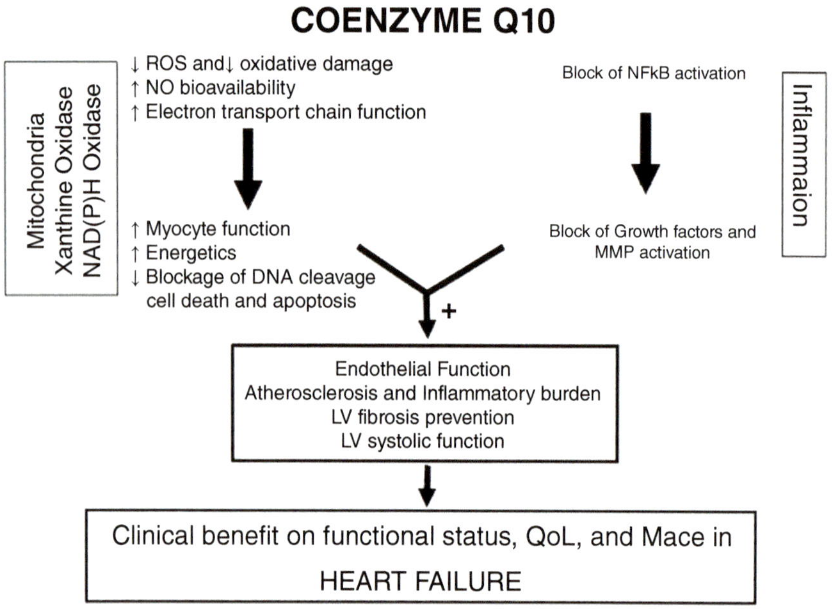

Figure 9.11 Pathophysiological mechanisms and potential clinical benefits of CoQ10 in heart failure. LV, left ventricle; MACE, major adverse cardiovascular events; MMP, matrix metalloproteinase; NAD(P)H, Nicotinamide adenine dinucleotide phosphate; NFkB, nuclear factor kappa-light-chain-enhancer of activated B cells; NO, nitric oxide; QoL, quality of life; ROS, reactive oxygen species. *Source:* [144]/ MDPI / CC BY 4.0.

9.4 Cancer Prevention

Cancer is a global health issue, and prevention strategies are growing [147]. Natural bioactive compounds, such as marine peptides, seaweed polysaccharides, and marine carotenoids, are being investigated for their potential in cancer prevention [33]. This report examines the scientific understanding of these marine-derived substances, with the intention of understanding their potential in preventing this complex disease (Fig. 9.12).

9.4.1 Omega-3 Fatty Acids (EPA and DHA)

Omega-3 fatty acids have potential anticancer effects through various mechanisms, including the inhibition of cancer cell proliferation, promotion of apoptosis, and the selective targeting of cancer cells [33, 148]. They also have anti-inflammatory properties, influencing the inflammatory response by modulating lipid mediators and regulating gene expression (Fig. 9.13). These properties may indirectly contribute to cancer prevention by counteracting the cellular and molecular environment that favors tumorigenesis, potentially minimizing the adverse effects of traditional cancer treatments [149].

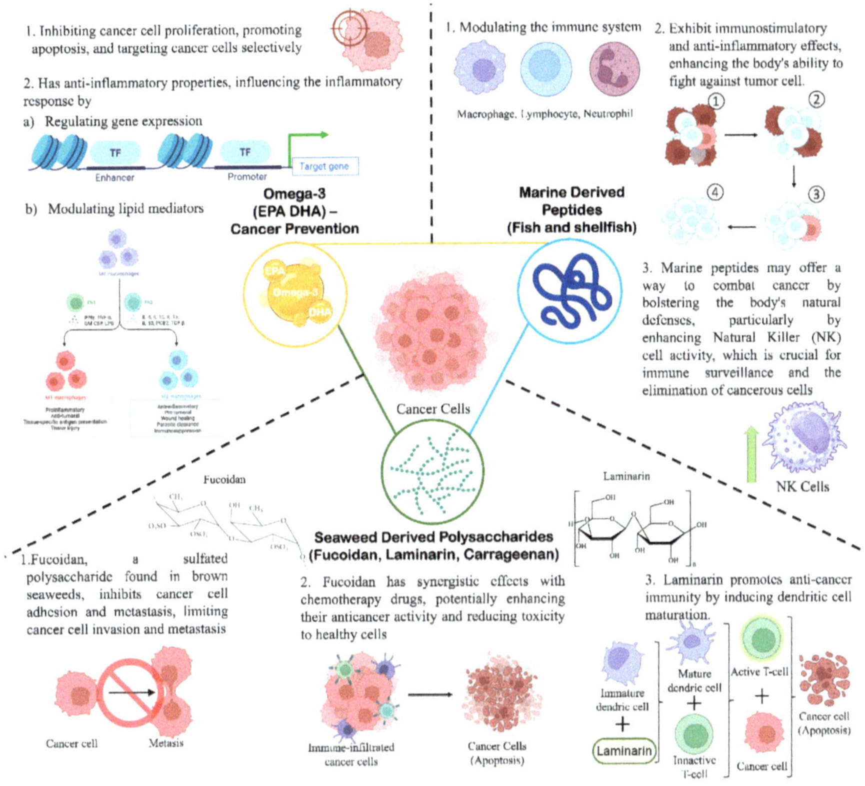

Figure 9.12 Marine-derived bioactive compounds—including omega-3 PUFAs, peptides, polysaccharides, and phospholipid derivatives—exert antitumor effects via modulation of membrane integrity, oxidative stress, inflammation, and oncogenic signaling. These mechanisms contribute to inhibition of cancer cell proliferation, immune evasion, and tumor progression, highlighting the therapeutic potential of aquatic nutraceuticals in cancer prevention and adjunctive care.

9.4.2 Marine Peptides (Derived from Fish and Shellfish)

Marine bioactive peptides, derived from various marine organisms like fish, shellfish, algae, and microorganisms, have shown potential as anticancer agents. These peptides, often extracted from marine proteins through enzymatic hydrolysis, have demonstrated anticancer properties through various mechanisms, including modulation of the immune system [150–152]. They can exhibit immunostimulatory and anti-inflammatory effects, enhancing the body's ability to fight against tumor cells (Fig. 9.14). Marine peptides may offer a potential approach to combating cancer by bolstering the body's natural defenses, particularly by enhancing natural killer (NK) cell activity, which is crucial for immune surveillance and the elimination of cancerous cells by inducing apoptosis in cancer cells and inhibiting cell migration [153, 154].

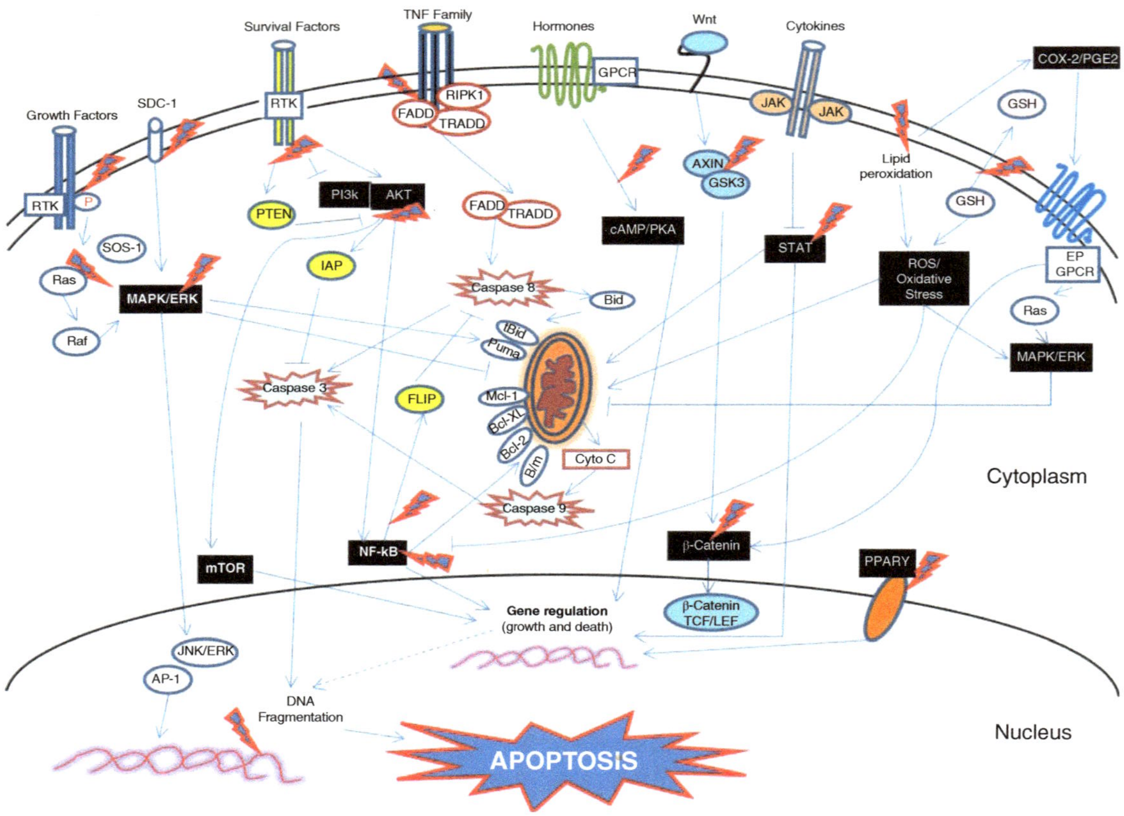

Figure 9.13 Multiple apoptotic molecular signals targeted by n−3 PUFAs in cancer cells. Abbreviations: RTK, protein tyrosine kinase; SOS-1, son of sevenless-1; Erk, extracellular-signal-regulated kinase; MAPK, mitogen-activated protein kinase; JNK, Jun N-terminal kinase; AP-1, activator protein-1; SDC-1, syndecan-1; PTEN, phosphatase and tensin homolog deleted on chromosome ten; PI3K, phosphatidylinositol-3-kinase; mTOR, mammalian target of rapamycin; IAP, inhibitor of apoptosis; NF-κB, nuclear factor-κB; FLIP, FLICE-like inhibitory protein; RIPK-1, receptor-like protein kinase-1; FADD, Fas-associated death domain; TRADD, TNF receptor-associated death domain; Bcl-2, B-cell lymphoma protein-2; Bim, Cyto C, cytochrome C; GPCR, G-protein coupled receptor; cAMP, cyclic adenosine monophosphate; PKA, protein kinase A; GSK, glycogen synthase kinase; TCF, T-cell factor; LEF, lymphoid enhancer-binding factor; JAK, Janus kinase; STAT, signal transducer and activator of transcription; ROS, reactive oxygen species; GSH, glutathione; COX-2, cyclooxygenase-2; PGE2, prostaglandin E2; PPAR-γ, peroxisome proliferator-activated receptor γ. arrows, activation; ⊥, inhibition; dashed arrows, indirect action; red/blue flash, targeted by n−3 PUFAs. *Source:* [148]/ MDPI / CC BY 4.0.

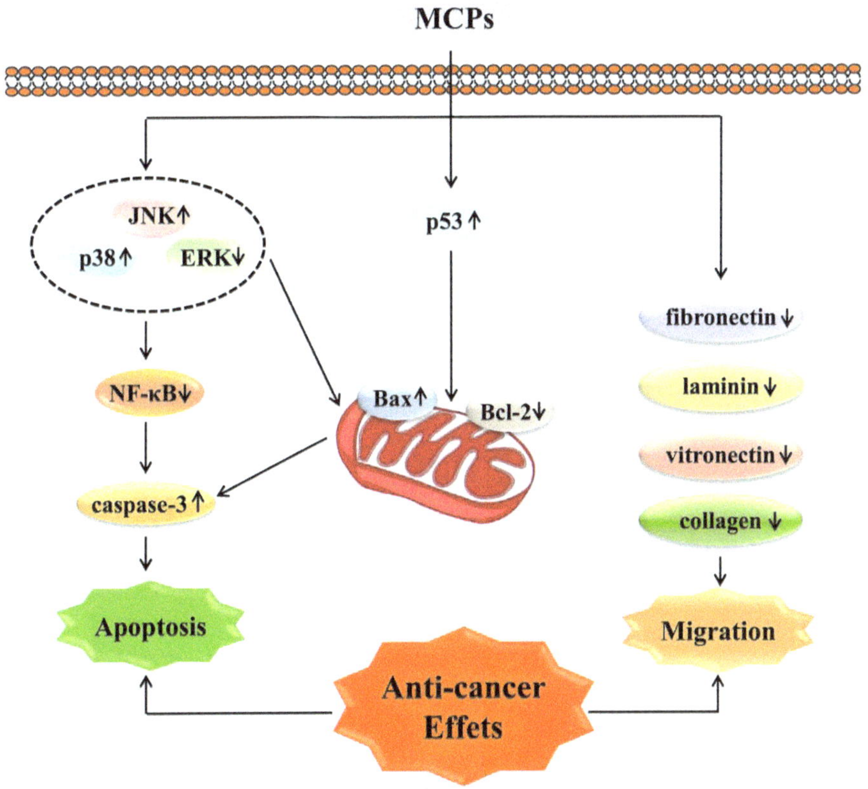

Figure 9.14 Marine collagen peptides (MCPs) exert anticancer activity by inducing cell apoptosis and inhibiting cell migration. *Source:* [154]/ Elsevier / CC BY 4.0.

9.4.3 Seaweed Polysaccharides (Fucoidan, Laminarin, and Carrageenan)

Seaweeds, including brown, red, and green algae, are rich in complex polysaccharides with diverse biological activities [155]. Fucoidan, a sulfated polysaccharide found in brown seaweeds, inhibits cancer cell adhesion and metastasis, limiting invasion and metastasis [156–158]. These polysaccharides activate the immune response against tumor cells by enhancing the activity of immune cells, such as NK and T cells. Laminarin promotes anticancer immunity by inducing the maturation of dendritic cells [159] (Fig. 9.15). Fucoidan has synergistic effects with chemotherapy drugs, potentially enhancing their anticancer activity and reducing toxicity to healthy cells. Marine-derived polysaccharide chitosan enhances NK cell activity, suggesting it could provide a proactive defense mechanism against cancer development and in its early stages. These polysaccharides have the potential to develop novel therapeutic strategies against cancer [160, 161].

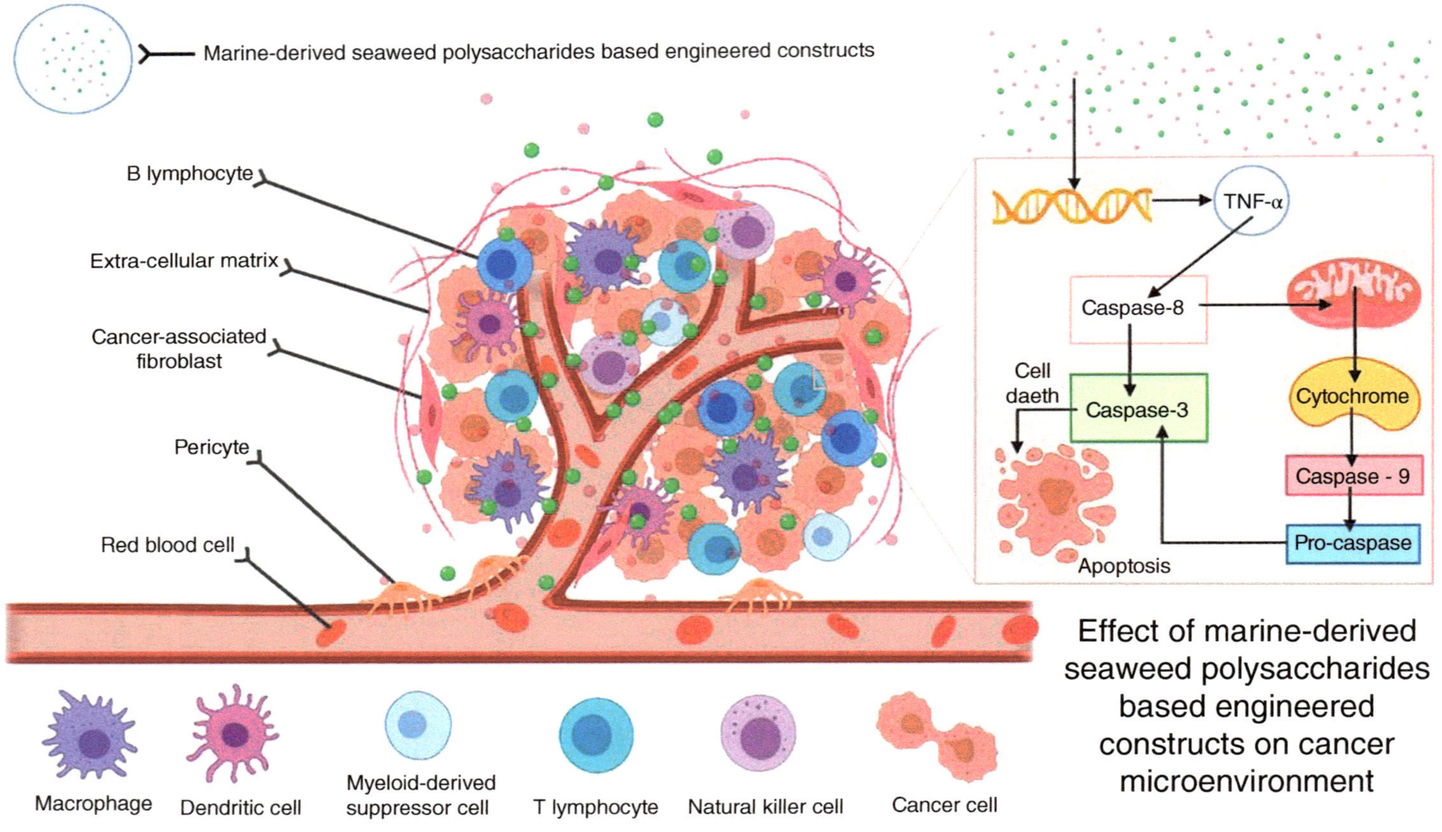

Figure 9.15 Anticancer potentialities of marine-derived seaweed polysaccharides-based engineered cues to destroy cancer cells and prevent metastasis by acting on the tumor cells' microenvironment via different mechanisms, such as cell death via apoptosis. *Source:* [158]/ MDPI / CC BY 4.0.

9.5 Weight Management

The global prevalence of overweight and obesity has reached alarming levels, posing significant risks to public health and increasing the burden on healthcare systems [162]. In response to this growing concern, there has been a surge in interest among consumers and researchers alike in exploring natural and dietary interventions for effective and sustainable weight management [163]. Aquatic organisms represent a rich and diverse source of bioactive compounds, holding considerable promise for addressing weight management due to their varied mechanisms of action (Fig. 9.16). This report aims to provide a comprehensive overview of the key bioactive components derived from aquatic products that have demonstrated potential for weight management.

9.5.1 Omega-3 Fatty Acids (EPA, DHA)

Omega-3 fatty acids, like EPA and DHA, are crucial for overall health and weight management. They enhance insulin sensitivity, reduce fat accumulation, and increase fat oxidation [103, 164]. They regulate gene expression in lipid metabolism, reducing lipogenesis and beta-oxidation [165]. They also reduce visceral fat, associated with metabolic syndrome, CVD, and type 2 diabetes. These benefits are most noticeable when combined with a balanced diet and regular exercise [166, 167].

9.5.2 High-Quality Protein and Marine Peptides from Fish and Shellfish

9.5.2.1 Higher Satiety Levels Compared to Other Protein Sources

Marine proteins and peptides, particularly those from fish and shellfish, have satiety-enhancing properties. MCPs promote a greater sense of fullness than other protein sources, such as whey or casein [168]. These peptides interact with the gut's endocrine system, stimulating the release of anorexigenic hormones, such as cholecystokinin, GLP-1, and PYY [169]. Consuming collagen increases satiety more effectively than consuming carbohydrates or fats, mediated by increased GLP-1 levels and decreased ghrelin levels [170–172] (Fig. 9.17).

9.5.2.2 Stimulation of Thermogenesis and Metabolic Rate

Marine peptides have a higher thermogenic effect than carbohydrates or fats, leading to increased energy expenditure and calorie burning [172, 173]. This property supports weight management efforts by promoting fat loss. Marine peptides may also regulate genes involved in energy metabolism, influencing metabolic rate [174, 175]. Research suggests that MCPs may help reduce body fat in overweight adults by promoting fat metabolism (Fig. 9.18). Marine peptides also support the breakdown and utilization of fats for energy production, regulating fat storage and utilization [176–178].

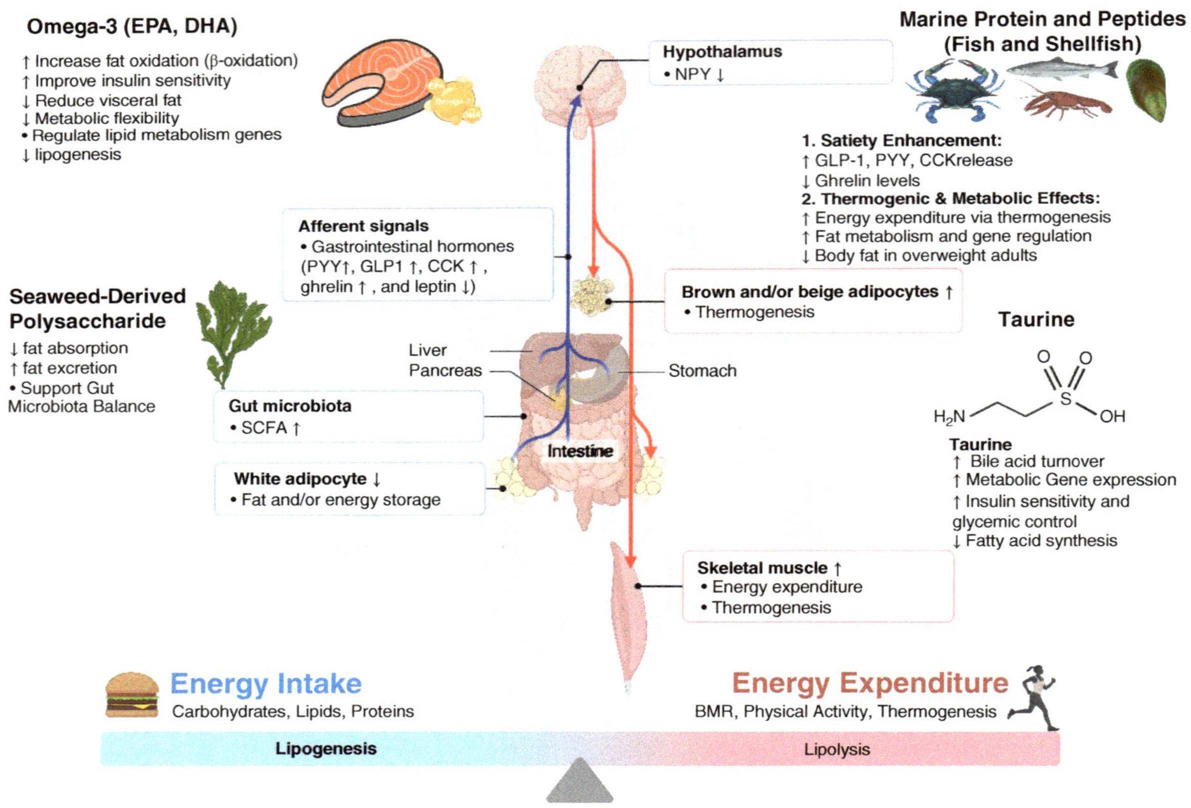

Figure 9.16 Marine-derived bioactive compounds—including proteins, peptides, omega-3 PUFAs, polysaccharides, and taurine—modulate energy metabolism by influencing central appetite regulation, gut hormone secretion, microbiota composition, thermogenesis, and lipid oxidation. These coordinated effects enhance energy expenditure, suppress energy intake, and rebalance lipogenesis/lipolysis, contributing to anti-obesity, anti-inflammatory, and insulin-sensitizing outcomes.

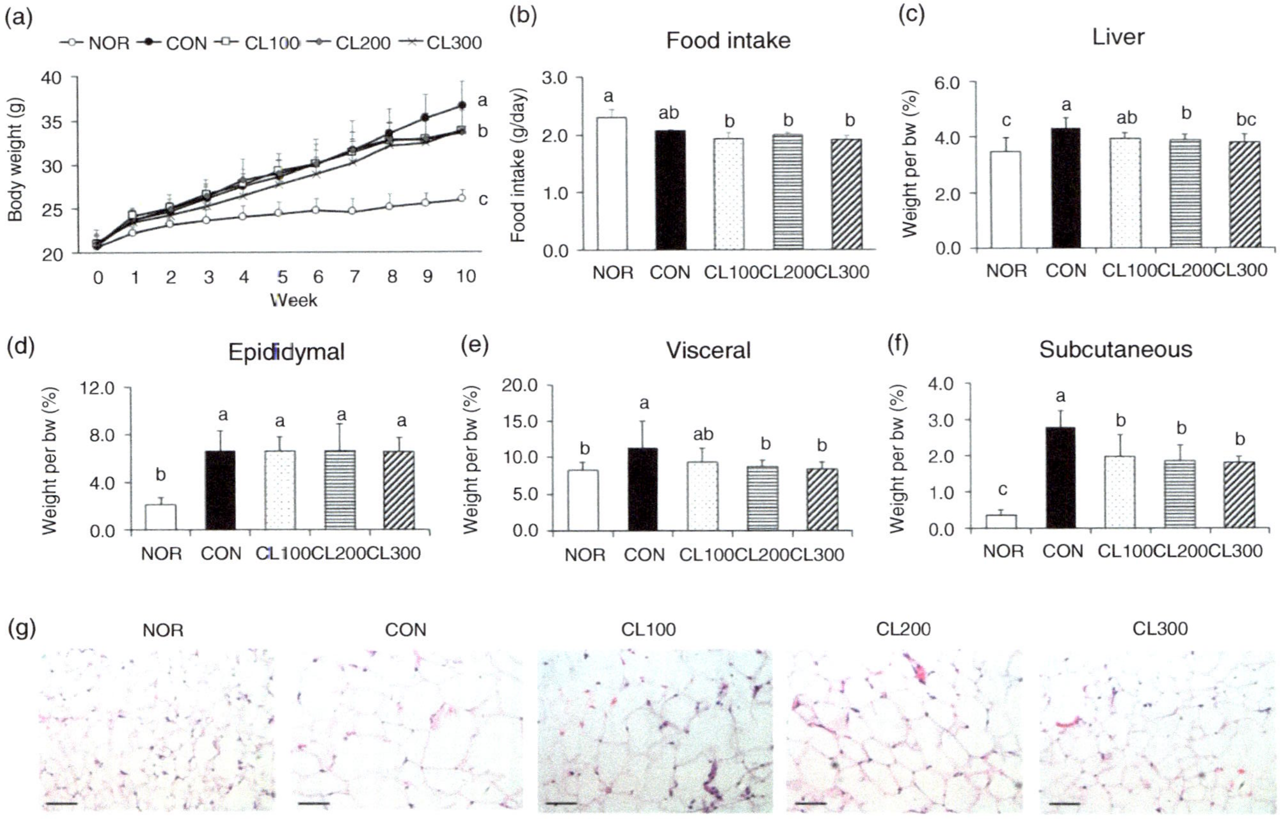

Figure 9.17 Effects of skate collagen peptide on body weight, food intake, and organ weight, and histological analysis of adipose tissue in high-fat diet-fed C57BL6/J mice for eight weeks. Data are mean ± standard deviation (SD) (n = 9 per group). Normal (NOR) C57BL/6J mice fed a chow diet with water; control (CON) C57BL/6J mice fed a high fat diet (HFD) with water; collagen 100 (CL100), collagen 200 (CL200), and collagen 300 (CL300) C57BL/6J mice fed a HFD with oral administration of skate collagen peptide at a concentration of 100, 200, and 300 mg/kg body weight per day, respectively. (A–C) Different letters mean significant differences to one-way analysis of variance (ANOVA), followed by Duncan's multiple-range test at $p < 0.05$. (a) change in body weight (bw) for 10 weeks; (b) food intake; (c) liver weight per bw; (d) epididymal adipose tissue weight per bw; (e) visceral adipose tissue weight per bw; (f) subcutaneous adipose tissue weight per bw; (g) hematoxylin and eosin staining, magnification: 200×, bar: 50 μm. *Source:* [172] / MDPI / CC BY 4.0.

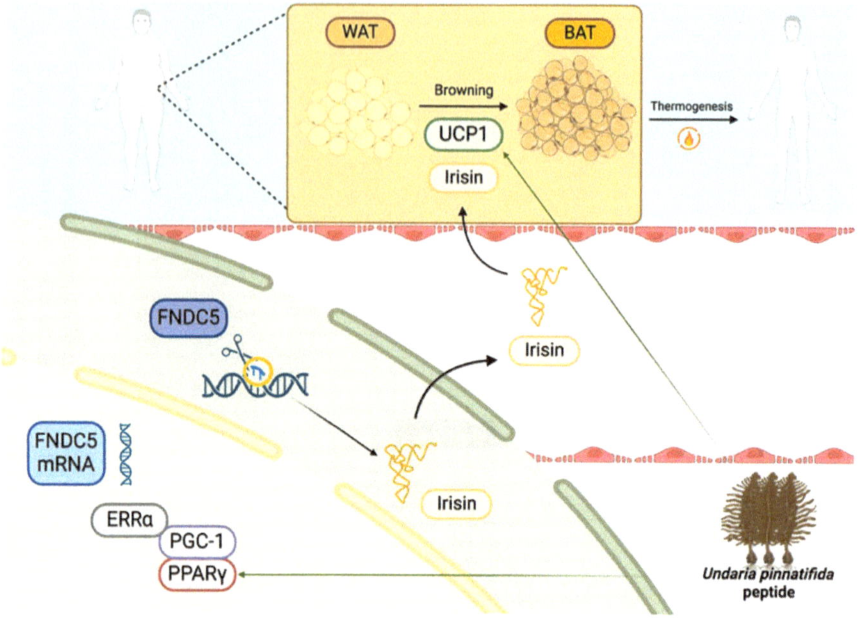

Figure 9.18 Functional food peptides enhance fat metabolism by promoting irisin release and BAT development, supporting anti-obesity strategies. [175]/ Elsevier / CC BY-NC-ND 4.0.

9.5.3 Seaweed-Derived Polysaccharides

Among seaweed-derived polysaccharides, alginate has strong evidence supporting its ability to reduce fat absorption in the intestine [155]. Found in brown seaweed, alginate forms a viscous gel in the acidic environment of the stomach and intestines. This gel binds to dietary fats, inhibiting their digestion and absorption, which results in increased fat excretion and lower calorie intake from fat [179]. Additionally, alginate can directly inhibit pancreatic lipase, the enzyme responsible for breaking down triglycerides into absorbable fatty acids, further reducing fat absorption. The effectiveness of alginate in this role appears to be related to its guluronate content, with higher guluronate alginates showing more potent lipase inhibition [180].

Chitosan is promoted for its ability to bind dietary fat in the digestive tract, which may reduce the absorption of fat and cholesterol from food. Often referred to as a "fat blocker," chitosan forms a gel in the stomach that binds to fat molecules, allowing them to be excreted rather than absorbed [43, 181, 182] (Fig. 9.19). Moreover, Chitosan may also have prebiotic properties, potentially supporting the growth of beneficial gut bacteria and contributing indirectly to weight management [183].

9.5.4 Other Key Bioactive Compounds for Weight Management

9.5.4.1 Taurine-Enhances Lipid Metabolism and Energy Expenditure

Taurine, a sulfur-containing amino acid, has been shown to improve metabolic health and reduce body weight (bw) in overweight adults [184]. It regulates lipid

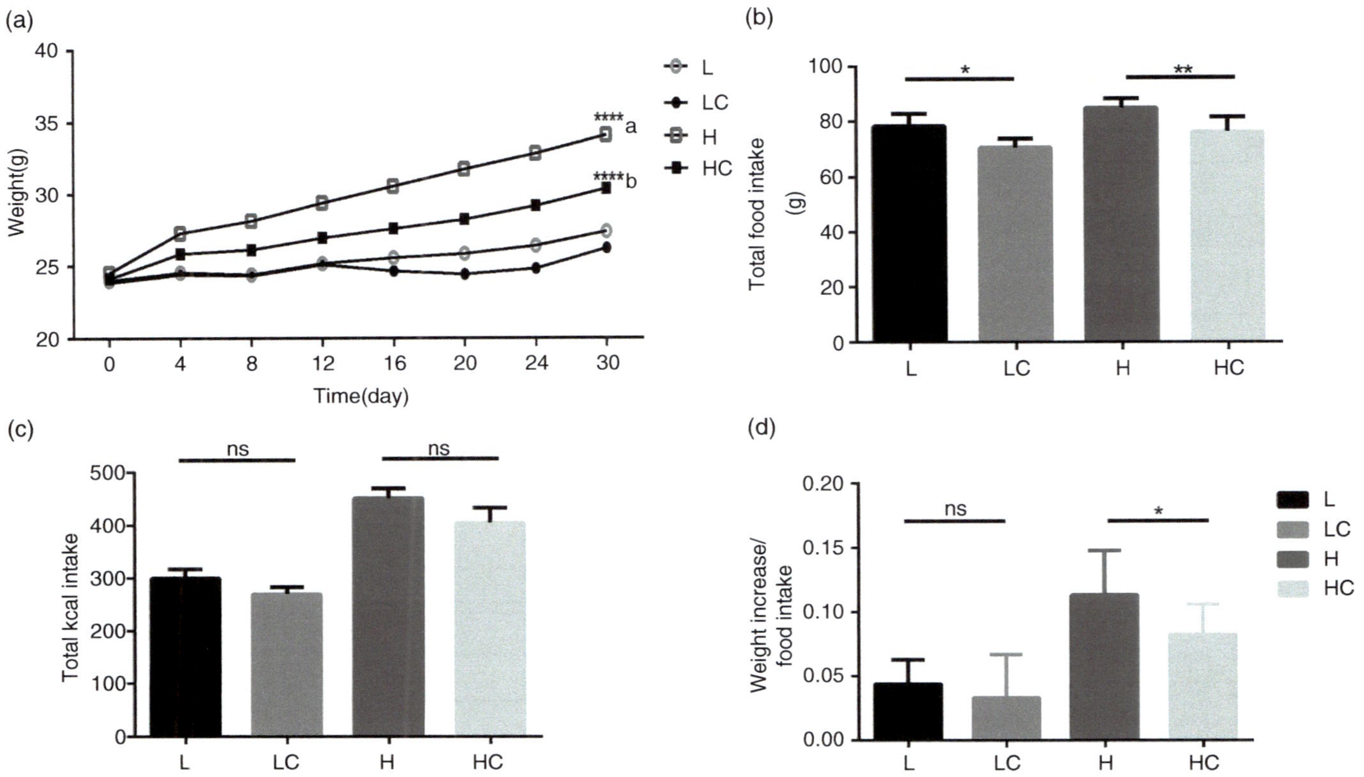

Figure 9.19 Effects of dietary chitosan on body weight, food consumption, energy intake, and feed efficiency in C57BL/6J mice fed low-fat or high-fat diets. Mice were allocated to four groups: L, low-fat diet (4.3% fat, 5% cellulose, w/w); LC, low-fat diet + 5% chitosan replacing cellulose; H, high-fat diet (35.4% fat, 5% cellulose, w/w); HC, high-fat diet + 5% chitosan replacing cellulose. (a) Body weight trajectories over the study per od (L: ○; LC: ●; H: □; HC: ■). (b) Cumulative food intake. (c) Cumulative caloric intake. (d) Feed efficiency is calculated as total weight gain divided by total food intake. Data are presented as mean ± SEM (n = 12 per group). Statistical comparisons were made by one-way ANOVA with Tukey's post hoc test: *$p < 0.05$, **$p < 0.01$ versus L (for LC) or versus H (for HC); ns, not significant. *Source:* [182]/ Elsevier / CC BY-NC-ND 4.0.

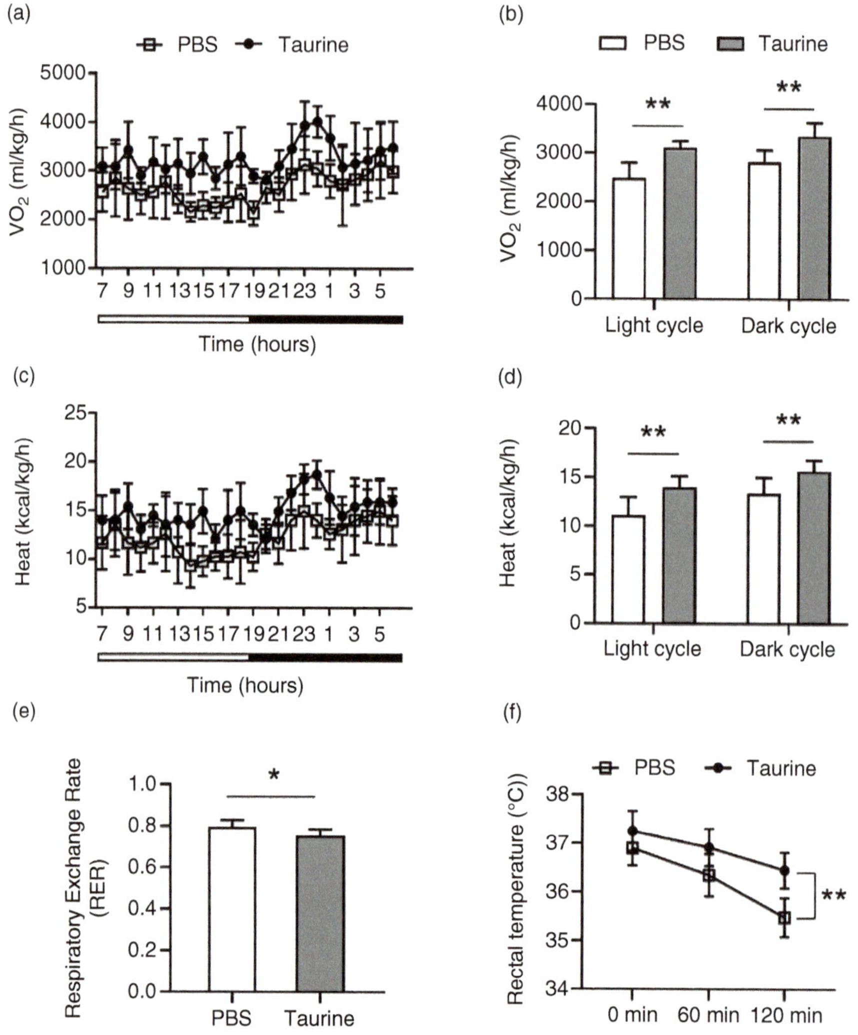

Figure 9.20 **Taurine increases energy expenditure and adaptive thermogenesis.** Mice were treated as indicated in (a) and (b). Whole-body oxygen consumption rate (VO_2) of the mice during a 12-h light/12-h dark cycle was measured in a metabolic cage (a), and the average values for 12-h light/12-h dark periods are shown in (b). (c, d) Heat production of the mice during a 12-h light/12-h dark cycle was calculated and shown (c), and the average values for 12-h light/12-h dark periods were shown in (d). (e) The average values of respiratory exchange ratio (RER) in the mice for a 24-h cycle were calculated from the metabolic cage data. (f) Rectal temperature of the mice was recorded at the indicated time points after cold exposure (4 °C). For statistical analysis, two-way analysis of variance and Bonferroni's post hoc. *Source:* [186]/ Elsevier / CC BY 4.0.

metabolism and promotes energy expenditure, reducing fatty acid synthesis and bile acid synthesis [185] (Fig. 9.20). Taurine also upregulates metabolic gene expression and enhances energy expenditure [186]. Studies suggest it may improve insulin sensitivity and glycemic control in overweight individuals [187].

9.6 Other Health Benefits

9.6.1 Immune System Support

Long-chain n-3 PUFA—notably EPA and DHA—abundant in mussels (Mytilus edulis), oysters, and Japanese anchovy (Engraulis japonicus), as well as in marine fish such as salmon, mackerel, sardine, and microalgae, modulate immune function by down-regulating nuclear factor-κB (NF-κB) and mitogen-activated protein kinase (MAPK) pathways, reducing pro-inflammatory cytokines (TNF-α, IL-1β, IL-6), and generating specialized pro-resolving mediators including resolvins, protectins, and maresins [188–190].

Marine-derived peptides—including collagen hydrolysates from fish skin/scales and from invertebrates such as sturgeon, seahorse (Hippocampus spp.), jellyfish (Rhopilema esculentum), blue mussel (M. edulis), and oyster—enhance lymphocyte proliferation, adjust Th1/Th2 balance, stimulate macrophage phagocytosis, and protect immune cells via antioxidant activity [191, 192].

Marine polysaccharides such as fucoidan (from brown seaweeds), carrageenan (red algae), alginate (kelp), and chitosan (from crustacean shells) activate macrophages, NK cells, and dendritic cells; increase IL-2 and IFN-γ secretion; and improve antigen presentation—with bioactivity influenced by sulfation and molecular weight [193, 194]. Other compounds, including astaxanthin (found in microalgae and krill) and phlorotannins (present in brown seaweeds), provide potent antioxidant and anti-inflammatory support, shielding immune cells from oxidative damage and contributing to balanced immune regulation [195, 196]. Collectively, these marine-derived foods and compounds provide a range of nutrients that play essential roles in maintaining and strengthening immune health.

9.6.2 Joint Health

9.6.2.1 Omega-3 for Joint Health

Blue mussels (M. edulis), salmon, mackerel, and Japanese anchovy (E. japonicus) are highly valued for their rich content of omega-3 fatty acids, specifically EPA and DHA. Other notable marine sources of omega-3 include sardines, herring, tuna, krill, cod liver, trout, and marine microalgae, which serve as the primary origin of EPA and DHA in the aquatic food chain [188] (Fig. 9.21).

Omega-3 supplementation has shown promise in supporting osteoarthritis (OA) management by reducing cartilage destruction, inhibiting pro-inflammatory cytokines, and promoting the production of specialized pro-resolving mediators, such as resolvins, protectins, and maresins. These bioactive compounds help shift immune responses toward anti-inflammatory pathways, decrease pain, stiffness, and swelling, and improve joint mobility and physical function. Studies indicate that omega-3 fatty acids, particularly when balanced against omega-6 intake, can inhibit key inflammatory signaling pathways (e.g., NF-κB, MAPK), reduce cartilage-degrading enzymes (matrix metalloproteinase (MMP)-13, ADAMTS-5), and enhance chondrocyte survival [188, 197, 198] (Fig. 9.22). Clinical and animal

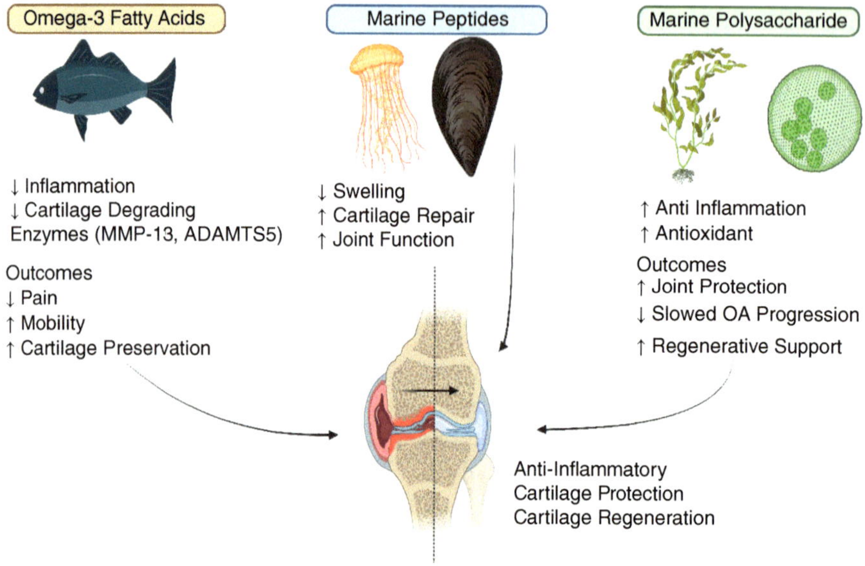

Figure 9.21 Bioactive molecules from fish, shellfish, and algae act on inflammatory and oxidative pathways to restore joint homeostasis and delay osteoarthritic degeneration.

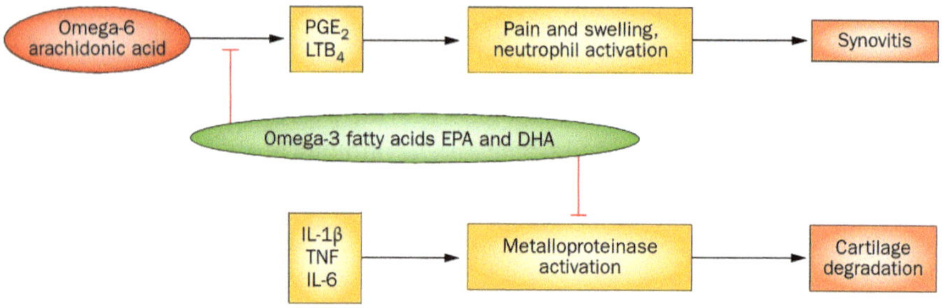

Figure 9.22 **Potential protective actions of omega-3 fats in OA.** The omega-3 fats in dietary fish oil, EPA and DHA, are homologues of arachidonic acid, the omega-6 fatty acid precursor of proinflammatory prostaglandins and leukotrienes, such as PGE2 and LTB4. Competitive inhibition of arachidonic acid metabolism by EPA and DHA could reduce inflammation, pain, and synovitis. EPA and DHA suppress chondrocyte metalloproteinases production in vitro, and dietary fish oil has a protective effect on cartilage and subchondral bone in animal models of OA. Abbreviations: DHA, docosahexaenoic acid; EPA, eicosapentaenoic acid; LTB4, leukotriene B4; OA, osteoarthritis; PGE2, prostaglandin E2. *Source:* [197]/ Springer Nature.

research also suggests benefits in slowing OA progression, improving quality of life (QoL), and potentially reducing the need for surgical intervention.

9.6.2.2 Marine Peptides

Marine-derived peptides, especially those from collagen-rich sources such as sturgeon skin, seahorse tissue, jellyfish, and blue mussel (M. edulis), are gaining attention as functional nutraceuticals for the management of OA. These bioactive peptides,

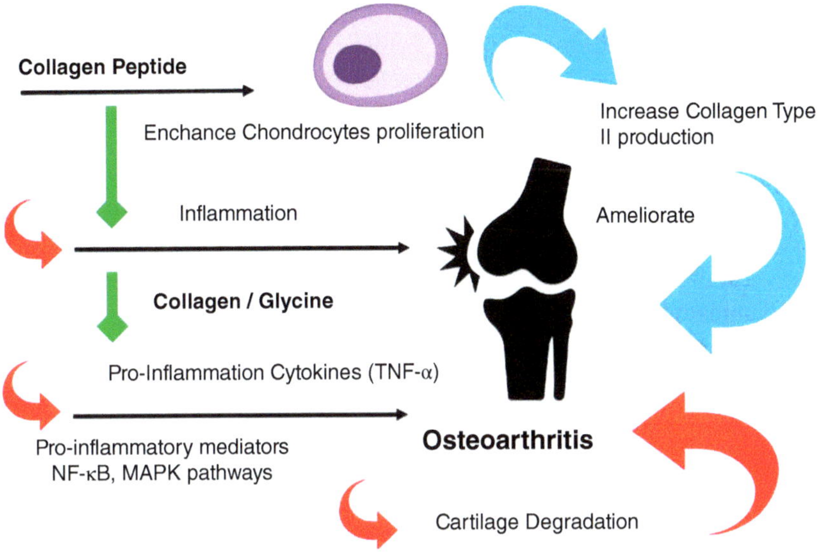

Figure 9.23 **Schematic overview of jellyfish-derived collagen peptide and collagen/glycine actions in osteoarthritis. Jellyfish collagen peptide promotes chondrocyte proliferation and increases collagen type II (Col II) synthesis.** In contrast, collagen/glycine attenuates inflammation by downregulating tumor necrosis factor-alpha (TNF-α) and inhibiting nuclear factor kappa B (NF-κB) and mitogen-activated protein kinase (MAPK) signaling, together preserving cartilage integrity in the osteoarthritic knee.
Source: [201]/ Frontiers Media SA / CC BY 4.0.

typically obtained through enzymatic hydrolysis, exert multiple joint-protective effects by modulating inflammation, oxidative stress, and cartilage metabolism.

Sturgeon collagen peptides supply type II collagen–building amino acids and can down-regulate cartilage-degrading enzymes, helping to preserve extracellular matrix structure [199]. Seahorse protein hydrolysates have been shown in obese post-traumatic OA models to suppress pro-inflammatory mediators, reduce proteoglycan loss, and improve joint swelling and pain. Jellyfish collagen peptides—particularly from Rhopilema esculentum enhance antioxidant capacity, inhibit inflammatory cytokines, and promote cartilage repair [191]. Blue mussel water extracts are rich in bioactive peptides that attenuate inflammatory responses, bolster antioxidant enzymes, and protect cartilage from degradation in OA, especially in obesity related models [200].

Collectively, these marine peptides act on key inflammatory pathways (e.g., NF-κB, MAPK), reduce MMPs such as MMP-3 and MMP-13, and support chondrocyte viability, making them promising adjuncts for slowing the progression of OA and improving joint function [201] (Fig. 9.23).

9.6.2.3 Polysaccharides

Marine-derived polysaccharides—found in seaweeds, shellfish, and certain marine plants—are increasingly recognized for their potential in OA management. Compounds such as fucoidan from brown algae, carrageenan from red algae, and alginate extracted from kelp display strong anti-inflammatory activity. They work by downregulating key pro-inflammatory cytokines, such as IL-1β and TNF-α, suppressing cartilage-degrading enzymes (MMP-3, MMP-13), and reducing nitric oxide production [202–205]. These actions help protect articular cartilage from degradation while also decreasing joint inflammation and swelling in experimental OA models.

Some marine polysaccharides also possess remarkable antioxidant capacity. β-1,3-Xylooligosaccharides from Caulerpa lentillifera, for example, have been shown to scavenge ROS and restore antioxidant enzyme activity in chondrocytes under inflammatory stress [206, 207]. Chitosan and its oligosaccharides—obtained from crustacean shells—can inhibit NF-κB signaling, limiting inflammation and oxidative damage. By mitigating oxidative stress, these compounds help preserve extracellular matrix integrity and enhance chondrocyte survival, both of which are crucial for slowing OA progression [208].

In addition to their systemic effects, specific marine polysaccharides serve as biomaterial scaffolds or delivery systems for joint-protective agents. Alginate hydrogels, for instance, can be used to deliver therapeutic molecules directly to the joint space, while low-molecular-weight carrageenan fractions may enhance cartilage regeneration [209]. Together, these anti-inflammatory, antioxidant, and structural-support functions make marine polysaccharides a versatile class of nutraceutical and biomedical candidates for complementary OA treatment strategies.

9.6.3 Gut Health

PUFAs, particularly omega-3 fatty acids such as EPA and DHA, found in fish oil and microalgae, play a crucial role in maintaining gut health. These lipids exert anti-inflammatory effects by modulating the NF-κB and PPARγ pathways, which help suppress pro-inflammatory cytokines such as TNF-α and IL-6 [210] (Figs. 9.24 and 9.25). Omega-3s also enhance intestinal barrier integrity by upregulating tight junction proteins and promote the growth of beneficial gut microbes, including Lactobacillus and Bifidobacterium. Through these mechanisms, omega-3s contribute to enhanced mucosal immunity and reduced endotoxemia, making them valuable in managing conditions such as inflammatory bowel disease and metabolic syndrome [211].

Marine-derived peptides, obtained from fish muscle, collagen hydrolysates, and seaweed proteins, offer a range of multifaceted benefits for gut health. These bioactive sequences exhibit potent antioxidant and anti-inflammatory properties, which help mitigate oxidative stress and support epithelial repair [212–215]. Some peptides interact directly with gut receptors, modulating immune responses and enhancing mucosal defense [216]. Additionally, emerging evidence suggests that marine peptides can influence gut microbiota composition, fostering a more diverse and resilient microbial ecosystem [217]. Their dual action—both biochemical and microbial—positions them as promising candidates for functional foods targeting gut integrity and immune modulation.

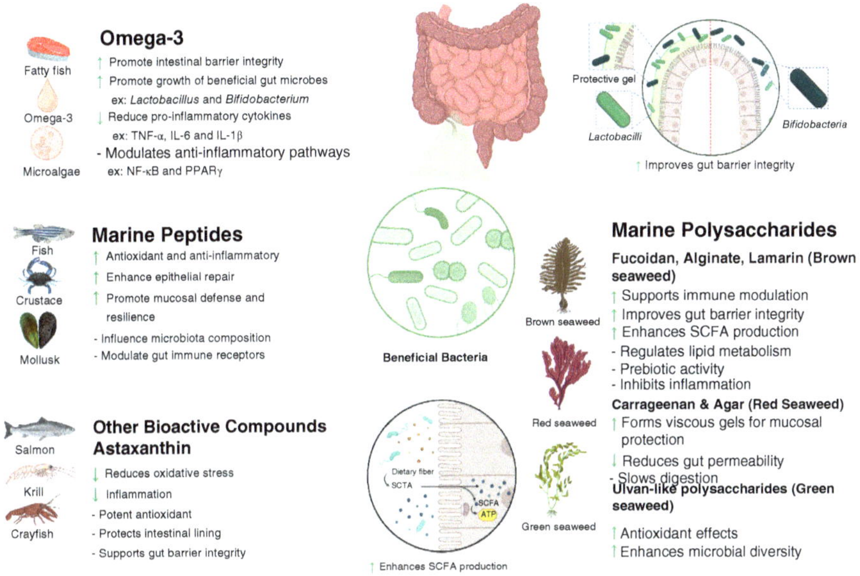

Figure 9.24 Aquatic-derived compounds influence gut microbial ecology and metabolite profiles, contributing to improved mucosal immunity and metabolic balance.

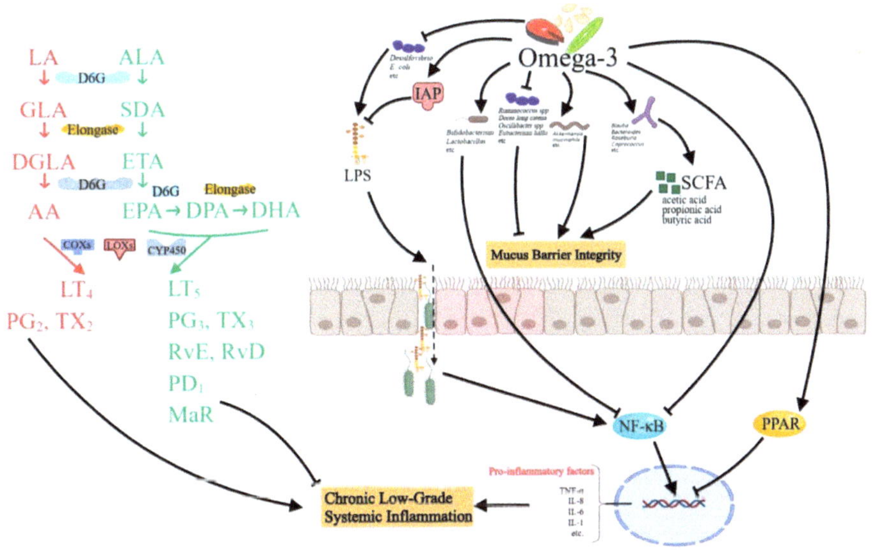

Figure 9.25 Omega-3 fatty acids (EPA/DHA) generate anti-inflammatory mediators (e.g., resolvins, protectins, maresins) via enzymatic pathways, modulate gut microbiota (e.g., ↑ Akkermansia, Bifidobacterium, Lactobacillus), enhance mucus barrier integrity, and promote SCFA production—ultimately suppressing NF-κB and PPAR-driven chronic low-grade inflammation. *Source:* [210]/ Frontiers Media SA / CC BY 4.0.

Seaweed-derived polysaccharides, including fucoidan, alginate, laminarin, carrageenan, and ulvan-like compounds, represent a rich source of dietary fiber with prebiotic potential [130, 218–220]. Brown seaweeds, such as Sargassum hemiphyllum and C. lentillifera, contain fucoidan and alginate, which are fermented by gut microbiota into short-chain fatty acids (SCFAs), including butyrate and acetate [221, 222]. These SCFAs regulate immune function, reinforce the gut barrier, and modulate lipid metabolism (Fig. 9.26). Red seaweeds, such as Eucheuma cottonii, provide carrageenan and agar, which form viscous gels that slow digestion and may reduce gut permeability [223]. Green seaweeds contribute antioxidant ulvan-like polysaccharides that support microbial diversity and immune balance [224]. Together, these compounds act along the microbiota–immune–metabolic axis, offering therapeutic potential for gut dysbiosis, obesity, and chronic inflammation.

Moreover, Astaxanthin, a potent antioxidant carotenoid found in microalgae, krill, and salmon, protects the intestinal lining against oxidative stress and inflammation, supporting gut barrier integrity [225, 226].

9.6.4 Skin Health

PUFAs, particularly omega-3 fatty acids such as EPA and DHA, are crucial for maintaining skin barrier integrity, hydration, and resilience. These fatty acids are incorporated into epidermal phospholipids and modulate inflammatory responses via eicosanoid signaling [227]. Omega-6 PUFAs such as linoleic acid support ceramide synthesis and barrier function, while omega-3s attenuate UV-induced photodamage and reduce inflammatory cytokines like IL-1β and TNF-α. Both oral and topical delivery of PUFAs can improve skin elasticity and reduce transepidermal water loss (TEWL), making them valuable in managing dry skin, photoaging, and inflammatory dermatoses [228] (Fig. 9.27).

Marine peptides, derived from fish skin, mollusks, and jellyfish, exhibit potent antioxidant, anti-inflammatory, and anti-photoaging properties [229]. These peptides inhibit MMPs, elastase, and hyaluronidase—enzymes that degrade collagen and elastin under UV stress [230, 231] (Fig. 9.28). MCPs, in particular, stimulate endogenous collagen synthesis, thereby improving skin elasticity and hydration and reducing fine lines [232, 233]. Their low molecular weight enhances bioavailability, and they are increasingly used in cosmeceuticals and nutraceuticals for skin rejuvenation and wound healing.

Marine polysaccharides, such as ulvan (from Ulva spp.), fucoidan (from brown seaweed), and carrageenan (from red algae), offer multifunctional skin benefits. Ulvan demonstrates excellent biocompatibility, antioxidant, immunomodulatory, and antimicrobial properties, making it suitable for use in hydrogels, films, and emulsions in skincare applications [234, 235]. Fucoidan and alginate contribute to skin hydration, barrier protection, and anti-inflammatory effects, while carrageenan forms protective gels that enhance moisture retention [236, 237] (Fig. 9.29). These polysaccharides also inhibit collagenase and elastase, support fibroblast proliferation, and reduce oxidative stress, positioning them as sustainable alternatives in antiaging and wound care formulations [238].

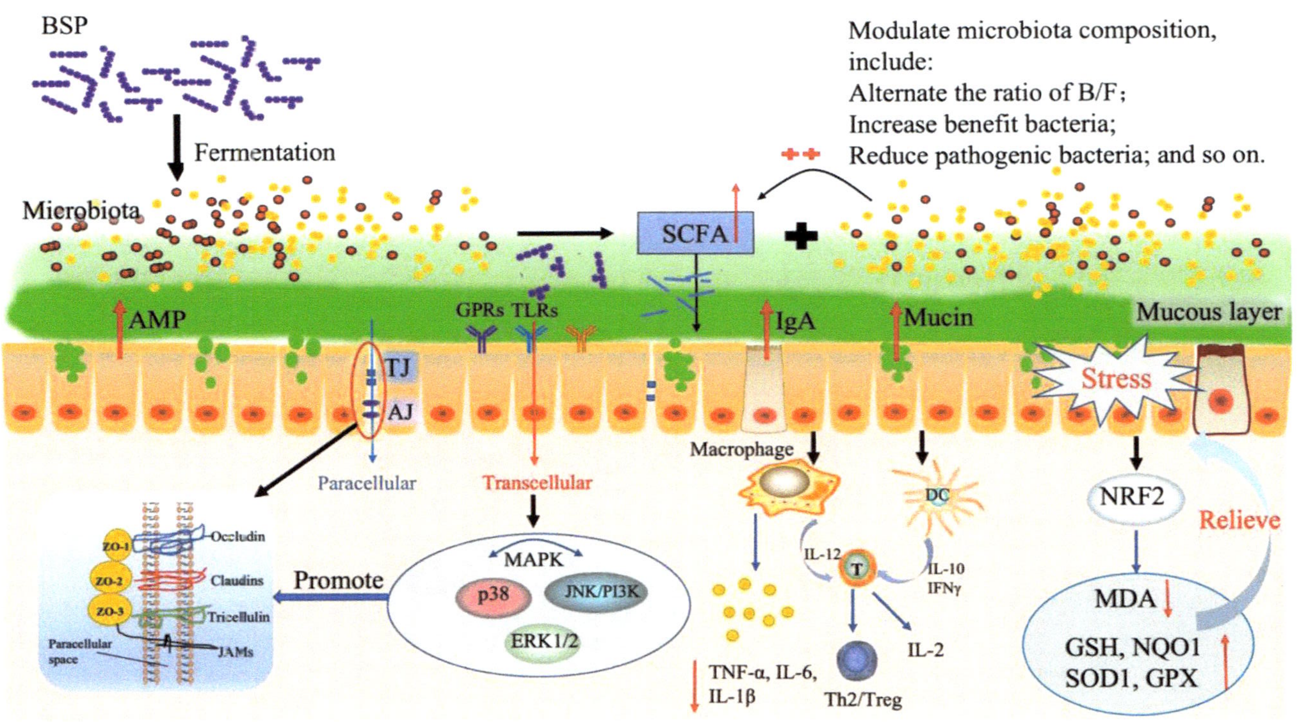

Figure 9.26 The role of brown seaweed polysaccharides (BSP) in the intestinal barrier. BSP that gastric and intestinal enzymes could not process was degraded and fermented by specific intestinal microbiota. BSP intervention modulated the composition of the microbiota. Fermentation of BSP produced SCFAs and other metabolites. SCFAs could be easily absorbed and promote the intestinal barrier function and the immune system. BSP protected intestinal integrity by preserving cellular antioxidant status, promoting cytokine production, and maintaining intestinal permeability. BSP, brown seaweed polysaccharide; SCFA, short chain fatty acid; B/F, Bacteroidetes/Firmicutes; AMP, antimicrobial peptide; IgA, immunoglobulin A; TJ, tight junction; AJ, adhesion junction; TLRs, Toll-like receptors; MDA, malonaldehyde; GSH, glutathione; NQO1, Quinone oxidoreductase 1; SOD, superoxide dismutase; GPX, glutathione peroxidase; DC, dendritic cell; Th2/Treg, T-helper cell/T-regulatory cells; IL, interleukin; IFN-γ, interferon-γ; TNF-α, tumor necrosis factor-α; MAPK, mitogen-activated protein kinase. *Source:* [222]/ Elsevier / CC BY-NC-ND 4.0.

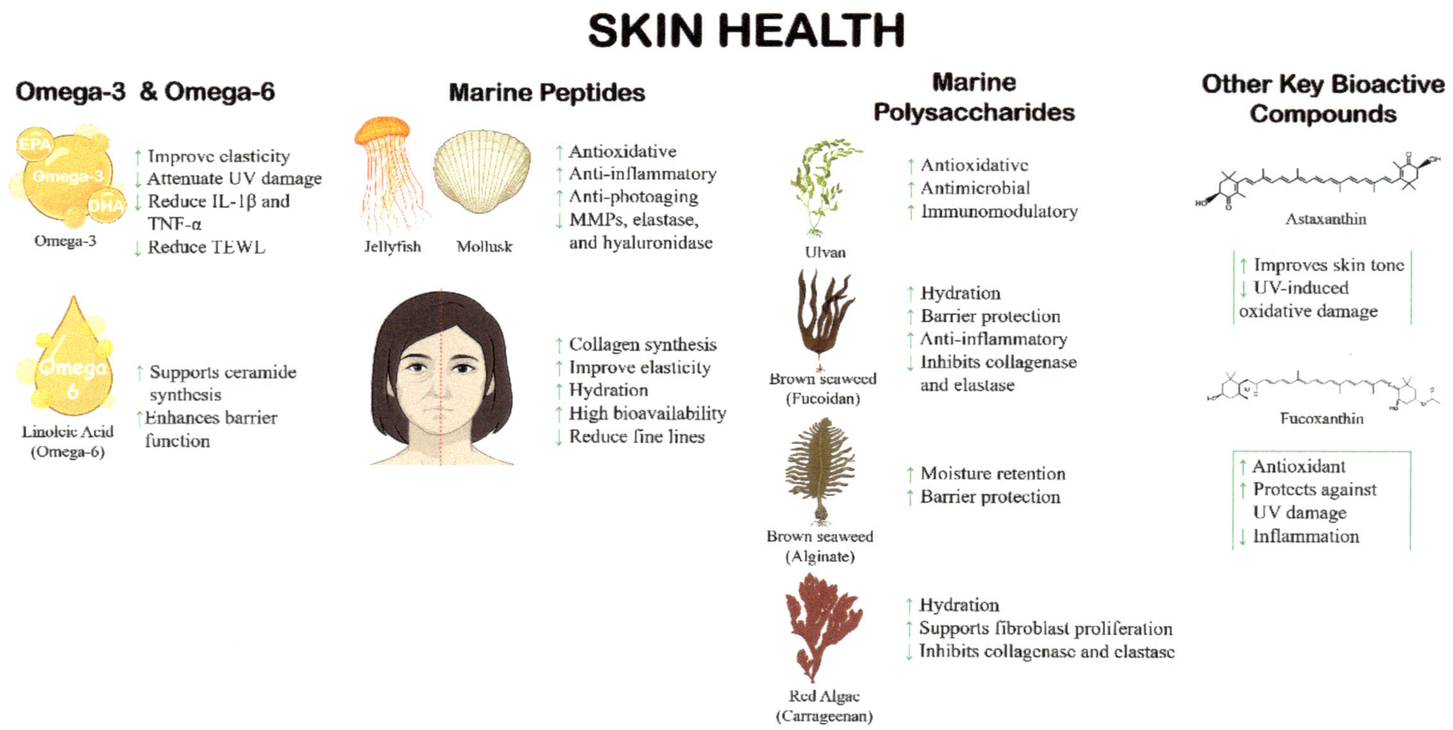

Figure 9.27 Marine bioactives target key pathways in skin physiology—including NF-κB, MAPK, and MMP regulation—resulting in improved dermal structure, reduced inflammation, and enhanced regeneration.

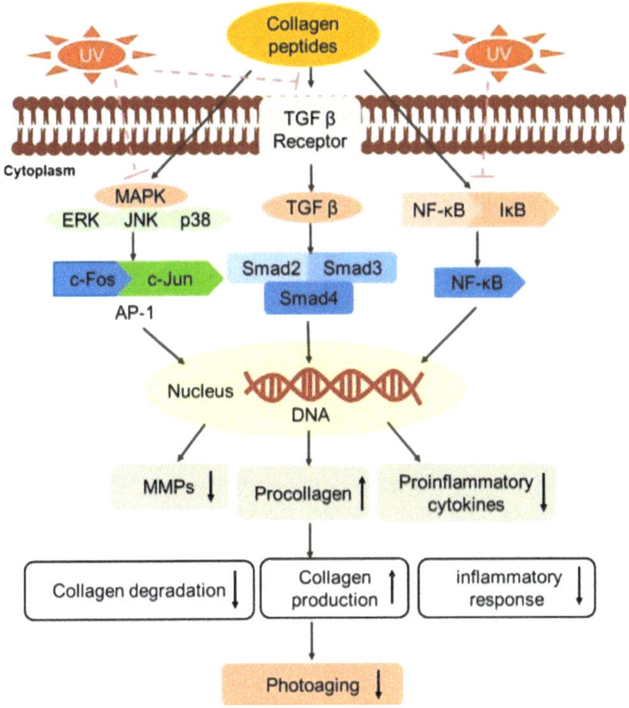

Figure 9.28 Collagen peptides inhibit UV-induced photoaging by blocking MAPK (ERK/JNK/p38) and NF-κB activation, reducing AP-1–driven MMP expression, and restoring TGF-β/Smad/Smad-mediated procollagen production to preserve dermal collagen and prevent inflammation. *Source:* [231]/ Elsevier / CC BY-NC-ND 4.0.

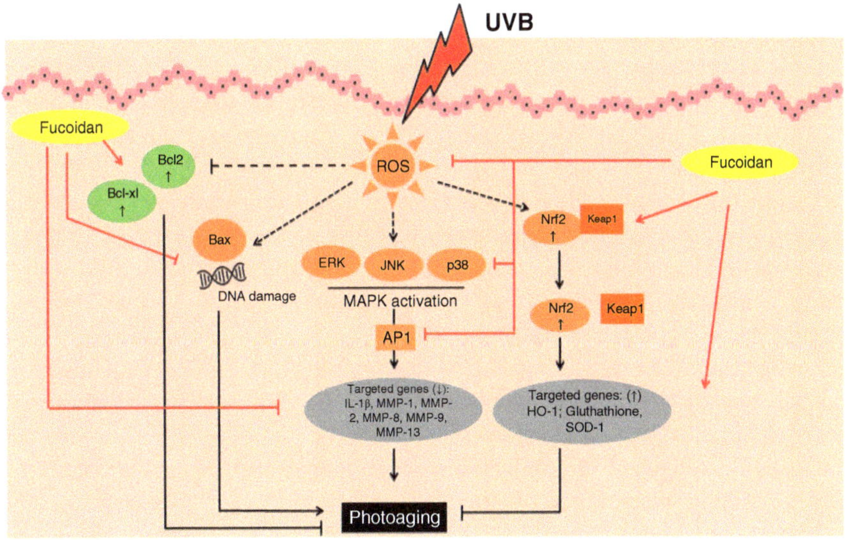

Figure 9.29 Fucoidan protects against UVB-induced photoaging by scavenging ROS and preventing DNA damage through upregulation of Bcl-xl/Bcl2 (inhibiting Bax), suppressing the MAPK (ERK/JNK/p38)/AP-1 axis to reduce MMP-1/2/8/9/13 and IL-1β expression, and activating the Nrf2 pathway (↑HO-1, glutathione, SOD-1; ↓Keap1) to bolster cellular antioxidant defenses. *Source:* [237]/ MDPI / CC BY 4.0.

Other marine bioactives include carotenoids, such as astaxanthin and fucoxanthin, which protect against UV-induced oxidative damage and improve skin tone [239]. Marine-derived polyphenols and sterols also contribute to anti-inflammatory and photoprotective effects [240]. Collectively, these compounds act synergistically to maintain skin homeostasis, prevent premature aging, and support dermal regeneration.

9.7 Conclusion and Future Research Directions

The field of aquatic product research is dynamic, with discoveries and emerging trends continuously shaping our understanding of their impact on human health. Aquatic products offer a wealth of health benefits, including improved brain and heart health, potential cancer prevention, and weight management, as well as other notable advantages related to immunity and inflammation. These benefits are primarily attributed to their rich content of omega-3 fatty acids, high-quality protein, and a diverse array of essential vitamins and minerals. However, the complex interplay of these nutrients, the long-term impacts of consumption, the sustainability and safety of aquaculture, the presence of environmental contaminants, and the specific needs of various demographic groups all necessitate further in-depth research. Future studies should focus on unraveling the synergistic effects of nutrients, conducting longitudinal trials, optimizing aquaculture practices, mitigating environmental risks, targeting research to specific populations, and exploring novel applications and emerging trends. By addressing these critical research directions, we can gain a more comprehensive understanding of the multifaceted role of aquatic products in promoting human health and develop evidence-based recommendations to maximize their benefits while ensuring their sustainable and safe consumption.

Future research should continue to explore and validate novel findings, such as the recently suggested link between seafood consumption in children and prosocial behavior. Understanding the mechanisms of such associations could have significant implications for dietary guidelines and public health initiatives. Furthermore, the potential of novel bioactive compounds derived from aquatic sources for promoting health and preventing disease represents an exciting area for future investigation. This includes exploring compounds with antioxidant, anti-inflammatory, and anticancer properties. Finally, the role of aquatic foods in shaping the gut microbiome and the subsequent impact on overall health is a promising avenue for future research. Understanding how different aquatic products influence the composition and function of the gut microbiota could reveal new pathways through which these foods exert their health benefits. Continued exploration of these novel applications and emerging trends will be crucial for fully harnessing the potential of aquatic products to improve human health and well-being.

References

1 Roos, N. (2025). Spotlight: aquaculture for human nutrition and health outcomes. *Frontiers in Aquaculture* 4, 1574847.

2 Action, S. (2020). World fisheries and aquaculture. *Food and Agriculture Organization* 2020: 1–244.

3 Arshad, N., Samat, N., and Lee, L.K. (2022). Insight into the relation between nutritional benefits of aquaculture products and its consumption hazards: a global viewpoint. *Frontiers in Marine Science* 9, 925463.

4 Bavinck, M., Ahern, M., Hapke, H. M., Johnson, D. S., Kjellevold, M., Kolding, J., ... & Franz, N. (2023). Small fish for food security and nutrition. *FAO fisheries and aquaculture technical paper*, (694), I-154.

5 Even, B., Thai, H. T. M., Pham, H. T. M., & Béné, C. (2024). Defining barriers to food systems sustainability: a novel conceptual framework. Frontiers in Sustainable Food Systems, 8, 1453999.

6 Ahern, M., Thilsted, S., Oenema, S., & Kühnhold, H. (2021). The role of aquatic foods in sustainable healthy diets. UN Nutrition Discussion Paper.

7 Mittal, R. K., Mishra, R., Sharma, V., & Purohit, P. (2024). Bioactive exploration in functional foods: unlocking nature's treasures. Current Pharmaceutical Biotechnology, 25(11), 1419–1435.

8 Pradeepkiran, J.A. (2019). Aquaculture role in global food security with nutritional value: a review. *Transl Anim Sci* 3(2): 903–910.

9 Tacon, A.G. and Metian, M. (2013). Fish matters: importance of aquatic foods in human nutrition and global food supply. *Reviews in Fisheries Science* 21(1): 22–38.

10 Agriculture Organization of the United Nations. Fisheries Department. (2018). The state of world fisheries and aquaculture. Food and Agriculture Organization of the United Nations.

11 Fan, J., Yang, Y., Yang, Y., Rui, D., Guo, S., Li, M., ... & Zhao, C. (2025). Predictive role of the dietary inflammatory index on stroke risk among hypertensive patients. Scientific Reports, 15(1), 13602.

12 Hosomi, R., Yoshida, M., and Fukunaga, K. (2012). Seafood consumption and components for health. *Global Journal of Health Science* 4 (3): 72.

13 Kris-Etherton, P.M., Harris, W.S., and Appel, L.J. (2002). Fish consumption, fish oil, omega-3 fatty acids, and cardiovascular disease. *Circulation* 106 (21): 2747–2757.

14 Reames, E. (2012). Nutritional benefits of seafood (SRAC Publication No. 7300). Southern Regional Aquaculture Center, LSU AgCenter.

15 Who, J. and Consultation, F.E. (2003). Diet, nutrition and the prevention of chronic diseases. *World Health Organ Tech Rep Ser* 916 (i-viii): 1–149.

16 Zhang, B., Xiong, K., Cai, J., & Ma, A. (2020). Fish consumption and coronary heart disease: a meta-analysis. Nutrients, 12(8), 2278.

17 Beal, T., Manohar, S., Miachon, L., & Fanzo, J. (2024). Nutrient-dense foods and diverse diets are important for ensuring adequate nutrition across the life course. *Proceedings of the National Academy of Sciences*, 121(50), e2319007121.

18 Troesch, B., Biesalski, H. K., Bos, R., Buskens, E., Calder, P. C., Saris, W. H., ... & Eggersdorfer, M. (2015). Increased intake of foods with high nutrient density can help to break the intergenerational cycle of malnutrition and obesity. *Nutrients*, 7(7), 6016–6037.

19 Drewnowski, A. and Fulgoni, V.L. 3rd (2014). Nutrient density: principles and evaluation tools. *Am J Clin Nutr* 99 (5 Suppl): 1223s–1228s.

20 Nicklas, T.A., Drewnowski, A., and O'Neil, C.E. (2014). The nutrient density approach to healthy eating: challenges and opportunities. *Public Health Nutr* 17 (12): 2626–2636.

21 Pateiro, M., Domínguez, R., Varzakas, T., Munekata, P. E., Movilla Fierro, E., & Lorenzo, J. M. (2021). Omega-3-rich oils from marine side streams and their potential application in food. *Marine Drugs*, 19(5), 233.

22 Saidaiah, P., Banu, Z., Khan, A. A., Geetha, A., & Somraj, B. (2024). A comprehensive review of Omega-3 fatty acids: Sources, industrial applications, and health benefits. *Ann Phytomed*, 13(1), 209–225.

23 Ucak, I., Afreen, M., Montesano, D., Carrillo, C., Tomasevic, I., Simal-Gandara, J., & Barba, F. J. (2021). Functional and bioactive properties of peptides derived from marine side streams. *Marine Drugs*, 19(2), 71.

24 Walquist, M. J., Eilertsen, K. E., Elvevoll, E. O., & Jensen, I. J. (2024). Marine-derived peptides with anti-hypertensive properties: Prospects for pharmaceuticals, supplements, and functional food. *Marine Drugs*, 22(4), 140.

25 Shahidi, F. and Saeid, A. (2025). Bioactivity of marine-derived peptides and proteins: a review. *Marine Drugs* 23(4): 157.

26 Das, B. K., Kumari, K., Ganguly, S., Talukder, A. K., Ray, A., Dutta, S., & Baisakhi, B. (2024). Unlocking the nutritional potential: amino acid profile of eight Indian food fishes and their role in meeting recommended dietary allowances. *Frontiers in Sustainable Food Systems*, 8, 1432034.

27 Balgir, P.P., Kaur, T., and Sharma, M. (2016). Antihypertensive peptides derived from food sources. *MOJ Food Processing & Technology* 2 (1): 1–6.

28 Nardo, A. E., Suárez, S., Quiroga, A. V., & Añón, M. C. (2020). Amaranth as a source of antihypertensive peptides. *Frontiers in Plant Science*, 11, 578631.

29 Pujiastuti, D. Y., Ghoyatul Amin, M. N., Alamsjah, M. A., & Hsu, J. L. (2019). Marine organisms as potential sources of bioactive peptides that inhibit the activity of angiotensin I-converting enzyme: a review. *Molecules*, 24(14), 2541.

30 Ryu, B., Shin, K.H., and Kim, S.K. (2021). Muscle protein hydrolysates and amino acid composition in fish. *Mar Drugs* 19(7), 377.

31 Meng, W., Zhang, J., Hou, H., Yu, L., & Dong, P. (2025). Exploring the structures and molecular mechanisms of bioactive compounds from marine foods for hyperuricemia prevention: a systematic review. *Critical Reviews in Food Science and Nutrition*, 65(30), 7251–7269.

32 Cadar, E., Popescu, A., Dragan, A. M. L., Pesterau, A. M., Pascale, C., Anuta, V., ... & Ionescu, A. M. (2025). Bioactive compounds of marine algae and their potential health and nutraceutical applications: a review. *Marine Drugs*, 23(4), 152.

33 Ghosh, S., Sarkar, T., Pati, S., Kari, Z. A., Edinur, H. A., & Chakraborty, R. (2022). Novel bioactive compounds from marine sources as a tool for functional food development. *Frontiers in Marine Science*, 9, 832957.

34 Lomartire, S. and Gonçalves, A.M.M. (2022). An overview of potential seaweed-derived bioactive compounds for pharmaceutical applications. *Mar Drugs* 20 (2): 141.

35 Lomartire, S., Marques, J.C., and Gonçalves, A.M.M. (2021). An overview to the health benefits of seaweeds consumption. *Mar Drugs* 19 (6): 341.

36 Lordan, S., Ross, R.P., and Stanton, C. (2011). Marine bioactives as functional food ingredients: potential to reduce the incidence of chronic diseases. *Mar Drugs* 9 (6): 1056–1100.

37 Carrasqueira, J., Bernardino, S., Bernardino, R., & Afonso, C. (2025). Marine-derived polysaccharides and their potential health benefits in nutraceutical applications. *Marine drugs*, 23(2), 60.

38 Dang, Y., Li, Z., and Yu, F. (2024). Recent advances in astaxanthin as an antioxidant in food applications. *Antioxidants* 13 (7): 879.

39 Lordan, R., Tsoupras, A., and Zabetakis, I. (2017). Phospholipids of animal and marine origin: Structure, function, and anti-inflammatory properties. *Molecules* 22 (11): 1964.

40 Wallace, T. C., Blusztajn, J. K., Caudill, M. A., Klatt, K. C., Natker, E., Zeisel, S. H., & Zelman, K. M. (2018). Choline: the underconsumed and underappreciated essential nutrient. *Nutrition today*, 53(6), 240–253.

41 Venugopal, V. and Gopakumar, K. (2017). Shellfish: nutritive value, health benefits, and consumer safety. *Comprehensive Reviews in Food Science and Food Safety* 16 (6): 1219–1242.

42 Saini, R. (2011). Coenzyme Q10: the essential nutrient. *J Pharm Bioallied Sci* 3 (3): 466–467.

43 Santisteban, V., Muñoz-Garcia, N., López-Yerena, A., Puntes, M., Badimon, L., & Padro, T. (2024). Efficacy of food Industry by-product β-glucan/chitin–chitosan on lipid profile of overweight and obese individuals: sustainability and nutraceuticals. *Nutrients*, 16(19), 3420.

44 Qaradakhi, T., Gadanec, L. K., McSweeney, K. R., Abraham, J. R., Apostolopoulos, V., & Zulli, A. (2020). The anti-inflammatory effect of taurine on cardiovascular disease. *Nutrients*, 12(9), 2847.

45 Akram, M., Munir, N., Daniyal, M., Egbuna, C., Găman, M. A., Onyekere, P. F., & Olatunde, A. (2020). Vitamins and Minerals: Types, sources and their functions. In *Functional foods and nutraceuticals: bioactive components, formulations and innovations* (pp. 149–172). Cham: Springer International Publishing.

46 Hoque, M. (2023). A review on different dietary sources of important vitamins and electrolytes. *International Journal of Research Publication and Reviews* 4 (8): 731–736.

47 Sarojnalini, C. and Hei, A. (2019). *Fish as an Important Functional Food for Quality Life. U: Functional Foods*, 77–97. Lagouri, V, Ured.

48 Singh, R. D., Arambam, K., Singh, A. M., & Singh, S. K. (2025). Health and nutritional benefits of fish as human food. In *Food Security, Nutrition and Sustainability Through Aquaculture Technologies* (pp. 43–59). Cham: Springer Nature Switzerland.

49 Vieira, B. M., de Mello e Silva, G. N., & Silva, M. I. (2025). Nutritional elements II: Vitamins and minerals. In *Fundamentals of drug and non-drug interactions: Physiopathological perspectives and clinical approaches* (pp. 57–86). Cham: Springer Nature Switzerland.

50 Puri, S., Shaheen, M., and Grover, B. (2023). Nutrition and cognitive health: a life course approach. *Front Public Health* 11: 1023907.

51 Fekete, M., Lehoczki, A., Tarantini, S., Fazekas-Pongor, V., Csípő, T., Csizmadia, Z., & Varga, J. T. (2023). Improving cognitive function with nutritional supplements in aging: a comprehensive narrative review of clinical studies investigating the effects of vitamins, minerals, antioxidants, and other dietary supplements. *Nutrients*, 15(24), 5116.

52 Mora, I., Arola, L., Caimari, A., Escoté, X., & Puiggròs, F. (2022). Structured long-chain omega-3 fatty acids for improvement of cognitive function during aging. *International journal of molecular sciences*, 23(7), 3472.

53 Sinclair, A.J. (2019). Docosahexaenoic acid and the brain-what is its role? *Asia Pacific Journal of Clinical Nutrition* 28 (4): 675–688.

54 Hashimoto, M., Hossain, S., Al Mamun, A., Matsuzaki, K., & Arai, H. (2017). Docosahexaenoic acid: one molecule diverse functions. *Critical reviews in biotechnology*, 37(5), 579–597.

55 von Schacky, C. (2021). Importance of EPA and DHA blood levels in brain structure and function. *Nutrients* 13 (4).

56 Dyall, S.C. (2015). Long-chain omega-3 fatty acids and the brain: a review of the independent and shared effects of EPA, DPA and DHA. *Front Aging Neurosci* 7: 52.

57 Bradbury, J. (2011). Docosahexaenoic acid (DHA): an ancient nutrient for the modern human brain. *Nutrients* 3 (5): 529–554.

58 DiNicolantonio, J.J. and O'Keefe, J.H. (2020). The importance of marine omega-3s for brain development and the prevention and treatment of behavior, mood, and other brain disorders. *Nutrients* 12 (8).

59 Zhou, L., Xiong, J. Y., Chai, Y. Q., Huang, L., Tang, Z. Y., Zhang, X. F., ... & Zhang, J. T (2022). Possible antidepressant mechanisms of omega-3 polyunsaturated fatty acids acting on the central nervous system. *Frontiers in Psychiatry* 13: 933704.

60 Cao, D., Kevala, K., Kim, J., Moon, H. S., Jun, S. B., Lovinger, D., & Kim, H. Y. (2009). Docosahexaenoic acid promotes hippocampal neuronal development and synaptic function. *Journal of Neurochemistry* 111 (2): 510–521.

61 Di Miceli, M., Bosch-Bouju, C., and Layé, S. (2020). PUFA and their derivatives in neurotransmission and synapses: a new hallmark of synaptopathies. *Proceedings of the Nutrition Society* 79 (4): 388–403.

62 Crupi, R., Marino, A., and Cuzzocrea, S. (2013). n-3 fatty acids: role in neurogenesis and neuroplasticity. *Current Medicinal Chemistry* 20 (24): 2953–2963.

63 Zinkow, A., Grodzicki, W., Czerwińska, M., & Dziendzikowska, K. (2025). Molecular mechanisms linking omega-3 fatty acids and the gut–brain axis. *Molecules* 30 (1): 71.

64 Kim, H.-Y., Huang, B.X., and Spector, A.A. (2022). Molecular and signaling mechanisms for docosahexaenoic acid-derived neurodevelopment and neuroprotection. *International Journal of Molecular Sciences* 23 (9): 4635.

65 Tian, J., Zhang, Y., and Zhao, X. (2025). The effects and mechanisms of n-3 and n-6 polyunsaturated fatty acids in the central nervous system. *Cell Mol Neurobiol* 45 (1): 25.

66 Weiser, M.J., Butt, C.M., and Mohajeri, M.H. (2016). Docosahexaenoic acid and cognition throughout the lifespan. *Nutrients* 8 (2): 99.

67 Yurko-Mauro, K., Alexander, D.D., and Van Elswyk, M.E. (2015). Docosahexaenoic acid and adult memory: a systematic review and meta-analysis. *PloS one* 10 (3): e0120391.

68 Devassy, J. G., Leng, S., Gabbs, M., Monirujjaman, M., & Aukema, H. M. (2016). Omega-3 polyunsaturated fatty acids and oxylipins in neuroinflammation and management of Alzheimer disease. *Adv Nutr* 7 (5): 905–916.

69 Joffre, C., Rey, C., and Layé, S. (2019). N-3 polyunsaturated fatty acids and the resolution of neuroinflammation. *Frontiers in pharmacology* 10: 1022.

70 Calder, P.C. (2020). n-3 PUFA and inflammation: from membrane to nucleus and from bench to bedside. *Proceedings of the Nutrition Society* 79 (4): 404–416.

71 Bălaşa, A.F., Chircov, C., and Grumezescu, A.M. (2020). Marine biocompounds for neuroprotection-a review. *Mar Drugs* 18 (6), 290.

72 Guimarães, P. L., Tavares, D. Q., Carrião, G. S., Oliveira, M. E. H., & Oliveira, C. R.(2023). Potential of marine compounds in the treatment of neurodegenerative diseases: a review. *Brazilian Journal of Biology* 83: e266795.

73 Silva, M., Seijas, P., and Otero, P. (2021). Exploitation of marine molecules to manage Alzheimer's disease. *Mar Drugs* 19 (7), 373.

74 Huang, C., Zhang, Z., and Cui, W. (2019). Marine-derived natural compounds for the treatment of Parkinson's disease. *Mar Drugs* 17 (4), 221.

75 Jia, C., Chai, J., Zhang, S., Sun, Y., He, L., Sang, Z., ... & Zheng, X. (2025). The advancements of marine natural products in the treatment of Alzheimer's disease: a study based on cell and animal experiments. *Marine Drugs* 23 (3): 91.

76 Faller, P. and Hureau, C. (2021). Reproducibility problems of amyloid-β self-assembly and how to deal with them. *Frontiers in Chemistry* 8: 611227.

77 Wang, Q., Yu, X., Li, L., & Zheng, J. (2014). Inhibition of amyloid-β aggregation in Alzheimer's disease. *Current pharmaceutical design*, 20(8), 1223–1243.

78 Tsoy, A., Umbayev, B., Kassenova, A., Kaupbayeva, B., & Askarova, S. (2024). Pathology of amyloid-β (Aβ) peptide peripheral clearance in Alzheimer's disease. *International Journal of Molecular Sciences* 25(20): 10964.

79 Cheung, R.C., Ng, T.B., and Wong, J.H. (2015). Marine peptides: bioactivities and applications. *Mar Drugs* 13 (7): 4006–4043.

80 Ochoa de la Paz, L., Zenteno, E., Gulias-Cañizo, R., & Quiroz-Mercado, H. (2019). Taurine and GABA neurotransmitter receptors, a relationship with therapeutic potential?. *Expert review of neurotherapeutics*, 19(4), 289–291.

81 Tang, C.-d., J.-H. Cheng, and D.-W. Sun, Structure-activity relationships and activity enhancement techniques of marine bioactive peptides (MBPs). *Critical Reviews in Food Science and Nutrition*, 2024: p. 1–23.

82 Wu, J.Y. and Prentice, H. (2010). Role of taurine in the central nervous system. *J Biomed Sci* 17 (Suppl 1): S1.

83 Xu LinLin, X. L., Dong WenHong, D. W., Zhao Jie, Z. J., & Xu YaJun, X. Y. (2015). Effect of marine collagen peptides on physiological and neurobehavioral development of male rats with perinatal asphyxia.

84 Zhu, L., Ren, Y., Zhang, S., Guo, Y., Zong, J., & Liu, Y. (2024). Marine-derived polysaccharides: the potential agents against neurodegenerative diseases. *Frontiers in Pharmacology* 15: 1506789.

85 Guo R, Pang J, Zhao J, Xiao X, Li J, Li J, Wang W, Zhou S, Zhao Y, Zhang Z, Chen H, Yuan T, Wu S & Liu Z. (2023). Unveiling the neuroprotective potential of dietary polysaccharides: a systematic review. *Frontiers in Nutrition* 10: 1299117.

86 Medoro, A., Davinelli, S., Milella, L., Willcox, B. J., Allsopp, R. C., Scapagnini, G., & Willcox, D. C. (2023). Dietary astaxanthin: a promising antioxidant and anti-inflammatory agent for brain aging and adult neurogenesis. *Marine Drugs* 21 (12): 643.

87 Wang, S. and Qi, X. (2022). The putative role of astaxanthin in neuroinflammation modulation: mechanisms and therapeutic potential. *Frontiers in Pharmacology* 13: 916653.

88 Choudhary, R. C., Kuschner, C. E., Kazmi, J., Mcdevitt, L., Espin, B. B., Essaihi, M., Nishikimi, M., Becker, L. B., & Kim, J. (2024). The Role of Phospholipid Alterations in Mitochondrial and Brain Dysfunction after Cardiac Arrest. *International Journal of Molecular Sciences,* 25(9), 4645.

89 Dai, Y., Tang, H., and Pang, S. (2021). The crucial roles of phospholipids in aging and lifespan regulation. *Frontiers in Physiology* 12: 775648.

90 Gimenez, M.S., Oliveros, L.B., and Gomez, N.N. (2011). Nutritional deficiencies and phospholipid metabolism. *Int J Mol Sci* 12 (4): 2408–2433.

91 Hasselmo, M.E. (2006). The role of acetylcholine in learning and memory. *Curr Opin Neurobiol* 16 (6): 710–715.

92 Chen, Z. R., Huang, J. B., Yang, S. L., & Hong, F. F. (2022). Role of cholinergic signaling in Alzheimer's disease. *Molecules* 27 (6): 1816.

93 Michaeloudes, C., Christodoulides, S., Christodoulou, P., Kyriakou, T. C., Patrikios, I., & Stephanou, A (2023). Variability in the clinical effects of the omega-3 polyunsaturated fatty acids DHA and EPA in cardiovascular disease—possible causes and future considerations. *Nutrients* 15 (22): 4830.

94 Fassett, R.G. and Coombes, J.S. (2012). Astaxanthin in cardiovascular health and disease. *Molecules* 17 (2): 2030–2048.

95 Jo, D.-M., Khan, F., Park, S.-K., Ko, S.-C., Kim, K. W., Yang, D., Kim, J.-Y., Oh, G.-W., Choi, G., Lee, D.-S., & Kim, Y.-M. (2024). From Sea to Lab: Angiotensin I-Converting Enzyme Inhibition by Marine Peptides—Mechanisms and Applications. Marine Drugs, 22(10), 449. https://doi.org/10.3390/md22100449

96 Kishimoto, Y., Yoshida, H., and Kondo, K. (2016). Potential anti-atherosclerotic properties of astaxanthin. *Marine Drugs* 14 (2): 35.

97 Patil, N. P., Le, V., Sligar, A. D., Mei, L., Chavarria, D., Yang, E. Y., & Baker, A. B. (2018). Algal polysaccharides as therapeutic agents for atherosclerosis. *Frontiers in cardiovascular medicine*, 5, 153.

98 Sheikh, O., Vande Hei, A. G., Battisha, A., Hammad, T., Pham, S., & Chilton, R. (2019). Cardiovascular, electrophysiologic, and hematologic effects of omega-3 fatty acids beyond reducing hypertriglyceridemia: as it pertains to the recently published REDUCE-IT trial. Cardiovascular diabetology, 18(1), 84.

99 Golanski, J., Szymanska, P., and Rozalski, M. (2021). Effects of omega-3 polyunsaturated fatty acids and their metabolites on haemostasis—current perspectives in cardiovascular disease. *International Journal of Molecular Sciences* 22 (5): 2394.

100 Wang, T., Zhang, X., Zhou, N., Shen, Y., Li, B., Chen, B. E., & Li, X. (2023). Association Between Omega-3 Fatty Acid Intake and Dyslipidemia: A Continuous Dose–Response Meta-Analysis of Randomized Controlled Trials. *Journal of the American Heart Association*, 12(11), e029512.

101 Covington, M.B. (2004). Omega-3 fatty acids. *American Family Physician* 70 (1): 133–140.

102 Skulas-Ray, A. C., Wilson, P. W., Harris, W. S., Brinton, E. A., Kris-Etherton, P. M., Richter, C. K., ... & Welty, F. K. (2019). Omega-3 fatty acids for the management of hypertriglyceridemia: a science advisory from the American Heart Association. *Circulation*, 140(12), e673–e691.

103 Bays, H. E., Tighe, A. P., Sadovsky, R., & Davidson, M. H. (2008). Prescription omega-3 fatty acids and their lipid effects: physiologic mechanisms of action and clinical implications. *Expert review of cardiovascular therapy*, 6(3), 391–409.

104 Liao, J., Xiong, Q., Yin, Y., Ling, Z., & Chen, S. (2022). The effects of fish oil on cardiovascular diseases: systematical evaluation and recent advance. *Frontiers in Cardiovascular Medicine*, 8, 802306.

105 Cui, X., Zhao, X., Li, J., Li, Z., Ren, X., Zhao, Y., ... & Li, X. (2021). Advances in the global distribution, chemical constituents, and pharmacology of hippocampus, a traditional marine Chinese medicine. *Frontiers in Marine Science*, 8, 774927.

106 Wijesekara, I. and Kim, S.-K. (2010). Angiotensin-I-converting enzyme (ACE) inhibitors from marine resources: prospects in the pharmaceutical industry. *Marine Drugs* 8 (4): 1080–1093.

107 Zhuang, Y., Sun, L., Zhang, Y., & Liu, G. (2012). Antihypertensive effect of long-term oral administration of jellyfish (Rhopilema esculentum) collagen peptides on renovascular hypertension. *Marine Drugs*, 10(2), 417–426.

108 CunhaNeves, A., Harnedy-Rothwell, P.A., and FitzGerald, R.J. (2022). In vitro angiotensin-converting enzyme and dipeptidyl peptidase-IV inhibitory, and antioxidant activity of blue mussel (*Mytilus edulis*) byssus collagen hydrolysates. *European Food Research and Technology* 248 (7): 1721–1732.

109 Do, J. R., Heo, I. S., Jo, J. H., Kim, D. S., Kim, H. K., Kim, S. S., & Han, C. K. (2006). Effect of antihypertensive peptides originated from various marine proteins on ACE inhibitory activity and systolic blood pressure in spontaneously hypertensive rats. *Korean Journal of Food Science and Technology*, 38(4).

110 Alteration of gene expression profile in kidney of spontaneously hypertensive rats treated with protein hydrolysate of blue mussel (*Mytilus edulis*) by DNA microarray analysis. *i StartPLoS* Onei End b Start10b End (10):

111 Heo, J. H., Kim, E. A., Kang, N., Heo, S. Y., Ahn, G., & Heo, S. J. (2024). The antioxidant effects of trypsin-hydrolysate derived from abalone viscera and fishery by-products, and the angiotensin-I converting enzyme (ACE) inhibitory activity of its purified bioactive peptides. *Marine Drugs*, 22(10), 461.

112 Korczek, K. R., Tkaczewska, J., Duda, I., & Migdał, W. (2020). Effect of heat treatment on the antioxidant and antihypertensive activity as well as in vitro digestion stability of mackerel (*Scomber scombrus*) protein hydrolysates. *Journal of Aquatic Food Product Technology* 29 (1): 73–89.

113 Li, J., Cai, D., Zhai, Y. N., Wu, C. X., Zhang, H. L., Zhang, J., ... & Qu, J. B. (2025). Two antihypertensive and antioxidant peptides derived from Alaska Pollack (*Theragra chalcograma*) skin: in silico, in vitro, and in vivo investigation. *Journal of Agricultural and Food Chemistry* 73 (16): 9932–9945.

114 Tsai, J.-S., Chen, J.-L., and Pan, B.S. (2008). ACE-inhibitory peptides identified from the muscle protein hydrolysate of hard clam (*Meretrix lusoria*). *Process Biochemistry* 43 (7): 743–747.

115 Tsai, J.-S. et al. (2006). The inhibitory effects of freshwater clam (*Corbicula fluminea*, Muller) muscle protein hydrolysates on angiotensin I converting enzyme. *Process Biochemistry* 41 (11): 2276–2281.

116 Wu, Q., Cai, Q. F., Tao, Z. P., Sun, L. C., Shen, J. D., Zhang, L. J., ... & Cao, M. J. (2015). Purification and characterization of a novel angiotensin I-converting enzyme inhibitory peptide derived from abalone (Haliotis discus hannai Ino) gonads. *European food research and technology*, 240(1), 137–145.

117 You, Q., Sun, X., Chen, J., Yu, J., & Wei, Y. (2023). Ameliorative effect of mussel-derived ACE inhibitory peptides on spontaneous hypertension rats. *European Journal of Nutrition* 62 (7): 3097–3111.

118 Tomosugi, N., Yamamoto, S., Takeuchi, M., Yonekura, H., Ishigaki, Y., Numata, N., ... & Sakai, Y.. et al. (2017). Effect of collagen tripeptide on atherosclerosis in healthy humans. *J Atheroscler Thromb* 24 (5): 530–538.

119 Oliveira, G. V. D., Volino-Souza, M., Cordeiro, E. M., Conte-Junior, C. A., & Alvares, T. S. (2019). Effects of fish protein hydrolysate ingestion on endothelial function compared to whey protein hydrolysate in humans. *International Journal of Food Sciences and Nutrition*.

120 Cardoso, S. M., Pereira, O. R., Seca, A. M., Pinto, D. C., & Silva, A. M. et al. (2015). Seaweeds as preventive agents for cardiovascular diseases: from nutrients to functional foods. *Mar Drugs* 13 (11): 6838–6865.

121 Dong, X., Zhou, M., Li, Y., Li, Y., Ji, H., & Hu, Q. (2021). Cardiovascular protective effects of plant polysaccharides: a review. *Frontiers in Pharmacology* 12,783641.

122 Jimenez-Escrig, A.B. and Sánchez-Muniz, F. (2000). Dietary fibre from edible seaweeds: chemical structure, physicochemical properties and effects on cholesterol metabolism. *Nutrition Research* 20: 585–598.

123 Kalisz, G. and Popiolek-Kalisz, J. (2025). Polysaccharides: the sweet and bitter impacts on cardiovascular risk. *Polymers* 17 (3): 405.

124 HATA, Y., NAKAJIMA, K., UCHIDA, J. I., HIDAKA, H., & NAKANO, T. (2001). Clinical effects of brown seaweed, *Undaria pinnatifida* (wakame), on blood pressure in hypertensive subjects. *Journal of Clinical Biochemistry and Nutrition* 30: 43–53.

125 Wada, K., Nakamura, K., Tamai, Y., Tsuji, M., Sahashi, Y., Watanabe, K., ... & Nagata, C. (2011). Seaweed intake and blood pressure levels in healthy pre-school Japanese children. *Nutr J* 10: 83.

126 de Jesus Raposo, M.F., de Morais, A.M., and de Morais, R.M. (2015). Marine polysaccharides from algae with potential biomedical applications. *Mar Drugs* 13 (5): 2967–3028.

127 da Silva Chagas, F. D., Lima, G. C., Dos Santos, V. I. N., Costa, L. E. C., de Sousa, W. M., Sombra, V. G., ... & Freitas, A. L. P. (2020). Sulfated polysaccharide from the red algae *Gelidiella acerosa*: anticoagulant, antiplatelet and antithrombotic effects. *International Journal of Biological Macromolecules* 159.

128 Chen, G. L., Cai, H. Y., Chen, J. P., Li, R., Zhong, S. Y., Jia, X. J., ... & Song, B. B. (2022). Chitosan/alginate nanoparticles for the enhanced oral antithrombotic activity of clam heparinoid from the clam coelomactra antiquata. *Marine Drugs* 20 (2): 136.

129 .Imran, M., Sajwan, M., Alsuwayt, B., & Asif, M. (2020). Synthesis, characterization and anticoagulant activity of chitosan derivatives. *Saudi Pharmaceutical Journal* 28 (1): 25–32.

130 Sudirman, S., Hsu, Y. H., He, J. L., & Kong, Z. L. (2018). Dietary polysaccharide-rich extract from *Eucheuma cottonii* modulates the inflammatory response and suppresses colonic injury on dextran sulfate sodium-induced colitis in mice. *PloS one* 13 (10): e0205252.

131 Carvalhal, F., Cristelo, R. R., Resende, D. I., Pinto, M. M., Sousa, E., & Correia-da-Silva, M. (2019). Antithrombotics from the sea: polysaccharides and beyond. *Marine drugs* 17 (3): 170.

132 Dwivedi, R. and Pomin, V.H. (2020). Marine antithrombotics. *Marine Drugs* 18 (10): 514.

133 Xie, L., Chen, M. H., Li, J., Yang, X. M., & Huang, Q. J. (2011). Antithrombotic effect of a polysaccharide fraction from *Laminaria japonica* from the South China Sea. *Phytotherapy Research* 25 (9): 1362–1366.

134 Serrano, G., Nishida, Y., and Wood, V. (2014). Natural astaxanthin improves blood flow and fights high blood pressure. *Agro Food Industry Hi-Tech* 25: 8–12.

135 Blas-Valdivia, V., Moran-Dorantes, D. N., Rojas-Franco, P., Franco-Colin, M., Mirhosseini, N., Davarnejad, R., ... & Cano-Europa, E. (2022). C-Phycocyanin prevents acute myocardial infarction-induced oxidative stress, inflammation and cardiac damage. *Pharmaceutical Biology* 60 (1): 755–763.

136 Cadar, E., Popescu, A., Dragan, A. M. L., Pesterau, A. M., Pascale, C., Anuta, V., ... & Ionescu, A. M., (2025). Bioactive compounds of marine algae and their potential health and nutraceutical applications: a review. *Marine Drugs* 23 (4): 152.

137 Baliou, S., Adamaki, M., Ioannou, P., Pappa, A., Panayiotidis, M. I., Spandidos, D. A., ... & Zoumpourlis, V, (2021). Protective role of taurine against oxidative stress. *Molecular Medicine Reports* 24 (2): 1–19.

138 Ibrahim, M.A., Eraqi, M.M., and Alfaiz, F.A. (2020). Therapeutic role of taurine as antioxidant in reducing hypertension risks in rats. *Heliyon* 6 (1): e03209.

139 Hernández-Camacho, J. D., Bernier, M., López-Lluch, G., & Navas, P. (2018). Coenzyme Q10 supplementation in aging and disease. *Frontiers in Physiology* 9, 316577.

140 Gutierrez-Mariscal, F. M., de la Cruz-Ares, S., Torres-Peña, J. D., Alcalá-Diaz, J. F., Yubero-Serrano, E. M., & López-Miranda, J (2021). Coenzyme Q10 and cardiovascular diseases. *Antioxidants* 10 (6): 906.

141 Martelli, A., Testai, L., Colletti, A., & Cicero, A. F. (2020). Coenzyme Q10: clinical applications in cardiovascular diseases. *Antioxidants* 9 (4): 341.

142 Rabanal-Ruiz, Y., Llanos-González, E., and Alcain, F.J. (2021). The use of coenzyme Q10 in cardiovascular diseases. *Antioxidants (Basel)* 10 (5),755.

143 Sue-Ling, C.B., Abel, W.M., and Sue-Ling, K. (2022). Coenzyme Q10 as adjunctive therapy for cardiovascular disease and hypertension: a systematic review. *The Journal of Nutrition* 152 (7): 1666–1674.

144 Di Lorenzo, A. et al. (2020). Clinical evidence for Q10 coenzyme supplementation in heart failure: from energetics to functional improvement. *Journal of Clinical Medicine* 9 (5): 1266.

145 Ibrahim, N. I., Fairus, S., Zulfarina, M. S., & Naina Mohamed, I. (2020). The efficacy of squalene in cardiovascular disease risk-a systematic review. *Nutrients*, 12(2), 414.

146 Ibrahim, N. and Naina Mohamed, I. (2021). Interdependence of anti-inflammatory and antioxidant properties of squalene-implication for cardiovascular health. *Life (Basel)* 11 (2),103.

147 Gopal, S. and Sharpless, N.E. (2021). Cancer as a global health priority. *JAMA* 326 (9): 809–810.

148 D'Eliseo, D. and Velotti, F. (2016). Omega-3 fatty acids and cancer cell cytotoxicity: implications for multi-targeted cancer therapy. *J Clin Med* 5 (2),15.

149 da Silva Batista, E., Nakandakari, S. C. B. R., Ramos da Silva, A. S., Pauli, J. R., Pereira de Moura, L., Ropelle, E. R., ... & Cintra, D. E. (2024). Omega-3 pleiad: The multipoint anti-inflammatory strategy. *Critical reviews in food science and nutrition*, 64(14), 4817–4832.

150 Chakniramol, S., Wierschem, A., Cho, M. G., & Bashir, K. M. I. (2022). Physiological and clinical aspects of bioactive peptides from marine animals. *Antioxidants*, 11(5), 1021.

151 Hasani, K., Ariaii, P., and Ahmadi, M. (2022). Antimicrobial, antioxidant and anti-cancer properties of protein hydrolysates from Indian mackerel (*Rastrelliger kanagurta*) waste prepared using commercial enzyme. *International Journal of Peptide Research and Therapeutics* 28 (3): 86.

152 Shin, J., Song, M. H., Yu, J. W., Ko, E. Y., Shang, X., Oh, J. W., ... & Saini, R. K. (2021). Anticancer potential of lipophilic constituents of eleven shellfish species commonly consumed in Korea. *Antioxidants*, 10(10), 1629.

153 Okeke, E. S., Okagu, I. U., Chukwudozie, K., Ezike, T. C., & Ezeorba, T. P. C. (2024). Marine-derived bioactive proteins and peptides: A review of current knowledge on anticancer potentials, clinical trials, and future prospects. *Natural Product Communications*, 19(3), 1934578X241239825.

154 Luo, Y., Zhang, Y., Zhang, T., Li, Y., Xue, H., Cao, J., ... & Jin, C. (2023). Review on marine collagen peptides induce cancer cell apoptosis, necrosis, and autophagy by reducing oxidized free radicals. *Biocell* 47 (5): 965–975.

155 Xie, C., Lee, Z. J., Ye, S., Barrow, C. J., Dunshea, F. R., & Suleria, H. A. (2024). A review on seaweeds and seaweed-derived polysaccharides: nutrition, chemistry, bioactivities, and applications. *Food Reviews International* 40 (5): 1312–1347.

156 Jin, J. O., Yadav, D., Madhwani, K., Puranik, N., Chavda, V., & Song, M. (2022). Seaweeds in the oncology arena: anti-cancer potential of fucoidan as a drug—a review. *Molecules* 27 (18): 6032.

157 Sanniyasi, E., Gopal, R. K., Damodharan, R., Arumugam, A., Sampath Kumar, M., Senthilkumar, N., & Anbalagan, M. (2023). In vitro anticancer potential of laminarin and fucoidan from Brown seaweeds. *Scientific Reports*, 13(1), 14452.

158 Bilal, M. and Iqbal, H.M.N. (2020). Marine seaweed polysaccharides-based engineered cues for the modern biomedical sector. *Marine Drugs* 18 (1): 7.

159 Karuppusamy, S., Rajauria, G., Fitzpatrick, S., Lyons, H., McMahon, H., Curtin, J., ... & O'Donnell, C. (2022). Biological properties and health-promoting functions of laminarin: A comprehensive review of preclinical and clinical studies. *Marine Drugs*, 20(12), 772.

160 Mawazi, S. M., Kumar, M., Ahmad, N., Ge, Y., & Mahmood, S. (2024). Recent applications of chitosan and its derivatives in antibacterial, anticancer, wound healing, and tissue engineering fields. Polymers, 16(10), 1351.

161 Baharlouei, P. and Rahman, A. (2022). Chitin and chitosan: prospective biomedical applications in drug delivery, cancer treatment, and wound healing. *Marine Drugs* 20 (7): 460.

162 Burra, P., Verduci, E., Dias, J. A., Buti, M., Carboni, A., Demirtas, C. O., ... & Group, U. P. A. (2025). The growing burden of obesity: addressing a global public health challenge. *United European Gastroenterology Journal*, 13(7), 1343.

163 Baker, J. S., Supriya, R., Dutheil, F., & Gao, Y. (2022). Obesity: treatments, conceptualizations, and future directions for a growing problem. *Biology*, 11(2), 160.

164 Albracht-Schulte, K., Kalupahana, N. S., Ramalingam, L., Wang, S., Rahman, S. M., Robert-McComb, J., & Moustaid-Moussa, N. (2018). Omega-3 fatty acids in obesity and metabolic syndrome: a mechanistic update. *J Nutr Biochem* 58: 1–16.

165 Nakamura, M.T., Yudell, B.E., and Loor, J.J. (2014). Regulation of energy metabolism by long-chain fatty acids. *Progress in Lipid Research* 53: 124–144.

166 Sureda, A., & Monserrat-Mesquida, M. (2025). Effects of Omega-3 Fatty Acids on Metabolic Syndrome and Fatty Liver Disease. In Omega-3 Fatty Acids: Keys to Nutritional Health and Disease (pp. 75–87). Cham: Springer Nature Switzerland.

167 Sinha, S., Haque, M., Lugova, H., & Kumar, S. (2023). The effect of omega-3 fatty acids on insulin resistance. *Life* 13 (6): 1322.

168 Ignot-Gutierrez, A., Serena-Romero, G., Guajardo-Flores, D., Alvarado-Olivarez, M., Martinez, A. J., & Cruz-Huerta, E. (2024). Proteins and peptides from food sources with effect on satiety and their role as anti-obesity agents: a narrative review. *Nutrients* 16 (20): 3560.

169 Huda, M., Wilding, J., and Pinkney, J. (2006). Gut peptides and the regulation of appetite. *Obesity Reviews* 7 (2): 163–182.

170 López-Yoldi, M., Riezu-Boj, J. I., Abete, I., Ibero-Baraibar, I., Aranaz, P., González-Salazar, I., ... & Zulet, M. A. (2024). Anti-obesity effects of a collagen with low digestibility and high swelling capacity: a human randomized control trial. *Nutrients* 16 (20): 3550.

171 Munekata, P. E., Pérez-Álvarez, J. Á., Pateiro, M., Viuda-Matos, M., Fernández-López, J., & Lorenzo, J. M. (2021). Satiety from healthier and functional foods. *Trends in Food Science & Technology* 113: 397–410.

172 Woo, M., Song, Y. O., Kang, K. H., & Noh, J. S. (2018). Anti-obesity effects of collagen peptide derived from skate (*Raja kenojei*) skin through regulation of lipid metabolism. *Marine Drugs* 16 (9): 306.

173 Manikkam, V., Vasiljevic, T., Donkor, O. N., & Mathai, M. L. (2016). A review of potential marine-derived hypotensive and anti-obesity peptides. *Critical Reviews in Food Science and Nutrition*, 56(1), 92–112.

174 Xing, L., Wang, Z., Hao, Y., & Zhang, W. (2022). Marine products as a promising resource of bioactive peptides: update of extraction strategies and their physiological regulatory effects. *Journal of Agricultural and Food Chemistry* 70 (10): 3081–3095.

175 Baravia, P. A., Qhabibi, F. R., Hadinata, E., Katuuk, S. H. H., Alfaray, R. I., Rampengan, D. D., ... & Nurkolis, F. (2025). Algal peptides for future food: key players and multifaceted agents in the fight against cardiometabolic syndrome. *Journal of Agriculture and Food Research* 20: 101786.

176 Kim, S.-K. (Ed.). (2011). Marine proteins and peptides: Biological activities and applications. Wiley-Blackwell. https://books.google.com.tw/books

177 Abachi, S., Pilon, G., Marette, A., Bazinet, L., & Beaulieu, L. (2023). Beneficial effects of fish and fish peptides on main metabolic syndrome associated risk factors: diabetes, obesity and lipemia. *Critical Reviews in Food Science and Nutrition* 63 (26): 7896–7944.

178 Liu, J., Zhen, D., Hu, C., Liu, Y., Shen, X., Fu, P., & He, Y. (2022). Reconfiguration of gut microbiota and reprogramming of liver metabolism with phycobiliproteins bioactive peptides to rehabilitate obese rats. *Nutrients* 14 (17): 3635.

179 Guo, L., Goff, H. D., Xu, F., Liu, F., Ma, J., Chen, M., & Zhong, F. (2020). The effect of sodium alginate on nutrient digestion and metabolic responses during both in vitro and in vivo digestion process. *Food Hydrocolloids* 107: 105304.

180 Houghton, D., Wilcox, M. D., Chater, P. I., Brownlee, I. A., Seal, C. J., & Pearson, J. P. (2015). Biological activity of alginate and its effect on pancreatic lipase inhibition as a potential treatment for obesity. *Food Hydrocolloids* 49: 18–24.

181 Panith, N., Wichaphon, J., Lertsiri, S., & Niamsiri, N (2016). Effect of physical and physicochemical characteristics of chitosan on fat-binding capacities under in vitro gastrointestinal conditions. *LWT - Food Science and Technology* 71: 25–32.

182 Tang, D., Wang, Y., Kang, W., Zhou, J., Dong, R., & Feng, Q. (2020). Chitosan attenuates obesity by modifying the intestinal microbiota and increasing serum leptin levels in mice. *Journal of Functional Foods* 64: 103659.

183 Uyanga, V. A., Ejeromedoghene, O., Lambo, M. T., Alowakennu, M., Alli, Y. A., Ere-Richard, A. A., ... & Lin, H. (2023). Chitosan and chitosan-based composites as beneficial compounds for animal health: Impact on gastrointestinal functions and biocarrier application. *Journal of Functional Foods* 104: 105520.

184 Bae, M., Ahmed, K., and Yim, J.E. (2022). Beneficial effects of taurine on metabolic parameters in animals and humans. *J Obes Metab Syndr* 31 (2): 134–146.

185 Kp, A.D. and Martin, A. (2023). Recent insights into the molecular regulators and mechanisms of taurine to modulate lipid metabolism: a review. *Critical Reviews in Food Science and Nutrition* 63 (23): 6005–6017.

186 Guo, Y. Y., Li, B. Y., Peng, W. Q., Guo, L., & Tang, Q. Q. (2019). Taurine-mediated browning of white adipose tissue is involved in its anti-obesity effect in mice. *Journal of Biological Chemistry* 294: jbc.RA119.009936.

187 Wen, C., Li, F., Zhang, L., Duan, Y., Guo, Q., Wang, W., ... & Yin, Y. (2019). Taurine is involved in energy metabolism in muscles, adipose tissue, and the liver. *Molecular Nutrition & Food Research* 63 (2): 1800536.

188 Shawl, M., Geetha, T., Burnett, D., & Babu, J. R. (2024). Omega-3 supplementation and its effects on osteoarthritis. *Nutrients* 16 (11): 1650.

189 Kong, Z. L., Gau, S. Y., Feng, K. K., & Chen, T. S. (2002). *Immue Bioactive Substances from Fish Larvae "Bull-Ary". In Animal Cell Technology: Basic & Applied Aspects: Proceedings of the Thirteenth Annual Meeting of the Japanese Association for Animal Cell Technology (JAACT), Fukuoka-Karatsu, November* 16–21, 2000 (pp. 449–453). Dordrecht: Springer Netherlands.

190 Gutiérrez, S., Svahn, S.L., and Johansson, M.E. (2019). Effects of omega-3 fatty acids on immune cells. *International Journal of Molecular Sciences* 20 (20): 5028.

191 Sudirman, S., Chen, C. Y., Chen, C. K., Felim, J., Kuo, H. P., & Kong, Z. L. (2023). Fermented jellyfish (*Rhopilema esculentum*) collagen enhances antioxidant activity and cartilage protection on surgically induced osteoarthritis in obese rats. *Front Pharmacol* 14: 1117893.

192 Abril, A.G., Carrera, M., and Pazos, M. (2024). Marine bioactive compounds with functional role in immunity and food allergy. *Nutrients* 16 (16): 2592.

193 Apostolova, E., Lukova, P., Baldzhieva, A., Katsarov, P., Nikolova, M., Iliev, I., ... & Kokova, V. (2020). Immunomodulatory and anti-inflammatory effects of fucoidan: a review. *Polymers* 12 (10): 2338.

194 Hwang, J., Yadav, D., Lee, P. C., & Jin, J. O. (2022). Immunomodulatory effects of polysaccharides from marine algae for treating cancer, infectious disease, and inflammation. *Phytotherapy Research* 36 (2): 761–777.

195 Park, J. S., Chyun, J. H., Kim, Y. K., Line, L. L., & Chew, B. P. (2010). Astaxanthin decreased oxidative stress and inflammation and enhanced immune response in humans. *Nutrition & Metabolism* 7 (1): 18.

196 Sugiura, Y., Matsuura, Y., Katsuzaki, H., Kakinuma, M., Amano, H., Usui, M., ... & Miyata, M. (2022). The immunomodulating effect of phlorotannins from a brown alga, *Eisenia nipponica*, on mice stimulated with ovalbumin through T cell regulation. *Plant Foods for Human Nutrition* 77 (2): 307–316.

197 Cleland, L.G. and James, M.J. (2012). Omega-3 fatty acids and synovitis in osteoarthritic knees. *Nature Reviews Rheumatology* 8 (6): 314–315.

198 Deng, W., Yi, Z., Yin, E., Lu, R., You, H., & Yuan, X. (2023). Effect of omega-3 polyunsaturated fatty acids supplementation for patients with osteoarthritis: a meta-analysis. *Journal of Orthopaedic Surgery and Research* 18 (1): 381.

199 Felim, J., Wang, P. Y., Pan, M. H., Kuo, H. P., & Kong, Z. L. (2025). Comparative study of the therapeutic effects of sturgeon collagen peptide chelate-zinc on osteoarthritis induced by anterior cruciate ligament transection and medial meniscectomy in obese male rats. *Biomedicine & Pharmacotherapy* 189: 118236.

200 Chang, H. W., Sudirman, S., Yen, Y. W., Mao, C. F., Ong, A. D., & Kong, Z. L. (2020). Blue mussel (*Mytilus edulis*) water extract ameliorates inflammatory responses and oxidative stress on osteoarthritis in obese rats. *Journal of Pain Research* 13: 1109–1119.

201 Felim, J., Chen, C. K., Tsou, D., Kuo, H. P., & Kong, Z. L. (2022). Effect of different collagen on anterior cruciate ligament transection and medial meniscectomy-induced osteoarthritis male rats. Frontiers in Bioengineering and Biotechnology, 10, 917474.

202 Sudirman, S., Ong, A. D., Chang, H. W., & Kong, Z. L. (2018). Effect of fucoidan on anterior cruciate ligament transection and medial meniscectomy induced osteoarthritis in high-fat diet-induced obese rats. *Nutrients* 10 (6): 686.

203 Sudirman, S., Chang, H. W., Chen, C. K., & Kong, Z. L. (2019). A dietary polysaccharide from Eucheuma cottonii downregulates proinflammatory cytokines and ameliorates osteoarthritis-associated cartilage degradation in obese rats. Food & function, 10(9), 5697–5706.

204 Ashraf, S., Mapp, P. I., Shahtaheri, S. M., & Walsh, D. A. (2018). Effects of carrageenan induced synovitis on joint damage and pain in a rat model of knee osteoarthritis. *Osteoarthritis and Cartilage* 26 (10): 1369–1378.

205 Lu, H. T., Hsieh, M. S., Cheng, C. W., Yao, L. F., Hsu, T. Y., Lan, J., ... & Chen, C. H. (2015). Alterative effects of an oral alginate extract on experimental rabbit osteoarthritis. Journal of Biomedical Science, 22(1), 64.

206 Cai, L., Zheng, J., Liu, L., Chen, X., & Wang, H. (2024). Unlocking the potential of β-1, 3-xylooligosaccharides from *Caulerpa lentillifera*: structural characterization, antioxidative and anti-osteoarthritis applications. *Chemical and Biological Technologies in Agriculture* 11 (1): 54.

207 Khairuddin, K., Sudirman, S., Huang, L., & Kong, Z. L. (2020). *Caulerpa lentillifera* polysaccharides-rich extract reduces oxidative stress and proinflammatory cytokines levels associated with male reproductive functions in diabetic mice. *Applied Sciences* 10 (24): 8768.

208 Zhang, C., Liao, Q., Ming, J. H., Hu, G. L., Chen, Q., Liu, S. Q., & Li, Y. M. (2017). The effects of chitosan oligosaccharides on OPG and RANKL expression in a rat osteoarthritis model. *Acta Cirurgica Brasileira* 32 (06): 418–428.

209 Oprenyeszk, F., Chausson, M., Maquet, V., Dubuc, J. E., & Henrotin, Y. (2013). Protective effect of a new biomaterial against the development of experimental osteoarthritis lesions in rabbit: a pilot study evaluating the intra-articular injection of alginate-chitosan beads dispersed in an hydrogel. *Osteoarthritis and Cartilage* 21 (8): 1099–1107.

210 Zou, B., Zhao, D., Zhou, S., Kang, J. X., & Wang, B. (2025). Insight into the effects of Omega-3 fatty acids on gut microbiota: impact of a balanced tissue Omega-6/Omega-3 ratio. Frontiers in Nutrition, 12, 1575323.

211 Seethaler, B., Lehnert, K., Yahiaoui-Doktor, M., Basrai, M., Vetter, W., Kiechle, M., & Bischoff, S. C. (2023). Omega-3 polyunsaturated fatty acids improve intestinal barrier integrity—albeit to a lesser degree than short-chain fatty acids: an exploratory analysis of the randomized controlled LIBRE trial. *European Journal of Nutrition* 62 (7): 2779–2791.

212 Gong, T., Fu, J., Shi, L., Chen, X., & Zong, X. (2021). antimicrobial peptides in gut health: a review. *Front Nutr* 8: 751010.

213 Mei, F., Duan, Z., Chen, M., Lu, J., Zhao, M., Li, L., ... & Chen, S. (2020). Effect of a high-collagen peptide diet on the gut microbiota and short-chain fatty acid metabolism. *Journal of Functional Foods* 75: 104278.

214 Xiang, X. W., Zhou, X. L., Wang, R., Shu, C. H., Zhou, Y. F., Ying, X. G., & Zheng, B. (2021). Protective effect of tuna bioactive peptide on dextran sulfate sodium-induced colitis in mice. *Marine Drugs* 19 (3): 127.

215 Yu, S., Guo, H., Ji, Z., Zheng, Y., Wang, B., Chen, Q., ... & Yuan, B. (2023). Sea cucumber peptides ameliorate DSS-induced ulcerative colitis: the role of the gut microbiota, the intestinal barrier, and macrophage polarization. *Nutrients*, 15(22), 4813.

216 Covasa, M., Stephens, R. W., Toderean, R., & Cobuz, C. (2019). Intestinal sensing by gut microbiota: targeting gut peptides. *Frontiers in endocrinology*, 10, 82.

217 Zhang, Z., Tang, S., Wang, Z., Han, J., Zhou, J., Lu, C., ... & Su, X. (2025). Protective effect of marine animal hydrolysis-peptides on adjuvant-induced arthritis in mice by gut microbiota and short-chain fatty acids. *Food Science and Human Wellness* 14: 9250089.

218 Mizuno, H., Bamba, S., Abe, N., & Sasaki, M. (2020). Effects of an alginate-containing variable-viscosity enteral nutrition formula on defecation, intestinal microbiota, and short-chain fatty acid production. *Journal of Functional Foods* 67: 103852.

219 Charoensiddhi, S., Abraham, R. E., Su, P., & Zhang, W. (2020). Seaweed and seaweed-derived metabolites as prebiotics. *Advances in Food and Nutrition Research* 91: 97–156.

220 Balasubramaniam, V., Aznyda, N., Hussin, M., Faradianna, L., Aswir, A. R., & Mohd Fairulnizal, M. N. (2020). Effect of red edible seaweed *Eucheuma denticulatum* on diet-induced obesity in vivo. *Journal of Applied Phycology* 32 (4): 2407–2417.

221 Wei, B., Zhang, B., Du, A. Q., Zhou, Z. Y., Lu, D. Z., Zhu, Z. H., ... & Wang, H. (2022). *Saccharina japonica* fucan suppresses high fat diet-induced obesity and enriches fucoidan-degrading gut bacteria. *Carbohydrate Polymers* 290: 119411.

222 Yao, W., Kong, Q., You, L., Zhong, S., & Hileuskaya, K. (2023). Polysaccharides from brown seaweed: physicochemical properties, absorption in the intestine, and beneficial effects on intestinal barrier. *Food Frontiers* 4 (4): 1547–1560.

223 Komisarska, P., Pinyosinwat, A., Saleem, M., & Szczuko, M. (2024). Carrageenan as a potential factor of inflammatory bowel diseases. *Nutrients*, 16(9), 1367.

224 Pari, R. F., Uju, U., Hardiningtyas, S. D., Ramadhan, W., Wakabayashi, R., Goto, M., & Kamiya, N. (2025). Ulva seaweed-derived ulvan: a promising marine polysaccharide as a sustainable resource for biomaterial design. *Mar Drugs* 23 (2).

224 Lin, X., Bo, H., Gu, J., Yi, X., Zhang, P., Liu, R., ... & Lin, C. H. (2022). Astaxanthin, a carotenoid antioxidant, pretreatment alleviates cognitive deficits in aircraft noised mice by attenuating inflammatory and oxidative damage to the gut, heart and hippocampus. *Biomedicine & Pharmacotherapy* 148: 112777.

226 Wu, L., Lyu, Y., Srinivasagan, R., Wu, J., Ojo, B., Tang, M., ... & Lin, D. (2020). Astaxanthin-shifted gut microbiota is associated with inflammation and metabolic homeostasis in mice. *The Journal of Nutrition* 150 (10): 2687–2698.

227 Vasconcelos, R. O., Bustos, S. O., Gonzalez, J. R., Soares, C. W., Barbosa, M. D., Chammas, R., ... & Trindade, G. S. (2016). Interaction between omega 3 PUFA and UVB radiation: photoprotective effect in normal and tumoral murine melanocytes? *Journal of Photochemistry and Photobiology B: Biology* 164: 361–368.

228 Harauma, A., Enomoto, Y., Endo, S., Hariya, H., & Moriguchi, T. (2024). Omega-3 fatty acids mitigate skin damage caused by ultraviolet-B radiation. *Prostaglandins, Leukotrienes and Essential Fatty Acids* 203: 102641.

229 Zhang, X., Zhuang, H., Wu, S., Mao, C., Dai, Y., & Yan, H. (2024). Marine bioactive peptides: anti-photoaging mechanisms and potential skin protective effects. *Curr Issues Mol Biol* 46 (2): 990–1009.

230 Wei, M., Qiu, H., Zhou, J., Yang, C., Chen, Y., & You, L. (2022). The anti-photoaging activity of peptides from *Pinctada martensii* meat. *Marine Drugs* 20 (12): 770.

231 Li, C., Fu, Y., Dai, H., Wang, Q., Gao, R., & Zhang, Y. (2022). Recent progress in preventive effect of collagen peptides on photoaging skin and action mechanism. *Food Science and Human Wellness* 11 (2): 218–229.

232 Rahman, A., Rehmani, R., Pirvu, D. G., Huang, S. M., Puri, S., & Arcos, M. (2024). Unlocking the therapeutic potential of marine collagen: a scientific exploration for delaying skin aging. *Mar Drugs* 22 (4).

233 Pu, S. Y., Huang, Y. L., Pu, C. M., Kang, Y. N., Hoang, K. D., Chen, K. H., & Chen, C. (2023). Effects of oral collagen for skin anti-aging: a systematic review and meta-analysis. *Nutrients* 15 (9).

234 Shen, S., Chen, X., Shen, Z., & Chen, H. (2021). Marine polysaccharides for wound dressings application: an overview. *Pharmaceutics* 13 (10).

235 Statha, D., Papaioannou, A., Kikionis, S., Kostaki, M., Sfiniadakis, I., Vitsos, A., ... & Rallis, M. C. (2024). Healing potential of the marine polysaccharides carrageenan and ulvan on second-degree burns. *Journal of Functional Biomaterials* 15 (9): 257.

236 Kang, J. W., Hyun, S. H., Kim, H. M., Park, S. Y., Lee, J. A., Lee, I. C., & Bae, J. S. (2024). The effects of fucoidan-rich polysaccharides extracted from *Sargassum horneri* on enhancing collagen-related skin barrier function as a potential cosmetic product. *J Cosmet Dermatol* **23** (4): 1365–1373.

237 Pangestuti, R., Shin, K.-H., and Kim, S.-K. (2021). Anti-photoaging and potential skin health benefits of seaweeds. *Marine Drugs* 19 (3): 172.

238 Campos, P. M. B. G. M., Melo, M. O., & Camargo Junior, F. B. (2014). Effects of polysaccharide-based formulations on human skin. *Polysaccharides*; Springer: Cham, Switzerland, 1–8.

239 Catanzaro, E., Bishayee, A., and Fimognari, C. (2020). On a beam of light: photoprotective activities of the marine carotenoids astaxanthin and fucoxanthin in suppression of inflammation and cancer. *Marine Drugs* 18: 544.

240 Pereira, L., & Cotas, J. (2023). Therapeutic potential of polyphenols and other micronutrients of marine origin. *Marine drugs*, 21(6), 323.

10

Proteins and Peptides for Health and Functionality

N. M. Liyanage[1], *Changliang Zhu*[2], *and You-Jin Jeon*[3]

[1] *University of Alberta, Edmonton, Canada*
[2] *Ocean University of China, Qingdao, China*
[3] *Jeju National University, Jeju, South Korea*

10.1 Introduction

As introduced in the previous chapter, aquatic ecosystems provide one of the richest sources of bioactive compounds, many of which have been shown to exert beneficial effects on human health and wellbeing. Among these, proteins and peptides derived from fish and shellfish are emerging as a unique class of natural molecules with applications that extend beyond conventional nutrition. Traditionally, seafood has been consumed for its high-quality protein and nutrient content. However, advances in food science, biotechnology, and nutraceutical development have revealed that these proteins are not just building blocks for human growth and metabolism but also serve as precursors to bioactive peptides capable of regulating key physiological processes. These discoveries have placed fish and shellfish-derived proteins and peptides at the forefront of research in functional foods and biomedical innovations [1]. Therefore, in this chapter, we further focus on exploring the proteins and peptides extracted from fish and shellfish for their health-promoting effects and functional characteristics.

Fish and shellfish-derived proteins are valuable primarily because they are both abundant and readily accessible. Fish and shellfish are among the most widely consumed foods worldwide. A sustainable supply of premium-quality proteins is ensured by these resources, driven by the development of aquaculture and optimized use of processing by-products [2]. In contrast to most terrestrial protein sources—including meat, dairy products, and plant-derived proteins—fish and shellfish-derived proteins often exhibit unique structural and compositional features that enhance their nutritional and functional value. In particular, fish and shellfish proteins are distinguished by a well-balanced and commonly superior amino acid profile, they contain higher concentrations of essential amino acids, including methionine, threonine, and lysine, in contrast to cereal proteins, which are frequently deficient in one or more such amino acids [3]. These amino acids are critical for protein synthesis, enzymatic functions, and metabolic regulation, especially in populations where plant-based proteins dominate diets.

Beyond their amino acid composition, these proteins also demonstrate favorable digestibility and bioavailability [4]. Myofibrillar proteins from fish are more easily hydrolyzed by gastrointestinal enzymes compared to red meat proteins, partly due to their lower connective tissue content and unique structural characteristics [5]. This not only guarantees enhanced efficiency of protein utilization by the human body but also facilitates the liberation of bioactive peptides during the digestive process, and these peptides can exert multiple physiological benefits, including antioxidant capacity, antihypertensive properties, and enhanced immune regulation [6]. Collagenous proteins from fish skin and bones are also notable for producing collagen peptides that are absorbed readily in small peptide forms, supporting skin, joint, and bone health [7, 8]. Collectively, these attributes underscore why fish and shellfish proteins stand out as both a sustainable and functionally superior resource for developing future health-oriented foods and nutraceuticals.

Peptides originating from these proteins hold comparable significance. These peptides usually contain between 2 and 20 amino acid residues and demonstrate a wide breadth of biological activities. [9]. Research has demonstrated their ability to function as antioxidants, antimicrobials, antihypertensives, immunomodulators, and anti-inflammatory agents. Unlike synthetic drugs, many of these bioactive peptides come with fewer side effects and are compatible with natural metabolic pathways [9]. Furthermore, the marine environment harbors species-specific proteins with significant biodiversity. Proteins derived from shrimp, crab, mollusks, and various finfish contain amino acid sequences that are often distinct from those found in terrestrial sources, making them a unique wellspring of novel and multifunctional peptides [10]. This ecological diversity provides opportunities to discover and optimize functional molecules. On the basis of newly elucidated mechanisms of action, this affirms the crucial role that marine proteins play in human health maintenance and the research and development of next-generation bioactive agents.

Classified as complete proteins, fish and shellfish proteins possess the full complement of nine essential amino acids in proportions that meet the nutritional needs of the human body. They also offer an excellent profile of branched-chain amino acids, known for their roles in muscle protein synthesis, recovery, and energy metabolism [11]. For communities with limited access to varied protein sources, they help to alleviate protein-energy malnutrition by serving as affordable and high-quality dietary protein.

This chapter assesses the nutritional efficacy and therapeutic capacity of fish- and shellfish-derived proteins and peptides. It reviews their sources, including edible species and fishery by-products, and explores extraction methods and their therapeutic potential. Overall, it provides a comprehensive view of the opportunities and challenges in advancing marine proteins for health and biomedical innovation.

10.2 Sources of Fish and Shellfish Proteins

Marine environments supply an extensive and varied protein resource derived from fish, crustaceans, and mollusks. These proteins differ in amino acid composition, structural features, and functional properties, offering valuable raw materials for nutritional applications and bioactive peptide development [12]. In addition, the growing emphasis on sustainability and circular economy approaches has placed seafood by-products and underutilized species into focus as untapped reservoirs of high-quality proteins. Understanding the sources of these proteins is crucial for harnessing their potential in health-promoting and functional applications.

10.2.1 Major Fish Species Utilized

Finfish constitute a major reservoir of premium-quality protein, furnishing essential amino acids that play a critical role in human metabolic function. Their proteins include a range of structural and functional types, such as collagen, myoglobin, and actin, exhibiting wide distribution throughout tissues comprising scales, bones, skin, and internal organs. Fish-derived proteins and peptides are sourced from both fillets and processing discards as well as coproducts [13].

Globally, whitefish species such as cod (*Gadus morhua*), pollock (*Theragra chalcogramma*), haddock (*Melanogrammus aeglefinus*), tilapia (*Oreochromis niloticus*), and catfish provide substantial amounts of muscle proteins that are widely used not only in diets but also in industrial protein extraction. These fish are particularly rich in myofibrillar proteins (actin and myosin), sarcoplasmic proteins, and collagen, each carrying potentially bioactive peptide sequences once enzymatically hydrolyzed [14]. Oily fish species like salmon, tuna, mackerel, and sardine also serve as significant protein resources alongside their well-known lipid benefits. Compared to whitefish, these oily variants often provide higher levels of certain sulfur-containing amino acids and histidine, which may contribute to anti-inflammatory and antioxidant activities when converted into peptides [15].

10.2.2 Shellfish Resources

Shrimp, crab, crayfish, and lobster all fall under the category of shellfish, which are notable sources of proteins and peptides. Bioactive peptides identified in shellfish species include the following categories: opioid peptides (mussels, clams); antimicrobial peptides (blue mussels, oysters, scallops, shrimp); angiotensin-converting enzyme (ACE) inhibitory peptides (shrimp, crabs); immunomodulatory peptides (fish, mussels, scallops); and antioxidant peptides (scallops). Within the shellfish opioid peptide subset, dynorphins are isolated from the blue mussel (*Mytilus edulis*), methionine-enkephalins from oysters and clams, and β-endorphins from shrimp [16, 17]. Mussel proteins are particularly valued for their anti-inflammatory peptide sequences, while oyster peptides have shown strong antioxidant and hepatoprotective functions [18]. Scallop and squid-derived proteins are under study for muscle health benefits, including anti-fatigue and recovery-promoting effects [19]. In Asia, mollusk proteins have been used traditionally in medicinal diets, highlighting their long-standing therapeutic associations.

Shellfish proteins often contain unique amino acid compositions, including high levels of taurine and glycine, supporting their role in maintaining cardiovascular and neurological health [20]. Taurine plays key roles in osmoregulation, antioxidant defense, and calcium signaling, all crucial for maintaining heart function and neural activity. Glycine, on the other hand, functions as an inhibitory neurotransmitter and modulates antioxidant systems by enhancing glutathione synthesis, supporting neuroprotection and immune responses. Shellfish muscles and shells contain high levels of these amino acids, along with arginine and proline, which also promote vascular health and collagen synthesis [21].

10.2.3 Fish and Shellfish By-products

A key promising pathway for sustainable protein production lies in harnessing seafood processing by-products. On a global scale, seafood processors generate massive volumes of residual materials (heads, skins, bones, scales, viscera, and shells), many of which are currently discarded or allocated to low-value animal feed applications. However, these materials often contain protein levels comparable to, or higher than, edible muscle tissue. Variations in the category and compositional characteristics of by-products have a direct bearing on the nutritional attributes and compositional profiles of derived protein hydrolysates [22]. Frames, trimmings, and heads are classified as clean fish processing by-streams with substantial potential for both direct human ingestion and deriving functional hydrolysates. A study documented that the proportions of Atlantic salmon by-products suitable for human consumption are 37.2% for heads, 56.7% for frames, and 81.0% for trimmings, based on total weight. In contrast, viscera are generally considered inedible (0% yield) because they contain lipids, bile acids, blood, and undigested feed, which give rise to oxidative processes and the synthesis of unfavorable metabolites during the hydrolysis reaction [23]. Accordingly, cleaner fish processing by-streams (frames, heads, and trimmings) with acceptable freshness profiles are ideal feedstocks for food applications, encompassing the manufacture of protein hydrolysates. Effective extraction of protein hydrolysates has been achieved using processing by-products (viscera, skin, and bones) from a broad spectrum of fish species, such as tilapia, channel catfish, Amur sturgeon, grass carp, cobia (*Rachycentron canadum*), *Lutjanus vitta*, sole fish, Alaska pollock (*T. chalcogramma*), and black scabbardfish (*Aphanopus carbo*) [24].

Shellfish processing residues, encompassing shells, cephalic parts, and visceral organs, act as abundant reservoirs of proteins and peptides endowed with excellent nutritional worth and functional potential. These processing residues are characterized by a protein content of 40%–60%, which encompasses myofibrillar proteins, sarcoplasmic proteins, and connective tissue-derived proteins such as collagen and gelatin. Proteins derived from the shells and heads of shrimp and crabs are distinguished not only by their favorable essential amino acid composition but also by their capacity to yield bioactive peptides with antioxidant, antimicrobial, and anti-inflammatory activities following enzymatic hydrolysis. Valorizing these by-products reduces environmental waste and provides sustainable raw materials for producing functional foods, nutraceuticals, and biomedical ingredients [25, 26]. The valorization of these resources not only confers economic benefits but also accords with sustainable development benchmarks, curbing waste generation and promoting the cyclic utilization of marine-derived biomaterials. The valorization of by-products through protein recovery and peptide development thus represents a critical intersection of nutrition science, sustainable food systems, and industrial biotechnology.

10.3 Extraction and Isolation Methods

The extraction of proteins and peptides from fish and shellfish is central to unlocking their functional and bioactive properties. Advances in both conventional and innovative techniques have enabled improved yields, purity, and bioactivity, tailoring marine-derived protein products for food, nutraceutical, and biomedical uses.

10.3.1 Fish and Shellfish Protein Hydrolysate Production

Several production methods for the protein hydrolysates are used, such as enzymatic hydrolysis, autolysis, chemical hydrolysis, and bacterial fermentation. Among them, enzymatic hydrolysis and chemical hydrolysis are the most common and widely used methods. Characterized by cost efficiency, operational speed, and high protein recovery yields, chemical hydrolysis is nonetheless deficient in product consistency regulation, with the non-specific cleavage of peptide bonds inducing considerable variability in the free amino acid spectrum [27].

Autolysis capitalizes on endogenous enzymes in fish muscle and visceral tissues, which catalyze the degradation of biomolecules into low-molecular-weight (LMW) peptides [28]. These endogenous enzyme levels, though, fluctuate significantly among individual fish of the same species and across different species, while also being highly contingent on seasonal changes and fish age. This, in turn, yields end products lacking consistent molecular profiles. Traditionally, autolysis has been utilized to manufacture fish sauce and silage for aquaculture and food applications. In contrast, utilizing commercially available enzymes for the enzymatic hydrolysis of proteins presents numerous benefits relative to autolysis or chemical hydrolysis, generating products with superior functional characteristics and enhanced bioactive properties. Autolysis may elicit an increase in the biosynthesis of unfavorable metabolites, the formation of nitrogenous species, and a reduction in sensory freshness when handling and storage protocols are suboptimal. Enzymatic Protein Hydrolysis(EPH)using commercially available proteases has been shown to preserve, and in some cases enhance, the functional properties of food proteins that are often diminished during conventional protein isolation [29].

Bacterial fermentation technology has long been used to retrieve and convert food constituents into high-value bio-based products, including organic acids and flavoring molecules, while enhancing nutritional quality, improving sensory characteristics, and reducing antinutritional factors [30]. However, it has been less frequently used for producing protein hydrolysates [31]. Bacterial fermentation can be divided into solid-state and submerged fermentation modalities. Solid-state fermentation is characterized by microbial growth and peptide production on a solid substrate in the absence of an aqueous phase, whereas submerged fermentation utilizes a liquid culture medium with incorporated protein substrates, where microbial metabolism drives the production of bioactive peptides [31]. As a result of the microbial proteolytic enzymes cleaving proteins according to their specificities, they generate peptides of varying sizes, sequences, and properties. Factors including the type of protein substrate, the enzymatic hydrolysis specificity, and parameters (pH, temperature, duration, and agitation) strongly influence the relative molecular weight and amino acid composition of peptides yielded from the process, which in turn determine their biological activities [32].

Enzymatic hydrolysis of protein substrates permits the fabrication of bioactive peptides, these bioactive molecules being acquirable through in vitro proteolysis of protein sources with appropriate proteolytic hydrolases. Targeted proteolytic enzymes catalyze the cleavage of fish muscle proteins, generating soluble and insoluble fractions simultaneously [33]. Enzymatic hydrolysis typically begins with homogenization, followed by heating at 85–90 °C for 20 minutes to inactivate endogenous enzymes, this ensures the precise evaluation of commercial enzyme effects on the resulting hydrolysates. During the hydrolysis reaction, critical parameters such as enzyme concentration, pH value, reaction time, and treatment temperature are the main factors influencing the characteristic properties of the hydrolysate [34]. Multiple proteolytic hydrolases comprising alcalase, neutrase, papain, pepsin, and trypsin are routinely applied for peptide production using protein raw materials. Jamil et al. (2016) reported that the degree of peptide bond hydrolysis elevates with increased enzyme concentration and thermal reaction conditions up to the enzyme's specific optimal parameters, beyond which the enzyme experiences structural denaturation [35]. The parameters of commonly used enzymes are given in Table 10.1.

Table 10.1 Commonly used enzymes and their parameters.

Sources	Proteolytic enzyme	pH	Temperature (°C)
Plants	Papain	6.5	50
Animals	Trypsin	8	37
	Pepsin	5	37
	α-Chymotrypsin	8	37
Microorganisms	Alcalase	9.5	50
	Neutrase	7	60
	Flavourzyme	7	55
	Protomax	6	40
	kojizyme	6	

10.3.2 Bioactive Peptide Generation from Fish and Shellfish Proteins

Bioactive peptides are generally categorized into two groups: naturally occurring bioactive peptides directly extractable from source organisms, and those generated via enzymatic hydrolysis of precursor proteins [10]. The extraction protocol for these distinct classes of bioactive peptides varies in the initial extraction stage, and this preliminary step serves to isolate high-molecular-weight proteins. Using ultrafiltration, microfiltration, nanofiltration, gel filtration, and ultracentrifugation [10]. Following the extraction and isolation of peptide fractions, a range of purification techniques comprising ion-exchange chromatography, gel permeation chromatography, and reversed-phase high-performance liquid chromatography are employed to yield highly purified peptide preparations [36].

Membrane separation typically serves as the primary purification step for enzymatic hydrolysates, allowing peptide fractionation based on molecular mass under applied pressure. Of all membrane separation methodologies, ultrafiltration is the most commonly employed owing to its cost efficiency and environmental compatibility [37]. Size-Exclusion Chromatography(SEC)and reverse-phase high-performance liquid chromatography (RP-HPLC), separating peptides on the basis of molecular size and hydrophobic interactions, are commonly utilized for peptide purification processes in hydrolysate processing. Capillary electrophoresis (CE) is an analytical approach for separating peptides according to their mass-to-charge (m/z) ratio by a fused quartz capillary. CE confers notable merits, including rapid analytical speed, minimal organic solvent consumption, economic efficiency, and superior efficiency, resolution, and selectivity. Nevertheless, its limited injection volume and overall efficiency make it less suitable for large-scale purification, so it is typically employed as a complementary method alongside chromatography [38]. After purification, peptides are identified. Cutting-edge mass spectrometry approaches, including Matrix-assisted laser desorption/ionization time-of-flight tandem mass spectrometry (MALDI–TOF–MS/MS) and Nuclear magnetic resonance(NMR)spectroscopy, are among the most widely utilized analytical tools for determining amino acid composition and their sequential arrangement in peptides [36]. Furthermore, cutting-edge in silico methods, including molecular docking and quantitative structure–activity relationship (QSAR) models, are employed in bioactive peptide generation, thereby circumventing the inherent constraints of conventional purification and identification approaches. [39]. The extraction and isolation procedure is summarized in Fig. 10.1.

10.4 Functional Properties of Protein Hydrolysates and Peptides

In food systems, the functional properties of proteins directly influence the quality and potential applications of the final product. During enzymatic hydrolysis, fish proteins are broken down into a mixture of free amino acids and short peptides

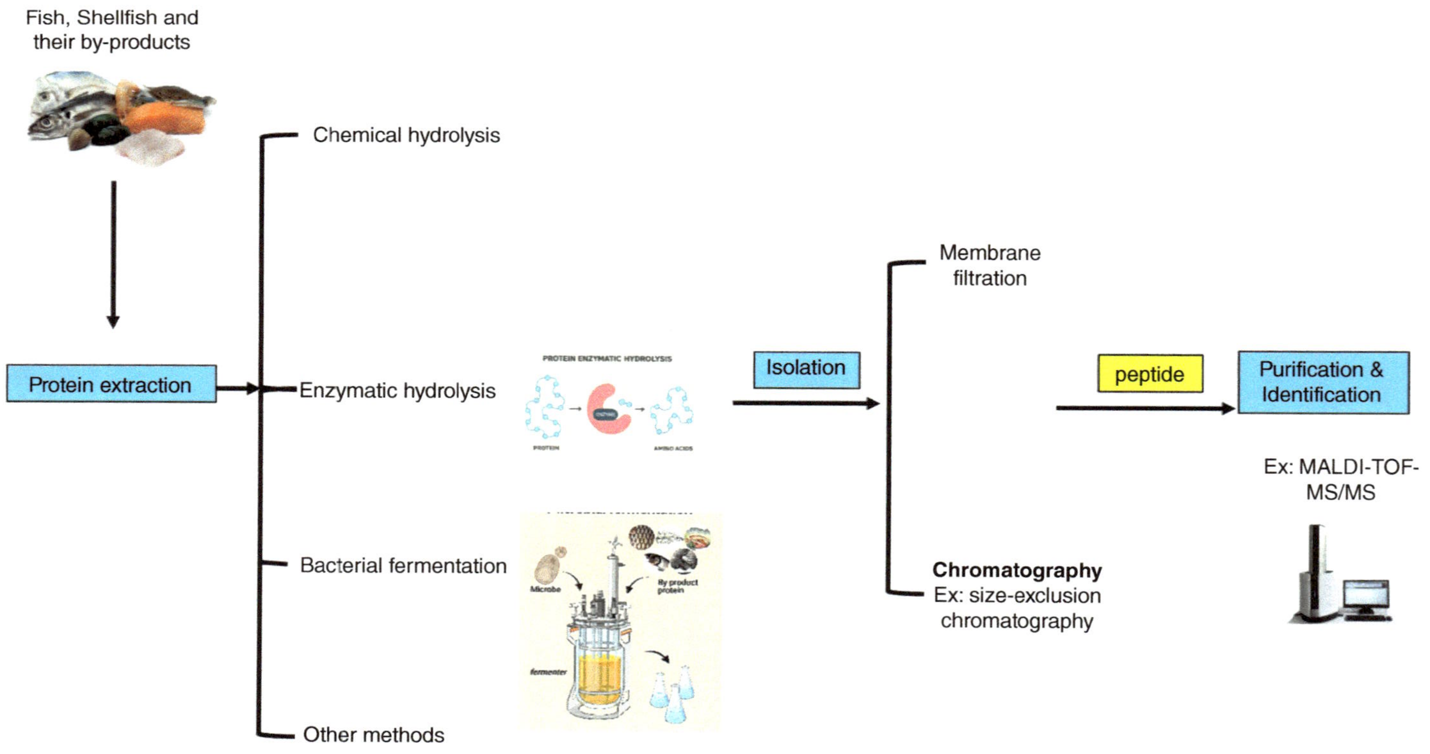

Figure 10.1 Schematic diagram of isolation and purification processes for fish and shellfish-derived peptides.

(di-, tri-, and oligopeptides). This process reduces peptide size and increases the number of amino acid carboxyl groups, simplifying the protein structure to enhance both functional quality and bioavailability. The important functional properties of proteins are solubility, thermal stability, gelation, emulsifying, foaming, fat binding, and water binding abilities [25].

10.4.1 Solubility

Increased solubility is regarded as both a favorable functional trait and a key quality index for fish protein hydrolysate (FPH), given that other functional properties, including emulsification and foaming capacity, are closely dependent on solubility levels. Proteins with high solubility disperse more readily, promoting better molecular distribution and resulting in the formation of superior colloidal systems. Solubility of protein is influenced by the basis of the degree of hydrolysis (DH) value, pH, incubation temperature, ionic strength, solvent type, and processing regimes [40]. It is postulated that extensive hydrolysis of large, complex protein macromolecules into LMW peptides—which possess an elevated number of surface ionizable polar moieties—thereby enhances their capacity to form hydrogen bonds with water molecules, and this elicits a significant enhancement in solubility. This phenomenon is further elucidated by the transformation of hydrophobic moieties to hydrophilic groups via the formation of terminal carbonyl and amino groups [41]. Moreover, the enhanced FPH solubility stems from the liberation of small soluble peptide fractions. This facilitates the enhanced exposure of additional charged and polar moieties to the surrounding aqueous environment. Additionally, high nitrogen solubility enhances the visual appeal and provides a smoother mouthfeel in foods containing protein hydrolysates [42].

10.4.2 Emulsifying Properties

The emulsification properties of enzymatically hydrolyzed FPHs bear a close relationship to their surface-active properties. Specifically, these hydrolysates are adept at lowering the interfacial tension between hydrophobic and hydrophilic constituents within food matrices. This enhanced ability to lower interfacial tension enables the formation and stabilization of emulsions, making FPH a valuable functional ingredient for improving texture and stability across diverse food products. Hydrolysis degree and the accompanying peptide size distribution represent primary factors modulating how effectively these hydrolysates perform at the oil–water interface, stabilizing dispersed droplets and averting phase separation in formulated food systems [43].

10.4.3 Foaming Ability

Foaming properties are commonly characterized in terms of foam capacity (FC) and foam stability (FS). A protein's FC measures the amount of interfacial area formed by whipping the protein, while its FS quantifies the time required for a foam to lose 50% of its initial volume. Dispersed proteins reduce interfacial tension at the air–water interface, thereby facilitating foam formation [44]. Foaming capacity is likewise influenced by peptide molecular weight. A higher DH elevates the content of LMW peptides in FPH, thereby facilitating the incorporation of air into aqueous solutions of small peptides. Such LMW peptides, which entrap a greater number of air cells, exhibit a reduced capacity to sustain stable foam structures [44].

10.4.4 Water Holding and Fat Binding Capacity

Water-holding capacity (WHC) refers to the capacity of a food product to retain and bind water during all stages of processing, and this constitutes a highly sought-after property within the food industry. The solubility of fish proteins has been associated with diminished WHC in such products [45]. Enzymatic hydrolysis boosts the abundance of polar moieties such as COOH and NH_2, which exert a marked influence on water adsorption capacity. LMW peptides derived from extended enzymatic hydrolysis tend to exhibit a higher WHC than larger-molecular-weight peptides, owing to the enhanced hydrophilicity of smaller peptide fragments [46]. Fat-binding capacity (FBC) denotes the capacity of proteins to absorb and retain oil, a key functional property that modulates the organoleptic attributes of food products. Key variables, including enzyme specificity, emulsification capacity, molecular density, and DH, exert a regulatory effect on FBC. Nonpolar amino acid residues of proteins facilitate hydrophobic interactions at the oil–water interface, whereas protein–lipid interactions are stabilized via multiple intermolecular bonding forces. FPH fractions prepared at low pH exhibit enhanced FBC, while those formulated at neutral pH possess superior water-retaining capacity [47]. Processing methodologies can exert a regulatory effect on the functional attributes of fish protein isolates (FPIs).

10.5 Bioactivities of Fish and Shellfish-Derived Peptides

Organisms like fish and shellfish are a leading source of bioactive proteins, which offer advantages like metabolic compatibility, absence of religious restrictions, and being free from terrestrial animal pathogens. However, the significance of these proteins extends far beyond basic nutrition. Once hydrolyzed into peptides, they exhibit therapeutic potential with diverse bioactivities. What sets these peptides apart is their multifunctionality. A single peptide sequence may simultaneously exert antioxidative, antihypertensive, anti-inflammatory, antidiabetic effects, etc. (Table 10.2). This versatility provides enormous potential in formulating multifunctional foods with synergistic health benefits, nutraceuticals, and pharmaceuticals that can address chronic, lifestyle-related diseases through integrative health support.

Table 10.2 Bioactivities of fish and shellfish-derived proteins and peptides.

Bioactivity	Source	Protein/peptide	Mode of action	References
Antioxidant	*Pinctada fucata*	Peptide	Hydroxyl and superoxide radical scavenging	[48]
	Antarctic krill (*Euphausia superba*)	SLPY, QYPPMQY, and EYEA LKPGN and LQP	Potent reducing capacity, protection against H_2O_2-induced DNA damage, ROS scavenging, and increased mitochondrial membrane potential	[49, 50]
	Siberian sturgeon (cartilage)	GEYGFE, PSVSLT, and IELFPGLP	DPPH and hydroxyl free radical scavenging activity	[51]
	Skipjack tuna	GHHAAA, PHPR, SVTEV, VRDQY, and SMDV	Protective effects on H_2O_2-induced oxidative damage, DPPH, ABTS, and metal chelating action	[52, 53]
	Stripped weakfish by-products (skin and bone)	IELIEKPMGIF and RADLSRELEEISERL	ABTS and hydroxyl radical scavenging	[54]
	Haliotis discus hannai	ATPGDEG	DPPH radical, hydroxyl, ROS scavenging	[55]
	By-product of serra Spanish mackerel	Protein hydrolysate	DPPH and ABTS radical scavenging, iron chelating	[56]
	Miiuy croaker (*Miichthys miiuy*)	FWKVV	DPPH, hydroxyl, and superoxide anion radical scavenging	[57]
Anti-inflammation	Pearl oyster (*Pinctada martensii*)	TWP, FPGA	NO production, pro-inflammatory cytokine production inhibition	[58]
	Baijiao sea bass	AADGPMKGILGY	COX1 and COX2 inhibition	[59]
	Herring (*Clupea harengus*)	IVPAS	iNOS activation reduction	[60]
	Green mussel (*Perna viridis*) foot	EGLLGDVF	Pro-inflammatory cytokines, NO, and COX-2 activation inhibition	[61]
	Sardine (*S. pilchardus*) by-products	Protein hydrolysate	Hindering the inflammation regulation in the endothelial cells	[62]

(*Continued*)

Table 10.2 (Continued)

Bioactivity	Source	Protein/peptide	Mode of action	References
Antihypertentsive	Pearl oyster (*Pinctada fucata*) shell	KKCHFWPFPW	ACE inhibitory activity	[63]
	Mackerel (*Scomber japonicas*)	PLITT	ACE inhibitory activity	[64]
	Shortfin scad (*Decapterus macrosoma*) waste	RGVGPVPAA	ACE inhibitory activity	[65]
	Ruditapes philippinarum	GRVSNCAA	ACE inhibitory activity	[49]
	Monkfish (*Lophius litulon*) swim bladders	SEGPK	Angiotensin-I-converting enzyme (ACE) inhibition	[66]
	Blue mussels (*Mytilus edulis*)	WF, and SWISS	ACE inhibitory activity	[67]
Antimicrobial	Half-Fin Anchovy (*Setipinna taty*)	FEDQLR, ALERTF, and RHPEYAVSVLLR	Increase H_2O_2 production and K^+ leakage, resulting in membrane damage	[68]
	Mollusk (*Cenchritis muricatus*)	SRSELIVHQRLF	Fungistatic effect	[69]
	Nerita versicolor	Nv-p1, Nv-p2, and Nv-p3	Antibiofilm action	[70]
	Yellow croaker (*Larimichthys crocea*),		Membrane damage, stimulation of flocculant secretion	[71]
Antiviral	Cod skin	Hot-water hydrolysate	Neuraminidase blocker	[72]
Anti-obesity	*Miichthys miiuy* muscle	VIAPW and IRWWW	Potent suppression of oleic acid-induced intracellular lipid accumulation	[73]
	Skate (*Raja kenojei*)	Collagen peptide	Suppression of fat accumulation	[74]
	Sphyrna mokarran skin	Collagen peptides	Ameliorating high-fat diet and alcohol-induced hyperlipidemia	[75]

10.5.1 Antioxidant Activity

Oxidation is an essential metabolic process in vertebrates and humans, facilitating numerous biochemical pathways required for energy production and cellular signaling. However, it likewise induces the production of reactive oxygen species (ROS) and free radicals, which, when unregulated, disrupt cellular homeostasis and trigger oxidative stress. Chronic oxidative stress could cause significant cellular damage and is implicated in the pathogenesis of various diseases, such as stroke, cardiovascular disease, diabetes, arteriosclerosis, and cancer [76, 77]. Recent research shows that antioxidant peptides from FPHs effectively scavenge free radicals and protect cells against oxidative injury, highlighting their role in disease prevention and overall health maintenance. The antioxidant capacity of these proteins and peptides has been evaluated via a series of in vitro assays; these include the scavenging of free radicals (e.g., 2,2-diphenyl-1-picrylhydrazyl (DPPH), hydroxyl, superoxide anion, and 2,2′-Azinobis-(3-ethylbenzothiazoline-6-sulfonic acid (ABTS)radicals) as well as metal ion chelation assays. The antioxidant potential of these molecules stems primarily from the capacity of such compounds to donate a hydrogen atom or electron, thus neutralizing free radicals and mitigating their harmful impacts on biological systems [78]. Hydrophobic amino acids, including phenylalanine (Phe, F), tryptophan (Trp, W), tyrosine (Tyr, Y), alanine (Ala, A), valine (Val, V), and leucine (Leu, L), boost the free radical neutralization capacity of peptides by facilitating their binding to and penetration of lipid bilayers in target cell membranes through hydrophobic interactions. By contrast, essential and acidic amino acids such as lysine (Lys, K), aspartic acid (Asp, D), and glutamic acid (Glu, E) act as efficient metal ion chelating agents. Additionally, aromatic amino acids (phenylalanine (Phe, F), tryptophan (Trp, W), tyrosine (Tyr, Y)) exert free radical scavenging activity through direct electron transfer pathways [79].

Antioxidant peptides derived from Antarctic krill (*Euphausia superba*) exhibited robust reducing power, protective capacity against plasmid DNA damage induced by H_2O_2, and inhibitory activity against lipid peroxidation [49]. A study investigating peptides extracted from the mantle type V of the pearl oyster (*Pinctada martensii*) demonstrated that such peptides possessed potent antioxidant activity, rendering them a valuable natural antioxidant source for the food processing industry [80]. Extensive investigations have centered on novel bioactive peptides derived from FPHs, especially those isolated from seafood processing by-products. For example, Najafian and Babji (2014) purified three distinct peptides from myofibrillar protein hydrolysates of *Pangasius sutchi*; subsequent purification and characterization assays revealed that these peptides exhibited outstanding antioxidant activity, which highlighted their potential as functional food ingredients with robust antioxidative properties [81]. Kim et al. (2021) identified a novel heptapeptide (TCGGQGR) from hydrolysates of mackerel processing by-products, and this peptide displayed strong antioxidant activity along with promising applicability for diverse industrial uses [82]. Similarly, protein hydrolysates derived from stripped weakfish by-products via Alcalase hydrolysis displayed potent DPPH radical scavenging activity, achieving scavenging levels of 60%–70%.

Although proteins and peptides of marine origin are widely recognized for their antioxidant properties, several knowledge gaps persist. Many studies confirm their radical-scavenging and metal-chelating activities, yet inconsistencies in extraction techniques, structural characterization, and *in vivo* validation remain unaddressed. Subsequent research should focus on standardizing methodologies, clarifying structure–activity relationships, and performing comprehensive *in vivo* investigations to confirm their therapeutic potential and practical applicability.

10.5.2 Anti-inflammatory Potential

Inflammation is the initial immune defense against toxins, pathogens, allergens, or injury in the body, characterized by intricate interactions between immune cells and soluble mediators. It triggers leukocyte recruitment and stimulates the release of tumor necrosis factor α (TNF-α), interleukin-1 (IL-1), prostaglandin E 2 (PGE_2), IL-6, and core enzymes, including cyclooxygenase-2 (COX-2)and inducible nitric oxide Synthase (iNOS), to enhance tissue repair. Pathogen recognition receptors, among which are Toll-like receptors (TLRs), recognize invading pathogenic microbes and trigger innate immune responses. Anti-inflammatory therapeutic strategies typically employ nonsteroidal anti-inflammatory drugs (NSAIDs) and corticosteroids, whereas NSAIDs can induce untoward effects, including gastrointestinal ulceration and, in infrequent instances, cardiovascular adverse events [78]. Numerous investigations have demonstrated that proteins and peptides derived from marine fish and shellfish modulate key inflammatory signaling pathways. One such study revealed that peptides TWP (402.19 Da), TAMY (484.19 Da), and FPGA (390.19 Da) isolated from meat hydrolysates of the pearl oyster (*P. martensii*) possess prominent anti-inflammatory activity. These peptides exert their anti-inflammatory effects by downregulating the production of nitric oxide (NO) and pro-inflammatory cytokines (TNF-α, IL-6, IL-1β) while upregulating the synthesis of the anti-inflammatory cytokine IL-10, thus rendering them promising therapeutic candidates for the management of inflammatory disorders [58]. A novel identified peptide EGLLGDVF (849.43 Da), extracted from the foot of the green mussel (*Perna viridis*), displays marked anti-inflammatory activity. It has been verified to downregulate pro-inflammatory cytokine production, lower NO levels, and inhibit COX-2 activation [61].

Two bioactive peptides isolated from the aqueous fraction of Baijiao sea bass (*Lateolabrax maculatus*) exerted anti-inflammatory activity by suppressing nitric oxide (NO) generation in lipopolysaccharide (LPS)-stimulated macrophages [83]. Additionally, an anti-inflammatory peptide LLFTTQ (721.80 Da), purified from a powdered extract of skipjack tuna (*Katsuwonus pelamis*), was found to lower the levels of IL-6, IL-10, and TNF-α—core cytokines associated with intestinal inflammatory responses [68]. Investigations have confirmed that marine mollusks possess anti-inflammatory properties stemming from their diverse peptide compositions. For instance, the peptide NPAQDC (647.55 Da), derived from the foot tissue of the Asiatic hard clam (*Meretrix meretrix*), has been demonstrated to downregulate the expression of pro-inflammatory cytokines and inhibit NO production, alongside suppressing COX-2 activation in LPS-challenged macrophages [61].

Peptides derived from marine fish and shellfish mediate anti-inflammatory activity via a multitude of interrelated molecular mechanisms. NF-κB constitutes a core molecular target in this process: this transcription factor modulates the synthesis of pro-inflammatory cytokines, including TNF-α, IL-6, and IL-1β, and numerous such marine-derived peptides inhibit its activation, thus downregulating the transcriptional expression of these pro-inflammatory cytokines. These peptides additionally regulate the MAPK signaling pathway, encompassing ERK, JNK, and p38, thereby attenuating inflammation-associated cellular responses. A further key mechanism entails reducing NO production via the downregulation of iNOS, a process that mitigates oxidative stress and curtails inflammatory tissue injury. In addition, these peptides frequently inhibit the COX-2/PGE_2 axis, decreasing prostaglandin-driven inflammation. Beyond direct pathway suppression, many marine bioactive peptides display immunomodulatory properties, simultaneously reducing pro-inflammatory mediators while upregulating the production of anti-inflammatory cytokines, including IL-10. Collectively, these interconnected mechanisms underscore the therapeutic potential of marine-derived proteins and peptides as viable therapeutic candidates for the management of inflammatory disorders [12].

10.5.3 Anticancer Potential

Cancer is defined as a group of disorders marked by the uncontrolled proliferation and division of aberrant cells. It arises when mutations in DNA interfere with the body's natural mechanisms for regulating cell growth and programmed cell death (PCD). Such disruptions can result in tumorigenesis and may prompt malignant cells to invade adjacent tissues or metastasize to distant organs via the circulatory (blood) and lymphatic systems. Anticarcinogenic, anticancer, and antiproliferative activities are defined as the capacity to neutralize carcinogenic agents or suppress the proliferation and metastasis of malignant cells. Agents with these properties known as anticarcinogens or anticancer compounds work by specifically identifying and either eliminating or suppressing malignant cells. Among these, bioactive peptides have emerged as promising therapeutic candidates. Their capacity to precisely target malignant cells while sparing normal somatic tissues confers the potential for effective therapeutic intervention with a reduced incidence of adverse effects [36].

Bioactive peptides derived from fish and shellfish exert anticancer activity through multiple molecular mechanisms, encompassing tumor growth inhibition and apoptotic induction in malignant cells. Apoptotic induction is primarily mediated via the intrinsic mitochondrial pathway—a core mechanistic route for apoptotic signaling. Within this signaling cascade, such peptides elicit the release of cytochrome c (Cyt c) from mitochondrial compartments and activate caspase-mediated signaling cascades, ultimately culminating in PCD. These peptides additionally modulate apoptosis-associated proteins through the upregulation of proapoptotic factors and downregulation of antiapoptotic factors. Early opening of the mitochondrial permeability transition pore (mPTP) at the inner mitochondrial membrane (IMM) represents a pivotal event in the initiation of primary necrosis. Such mPTP-mediated events abrogate ATP synthesis and induce the influx of large volumes of water and small molecular solutes into the mitochondrial matrix down their electrochemical gradients, leading to profound osmotic swelling of mitochondrial structures and subsequent necrotic cell death. Moreover, mitochondrial outer membrane permeabilization (MOMP) facilitates the release of proapoptotic mediators, including Cyt c, that drive the activation of caspases, apoptosis-inducing factor (AIF), second mitochondria-derived activator of caspase (Smac), and endonucleases (as illustrated in Fig. 10.2).

Moreover, such peptides can abrogate angiogenesis via the downregulation of vascular endothelial growth factor (VEGF), resulting in deprived tumors due to a lack of nutrient and oxygen supply [84]. Bioactive peptides derived from FPHs have exhibited considerable potential in alleviating oxidative stress by decreasing intracellular ROS levels. This reduction helps prevent genetic damage such as mutations and chromosomal aberrations, essential factors in cancer progression. Studies exploring the antiproliferative activity of protein hydrolysates obtained from fish by-products demonstrated that FPHs derived from the skin, bone, cephalic tissue, and visceral organs of diverse fish species potently inhibited the proliferation of human colorectal and breast cancer cells. These observations underscore the therapeutic potential of FPH as natural bioactive agents for the prevention and therapeutic management of cancer [85].

Yu et al. (2018) investigated the prospect of *Cyclina sinensis* protein hydrolysates (CSPs) for the development of novel antiproliferative pentapeptides that selectively target prostate cancer cells. Their experimental results demonstrated that CSPs exerted a significant inhibitory effect on the proliferation of DU-145 prostate cancer cells via apoptotic induction [86]. Hsu et al. (2011) assessed the antiproliferative activity of bioactive peptides obtained from by-products of tuna dark muscle, which were enzymatically hydrolyzed with papain and protease XXIII. The resulting peptide fractions exhibited strong antiproliferative activity. Two specific peptides isolated, Leu–Pro–His–Val–Leu–Thr–Pro–Glu–Ala–Gly–Ala–Thr (1.21 kDa) and Glu–Gly–Gly–Val–Tyr–Met–Val–Thr (1.12 kDa) demonstrated a dose-dependent inhibitory effect on MCF-7 human breast cancer cells, highlighting their potential as bioactive agents in cancer therapy [87]. While peptides derived from fish and shellfish exhibit considerable anticancer potential, several critical knowledge gaps remain to be elucidated. Such gaps primarily encompass comprehensive *in vivo* investigations, the precise characterization of their mechanisms of action, and the establishment of clinical translational relevance.

10.5.4 Antihypertensive Effect

Hypertension represents a severe chronic disorder and a predominant global public health challenge, especially in adult populations. While a broad range of angiotensin I-converting enzyme (ACE) inhibitors are clinically available, hypertension-associated mortality rates continue to rise. ACE exerts a crucial regulatory role in

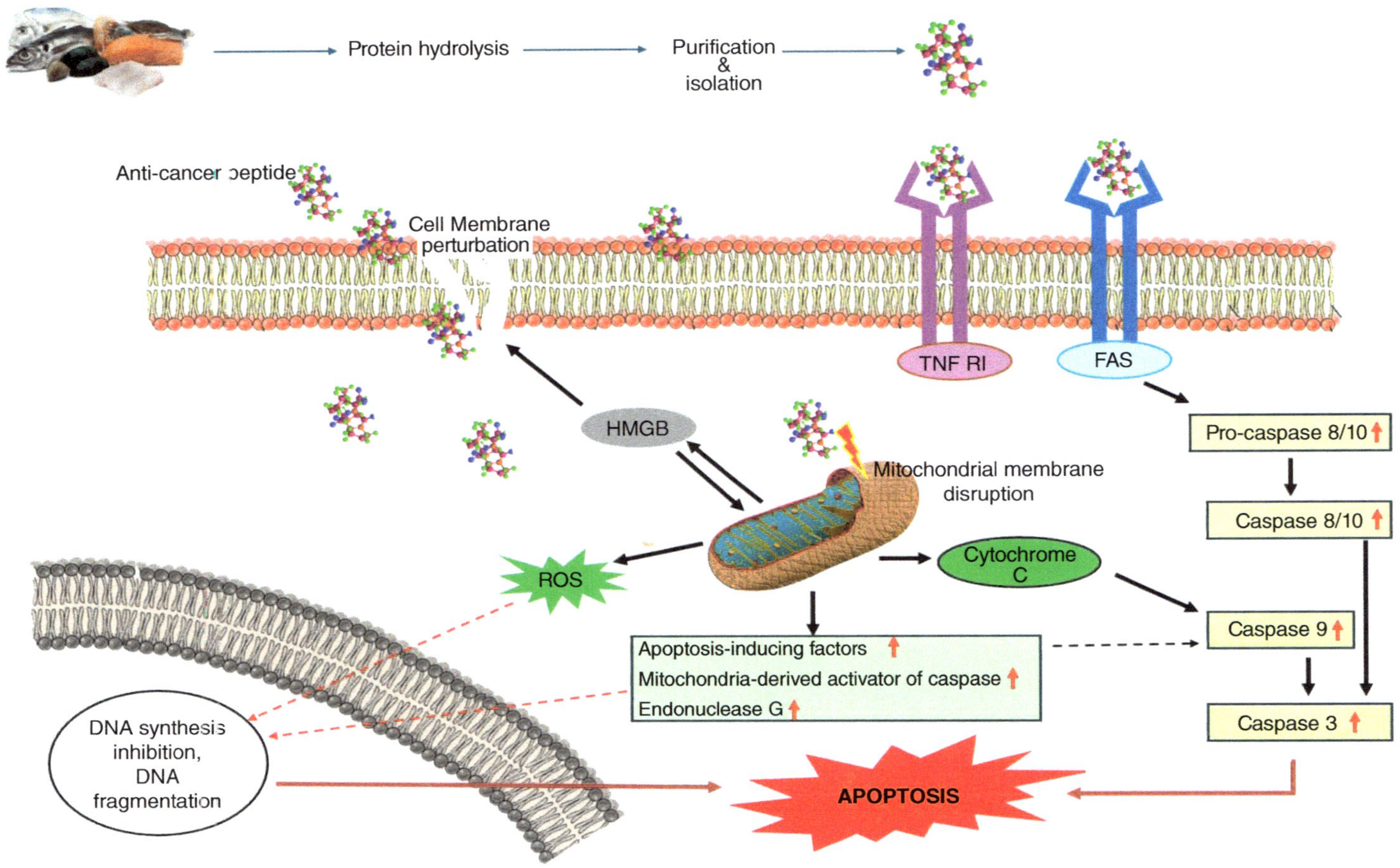

Figure 10.2 Anticancer mechanism of peptides isolated from fish and shellfish.

the modulation of blood pressure via catalyzing the conversion of angiotensin I to angiotensin II. Angiotensin II not only narrows blood vessels but also suppresses the vasodilatory effects of bradykinin, jointly contributing to raised blood pressure. Therefore, inhibiting ACE activity remains a central therapeutic approach for lowering hypertension risk and effectively managing high blood pressure [88]. Numerous studies have underscored the potential of aquatic resource-derived proteins and peptides as natural ACE inhibitors. A notable illustration is the IK, YEGDP, WF, and SWISS peptides, isolated from the blue mussel (*M. edulis*) via enzymatic hydrolysis, which displayed the most potent ACE inhibitory activity. These bioactive peptides effectively interfere with the ACE pathway, suggesting strong potential for managing hypertension through natural, marine-derived compounds [67].

Bioactive peptides isolated from the swim bladders of monkfish (*Lophius litulon*) via combined acid and enzymatic hydrolysis have exhibited considerable ACE inhibitory activity. In a similar vein, peptides extracted from processing by-products of shortfin scad (*Decapterus macrosoma*) through alcalase-mediated enzymatic hydrolysis demonstrated potent ACE inhibitory activity. Such observations highlight the clinical potential of marine-derived peptides as natural bioactive agents for the clinical management of hypertension, while also promoting sustainable use of seafood by-products [66]. According to Wijesekara and Kim (2010), distinct amino acid residues exert a pivotal regulatory role in the ACE inhibitory activity of bioactive peptides. Peptides harboring hydrophobic or aromatic amino acid residues—including tryptophan, tyrosine, proline, and phenylalanine—at their N-terminal positions exhibit such inhibitory properties and, along with particular residues at the C-terminal, are effective competitive inhibitors. Such structural characteristics facilitate the direct binding of these peptides to the catalytic active site of ACE, thus inhibiting its enzymatic activity and thereby supporting the modulation of systemic blood pressure [89]. Shrimp shell waste was reported as a valuable natural source of ACE-inhibitory peptides. Specifically, shrimp shell-derived peptides isolated from neutral proteases exhibit higher ACE inhibitory activity [88].

10.5.5 Antidiabetic Potential

Diabetes mellitus (DM) represents a major global health challenge. Type 2 diabetes mellitus (T2DM), with epidemiological estimates suggesting it will afflict approximately 700 million individuals worldwide by 2045, which constitutes roughly 90% of all diabetic cases, is a metabolic disorder predominantly linked to insulin resistance and strongly modulated by dietary patterns and lifestyle determinants. While conventional treatments exist, many are accompanied by undesirable side effects, prompting growing interest in alternative therapies. Among these, dietary proteins and bioactive peptides obtained from natural origins have emerged as viable candidates due to their ability to modulate glucose metabolism and boost insulin sensitivity with minimal side effects [79]. Dipeptidyl peptidase IV (DPP-IV) represents a pivotal therapeutic target in the clinical management of T2DM, given that it inactivates the incretin hormones glucagon-like peptide-1 (GLP-1) and glucose-dependent insulinotropic polypeptide (GIP). Such incretin hormones exert an essential regulatory role in modulating postprandial insulin secretion and sustaining systemic glycemic levels. By abrogating DPP-IV activity, the activity of GLP-1 and GIP is prolonged, enhancing insulin release and improving glycemic control. Beyond DPP-IV inhibition, targeting carbohydrate-digesting enzymes such as α-amylase and α-glucosidase can effectively decrease intestinal glucose uptake, offering an additional strategy to curb post-meal glycemic spikes [79]. Comprehensive research efforts have investigated the anti-diabetic potential of fish-derived peptides. The peptides IPV and IPVDM isolated from *Capros aper* have shown potential antidiabetic activity through DPP-IV inhibitory activity [90]. Moreover, Harnedy et al. (2018) showed that Atlantic salmon skin hydrolysate exhibited potent antidiabetic activity in vitro. The hydrolysate was found to stimulate the GLP-1 and contribute to glucose homeostasis [90]. Additionally, *Sardina pilchardus* peptides NAPNPR and YECSVR inhibited DPP_IV [90].

Proteins sourced from mollusks have exhibited notable antidiabetic potential, particularly through their influence on key metabolic enzymes. Among them, protein hydrolysates derived from blue mussels (*M. edulis*) have shown significant inhibitory activity against DPP-IV, an enzyme involved in glucose metabolism. These hydrolysates effectively prolong the activity of incretin hormones, thereby enhancing insulin secretion and improving postprandial glycemic control. These findings emphasize the clinical value of marine-derived peptides as natural therapeutic agents in the intervention of T2DM [91].

10.6 Challenges and Future Direction

Despite the considerable therapeutic potential of fish and shellfish-derived bioactive proteins and peptides, their widespread commercialization and industrial-scale application remain hindered by a range of scientific, technological, and economic challenges. Traditional extraction techniques, such as solvent-based acetic acid treatments, are particularly ineffective for fish collagens due to their inherent insolubility and extensive crosslinking, which result in low yields and inefficient workflows. Although alkaline solubilization methods can improve the extraction efficiency of FPIs, they often compromise amino acid integrity, disrupt secondary protein structures, and diminish both solubility and bioactivity, thereby limiting their functional and therapeutic value. Enzymatic extraction has emerged as a promising alternative, offering enhanced collagen recovery and the release of targeted bioactive peptide sequences. Nevertheless, the elevated cost of industrial-grade enzymes constitutes a substantial impediment to large-scale scalability, restricting the feasibility of these methods beyond specialized or niche markets. While microbial fermentation is a widely adopted and cost-effective method for protein hydrolysis, its industrial application is constrained by low overall yields and limited control over peptide specificity. This lack of selectivity hinders the utility of this approach in the targeted biosynthesis of high-value bioactive substances.

On the downstream end, purification of target peptides at the laboratory scale continues to depend on conventional separation techniques, including membrane filtration, dialysis, gel filtration chromatography, ion-exchange chromatography, and RP-HPLC. While effective for research, these approaches are difficult and expensive to translate to industrial scales, especially given the often-low concentrations of the most active peptide fractions in crude extracts. Even advanced separation and purification systems, such as ultrafiltration and sophisticated chromatographic methods, remain costly and require complex optimization to be viable in large-scale settings [48]. These fish and shellfish-derived peptides are generally recognized for their favorable safety profile, exhibiting lower allergenicity and minimal toxicity compared to many other bioactive compounds. A major barrier to their application is peptide stability and bioavailability, given that these biomolecules are vulnerable to breakdown during food processing and gastrointestinal digestion. In order to overcome this challenge, innovative delivery systems such as nanoemulsions, nanoliposomes, and polymeric nanoparticles are being vigorously explored to enhance oral bioavailability and improve the in vivo stability of target peptides. However, the successful commercialization of nanoencapsulated peptides demands rigorous investigation into their behavior, fate, and safety within both food matrices and the human body to ensure efficacy and consumer protection.

Furthermore, the bioactivity of the peptides is mostly documented in vitro. Their molecular targets, mechanisms of action, and metabolic pathways are not yet fully understood. The rapid advancement of bioinformatics, molecular docking, and computational modeling is revolutionizing the isolation and targeted design of bioactive peptides with targeted therapeutic functions. These technologies enable high-throughput screening and predictive analysis, significantly accelerating the identification of promising candidates. To fully harness the potential of bioactive proteins and peptides, future strategies ought to prioritize comprehensive human studies to validate therapeutic claims, alongside robust safety assessments.

10.7 Conclusion

Fish and shellfish proteins and peptides represent one of the most promising classes of marine bioactives for nutrition, health promotion, and therapeutic development. Their unique amino acid composition, high digestibility, and diverse bioactivities, including antioxidative, anti-inflammatory, antihypertensive, antidiabetic, and anticancer effects, make them valuable candidates for functional food, nutraceutical, and pharmaceutical applications. Advances in extraction and production methods, especially enzymatic hydrolysis and emerging green technologies, have markedly improved the yield and functional properties of marine bioactive compounds. Nevertheless, large-scale application and commercialization are challenged by factors such as cost, optimization of processing, product stability, bioavailability, and rigorous safety evaluations. Innovative strategies, including nanotechnology-enabled delivery systems, targeted purification, and AI-driven peptide discovery, are paving the way for next-generation marine functional ingredients. For the marine peptide sector to realize its full potential, future research must prioritize human clinical validation, mechanistic studies, and alignment with regulatory frameworks. Integrative efforts across food science, biotechnology, and medicine will ultimately enable the sustainable valorization of fish and shellfish proteins and peptides, positioning them at the forefront of solutions for modern health and nutrition challenges.

References

1 Ochiai, Y. and Ozawa, H. (2020). Biochemical and physicochemical characteristics of the major muscle proteins from fish and shellfish. *Fisheries Science* 86 (5): 729–740. https://doi.org/10.1007/s12562-020-01444-y.

2 Kemp, D.C. and Kwon, J.Y. (2021). Fish and shellfish-derived anti-inflammatory protein products: properties and mechanisms. *Molecules* 26: 3225.

3 Liu, R., Xing, L., Fu, Q. et al. (2016). A review of antioxidant peptides derived from meat muscle and by-products. *Antioxidants* 5: 32. https://doi.org/10.3390/antiox5030032.

4 Sarwar Gilani, G., Wu Xiao, C., and Cockell, K.A. (2012). Impact of antinutritional factors in food proteins on the digestibility of protein and the bioavailability of amino acids and on protein quality. *British Journal of Nutrition* 108 (S2): S315–S332. https://doi.org/10.1017/S0007114512002371.

5 Xu, Y. and Xu, X. (2021). Modification of myofibrillar protein functional properties prepared by various strategies: A comprehensive review. *Comprehensive Reviews in Food Science and Food Safety* 20 (1): 458–500. https://doi.org/10.1111/1541-4337.12665.

6 Kim, S.-K. and Wijesekara, I. (2010). Development and biological activities of marine-derived bioactive peptides: a review. *Journal of Functional Foods* 2 (1): 1–9. https://doi.org/10.1016/j.jff.2010.01.003.

7 Subhan, F., Hussain, Z., Tauseef, I. et al. (2021). A review on recent advances and applications of fish collagen. *Critical Reviews in Food Science and Nutrition* 61 (6): 1027–1037. https://doi.org/10.1080/10408398.2020.1751585.

8 Xu, S., Zhao, Y., Song, W. et al. (2023). Improving the sustainability of processing by-products: extraction and recent biological activities of collagen peptides. *Foods* 12 (10): 1965. https://doi.org/10.3390/foods12101965.

9 Ruiz-Ruiz, F., Mancera-Andrade, E.I., and Iqbal, H.M.N. (2017). Marine-derived bioactive peptides for biomedical sectors: a review. *Protein & Peptide Letters* 24 (2): 109–117.

10 Sable, R., Parajuli, P., and Jois, S. (2017). Peptides, peptidomimetics, and polypeptides from marine sources: A wealth of natural sources for pharmaceutical applications. *Marine Drugs* 15 (4): 124. https://doi.org/10.3390/md15040124.

11 Shi, J.M., Ren, F.-I., Zhang, X. et al. (2025). A comprehensive review on biological properties of fish and shellfish allergens. *European Food Research and Technology* 251 (11): 3417–3444. https://doi.org/10.1007/s00217-025-04848-1.

12 Shahidi, F. and Saeid, A. (2025). Bioactivity of marine-derived peptides and proteins: a review. *Mar Drugs* 23 (4): 157. https://doi.org/10.3390/md23040157.

13 Fereidoon, S. and Sarusha, S. (2022). Novel marine bioactives: application in functional foods, nutraceuticals, and pharmaceuticals. *Journal of Food Bioactives* 19 (0). https://doi.org/10.31665/JFB.2022.18316.

14 Remme, J.F., Korsnes, S., Steen, S. et al. (2023). The effects of enzymes, species, and storage of raw material on physicochemical properties of protein hydrolysates from whitefish heads. *Marine Drugs* 13 (21): 12088. https://doi.org/10.3390/md21110587.

15 Al Solami, L. and Korish, M. (2024). Proximate composition, fatty acid characteristics, amino acid profile and mineral content of fish *Acanthurus sohal*. *Heliyon* 10 (16): e36474. https://doi.org/10.1016/j.heliyon.2024.e36474.

16 Zheng, S.-L., Wang, Y.-Z., Zhao, Y.-Q. et al. (2023). High Fischer ratio oligopeptides from hard-shelled mussel: preparation and hepatoprotective effect against acetaminophen-induced liver injury in mice. *Food Bioscience* 53: 102638. https://doi.org/10.1016/j.fbio.2023.102638.

17 Liu, Z., Li, M., Yi, Q. et al. (2018). The neuroendocrine-immune regulation in response to environmental stress in marine bivalves. *Frontiers in Physiology* 9: https://doi.org/10.3389/fphys.2018.01456.

18 Wang, J. and Scheibel, T. (2018). Recombinant production of mussel byssus inspired proteins. *Biotechnology Journal* 13 (12): 1800146. https://doi.org/10.1002/biot.201800146.

19 Chai, T.-T., Law, Y.-C., Wong, F.-C., and Kim, S.-K. (2017). Enzyme-assisted discovery of antioxidant peptides from edible marine invertebrates: a review. *Marine Drugs* 15.

20 Huang, Z., Aweya, J.J., Zhu, C. et al. (2020). Modulation of crustacean innate immune response by amino acids and their metabolites: inferences from other species. *Frontiers in Immunology* 11: 574721. https://doi.org/10.3389/fimmu.2020.574721.

21 Hossain, A. and Shahidi, F. (2024). Upcycling shellfish waste: distribution of amino acids, minerals, and carotenoids in body parts of north Atlantic crab and shrimp. *Foods* 13 (17): 2700. https://doi.org/10.3390/foods13172700.

22 Vázquez, J.A., Valcarcel, J., Sapatinha, M. et al. (2023). Effect of the season on the production and chemical properties of fish protein hydrolysates and high-quality oils obtained from gurnard (*Trigla* spp.) by-products. *LWT* 177: 114576. https://doi.org/10.1016/j.lwt.2023.114576.

23 Nikoo, M., Regenstein, J.M., Haghi Vayghan, A., and Walayat, N. (2023). Formation of oxidative compounds during enzymatic hydrolysis of byproducts of the seafood industry. *Processes* 11 (2): 543. https://doi.org/10.3390/pr11020543.

24 Wangkheirakpam, M., Mahanand, S., Majumdar, R. et al. (2019). Fish waste utilization with reference to fish protein hydrolisate-a review. *Fish Technology* 56 (3): 169–178.

25 Venugopal, V. and Sasidharan, A. (2022). Functional proteins through green refining of seafood side streams. *Frontiers in Nutrition* 9: 974447. https://doi.org/10.3389/fnut.2022.974447.

26 Kim, S.-K. and Mendis, E. (2006). Bioactive compounds from marine processing byproducts – a review. *Food Research International* 39 (4): 383–393. https://doi.org/10.1016/j.foodres.2005.10.010.

27 Siddik, M.A.B., Howieson, J., Fotedar, R., and Partridge, G.J. (2021). Enzymatic fish protein hydrolysates in finfish aquaculture: a review. *Reviews in Aquaculture* 13 (1): 406–430. https://doi.org/10.1111/raq.12481.

28 Nikoo, M., Benjakul, S., and Ahmadi Gavlighi, H. (2022). Protein hydrolysates derived from aquaculture and marine byproducts through autolytic hydrolysis. *Comprehensive Reviews in Food Science and Food Safety* 21 (6): 4872–4899. https://doi.org/10.1111/1541-4337.13060.

29 Dent, T. and Maleky, F. (2023). Pulse protein processing: the effect of processing choices and enzymatic hydrolysis on ingredient functionality. *Critical Reviews in Food Science and Nutrition* 63 (29): 9914–9925. https://doi.org/10.1080/10408398.2022.2070723.

30 Tan, Y.X., Mok, W.K., Lee, J. et al. (2019). Solid state fermentation of brewers' spent grains for improved nutritional profile using *Bacillus subtilis* WX-17. *Fermentation* 5: 52. https://doi.org/10.3390/fermentation5030052.

31 Nasri, R., Abdelhedi, O., Nasri, M., and Jridi, M. (2022). Fermented protein hydrolysates: biological activities and applications. *Current Opinion in Food Science* 43: 120–127. https://doi.org/10.1016/j.cofs.2021.11.006.

32 Abd Rashid, N.Y., Manan, M.A., Pa'ee, K.F. et al. (2022). Evaluation of antioxidant and antibacterial activities of fish protein hydrolysate produced from Malaysian fish sausage (*Keropok lekor*) by-products by indigenous *Lactobacillus casei* fermentation. *Journal of Cleaner Production* 347: 131303. https://doi.org/10.1016/j.jclepro.2022.131303.

33 Molla, A.E. and Hovannisyan, H.G. (2011). Optimization of enzymatic hydrolysis of visceral waste proteins of beluga *Huso huso* using Protamex. *International Aquatic Research* 3 (2): 93–99.

34 Ortizo, R.G., Sharma, V., Tsai, M.-L. et al. (2023). Extraction of novel bioactive peptides from fish protein hydrolysates by enzymatic reactions. *Applied Sciences* 13 (9): 5768. https://doi.org/10.3390/app13095768.

35 Jamil, N.H., Halim, N.R.A., and Sarbon, N.M. (2016). Optimization of enzymatic hydrolysis condition and functional properties of eel (*Monopterus* sp.) protein using response surface methodology (RSM). *International Food Research Journal* 23 (1): 1–9.

36 Shaik, M.I. and Sarbon, N.M. (2022). A review on purification and characterization of anti-proliferative peptides derived from fish protein hydrolysate. *Food Reviews International* 38 (7): 1389–1409. https://doi.org/10.1080/87559129.2020.1812634.

37 Lee, H.-G., Kim, H.-S., Oh, J.-Y. et al. (2021). Potential antioxidant properties of enzymatic hydrolysates from *Stichopus japonicus* against hydrogen peroxide-induced oxidative stress. *Antioxidants* 10 (1): 110. https://doi.org/10.3390/antiox10010110.

38 Sridhar, K., Inbaraj, B.S., and Chen, B.-H. (2021). Recent developments on production, purification and biological activity of marine peptides. *Food Research International* 147: 110468. https://doi.org/10.1016/j.foodres.2021.110468.

39 Bo, W., Chen, L., Qin, D. et al. (2021). Application of quantitative structure-activity relationship to food-derived peptides: methods, situations, challenges and prospects. *Trends in Food Science & Technology* 114: 176–188. https://doi.org/10.1016/j.tifs.2021.05.031.

40 Nugraha, R., Ruethers, T., Johnston, E.B. et al. (2021). Effects of extraction buffer on the solubility and immunoreactivity of the Pacific Oyster Allergens. *In Foods* 10 (2): 409. https://doi.org/10.3390/foods10020409.

41 Ghelichi, S., Shabanpour, B., Pourashouri, P. et al. (2018). Extraction of unsaturated fatty acid-rich oil from common carp (*Cyprinus carpio*) roe and production of defatted roe hydrolysates with functional, antioxidant, and antibacterial properties. *Journal of the Science of Food and Agriculture* 98 (4): 1407–1415. https://doi.org/10.1002/jsfa.8608.

42 Tanuja, S., Viji, P., Zynudheen, A., and Joshy, C. (2012). Composition, functional properties and antioxidative activity of hydrolysates prepared from the frame meat of Striped catfish (*Pangasianodon hypophthalmus*). *Egyptian Journal of Biology* 14: 27–35. https://doi.org/10.4314/ejb.v14i1.3.

43 Naghdi, S., Rezaei, M., Tabarsa, M., and Abdollahi, M. (2023). Fish protein hydrolysate from sulfated polysaccharides extraction residue of tuna processing by-products with bioactive and functional properties. *Global Challenges* 7 (4): 2200214. https://doi.org/10.1002/gch2.202200214.

44 Abd El-Rady, T.K., Tahoun, A.-A.M., Abdin, M., and Amin, H.F. (2023). Effect of different hydrolysis methods on composition and functional properties of fish protein hydrolysate obtained from Bigeye tuna waste. *International Journal of Food Science and Technology* 58 (12): 6552–6562. https://doi.org/10.1111/ijfs.16769.

45 Ucak, I., Afreen, M., Montesano, D. et al. (2021). Functional and bioactive properties of peptides derived from marine side streams. *Marine Drugs* 19.

46 Halim, N.R.A., Yusof, H.M., and Sarbon, N.M. (2016). Functional and bioactive properties of fish protein hydolysates and peptides: a comprehensive review. *Trends in Food Science & Technology* 51: 24–33. https://doi.org/10.1016/j.tifs.2016.02.007.

47 Ghalamara, S., Brazinha, C., Silva, S., and Pintado, M. (2024). Valorization of fish processing by-products: biological and functional properties of bioactive peptides. *Current Food Science and Technology Reports* 2 (4): 393–409. https://doi.org/10.1007/s43555-024-00045-5.

48 Whitaker, R.D., Altintzoglou, T., Lian, K., and Fernandez, E.N. (2021). Marine bioactive peptides in supplements and functional foods - a commercial perspective. *Current Pharmaceutical Design* 27 (11): 1353–1364. https://doi.org/10.2174/1381612824999201105164000.

49 Zhang, S.-Y., Zhao, G.-X., Suo, S.-K. et al. (2021). Purification, identification, activity evaluation, and stability of antioxidant peptides from alcalase hydrolysate of Antarctic krill (*Euphausia superba*) proteins. *Marine Drugs* 19.

50 Wang, Q., Shi, J., Zhong, H.A. et al. (2021). High-degree hydrolysis sea cucumber peptides improve exercise performance and exert antifatigue effect via activating the NRF2 and AMPK signaling pathways in mice. *Journal of Functional Foods* 86: 104677. https://doi.org/10.1016/j.jff.2021.104677.

51 Sheng, Y., Qiu, Y.-T., Wang, Y.-M. et al. (2022). Novel antioxidant collagen peptides of *Siberian sturgeon* (acipenserbaerii) cartilages: the preparation, characterization, and cytoprotection of H2O2-damaged human umbilical vein endothelial cells (HUVECs). *Marine Drugs* 20.

52 Wang, Y.-M., Li, X.-Y., Wang, J. et al. (2022). Antioxidant peptides from protein hydrolysate of skipjack tuna milt: purification, identification, and cytoprotection on H2O2 damaged human umbilical vein endothelial cells. *Process Biochemistry* 113: 258–269. https://doi.org/10.1016/j.procbio.2022.01.008.

53 Phetchthumrongchai, T., Tachapuripunya, V., Chintong, S. et al. (2022). Properties of protein hydrolysates and bioinformatics prediction of peptides derived from thermal and enzymatic process of skipjack tuna (*Katsuwonus pelamis*) roe. *Fishes* 7 (5): 255. https://doi.org/10.3390/fishes7050255.

54 Lima, K., de Quadros, C., da Rocha, M. et al. (2019). Bioactivity and bioaccessibility of protein hydrolyzates from industrial byproducts of Stripped weakfish (*Cynoscion guatucupa*). *LWT* 111: 408–413.

55 Qian, Z.-J., Zhang, Y.Y., Oh, G.-W. et al. (2018). Antioxidant and angiotensin I converting enzyme inhibition effects and antihypertensive effect in spontaneously hypertensive rats of peptide isolated from boiled abalone by-products, *Hallotis discus hannai*. *Journal of Aquatic Food Product Technology* 27 (9): 946–960. https://doi.org/10.1080/10498850.2018.1518361.

56 Lima, D.A.S., Santos, M.M.F., Duvale, R.L.F. et al. (2021). Technological properties of protein hydrolysate from the cutting byproduct of serra Spanish mackerel (*Scomberomorus brasiliensis*). *Journal of Food Science and Technology* 58 (8): 2952–2962. https://doi.org/10.1007/s13197-020-04797-5.

57 He, Y., Pan, X., Chi, C.-F. et al. (2019). Ten new pentapeptides from protein hydrolysate of miiuy croaker (*Miichthys miiuy*) muscle: Preparation, identification, and antioxidant activity evaluation. *LWT* 105: 1–8. https://doi.org/10.1016/j.lwt.2019.01.054.

58 Shen, J., Zeng, M., Huang, P. et al. (2023). Purification and activity evaluation of novel anti-inflammatory peptides from pearl oyster (*Pinctada martensii*) hydrolysates. *Food & Function* 14 (9): 4242–4253. https://doi.org/10.1039/D2FO04046H.

59 Hayes, M., Aluko, R.E., Aurino, E., and Mora, L. (2030). Generation of bioactive peptides from porphyridium sp. and assessment of their potential for use in the prevention of hypertension, inflammation and pain. *Marine Drugs* 21.

60 Durand, R., Pellerin, G., Thibodeau, J. et al. (2020). Screening for metabolic syndrome application of a herring by-product hydrolysate after its separation by electrodialysis with ultrafiltration membrane and identification of novel anti-inflammatory peptides. *Separation and Purification Technology* 235: 116205. https://doi.org/10.1016/j.seppur.2019.116205.

61 Joshi, I. and Nazeer, R.A. (2020). EGLLGDVF: a novel peptide from green mussel *Perna viridis* foot exerts stability and anti-inflammatory effects on LPS-stimulated RAW264.7 cells. *Protein and Peptide Letters* 27 (9): 851–859. https://doi.org/10.2174/0929866527666200224111832.

62 Vieira, E.F., Van Camp, J., Ferreira, I.M.P.L.V.O., and Grootaert, C. (2018). Protein hydrolysate from canned sardine and brewing by-products improves TNF-α-induced inflammation in an intestinal–endothelial co-culture cell model. *European Journal of Nutrition* 57 (6): 2275–2286. https://doi.org/10.1007/s00394-017-1503-2.

63 Solstad, R.G., Johansen, C., Stensvåg, K. et al. (2020). Structure-activity relationship studies of shortened analogues of the antimicrobial peptide EeCentrocin 1 from the sea urchin *Echinus esculentus*. *Journal of Peptide Science* 26 (2): e3233. https://doi.org/10.1002/psc.3233.

64 Wang, B., Zhang, H., Wen, Y. et al. (2024). The novel angiotensin-I-converting enzyme inhibitory peptides from *Scomber japonicus* muscle protein hydrolysates: QSAR-based screening, molecular docking, kinetic and stability studies. *Food Chemistry* 447: 138873. https://doi.org/10.1016/j.foodchem.2024.138873.

65 Ishak, N.H., Shaik, M.I., Yellapu, N.K. et al. (2021). Purification, characterization and molecular docking study of angiotensin-I converting enzyme (ACE) inhibitory peptide from shortfin scad (*Decapterus macrosoma*) protein hydrolysate. *Journal of Food Science and Technology* 58 (12): 4567–4577. https://doi.org/10.1007/s13197-020-04944-y.

66 Hu, Y.-D., Xi, Q.-H., Kong, J. et al. (2023). Angiotensin-I-converting enzyme (ACE)-inhibitory peptides from the collagens of monkfish (*Lophius litulon*) swim bladders: isolation, characterization, molecular docking analysis and activity evaluation. *Marine Drugs* 21 (10): 516. https://doi.org/10.3390/md21100516.

67 Suo, S.-K., Zhao, Y.-Q., Wang, Y.-M. et al. (2022). Seventeen novel angiotensin converting enzyme (ACE) inhibitory peptides from the protein hydrolysate of *Mytilus edulis*: isolation, identification, molecular docking study, and protective function on HUVECs. *Food & Function* 13 (14): 7831–7846. https://doi.org/10.1039/D2FO00275B.

68 Wang, J., Wei, R., and Song, R. (2019). Novel antibacterial peptides isolated from the Maillard reaction products of half-fin anchovy (*Setipinna taty*) hydrolysates/glucose and their mode of action in *Escherichia coli*. *Marine Drugs* 17 (1): 47. https://doi.org/10.3390/md17010047.

69 López-Abarrategui, C., McBeth, C., Mandal, S.M. et al. (2015). Cm-p5: an antifungal hydrophilic peptide derived from the coastal mollusk *Cenchritis muricatus* (Gastropoda: Littorinidae). *Faseb Journal* 29 (8): 3315–3325. https://doi.org/10.1096/fj.14-269860.

70 Rodriguez, A., Martell-Huguet, E.M., González-García, M. et al. (2023). Identification and characterization of three new antimicrobial peptides from the marine mollusk Nerita versicolor (Gmelin, 1791). *International Journal of Molecular Sciences* 24 (4): 3852. https://doi.org/10.3390/ijms24043852.

71 Zheng, L., Li, Y., Wang, J. et al. (2020). Antibacterial and antiparasitic activities analysis of a hepcidin-like antimicrobial peptide from *Larimichthys crocea*. *Acta Oceanologica Sinica* 39 (10): 129–139. https://doi.org/10.1007/s13131-020-1580-6.

72 Li, J., Chen, Y., Yuan, N. et al. (2018). A novel natural influenza A H1N1 virus neuraminidase inhibitory peptide derived from cod skin hydrolysates and its antiviral mechanism. *Marine Drugs* 16 (10): 377. https://doi.org/10.3390/md16100377.

73 Wang, K., Siddanakoppalu, P.N., Ahmed, I. et al. (2020). Purification and identification of anti-allergic peptide from Atlantic Salmon (*Salmo salar*) byproduct enzymatic hydrolysates. *Journal of Functional Foods* 72: 104084. https://doi.org/10.1016/j.jff.2020.104084.

74 Woo, M., Song, Y.O., Kang, K.-H., and Noh, J.S. (2018). Anti-obesity effects of collagen peptide derived from skate (*Raja kenojei*) skin through regulation of lipid metabolism. *Marine Drugs* 16 (9): 306. https://doi.org/10.3390/md16090306.

75 Vijayan, D.K., Perumcherry Raman, S., Dara, P.K. et al. (2022). In vivo anti-lipidemic and antioxidant potential of collagen peptides obtained from great hammerhead shark skin waste. *Journal of Food Science and Technology* 59 (3): 1140–1151. https://doi.org/10.1007/s13197-021-05118-0.

76 Pisoschi, A.M., Pop, A., Iordache, F. et al. (2021). Oxidative stress mitigation by antioxidants - an overview on their chemistry and influences on health status. *European Journal of Medicinal Chemistry* 209: 112891. https://doi.org/10.1016/j.ejmech.2020.112891.

77 Taira, Y., Yamashita, T., Bian, Y. et al. (2020). Antioxidative effects of a novel dietary supplement Neumentix in a mouse stroke model. *Journal of Stroke and Cerebrovascular Diseases* 29 (8): 104818. https://doi.org/10.1016/j.jstrokecerebrovasdis.2020.104818.

78 Jegani, K.T., Balde, A., and Nazeer, R.A. (2025). A review on anti-inflammatory and antioxidant peptides derived from marine organisms: mechanism of action and therapeutic applications. *Food Bioscience* 63: 105745. https://doi.org/10.1016/j.fbio.2024.105745.

79 Leong, Y.K. and Chang, J.-S. (2024). Proteins and bioactive peptides from algae: Insights into antioxidant, anti-hypertensive, anti-diabetic and anti-cancer activities. *Trends in Food Science & Technology* 145: 104352. https://doi.org/10.1016/j.tifs.2024.104352.

80 Xia, G., Zhang, X., Dong, Z., and Shen, X. (2017). Comparative study on the antioxidant activity of peptides from pearl oyster (*Pinctada martensii*) mantle type V collagen and tilapia (*Oreochromis niloticus*) scale type I collagen. *Journal of Ocean University of China* 16 (6): 1175–1182. https://doi.org/10.1007/s11802-017-3323-7.

81 Najafian, L. and Babji, A.S. (2015). Isolation, purification and identification of three novel antioxidative peptides from patin (*Pangasius sutchi*) myofibrillar protein hydrolysates. *LWT - Food Science and Technology* 60 (1): 452–461. https://doi.org/10.1016/j.lwt.2014.07.046.

82 Kim, N.Y., Jung, H.Y., and Kim, J.K. (2021). Identification and characterisation of a novel heptapeptide mackerel by-product hydrolysate, and its potential as a functional fertiliser component. *Journal of Chromatography B* 1180: 122881. https://doi.org/10.1016/j.jchromb.2021.122881.

83 Chen, J., Bai, W., Cai, D. et al. (2021). Characterization and identification of novel anti-inflammatory peptides from Baijiao sea bass (*Lateolabrax maculatus*). *LWT* 147: 111521. https://doi.org/10.1016/j.lwt.2021.111521.

84 de la Fuente, B., Berrada, H., and Barba, F.J. (2022). Marine resources and cancer therapy: from current evidence to challenges for functional foods development. *Current Opinion in Food Science* 44: 100805. https://doi.org/10.1016/j.cofs.2022.01.001.

85 Kandyliari, A., Golla, J.P., Chen, Y. et al. (2020). Antiproliferative activity of protein hydrolysates derived from fish by-products on human colon and breast cancer cells. *Proceedings of the Nutrition Society* 79 (OCE2): E282. https://doi.org/10.1017/S002966512000230X.

86 Yu, F., Zhang, Y., Ye, L. et al. (2018). A novel anti-proliferative pentapeptide (ILYMP) isolated from *Cyclina sinensis* protein hydrolysate induces apoptosis of DU-145 prostate cancer cells. *Molecular Medicine Reports* 18 (1): 771–778. https://doi.org/10.3892/mmr.2018.9019.

87 Hsu, K.-C., Li-Chan, E.C.Y., and Jao, C.-L. (2011). Antiproliferative activity of peptides prepared from enzymatic hydrolysates of tuna dark muscle on human breast cancer cell line MCF-7. *Food Chemistry* 126 (2): 617–622. https://doi.org/10.1016/j.foodchem.2010.11.066.

88 Feng, S., Limwachiranon, J., Luo, Z. et al. (2016). Preparation and purification of angiotensin-converting enzyme inhibitory peptides from hydrolysate of shrimp (*Litopenaeus vannamei*) shell waste. *International Journal of Food Science and Technology* 51 (7): 1610–1617. https://doi.org/10.1111/ijfs.13131.

89 Wijesekara, I. and Kim, S.-K. (2010). Angiotensin-I-converting enzyme (ACE) inhibitors from marine resources: prospects in the pharmaceutical industry. *Marine Drugs* 8: 1080–1093.

90 Harnedy-Rothwell, P.A., McLaughlin, C.M., O'Keeffe, M.B. et al. (2020). Identification and characterisation of peptides from a boarfish (*Capros aper*) protein hydrolysate displaying in vitro dipeptidyl peptidase-IV (DPP-IV) inhibitory and insulinotropic activity. *Food Research International* 131: 108989. https://doi.org/10.1016/j.foodres.2020.108989.

91 Neves, A.C., Harnedy, P.A., and FitzGerald, R.J. (2016). Angiotensin converting enzyme and dipeptidyl peptidase-IV inhibitory, and antioxidant activities of a blue mussel (*Mytilus edulis*) meat protein extract and its hydrolysates. *Journal of Aquatic Food Product Technology* 25 (8): 1221–1233. https://doi.org/10.1080/10498850.2015.1051259.

11

Quality and Safety of Aquatic Products

Zhijun Tan[1], Dongyu Li[2], and Guanchao Zheng[1]

[1] *National Aquatic Product Quality Inspection and Testing Center, Qingdao, China*
[2] *Ocean University of China, Qingdao, China*

Food safety is defined by the World Health Organization (WHO) as "the assurance that food will not cause harm to the consumer when it is prepared and eaten according to its intended use." In fact, the definition is just a narrow concept of food safety, and the wider one also covers nutrition and food security. That means food safety, food nutrition, and food security are inextricably linked. By the historical development of human demand for food, food security—meaning enough food supply—is the first layer demand, food safety—meaning no-harm food supply—is the second layer demand, and food nutrition—meaning nutritious food supply—is the third and also the highest layer demand. Until now, food safety has been a threat to most countries worldwide. According to the WHO, an estimated 600 million people worldwide—almost 1 in 10 of the global population—fall ill each year after consuming contaminated food, resulting in 420,000 deaths annually. Additionally, unsafe food causes economic losses of US$ 110 billion yearly in low- and middle-income countries due to reduced productivity and increased medical expenses. Therefore, food safety is not only a matter of human health, but also an important economic and political issue. Compared with terrestrial food, the safety issues of aquatic products are more severe and prominent, primarily because of the complex factors associated with water's openness and fluidity, as well as the unique physiological characteristics of aquatic organisms (Fig. 11.1).

11.1 Overview of Aquatic Product Quality and Safety

11.1.1 International Evolution of Aquatic Product Safety and Quality Research

Throughout the development of global food quality and safety (including aquatic product safety), public attention to food safety has gradually grown alongside economic progress. Although early human societies also prioritized food safety, the

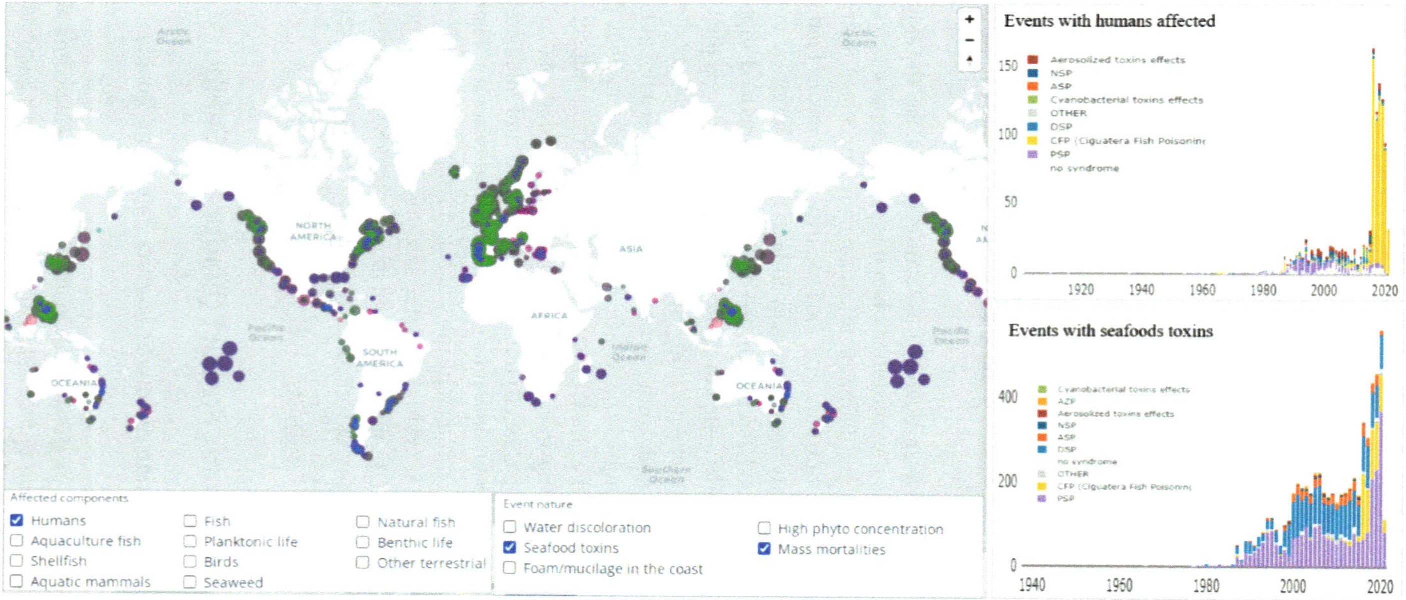

Figure 11.1 Global distribution map of shellfish toxin hazards (1800–2022). Note: the map and data from UNESCO / https://data.hais.ioc-unesco.org/ / last accessed Jan. 9, 2026.

modern concept of food safety is generally traced back to the establishment of the US Food and Drug Administration (FDA) in 1930, widely recognized as a landmark event. The founding of the European Food Safety Authority (EFSA) in 2002 further elevated the significance of food safety in developed countries. Beyond these two key entities—the US FDA and the EU's EFSA—other developed nations have established specialized food safety institutions, while global bodies such as the United Nations Food and Agriculture Organization (FAO), WHO, Codex Alimentarius Commission (CAC), and World Trade Organization (WTO) also play critical roles. In research depth and control technology, developed countries have long led global advancements, surpassing developing nations across all domains: detection technology, risk assessment, theoretical research, evaluation methodologies, prevention and control techniques, and policy frameworks. Relying on their advanced technological systems and robust theoretical foundations, they have formulated a range of trade barriers and maintained prolonged restrictive measures against developing countries, including China, causing substantial economic losses. For instance, China's marine shellfish products faced European Union (EU) restrictions for 19 consecutive years.

In scientific research, developed countries consistently prioritize fundamental studies, continuously innovating in new technologies, methods, and concepts. They have not only generated numerous original and highly promising research outcomes but also long guided global scientific trends, setting benchmarks for both technological innovation and theoretical development. Internationally, research advancements in aquatic product safety and quality management are converging on three strategic directions: (1) Interdisciplinary integration leveraging biology, oceanography, chemistry, and food science to advance systematic investigations through full-chain analysis and multi-omics integration. (2) Holistic risk-nutrition paradigm shift moving beyond traditional fragmented approaches to address co-formation mechanisms of safety hazards and nutritional attributes across the production-to-consumption continuum. (3) Smart risk-mitigation technologies combining predictive analytics for hazard detection, artificial intelligence (AI)-driven nutritional grading systems, and blockchain-enabled traceability platforms to elevate industry value chains.

11.1.2 International Aquatic Products Quality and Safety Supervision

Food quality and safety have become an economic, social, and national security issue worldwide, which has become a serious challenge to both developed and developing countries. Therefore, food safety has not only become a country or a region, but has even become one of the key factors affecting the future destiny of human beings. In response to food quality and safety issues, most countries in the world have established corresponding laws and regulations, especially in developed countries.

In general, the world food safety regulatory system can be divided into three modes: the first is a multi-model, the food control system based on the responsibility of multiple departments. The second is a comprehensive system, with one department responsible for formulating policies, standards, risk assessment, coordination guidance, and supervision; other departments are the basis for implementation. The third is a single system that combines all responsibility for protecting public health and food safety into a single food safety agency. The priority system advocated by the FAO and the WHO is a single system, and the integrated system is considered to be an intermediate form of transition from a distant system to a single system.

11.1.2.1 World Food Safety Organizations

11.1.2.1.1 United Nations Food and Agriculture Organization (FAO) The FAO was founded in 1945. It has 191 member states and 1 member organization (EU). The aim of food safety of FAO is to guarantee the basic food demands of all people in the world, which includes improving the efficiency of production and distribution of all food and agricultural products, improving the living conditions of the rural population, promoting the development of the rural economy, and ultimately eliminating hunger and poverty. The main functions of FAO are: (1) collecting, collating, analyzing, and disseminating information on world food production and trade; (2) providing technical assistance to member states, mobilizing the international community to invest, and implementing agricultural development projects of international development and financial institutions; (3) member states provide advisory services on food and agriculture policies and plans; (4) discuss major issues in the international food and agriculture sector, formulate relevant international codes of conduct and regulations, negotiate international standards and agreements in the field of food and agriculture, and strengthen consultation and cooperation among member states. In FAO, the Joint FAO/WHO Expert Committee on Food Additives (JECFA) major play the role in food safety.

11.1.2.1.2 World Health Organization (WHO) WHO was founded in 1948. The WHO Global Food Safety Strategy aims to reduce the health and social burden of foodborne diseases and to improve food safety capabilities worldwide. Mainly through the implementation of (1) the development of a risk-based, continuous integrated food safety system to promote and support; (2) the design of science-based measures throughout the food production chain, these measures will prevent unacceptable foods contact with microbiological and chemical levels; (3) collaboration with other sectors and partners, assessment and management of foodborne risks, and exchange of information, and three action lines to achieve overall strategic objectives.

11.1.2.1.3 Codex Alimentarius Commission (CAC) The CAC is an intergovernmental international organization jointly established by the FAO of the United Nations and the WHO in 1963. It is responsible for coordinating intergovernmental food standards and establishing a complete international food standard system. The CAC currently has 180 member countries, covering 98% of the world's population. The main work is carried out through the three Codex committees and their branches under the Executive Committee.

11.1.2.1.4 International Organization for Standardization (ISO) The ISO traces its origins to the International Federation of National Standardization Associations (ISA) and the United Nations Standards Coordinating Committee (UNSCC). In October 1946, 25 national standardization bodies convened a historic conference in London, resolving to establish a unified international body. This assembly drafted and ratified ISO's foundational constitution and procedural framework, culminating in its official establishment on February 23, 1947, headquartered in Geneva, Switzerland. Mandated to advance global standardization, ISO facilitates international trade and fosters cross-border collaboration in science, technology, and economic development through consensus-driven technical specifications.

11.1.2.2 Regional Food Safety Organization – European Union

The EU has established the world's most stringent food regulatory systems and standards, and has also established food safety regulatory agencies in the EU and its member states to address issues such as food safety, consumer health, and international relations related to agriculture. It mainly includes the European Food Regulatory Authority, the EU Public Health Consumer Protection Department, and the EU Food and Veterinary Office. At the level of member states, such as Britain, France, Germany, Holland and other countries, the relevant national food safety administration has been set up respectively, with emphasis on responsibilities and obligations. The EFSA, established in 2002 under EU Regulation 178/2002, serves as the cornerstone of the EU's food safety governance system. Its core mandate encompasses conducting independent scientific assessments and integrated risk analyses to inform EU regulatory frameworks on food chain hazards—from chemical contaminants to emerging technologies like gene editing. EFSA's evidence-based recommendations directly underpin EU policies on food safety, consumer protection, and international trade safeguards, ensuring alignment with global standards such as Codex Alimentarius. While EFSA does not directly enforce regulations, its scientific authority enables the EU to implement science-backed trade measures, such as import restrictions for non-compliant products, thereby balancing risk mitigation with market access objectives.

11.1.2.3 National Food Safety Organization

11.1.2.3.1 United States The US food safety regulatory system is generally divided into three basic levels: federal, state, and regional. Take the federal network as an example. Relevant regulatory agencies include FDA and US Centers for Disease Control (CDC) under the US Department of Health and Human Services (DHHS); the Food Safety and Inspection Service (FSIS), the Animal and Plant Health Inspection Service (APHIS), and Agricultural Marketing Service (AMS) under the US Department of Agriculture (USDA); Environmental Protection Agency (EPA); the Customs and Border Protection (CBP) under the Department of Homeland Security (DHS).

11.1.2.3.2 Canada Canada adopts a federal system and implements a federal, provincial, and municipal administrative system. In food safety management, the federal, provincial, and municipal authorities have a corresponding responsibility to manage food safety in a model of hierarchical management, mutual cooperation, and broad participation. The main governing bodies at the federal level are the Department of Health and the Department of Agriculture's Canadian Food Inspection Agency (CFIA). The Ministry of Health is responsible for the development of safety and nutritional quality standards for all foods sold in the country, as well as policies related to food safety. The CFIA is responsible for implementing these regulations and standards and overseeing the implementation of relevant regulations and standards. The provincial government's food safety agency provides inspections of small food companies that sell their products locally within their jurisdiction. The municipality is responsible for providing and supervising public health standards for restaurants that operate finished food products.

11.1.2.3.3 United Kingdom The food safety regulation in the United Kingdom is jointly undertaken by the federal government, local authorities, and various organizations to strengthen supervision. In the United Kingdom, the Food Standards Agency (FSA) is an independent monitoring agency not affiliated with any government department. It is responsible for the overall food safety affairs and the development of various standards. It submits an annual food inspection report to Congress. The FSA's governing body is based on the Board of Directors. Its members are appointed by the Minister of Health of the Central Government, the Ministers of Health of the local governments of Scotland, Wales, and Northern Ireland, to discuss the basic operational strategies and policy formulation of the Food Standards Bureau. A CEO is responsible for the day-to-day affairs, based in London, with branches in Scotland, Wales, and Northern Ireland, as well as related special working groups or committees, such as food additives, food hygiene, food emergency, and food pollution, to strengthen the monitoring of all aspects of the food chain.

11.1.2.3.4 China China's food safety management adopts a multi-modal model, which implements a segmented comanagement model in various government departments. The food safety management work at the central government level is mainly handled by the FDA, the Ministry of Agriculture and Rural Affairs, the Health Planning Commission, the General Administration of Quality Supervision, Inspection and Quarantine, the Ministry of Industry and Information Technology, and the Ministry of Commerce, and reports to the State Food Safety Office. These institutions are all formed into systems. There are corresponding extension institutions at the provincial, city, and county levels, and each organization has its own specific structure and management scope. These departments mainly supervise and manage food according to the principle of subsector supervision. The process of food from raw materials to processing to finished products is handled by one department. The General Administration of Food and Drugs is responsible for the unified supervision of production, circulation, and consumption. The Ministry of Agriculture and Rural Affairs is responsible for the quality and safety supervision and management of agricultural products and the management of genetically modified organisms. The Health Planning Commission is responsible for food safety risk assessment and food safety standards. The General Administration of Quality Supervision, Inspection, and Quarantine is responsible for food import and export management. And food-related product management, the Ministry of Commerce is responsible for the supervision of the circulation industry.

11.1.3 Main Features of Aquatic Product Quality and Safety

As mentioned above, the safety issues inherent to aquatic products present a significantly greater challenge compared to terrestrial foods. Aquatic organisms are highly dependent on the watery environment, which not only serves as a medium for various pollutants but also exhibits dimensional complexity and dynamic variability in its pollution characteristics. Furthermore, water-living organisms possess unique physiological and ecological traits—such as high-protein content—that make them particularly susceptible to contamination risks related to safety and quality. Additionally, their small size and diverse population further complicate the prevention and control of these safety and quality issues in aquatic products. According to recent reports, hazards identified in aquatic products can be categorized into external risk factors (e.g., biological, chemical, and physical contaminants) as well as emerging contaminants, alongside internal risk factors such as allergens and biogenic amines [1, 2].

11.1.3.1 Many Safety Hazards with a Wide Range of Sources

As the final environmental destination of pollutants, the aquatic environment contains a large number of toxic and harmful risk factors, such as microorganisms, drugs, toxins, organic pollutants, heavy metals, and microplastics, as well as formaldehyde, biogenic amines, and allergens produced by water-derived products. In comparison, the terrestrial animal food safety hazard factors are mostly present in fixed environmental units, such as soil or feed, while the aquatic product safety risk factors may come from various environmental units, such as the surface layer, the middle layer, the bottom layer, and the sediment of the water environment. In addition, physiology and ecology are significant differences between different aquatic organisms, which comprehensively form two major ways of exogenous (environmental transmission) or endogenous (biological self-production) of aquatic product safety risks. Also, the food chain(net) in a watery environment is more complicated. Therefore, the risk of aquatic products is more complicated, and the controllability is less feasible than that of terrestrial products [3, 4].

11.1.3.2 High Variety of Aquatic Organisms with Complex Potential Risk

Many kinds of watery organisms are cultured in China, such as fish, shrimp, shellfish, algae, and other catches, by a series of culture modes, including factory, pond, bottom sowing, shovel, and fishing. Due to the huge differences between biological species and their respective production models and living environments, the comprehensive safety risk difference among different varieties of aquatic products is extremely significant. For example, farmed fish, shrimp, and other products are mainly drug residues and allergens, algae are heavy metals and herbicides, while shellfish may have microorganisms, toxins, persistent organic pollutants (POPs), heavy metals, and microplastics. Even offshore catches such as tuna, squid, and krill are at risk of heavy metals such as mercury, fluorine, and endogenous formaldehyde. Particularly serious is that the three-dimensional nature of water pollution can easily cause multi-pollution characteristics in aquatic products, which brings serious compound risks to consumers.

11.1.3.3 Long Production Cycle with Absent Effective Control Technology

Whether it is planting or breeding, the production cycle of terrestrial food is mostly no more than one year; some kinds are only several months. The food chain only has "environment-bait (input)-biological," and the safety control point is certified. However, the production cycle of aquatic products is more than 6 months, and the longest is even over several years. Because the aquatic environment is far more complex and difficult to control than the terrestrial environment, the living environment of aquatic organisms is widely distributed in estuaries, bays, offshore, oceans, and even polar regions. In water, organisms may distribute in the surface, middle, and bottom layers, and the food chain includes two complex ones: that is, "environment-ecosystem food chain-biological" and "environment-bait (input)-biological," respectively. Therefore, the production cycle and chain of aquatic products are significantly longer than that of terrestrial food. Combined with the complicated processing methods of aquatic products, the risk of aquatic products is difficult to control and lacks effective control technology.

11.1.4 Risk Assessment of Aquatic Products

Risk assessment refers to the process of scientific assessment of the possible adverse effects and extent of biological, chemical, and physical hazards in aquatic products and additives on human health. The purpose of the risk assessment is to provide a description of the health risks associated with food/hazards based on scientific evidence. The ultimate goal include ① protect public health and enhance consumer confidence; ② promote international trade. Risk assessment is the scientific basis for formulating aquatic product safety standards, responding to aquatic product safety emergencies, issuing warnings, and preventing foodborne diseases.

The risk assessment process can be divided into four phases: hazard identification, hazard characterization, exposure assessment, and risk description. ① Hazard identification: According to scientific data and literature information, such as epidemiology, animal tests, in vitro tests, and structure-activity relationships, whether human body exposure to a certain hazard will adversely affect health and cause adverse effects. Possibilities, and the population and scope that may be at risk. ② Hazard characterization: Qualitative or quantitative description of the adverse health effects associated with hazards. Animal testing, clinical studies, and epidemiological studies can be used to determine dose–response relationships, mechanisms of action, and other mechanisms of adverse health effects. If possible, a threshold for toxic effects should establish a safe intake level. ③ Exposure assessment: Describe the pathways of harm entering the human body and estimate the level of hazard in different populations. According to the level of the hazard in the diet and the dietary consumption of the population, the total dietary intake of the hazard is estimated initially, and other non-meal access to the human body is considered, and the total intake of the human body is estimated and compared with the safe intake. ④ Risk characterization: Based on hazard identification, hazard characterization, and exposure assessment, a comprehensive analysis of the risk and extent of adverse effects on the health of the population should be conducted, and the uncertainty in the risk assessment process should be described and explained.

According to the evaluation method adopted, risk assessment can be generally divided into two types: deterministic assessment and probabilistic assessment. The probabilistic assessment process is relatively complex compared to deterministic assessments, but the assessment results provide more information for risk management decisions.

Depending on the outcome of the outcome, risk assessment can be divided into qualitative assessment and quantitative assessment. Qualitative assessments use high, medium, and low descriptive words to represent risk; quantitative assessments use quantitative values to represent the magnitude of the risk and its accompanying uncertainty.

11.2 Aquatic Product Safety (Pollution Risk) Category

The main factors affecting the safety of aquatic products can be divided into two categories. One is inherent in aquatic products, and is also known as endogenous influencing factors, such as formaldehyde, histamine, and endogenous organisms decomposed in aquatic products. The other is caused by external factors, also collectively referred to as exogenous factors, the accumulation of harmful substances caused by environmental pollution in aquatic products, and the accumulation of seaweed toxins in shellfish. Shellfish toxins, pesticide residues, and veterinary drug residues are all such. The safety and health factors in aquatic products include the following aspects.

11.2.1 Biological Pollutants

This section focuses on biological pollutants in aquatic products, including bacteria, viruses, parasites, and toxins produced by microorganisms, which pose significant food safety risks and can cause foodborne diseases [5, 6]. These contaminants may originate naturally from the aquatic environment or be introduced through contamination during handling and processing [7]. The discussion will center on their sources, transmission pathways, pathogenic mechanisms, and associated human health impacts.

11.2.1.1 Indigenous Pathogens: The *Vibrio* Paradigm

Vibrio is a genus of Gram-negative bacteria that are natural inhabitants of marine and estuarine environments worldwide. Unlike many foodborne pathogens that indicate fecal contamination, the presence of *Vibrio* species in seafood is often natural and environmentally driven. Certain species pose significant human health risks, causing illnesses ranging from gastroenteritis to life-threatening septicemia, primarily through the consumption of raw or undercooked seafood such as oysters and clams[8].

Vibrio species occupy specific ecological niches as free-living organisms in coastal waters, sediments, and estuaries, often associated with zooplankton such as copepods and the surfaces of aquatic animals. Their abundance is strongly correlated with environmental conditions, particularly water temperature (nearly 15 °C/59 °F) and salinity (0.5–4%). This environmental dependence makes *Vibrio*-related risks seasonal and geographically variable. Notably, rising sea temperatures due to climate change are linked to the expanded geographic range and longer seasonal prevalence of pathogenic vibrios, representing a growing public health concern [9].

Vibrio parahaemolyticus is a significant foodborne pathogen worldwide and a leading cause of seafood-associated bacterial gastroenteritis. The pathogenicity of *V. parahaemolyticus* is primarily attributed to the production of key virulence factors. Most clinical strains produce the thermostable direct hemolysin (TDH) and/or the TDH-related hemolysin (TRH), which act as enterotoxins and cause cytotoxicity and fluid secretion in the intestines.

Vibrio cholerae is a diverse bacterial species inhabiting estuarine and coastal environments worldwide, where it frequently associates with zooplankton and other aquatic organisms. Pathogenic strains of *V. cholerae* produce several key virulence factors that enable colonization and disease progression. The cholera toxin (CT), encoded by the ctxAB genes carried on a lysogenic bacteriophage (CTX), is the primary virulence determinant responsible for the severe watery diarrhea characteristic of cholera. Another primary virulence factor is toxin-co-regulated pilus (TCP). Bacterial colonization of the small intestine is facilitated by the TCP.

Vibrio vulnificus is a highly invasive and opportunistic bacterial pathogen found in warm coastal marine environments, particularly in temperate zones such as the Gulf of Mexico. It is recognized as one of the most serious foodborne pathogens due to its high fatality rates in susceptible individuals. The virulence of *V. vulnificus* is attributed to several key factors. The capsular polysaccharide is a major antiphagocytic factor that helps the bacterium evade host immune defenses. Additionally, it produces potent cytolysin/hemolysin (VvhA) and metalloproteases, which contribute to tissue damage, vascular permeability, and systemic invasion. Infections with *V. vulnificus* exhibit a strong epidemiological profile linked to underlying health conditions.

11.2.1.2 Contaminating Pathogens: The Cross-Contamination Threat

In contrast to indigenous pathogens, contaminating pathogens are not naturally present in the aquatic environment but are introduced into seafood products during harvesting, handling, processing, or distribution. Their presence typically indicates inadequate hygiene practices, poor sanitation, or cross-contamination from other sources. Unlike Vibrio species, whose growth is largely influenced by environmental conditions, the proliferation of these contaminants is primarily governed by time and temperature abuse during the postharvest phase. Controlling these hazards, therefore, relies heavily on strict adherence to Good Hygiene Practices (GHP) and Hazard Analysis Critical Control Point (HACCP) principles throughout the supply chain.

Staphylococcus aureus is a Gram-positive bacterium commonly associated with human skin and nasal passages. Its introduction into seafood is almost exclusively due to poor personal hygiene among food handlers [10]. The unique risk it poses stems from its ability to produce highly heat-stable enterotoxins. If contaminated products are held in the temperature danger zone (4 °C–60 °C) for an extended period, the bacteria can multiply and produce these toxins. Subsequent cooking may destroy the bacteria but will not inactivate the toxins, leading to staphylococcal food poisoning upon consumption.

Escherichia coli, particularly shiga toxin-producing strains (STEC) such as O157:H7, serves as a key indicator of fecal contamination [11]. While most E. coli strains are harmless commensals of the human and animal gut, pathogenic varieties can cause severe gastrointestinal illness. Their presence in seafood is unequivocal evidence of direct or indirect contact with fecal matter, typically through polluted harvest waters or improper handling during processing. Infection can result in bloody diarrhea and, in severe cases, hemolytic uremic syndrome (HUS).

Clostridium botulinum, a Gram-positive, spore-forming, anaerobic bacterium, is found in sediments and the intestinal tracts of fish [12]. The hazard materializes when spores, which are highly resistant to environmental stress, survive and germinate in anaerobic conditions, leading to the production of a potent neurotoxin. Seafood products packaged using modified atmospheres (MAP) or vacuum packaging are of particular concern, as these processes remove oxygen and create an ideal environment for toxin production. The risk is especially pronounced for psychrotrophic strains (Type E) that can grow and produce toxin at temperatures as low as 3 °C, underscoring the critical need for stringent, multi-factorial control measures.

11.2.2 Viruses

In terms of viruses, aquatic products are mainly contaminated with norovirus, hepatitis A virus (HAV), rotavirus, and astrovirus, which are also the main types of foodborne diseases that lead to human beings [13]. Unlike bacteria, viruses do not multiply in food or the environment; however, they remain highly infectious and can cause disease even at very low exposure levels. Viral contamination almost exclusively originates from human fecal matter, entering the aquatic environment through sewage discharge, contaminated runoff, or improper waste management. Bivalve mollusks, such as oysters, clams, and mussels, are particularly high-risk vehicles for viral transmission due to their filter-feeding behavior, which concentrates viral particles from surrounding water.

Norovirus is the most common cause of viral gastroenteritis worldwide and is frequently linked to shellfish consumption [14]. It is highly contagious, environmentally stable, and resistant to many common disinfectants and temperature fluctuations. Symptoms include acute vomiting, diarrhea, and abdominal pain.

HAV, though less common, poses a more serious public health threat due to its ability to cause severe liver infection, resulting in symptoms such as jaundice, fatigue, and abdominal pain. HAV can persist in the environment for extended periods and remains infectious in seawater and seafood.

Rotavirus and Astrovirus are also recognized as important viral pathogens in aquatic products, particularly in regions with inadequate sanitation infrastructure. Rotavirus has been a major cause of severe dehydrating diarrhea in children, though vaccination programs have reduced its prevalence in many countries. Astrovirus often leads to milder gastroenteritis but remains a concern in immunocompromised populations.

11.2.3 Parasites

Harmful organisms and parasites in aquatic products have now attracted much attention due to their serious damage to human health [15]. Among the more common aquatic products are parasitic nematodes such as *Angiostrongylus cantonensis*, as well as schistosomiasis, trematodes, mites, and others including Capillaria philippinensis, paragonimiasis, heterophyid trematodes, clonorchiasis, etc. Many parasites have complex life cycles involving multiple hosts, with aquatic organisms often serving as intermediate carriers. Human infection occurs through the ingestion of viable parasitic larvae or cysts present in fish, crustaceans, or mollusks.

11.2.4 Toxins Produced by Microorganisms

Toxins produced by microorganisms represent a unique and significant hazard in aquatic products, as they are not destroyed by cooking or standard processing methods and can cause severe, often rapid-onset, illnesses in consumers. These toxins originate from various biological sources, including bacteria, microalgae, and other marine organisms, and accumulate in seafood through the food web. Unlike microbial pathogens, the presence of these toxins is not correlated with hygiene or sanitation practices but is instead influenced by ecological and environmental conditions such as algal blooms, water temperature, and nutrient availability [16, 17].

11.2.5 Chemical Pollutants

11.2.5.1 Pesticides

Chemical pesticides are an important factor affecting the growth and safety of aquatic animals and consumer safety [18]. With the continuous improvement of the development speed of pesticide technology, there are more and more pesticides in the world. More than 1,500 kinds of commercial pesticides have been circulating in the market, and there are also three or four hundred kinds commonly used. The living environment of aquatic animals is greatly affected by the surrounding environment. Chemical pesticides are brought into the water due to rainfall and other forms, resulting in excessive pesticide residues in farmed fish, which poses a food safety hazard. The global prevalence of pesticide residues in aquatic products poses a significant public health concern, with six major chemical classes—organochlorines (OCs), organophosphates (OPs), pyrethroids, carbamates, neonicotinoids, and endocrine-disrupting pesticides—demonstrating persistent contamination patterns. These compounds exhibit dual environmental behaviors: (1) multi-residue accumulation through trophic transfer, and (2) biomagnification across aquatic food webs, resulting in concentration levels exceeding safety thresholds. Their dual toxicity profiles—causing both acute toxicity (e.g., neurological damage) and chronic endocrine disruption—have triggered extensive multidisciplinary research efforts to develop effective monitoring and mitigation strategies.

OCs, typical residual pollutants detected in aquatic products, such as hexachlorocyclohexane (HCH), dichlorodiphenyltrichloroethane (DDT) and its metabolites, and sodium pentachlorophenolate (PCP-Na), are classified as highly persistent and lipophilic organic compounds [19]. These substances can be transferred through food chains via trophic transfer and undergo biomagnification. They are significantly accumulated in organisms, particularly in lipid-rich tissues such as adipose and liver, which constitute the primary reservoirs. In addition to inducing acute neurotoxicity, these compounds pose potential chronic health risks, notably by disrupting the normal function of the endocrine system, thereby presenting long-term threats to ecosystem integrity and food safety.

(OPs, such as dichlorvos and dimethoate, are frequently detected as residues in aquatic products. The core toxic mechanism of OPs involves the irreversible inhibition of acetylcholinesterase (AChE) activity at synaptic clefts, leading to the excessive accumulation of acetylcholine [20]. This accumulation results in the overstimulation of cholinergic nerves. OPs do not undergo spontaneous degradation and are found as residues in water resources, fruits, vegetables, and processed foods. Repeated low-level exposure can lead to their accumulation in organisms, inducing a range of muscarinic and nicotinic effects such as salivation, muscle tremors, and respiratory distress. In severe cases, respiratory failure and death may occur.

Pyrethroids, common varieties such as deltamethrin, permethrin, and cypermethrin, exhibit relatively low toxicity to mammals but are highly toxic to aquatic organisms (e.g., fish and shrimp). Their mechanism of action involves targeting voltage-gated sodium channels in nerve axons, leading to delayed channel closure, sustained neuronal excitation, and consequent convulsions, paralysis, and death. These pesticides are characterized by high lipophilicity, which facilitates their accumulation in aquatic sediments and biological tissues. Through trophic transfer, they pose potential risks to fishery ecosystems and food safety [21].

Carbamates pose significant risks to aquatic organisms due to their high water solubility and mobility. The mechanism of action involves reversible inhibition of AChE activity, leading to excessive accumulation of the neurotransmitter acetylcholine and resulting in sustained excitation and functional disruption of the nervous system. These pesticides exhibit strong acute toxicity to fish and crustaceans, among other aquatic invertebrates, causing behavioral abnormalities, growth inhibition, and even mortality [22]. Furthermore, their transmission through the food chain may potentially compromise the stability and health of marine ecosystems.

Neonicotinoids are characterized by high water solubility and moderate environmental persistence. Their mechanism of action involves selective agonism of nicotinic acetylcholine receptors in the insect nervous system, leading to sustained receptor activation, neuronal hyperexcitation, paralysis, and mortality [23]. These pesticides are highly toxic to aquatic invertebrates (e.g., mayflies), severely disrupting the foundation of aquatic food webs, and inducing chronic adverse effects in crustaceans. Although exhibiting relatively low acute toxicity to mammals, chronic exposure may potentially cause neurodevelopmental toxicity and endocrine-disrupting effects.

Endocrine-disrupting Pesticides in aquatic products can disrupt the endocrine system functions of both aquatic organisms and humans through mechanisms such as mimicking/antagonizing endogenous hormones (e.g., binding to estrogen or androgen receptors) and interfering with hormone synthesis and metabolism. The hazards are manifested as reproductive impairments, developmental malformations, and abnormal sexual differentiation in aquatic organisms. Through trophic transfer, these compounds pose potential risks of reproductive toxicity, developmental defects, and hormone-related cancers to humans.

11.2.5.2 Drugs

Residues of fishery drugs refer to drugs that are artificially used for disease prevention and treatment in the cultivation process, and which are accumulated or incompletely metabolized in the living body. The use of illicit drugs or aquatic products that exceed the safety limit will cause potential risks to human health after ingestion. Drug residues in aquatic products can cause allergic reactions, toxic effects, hormone effects, and other effects, and resistance will occur to human body with long-term exposure [24]. Aquatic products may contain hundreds of veterinary drug residues, primarily encompassing antibiotics, hormones, synthetic antimicrobials, and insecticides [25]. These residues exhibit varied toxicological effects, with particular concern arising from persistent compounds that accumulate in aquatic ecosystems, thereby threatening both environmental integrity and human health through prolonged exposure.

Antibiotics and antibiotic residues are among the most common contaminants in aquaculture [26], primarily used to prevent and treat bacterial diseases. Their hazards are primarily manifested in driving the selection and dissemination of resistant bacteria and antibiotic resistance genes in the environment, while also potentially disrupting the balance of microbial communities in aquatic ecosystems and the human gut. Some antibiotics (e.g., fluoroquinolones) may even possess direct toxicity, potentially causing allergic reactions or latent organ damage in consumers.

Hormones and hormonal drugs are frequently employed in aquaculture to promote growth or regulate reproductive cycles. Their residues pose profound hazards to aquatic ecosystems and human health, primarily characterized by endocrine-disrupting effects. Even exposure to very low concentrations can lead to sex ratio imbalance, abnormal gonad development, and reproductive dysfunction in aquatic organisms.

Synthetic antimicrobials, synthetic antimicrobials (e.g., sulfonamides, nitrofurans) are widely used alongside antibiotics to control diseases in aquaculture.

Many of these compounds exhibit high environmental persistence and are resistant to natural degradation. Beyond contributing to antimicrobial resistance [27], certain types (e.g., the banned nitrofurazone) have been confirmed to have metabolites with genotoxic and carcinogenic properties, posing long-term risks to human health. Their mechanisms of action differ from conventional antibiotics; for instance, sulfonamides act as competitive antagonists of para-aminobenzoic acid, inhibiting bacterial folic acid synthesis, while nitrofurans are reduced to active compounds within bacterial cells, damaging DNA and disrupting its replication.

Insecticides, pesticides used to control aquatic parasites (e.g., fish lice and anchor worms), including OPs and pyrethroids, are generally highly toxic to nontarget organisms. Their hazard lies not only in their efficacy against target parasites but also in their potential to cause significant mortality among keystone species such as crustaceans and plankton, thereby destabilizing aquatic food webs. Although the risk of acute poisoning to human consumers is relatively low, chronic exposure may have potential impacts on the nervous and immune systems. Their core mechanism of action involves targeting key sites in the nervous system; for example, OPs and carbamates irreversibly or reversibly inhibit AChE activity, while pyrethroids act on voltage-gated sodium channels in nerve axons, leading to sustained neuronal excitation.

11.2.5.3 Biotoxins

Biotoxins can be produced by aquatic animals or be accumulated and transferred through the food chain in fish or shellfish. The most dangerous biotoxins in aquatic products in China mainly include shellfish toxins, tetrodotoxin, and ciguatoxin. These three kinds of biotoxins have become one of the most dangerous pollutants endangering human health, especially in some special aquatic products, such as bivalve shellfish, puffer fish, and nassariidae. Another biotoxins microcystin, also distributes widely in China and other countries, mainly affecting humans by drinking water. Biotoxins exhibit significantly higher lethality compared to conventional hazardous substances, characterized by their abrupt onset and resistance to control mechanisms. This dual attribute elevates their potential to trigger large-scale foodborne poisoning events, positioning them as a globally significant public health threat requiring urgent regulatory intervention.

Tetrodotoxin, a potent neurotoxin, is primarily found in the ovaries, liver, and skin of pufferfish. Its hazard is extreme, characterized by a short latency period and high mortality following human ingestion. Clinical symptoms include perioral numbness, motor dysfunction, and respiratory distress, with severe cases leading to death due to respiratory muscle paralysis. The mechanism of action involves highly selective blockade of voltage-gated sodium channels on nerve and muscle cells, thereby inhibiting action potential generation and nerve impulse conduction, resulting in progressive neuromuscular paralysis [28].

Shellfish toxins (e.g., paralytic, diarrhetic, neurotoxic, amnesic shellfish poisons) are not produced by the shellfish themselves but are synthesized by marine algae (e.g., dinoflagellates) and bioaccumulated in filter-feeding shellfish (e.g., oysters, scallops, and mussels) through the food chain. Their hazards vary by type: paralytic shellfish poison can cause respiratory failure and death [29]; diarrhetic shellfish poison induces gastrointestinal illness; neurotoxic shellfish poison triggers neurological symptoms; and amnesic shellfish poison may damage brain cells. The mechanisms of action are diverse; for instance, paralytic shellfish poison functions by blocking sodium channels, while amnesic shellfish poison acts as an agonist for excitatory neurotransmitter receptors, leading to neuronal damage [30].

Ciguatoxin is primarily produced by benthic microalgae in tropical and subtropical seas and bioaccumulates in the muscle, viscera, and skin of large coral reef fish (e.g., groupers, barracudas) through the algae-detritus-fish food chain. Its hazard is unique, with poisoning symptoms being complex and involving gastrointestinal, neurological, and cardiovascular abnormalities. The toxin's mechanism of action involves persistent activation of voltage-gated sodium channels and potential effects on potassium and calcium channels, leading to sustained cellular excitation and disruption of neuronal signaling [31].

11.2.5.4 Persistent Organic Pollutants (POPs)

POPs mainly refer to some polycyclic aromatic hydrocarbon (PAH) harmful compounds, such as polychlorinated biphenyls (PCBs) (including dioxins) and other halogenated hydrocarbons, biphenyl organic hydrocarbons, etc., mainly from pesticides and industrial chemistry. POPs are chemically stable, difficult to decompose, and have several toxicities, such as immunotoxicity, endocrine, reproductive, and even mutagenic, teratogenic, and carcinogenic toxicity. POPs are accumulative and can be scaled up through the food chain, causing harm to humans. Aquatic animals are the most affected category, and their impact on ecology and human health is very serious [32]. POPs, classified as emerging contaminants of global environmental concern, exhibit unique environmental persistence and bioaccumulative potential in aquatic ecosystems. Their prolonged environmental residence times and transboundary mobility have elevated them to a critical nexus in international efforts for food safety risk mitigation, while simultaneously triggering geopolitical tensions over divergent regulatory frameworks and enforcement capacities.

OC pesticides represent the earliest category of POPs to be recognized and subjected to global action, primarily consisting of synthetic insecticides. Representative substances include DDT, a globally infamous insecticide whose agricultural use is now prohibited worldwide; the highly effective soil insecticides aldrin, dieldrin, and endrin, used against termites, grasshoppers, and corn rootworms; as well as chlordane, heptachlor, mirex, toxaphene, hexachlorobenzene, endosulfan, and lindane. These compounds are recognized as probable human carcinogens and endocrine disruptors. They can be absorbed through the skin, exhibit neurotoxicity in humans, and cause skin conditions while damaging the liver, immune system, and nervous system. They pose severe risks to newborns.

Another major category of POPs comprises industrial chemicals, which were not used as pesticides but served specific functions in industrial production and products. Among these, PCBs, once widely employed as dielectric fluids in transformers and capacitors, as well as in heat exchangers and paints due to their exceptional insulating properties and stability, can cause cancer and damage the immune, reproductive, and nervous systems [33]. Hexabromocyclododecane (HBCDD) is an additive flame retardant primarily used in building insulation, while short-chain chlorinated paraffins (SCCPs) are used as additives in metalworking fluids and as plasticizers in plastics. These substances are characterized by their persistence, bioaccumulative potential, and endocrine-disrupting properties.

Furthermore, POPs include unintentional by-products, exemplified by dioxins and furans [34], which originate primarily from waste incineration, industrial processes, and incomplete combustion. International efforts under the Stockholm Convention enforce strict controls on these substances, including prohibitions and restrictions on their production, use, and emissions, alongside the sound management of wastes and stockpiles.

The list of POPs has been expanded as scientific understanding has advanced. Brominated flame retardants, such as polybrominated diphenyl ethers (PBDEs) [35], formerly used in furniture foam, carpets, and electronics, can leach from products into the environment and are associated with endocrine disruption, particularly of thyroid hormones, and potential neurodevelopmental effects. Another group, perfluorinated compounds, is noted for their unique water- and grease-repellent properties. The representative substance perfluorooctane sulfonic acid (PFOS) and its salts, previously used in textile treatments, fire-fighting foam, and food packaging, are highly persistent in humans and wildlife. Exposure is linked to thyroid disease, high cholesterol, reproductive issues, and cancer. The safety of their replacements is now under increasing scrutiny.

11.2.5.5 Chemical Additives Such As Processing Aids and Additives Used in Violation of Regulations

The application of processing chemicals and additives in food and aquatic product manufacturing is subject to stringent regulatory oversight; however, noncompliant usage introduces significant food safety hazards. Authorized functional additives—including antioxidants, colorants, preservatives, and texturizers—are indispensable for achieving technological objectives (e.g., sensory enhancement, shelf-life extension). Nevertheless, even ostensibly benign substances may undergo unintended chemical transformations when exposed to anti-nutritional factors or interact with food matrices, potentially generating toxic by-products. Such scenarios are exacerbated by the deliberate misuse of non-permitted chemical agents, which constitute a critical pathway for foodborne intoxication and regulatory noncompliance.

Antioxidants are primarily used to prevent or delay the oxidative deterioration of fats and oils in food. Butylated hydroxyanisole (BHA) [36] exhibits good stability after heating and is widely used internationally. Its antioxidant effect becomes more significant when combined with synergists such as citric acid, and it is generally considered safe. Butylated hydroxytoluene (BHT) demonstrates high stability and heat resistance, making it an effective and economical antioxidant widely applied in aquatic product processing. It is often used in combination with BHA, though it has slightly higher toxicity. Propyl gallate (PG), which is heat-stable, shows stronger antioxidant effects on lard compared to BHA and BHT, with relatively low toxicity.

Tert-butylhydroquinone (TBHQ) [37] is a newer phenolic antioxidant known for its superior antioxidant performance.

Preservatives (e.g., sodium benzoate and potassium sorbate) inhibit microbial growth to extend the shelf life of foods such as jams and candied fruits. Common violations include their unauthorized use in fresh wet noodles, raw meat and poultry, steamed buns, and cooked aquatic products (non-canned) [38]. Bleaching agents (e.g., sulfur dioxide and sodium metabisulfite) remove color or prevent browning through redox reactions. These compounds also exhibit preservative properties and help prevent the oxidation of vitamin C in fruits and vegetables. Illegal practices involve the use of bleaching agents in fresh shrimp, steamed buns (via sulfur fumigation), and fresh vegetables like bamboo shoots and bean sprouts to enhance appearance.

Color fixatives are used to improve or preserve food color. For instance, nitrates or nitrites are added in meat processing to impart a bright red color. However, nitrites are highly toxic and can cause methemoglobinemia, reducing the blood's oxygen-carrying capacity. Moreover, they are precursors to carcinogenic N-nitroso compounds, necessitating strict control of their usage levels. Colorants (or pigments) are applied to enhance the sensory properties of food. However, most synthetic colorants pose risks due to inherent chemical toxicity, harmful metabolites, or potential contamination with arsenic, lead, or other hazardous compounds during production.

Moisture retention agents and high-risk categories. In meat products and prepared foods, moisture retention agents (e.g., phosphates) are frequently overused alongside color fixatives (nitrites) and preservatives (e.g., potassium sorbate) to improve water retention, tenderness, yield, color, and microbial inhibition. Additionally, candied fruits, preserves, and dried fruit products represent a high-risk category for excessive additive use. To prolong shelf life under acidic and sweet conditions and enhance sensory appeal, preservatives (e.g., benzoic acid and sorbic acid), sweeteners (e.g., sodium saccharin and cyclamate), and colorants (e.g., amaranth and ponceau 4R) are often applied beyond permissible limits.

Heavy metals. When the pollution brings harmful heavy metals to the growing waters of aquatic animals due to rainfall, water migration, and other reasons, the plankton in the water filter and absorb higher levels of heavy metals. Aquatic animals are rich in heavy metals and can be enriched in the body through the food chain, which is an important safety hazard. For example, seawater algae, filter-feeding bivalve molluscs, and harmful heavy metals in large carnivorous fish have always been the focus of attention at home and abroad. The heavy metals in the aquatic products that are mainly concerned include lead (Pb), cadmium (Cd), methylmercury, and inorganic arsenic [39]. Overall, heavy metals primarily pose chronic toxicity risks to consumers; however, certain aquatic species exhibit selective bioaccumulation capacities for specific heavy metals. Notably, cadmium concentrations in scallops can reach levels dozens of times higher than those in other products, potentially inducing acute poisoning symptoms in consumers.

Lead is a toxic metal that can accumulate in human and animal tissues. Its primary sources include paint, batteries, smelting, metal products, electroplating, cosmetics, coal combustion, and aging water pipes. Lead enters the human body through the skin, digestive tract, and respiratory system, binding to various organs and causing toxic effects such as anemia, neurological dysfunction, and kidney damage. Children, the elderly, and immunocompromised individuals are particularly vulnerable. Aquatic products from contaminated waters (e.g., near industrial areas or ports) pose elevated risks, with bivalves and crustaceans readily accumulating lead. Irrigation with water containing 0.1–4.4 mg/L of lead significantly increases lead concentrations in crops. The normal blood lead level in humans should remain below 0.1 mg/L; exceeding this threshold may cause anemia and neurological damage. Children's developing brains are especially susceptible, necessitating active interventions, such as chelation therapy, when blood lead levels exceed safety standards.

Cadmium is a nonessential element with high toxicity and a strong tendency to accumulate in the body, primarily damaging the kidneys. Major sources include wastewater from electroplating, mining, smelting, battery production, and the chemical industry, as well as discarded batteries. Cadmium is also found in foods such as mushrooms, dairy products, and grains. Due to their filter-feeding behavior, bivalves such as scallops and oysters exhibit high cadmium concentrations [40]. Cadmium displaces calcium in bones, leading to severe softening and fractures, disrupts gastrointestinal function, interferes with zinc-dependent enzyme systems, and can elevate blood pressure. Miners and immunocompromised individuals are at heightened risk.

Methylmercury primarily originates from volcanic activity, coal combustion, mining, and industrial wastewater. It is formed from inorganic mercury converted by microorganisms in aquatic sediments. Methylmercury exhibits strong bioaccumulation and biomagnification, persisting in the muscle tissues of aquatic products and increasing in concentration up the food chain [41]. Large predatory fish (e.g., bigeye tuna, marlin) at the top of the food chain contain the highest levels. Methylmercury is highly neurotoxic, particularly to the developing brain, and exposure in fetuses and infants can cause cognitive and motor deficits. Pregnant women should exercise caution, as mercury crosses the placental barrier, impairing fetal neurological development. In adults, long-term exposure may lead to sensory disturbances, loss of coordination, and hearing or vision impairment.

Arsenic contamination stems from natural weathering, mining, smelting, and pesticide use. In aquatic products, arsenic is primarily present in the less toxic organic form (e.g., arsenobetaine), which is rapidly excreted by humans. However, inorganic arsenic—a potent carcinogen—may be present in certain seaweed species (e.g., hijiki) and heavily contaminated waters. While all seafood contains high levels of organic arsenic, hijiki has been flagged in several countries for potentially elevated inorganic arsenic content. Long-term ingestion of inorganic arsenic increases the risk of skin, lung, and bladder cancers, and may also cause skin lesions and cardiovascular diseases.

11.2.6 Physically Derived Hazards

Solid heterogeneous substances in aquatic products mainly contain their own fish bone spurs or metal-like heterogeneous substances introduced during processing. The bone spur is a general term for fish vertebrae and intermuscular spurs. Its shape and size vary with fish. It is the main heterosis in fish products. It may cause certain harm to the human body after being eaten. These pose a particular risk of throat or gastrointestinal perforation, choking, or dental injury.

Metal products are mixed with aquatic products. These substances may cause harm to consumers when they are consumed. It has always been one of the key concerns of aquatic product safety. The metal to be mixed is roughly classified into three types, namely, iron metal, nonferrous metal, and stainless steel [42].

11.2.7 Self-Sourced Hazards

The self-borne hazards in aquatic products are mainly toxic and harmful substances produced by aquatic products, mainly including allergens, biogenic amines, and formaldehyde substances.

Allergens in aquatic products are substances that can produce immune responses in the form of immune damage in some sensitive people, mostly proteinaceous substances. Aquatic food allergens can cause allergic reactions in most sensitive patients. Most allergens are stable to heat and have certain resistance to hydrolase. Many aquatic food allergies have common antigenic determinants, which make aquatic products cross-reactive [43].

Biogenic amines are a general term for a class of nitrogen-containing biologically active low molecular weight organic compounds, usually formed by amino acid decarboxylase secreted by active cells to specifically catalyze the decarboxylation of the corresponding free amino acids. The decarboxylation of an amino acid is the transfer of the a-carboxy group to form the corresponding biogenic amine compound [44].

The self-derived formaldehyde in aquatic products refers to the formaldehyde that is produced and metabolized by itself in the process of storage and processing of aquatic products and products, and does not include artificially added or externally contaminated and residual formaldehyde[45, 46].

11.3 Detection Technology

In the future, detection technology for quality and safety of aquatic products will focus on the precision detection, nondirectional screening, metabolitc structure analysis, and high-resolution mass spectrometry (HRMS) technology in known/unknown groups. The field of high-throughput, high-sensitivity rapid screening and confirmatory analysis of trace chemical contaminants and metabolites will become a hot research and application [47].

11.3.1 Conventional Detection Technologies

The detection of pollutants in aquatic products relies on a diverse array of analytical technologies tailored to the specific nature of the hazard—whether biological, physical, or endogenous. For microbiological contaminants such as pathogenic bacteria

(Vibrio, Salmonella, Listeria), conventional culture-based methods remain foundational, employing selective media and biochemical tests for isolation and identification. These are increasingly supplemented with rapid and highly sensitive molecular techniques, including polymerase chain reaction (PCR) and real-time PCR, which allow for the detection of specific pathogens and virulence genes even at low levels. For viruses (e.g., norovirus, hepatitis A), reverse transcription PCR (RT-PCR) is the method of choice due to its ability to detect viral RNA. Parasites such as Anisakis are often identified through visual inspection (e.g., candling) or pepsin digestion assays, though molecular methods are also gaining traction for species-specific confirmation.

In the case of physical hazards such as bone spurs and metal fragments, X-ray imaging is widely used in industrial settings for its nondestructive ability to detect dense foreign materials embedded in fillets or processed products. Metal detectors are routinely deployed along processing lines to identify ferrous and nonferrous metallic contaminants.

For endogenous contaminants, specific and quantitative approaches are required. Histamine and other biogenic amines—often indicative of spoilage or temperature abuse—are commonly analyzed using high-performance liquid chromatography (HPLC) or enzyme-linked immunosorbent assay (ELISA) kits, which offer high sensitivity and suitability for quality control screening. Allergens such as tropomyosin, a major allergen in crustaceans and mollusks, are typically detected using immunoassays (e.g., dipstick tests or ELISA) or mass spectrometry (MS)-based proteomics, enabling compliance with labeling regulations and prevention of cross-contact.

For the accurate quantification of chemical source pollutants, chromatographic techniques and their hyphenation with MS remain the internationally recognized reference methods and "gold standard." [48] Gas chromatography-mass spectrometry (GC-MS) is the core technology for detecting volatile and semi-volatile organic compounds, widely applied in the qualitative and quantitative analysis of pesticide residues, PCBs, PAHs (such as benzo[*a*]pyrene), and plasticizers. Its high separation efficiency and sensitivity provide a reliable basis for identifying trace pollutants in complex matrices. Liquid chromatography-mass spectrometry (LC-MS) is suitable for analyzing thermally labile and highly polar compounds, serving as the preferred method for detecting veterinary drug residues (e.g., antibiotics and hormones), certain pesticides and their metabolites, biotoxins (e.g., microcystins), and illicit additives (e.g., malachite green and nitrofuran metabolites). This technique combines exceptional separation capabilities with robust structural elucidation, making it indispensable for precise regulatory oversight.

The detection of chemical source pollutants in aquatic products—such as heavy metals (e.g., mercury, cadmium, and lead); POPs, including PCBs and dioxins; pesticide residues; veterinary drug residues (e.g., tetracyclines and sulfonamides); PAHs; and plasticizers (e.g., phthalates)—relies on a suite of mature analytical techniques. These methods are designed to address the chemical diversity and low concentrations of pollutants, typically requiring sample pretreatment steps such as extraction, purification, and concentration to minimize interference from matrix components in aquatic products (e.g., fish muscle and shellfish tissues), including lipids, proteins, and salts [49]. Conventional techniques emphasize accuracy, precision, and compliance with international standards, such as those from the Codex Alimentarius Commission, EU Regulation (EC) No 1881/2006, and China's GB 2762-2017 maximum residue limits (MRLs).

For heavy metals, atomic absorption spectrometry (AAS) and inductively coupled plasma optical emission spectrometry (ICP-OES) are foundational methods [50]. AAS, particularly graphite furnace AAS (GF-AAS), measures the absorption of light by vaporized metal atoms in samples, achieving detection limits at the ppb level. ICP-OES excites atoms and ions in a high-temperature plasma and detects emission wavelengths for multi-element analysis. Sample preparation typically involves acid digestion (e.g., microwave-assisted nitric acid digestion) to solubilize metals.

Organic chemical pollutants such as POPs and PAHs are primarily analyzed using GC coupled with electron capture detection (ECD) or MS. GC-ECD separates volatile compounds on capillary columns and detects halogenated compounds like PCBs via electron capture [51]. For nonvolatile or polar residues such as antibiotics, HPLC combined with ultraviolet (UV) or fluorescence detection is employed, separating analytes under high pressure based on interactions with the stationary phase.

These conventional methods form the core of routine monitoring in food safety laboratories, providing reliable quantitative data for regulatory enforcement. However, they require skilled operators, time-consuming sample preparation (e.g., QuEChERS—Quick, Easy, Cheap, Effective, Rugged, and Safe extraction), and complex instrumentation, rendering them unsuitable for on-site applications.

11.3.2 Rapid Detection Technology

Rapid detection technology is a very important part of efficient detection and identification technology. In recent years, rapid detection methods such as biosensors, visualization technology, and immunoassay have developed rapidly and become a very important part of efficient detection and identification technology. Biosensors mainly include enzyme biosensors (including inhibitory biosensors and catalytic biosensors) and immunobiosensors (including electrochemical immunosensors, optical immunosensors, piezoelectric immunosensors, mechanical immunosensors, and microcantilever sensors). Visualization technology is a rapid detection technology for rapid qualitative and quantitative analysis based on specific color changes caused by the interaction between the medium and the target. Commonly used visual rapid detection techniques include color reaction, test strips, and visual fluorescence. An immunoassay is the use of an antigen. Antibody-specific binding assays are used to detect various substances (drugs, hormones, proteins, microorganisms, etc.). Immuno chromatography is a widely recognized and applicable rapid test method, but due to the complexity of aquatic products, false positive results are prone to occur. At present, the existing technology is being improved, and colloidal gold immune chromatography is continuously optimized to meet the rapid inspection requirements of aquatic products.

Biosensors integrate biological recognition elements (e.g., enzymes and antibodies) with transducers for signal amplification [52]. For heavy metals, AChE inhibition-based biosensors detect OPs or mercury by measuring colorimetric or electrochemical changes in enzyme activity inhibition. Immunosensors, such as surface plasmon resonance (SPR) devices, enable real-time detection of antibiotics like chloramphenicol via antigen-antibody binding, with limits of detection (LODs) as low as 0.1 ng/mL. Optical biosensors, including quantum dot-based fluorescence quenching, offer high specificity for PAHs.

Immunoassay techniques, particularly ELISA and colloidal gold lateral flow assays, are widely used for pesticide and veterinary drug residues [53]. ELISA kits quantify antibiotics through competitive binding, providing semiquantitative results via visual or photometric readout. Colloidal gold test strips enable visual screening for sulfonamides or nitrofurans, where color lines indicate exceedance of thresholds (e.g., 10 ppb), although matrix components in fatty aquatic samples may cause false positives.

Electrochemical rapid methods, such as anodic stripping voltammetry for heavy metals [54], preconcentrate analytes on electrodes and measure redox currents, suitable for portable devices in aquaculture settings. Visualization techniques, like paper-based microfluidics with colorimetric reactions, facilitate spot testing for plasticizers via dye displacement. These rapid tools are integrated into Hazard Analysis and Critical Control Points (HACCP) systems for preliminary screening, with positive results requiring laboratory confirmation. Challenges include matrix interference from aquatic lipids, necessitating simple cleanup steps such as solvent extraction. Ongoing advancements, such as smartphone-integrated biosensors, enhance data portability and accuracy.

11.3.3 High-Throughput Analysis Technology

High-throughput screening (HTS) refers to a new type of simultaneous analysis of multiple indicators in a sample, or simultaneous analysis of one of multiple samples, to obtain the most biological information in the shortest time. Compared with traditional analytical methods, HTS has the advantages of high throughput, parallelism, microchemical, and automation. According to the detection principle, the HTS mainly includes biochip, pyrosequencing technology, and fluorescence polarization immunoassay technology. The microarray chip mainly includes a gene chip, a protein chip, a cell chip, and a tissue chip; the microfluidic chip is a technology for processing a small amount of liquid in a micro flow path system, and can be completed with a very small amount of samples and reagents, and has been widely used. In the field of biomedicine, it is especially suitable for rapid detection in countries or regions suffering from natural disasters and a lack of resources, including PCR chips, flow chips, capillary electrophoresis chips, microelectronic chips, chromatography chips, and various sample preparation chips. In addition, pyrosequencing technology, fluorescence polarization immunoassay technology, quantum dot fluorescence immunoassay technology, etc., all show advantages in high-throughput analysis [55].

In response to increasingly complex chemical pollution, nontargeted screening and high-throughput analytical capabilities have become core competencies in modern laboratories [56]. The application of liquid chromatography-quadrupole-time-of-flight mass spectrometry (LC-Q-TOF-MS) and Orbitrap-series HRMS has propelled the detection of chemical source pollutants from "limited target monitoring" to a new era of "comprehensive untargeted screening." These technologies enable the unbiased acquisition of precise molecular mass information for nearly all compounds in a sample at once, facilitating the discovery and identification of non-target compounds (e.g., novel transformation products, unknown illicit additives) through database matching, thereby providing critical technical support for risk early warning.

Liquid chromatography-tandem mass spectrometry (LC-MS/MS) and gas chromatography-tandem mass spectrometry (GC-MS/MS) dominate high-throughput workflows [57]. LC-MS/MS separates polar and semi-polar compounds (e.g., antibiotics, mycotoxins) on reversed-phase columns, followed by electrospray ionization (ESI) and tandem MS for structural confirmation. Multiple reaction monitoring (MRM) enables quantification of over 100 residues with LODs below 1 ppb, using isotope dilution for accuracy. GC-MS/MS targets volatile organics like PAHs and PCBs, employing electron impact ionization for library matching.

Microfluidic chips and multi-well plate systems facilitate automation; for instance, digital microfluidics (DMF) chips automate solid-phase extraction (SPE) and LC separation in nL-scale volumes, enabling parallel analysis of 96 samples [58]. Biochip arrays, such as those with immobilized antibodies or aptamers, support multiplexed immunoassays for heavy metals and pesticides, coupled with fluorescence detection for high sensitivity [59]. Pyrosequencing and next-generation sequencing (NGS) variants adapt to metabolomics profiling of pollutant transformation products, elucidating exposure pathways.

HRMS [60], including time-of-flight (TOF) and Orbitrap systems, excels in non-targeted screening of unknown pollutants. HRMS delivers precise mass measurements (resolution >50,000), permitting retrospective data analysis for emerging risks like microplastic-derived chemicals. Software tools such as MZmine or XCMS process spectral data for peak annotation against databases (e.g., NIST and METLIN). These technologies reduce analysis time from days to hours, with throughput exceeding 1,000 samples per day. Advantages include broad coverage and minimal false negatives, though high costs and data complexity demand specialized expertise. In aquatic products, they support "non-targeted metabolomics" to track pollutant bioaccumulation.

11.3.4 Sample Preparation Technology

Matrix interference is a bottleneck to the rapid detection of food safety. At present, various new SPE technologies based on SPE are the main research hotspots for sample pretreatment, such as online multi-channel solid-phase extraction, matrix-dispersed solid-phase extraction, dispersive solid-phase extraction, magnetic solid-phase extraction, and chip solid-phase extraction. Combined with rapid analysis technology, SPE can reduce sample processing time and improve analysis efficiency. Other rapid pretreatment techniques, such as liquid-phase microextraction, solid-phase microextraction (SPME), supercritical fluid extraction, stir bar extraction, membrane extraction, salting-out assisted liquid–liquid extraction, are also good choices for sample preparation.

Efficient and precise detection critically depends on high-quality sample pretreatment. For chemical source pollutants, sample preparation techniques are evolving toward rapidity, environmental friendliness, and miniaturization. QuEChERS, as a quick, easy, cheap, effective, rugged, and safe pretreatment method, has become mainstream for analyzing multi-class chemical pollutants like pesticide residues [61]. SPE and its novel variants, such as magnetic SPE and dispersive SPE, enable targeted enrichment and purification of analytes, effectively removing matrix interferences like lipids and proteins, which are crucial for ensuring accuracy in downstream instrumental analysis. Other miniaturized techniques, including SPME and liquid-phase microextraction (LPME), are gaining prominence in environmental and food sample analysis due to their low solvent consumption, operational simplicity, and ease of online coupling with analytical instruments.

In China, a standardized methodology system aligned with international practices has been established for detecting chemical source pollutants in aquatic products, with active promotion of frontier technologies [62]. For instance, in key marine areas and aquaculture zones, high-end equipment such as LC–MS/MS and ICP-MS is deployed for routine monitoring and risk assessment of fishery veterinary drugs, environmental persistent pollutants, and heavy metals. These systematic efforts not only provide data foundations for formulating scientific limit standards and regulatory measures but also integrate into full-chain HACCP control systems, implementing effective risk interventions at critical nodes, including preproduction environmental assessment, in-production input control, and postproduction market surveillance, significantly elevating the quality and safety levels of Chinese aquatic products.

Looking to the future, detection technologies for chemical pollutants in aquatic products will emphasize precision, nontargeted screening, and metabolite profiling, driven by HRMS for both known and unknown hazards. Focus will shift to high-throughput, ultra-sensitive platforms for trace multi-residue analysis, integrating AI for data interpretation.

Nontargeted HRMS workflows, utilizing Q-TOF or FT-ICR MS, will enable suspect and non-suspect screening without prior standards, identifying novel pollutants such as nanoplastics or transformation products. AI-enhanced spectral libraries and machine learning algorithms will automate peak identification, with in vitro models predicting toxicity. Portable HRMS devices, combined with ambient ionization (e.g., DART-MS), will facilitate real-time on-site detection on fishing vessels or markets [63].

Nanomaterial-based (e.g., graphene) biosensor arrays will advance rapid screening, achieving femtomolar LODs for antibiotics through enhanced signal transduction. Integrated LC-HRMS metabolomics will map bioaccumulation pathways, linking environmental exposure to product safety [64]. Alignment with global standards will promote standardized HRMS databases for international trade.

11.4 Aquatic Product Quality and Safety Control Technology

11.4.1 Environmental Applicability Evaluation Technology

Environmental pollution in agricultural production areas is a three-dimensional type of pollution. The source of pollution may come from water, soil, organisms, or the atmosphere. The types of pollutants may be chemical fertilizers, drugs, heavy metals, organic pollutants, microorganisms, or toxins. Therefore, the environmental control technology of the production area must be based on the risk assessment of environmental pollutants. At present, the international community mainly uses two methods to carry out related research, namely the three-step method of the United States and the four-step method of the EU. China generally adopts the three-step method of the United States, namely, problem description, analysis stage, and risk characterization. The four-step rule includes hazard identification, dose–response evaluation, exposure evaluation, and risk characterization. More specifically, some scholars have divided the risk assessment of agricultural water environment into five steps: source analysis, hazard determination, dose–response evaluation, exposure evaluation, and risk characterization.

The environmental risk assessment of agricultural production areas is combined with the ecological and environmental principles of pollutants and the physiological ecology of agricultural organisms. The HACCP principle is used to divide the risk of food safety into three key points: ①screening and safety of environmental pollutants in the producing area; ②ecological behavior and environmental fate of environmental pollutants; and ③formation process and regulation mechanism of environmental pollutants. Finally, the monitoring and controlling technologies are comprehensively established to ensure original control quality and safety.

11.4.2 Production Process Control Technology

All the potential hazards affecting the quality and safety of adult fish, such as feed and its additives, fertilizer and fishery inputs, pathogens, and parasites, are analyzed, and the probability and severity of each potential hazard are evaluated to find out the impact. The significant hazard factors will be incorporated into the safe production operation plan to ensure that all hazards affecting the quality and safety of aquatic products are controlled, and that the produced aquatic products meet the national food safety standards.

11.4.2.1 Monitoring of the Production Environment

Before selecting aquaculture land, the ecology, the pollution, and the sediments of the farm area must be evaluated; during the breeding process, all potential factors described above are closely monitored. By monitoring the ecological environment of the farm, the pollution sources in and around the farm, and the harmful toxic substances in the pond bottom, it is ensured that the indicators of the production environment meet the critical limits and prevent aquatic products from causing significant biological and chemical hazards associated with the environment. When it is found that the indicators of the production environment do not meet the critical limits, investigate and monitor the geological and surrounding pollution sources, find out the reasons for exceeding the standard, and take effective measures to eliminate the pollution; when the pollution cannot be eliminated, the aquatic products are transferred to the farms that meet the requirements. All aquatic product certification certificates obtained by the farm before it is put into production, and all environmental testing reports of qualified testing institutions shall be archived. The environmental monitoring of the production areas in the breeding process must also be fully recorded and filed to ensure the data in the production environment and aquatic products.

11.4.2.2 Monitoring of Aquaculture Water

The indicators of water quality are the key limits for monitoring in accordance with the provisions of. Before the site selection, the water quality of the water source must be tested and qualified by qualified testing institutions; the aquatic products should be cultured in strict accordance with the aquaculture technical operation rules, and the inputs should be used reasonably to ensure that the water quality of the culture is good and that the water pollution of the breeding itself is prevented. By monitoring changes in aquaculture water quality indicators, it is ensured that the indicators of aquaculture water meet critical limits and that aquatic products are exposed to significant biological and chemical hazards associated with aquaculture water. Before stocking seedlings, the farm must obtain the certificate of origin of aquatic products or entrust a qualified testing institution to test and evaluate the water quality of the water source. During the production process, the quality control personnel entrust a qualified testing institution to check the water quality of the water source and monitor whether the upstream of the water source has increased. The water quality of the water source poses a threatening source of pollution. Before and during the breeding process, the aquaculture technician or laboratory regularly tests the water quality of the culture in accordance with the requirements of the aquaculture technical operation rules.

11.4.2.3 Monitoring of Seedlings

The stocking of seedlings does not carry pathogens, and the absence of drug residues is a critical limit for monitoring. Choose a seed farm with a good reputation and an effective seedling inspection certificate. By monitoring specific pathogens and drug residues carried by stocking seedlings, it is ensured that stocked seedlings meet critical limits and prevent aquatic products from showing significant biological and chemical hazards associated with seedlings. Before the aquatic products are put on the market, if the species carrying specific pathogens or residual drugs are mixed into the farm, they must be tested and not carried with specific pathogens and residual drugs before being used for breeding. During the breeding process, special attention should be paid to whether the aquatic products are still carried with specific pathogens.

11.4.2.4 Monitoring of Aquaculture Inputs

The use of feed and its additives is based on the use of fishery drugs, which is < the MRL for veterinary drugs in animal foods> announced by the Ministry of Agriculture No. 235. By monitoring the quality, storage, and use of inputs such as fertilizers, feeds, and their additives and fishery drugs, it is ensured that the use of inputs meets relevant regulatory requirements and prevents aquatic products from exhibiting significant biological and chemical hazards associated with farmed inputs.

11.4.2.5 Monitoring of Breeding Diseases

Before the seedlings are stocked, the clearing ponds are strictly eliminated, the seedlings are strictly quarantined, and the drugs are disinfected before entering the pond; the water quality of the ponds is kept good and stable; the water bodies are disinfected regularly, and the baits are fed; the feeding of high-quality compound feeds is prevented. Fish disease. When fish disease occurs, isolation measures should be taken. Dead fish should be promptly removed and harmlessly treated. The used fishing gear in the sick fish pond should be immersed and disinfected. The diseased fish pond water should not be discharged without disinfection. Scientific treatment of fish disease. By monitoring the diseases that occur during aquaculture, ensuring that the microbes carried by aquatic products meet the critical limits, and preventing aquatic products from causing significant biological hazards associated with cultured diseases.

11.4.3 Logistics Process Control Technology

Aquatic products have many circulation links and long processes in the logistics process, which inevitably lead to difficulty to control quality and safety. This requires manufacturers to adopt strict quality assurance measures in all aspects of the cold chain. Cold chain logistics (CCL) refers to a systematic project in which frozen and frozen foods are always in the prescribed low temperature environment during production, storage, transportation, and consumption, to ensure food quality and reduce food loss. CCL is gradually established with the advancement of science and technology and the development of refrigeration technology. It is a low-temperature logistics process based on the refrigeration process and using refrigeration technology as a means. Based on the biological characteristics of perishable and rotten seafood, CCL is widely used in all aspects of seafood logistics to ensure food quality and safety during processing. The CCL of seafood refers to the logistics form with comprehensive economic benefits and social benefits based on cold chain technology and information network as the means in the procurement, processing, storage, transportation, and sales of seafood. The seafood cold chain is mainly composed of four parts: frozen processing, frozen storage, refrigerated transportation and distribution, and frozen sales. From the aspects of biological pollution, chemical pollution, and physical pollution, the key control points and key limits in the CCL of fish and shellfish supply chain were determined, and the main factors, such as pathogenic bacteria, heavy metals, and chemical drug residues, were monitored.

11.4.4 Certification Technology of Aquatic Products

Product certification is one of the important measures to ensure the quality and safety of aquatic products. In China, the main certified products of aquatic products mainly include four kinds of products, namely green food, pollution-free aquatic products, organic aquatic products, and geographical indication aquatic products. These four certified products are called "three products and one standard" in China. Aquatic product certification is based on standardized production.. In particular, China's economic development has now entered a new normal, paying more attention to the scale and industrialization of aquatic products, and at the same time, paying more attention to its high-quality period, and has entered a new stage in the establishment of its brand.

11.4.4.1 Pollution-Free Aquatic Products

Non-polluted aquatic products refer to unprocessed or primary processed edible aquatic products that meet the requirements of relevant national standards and norms, meet the requirements of relevant national standards and norms, and are certified to obtain certified certificates and permit the use of pollution-free aquatic products. The following conditions must be met: ① the production process meets the standards and requirements; ② the corresponding professional technology and management personnel; ③ control measures; ④ input meet the requirements; and ⑤ complete production and sales files.

11.4.4.2 Green Aquatic Products

Green aquatic products refer to safe, high-quality, nutritious aquatic products that are produced according to the principle of sustainable development, produced according to specific production methods, and approved by specialized agencies. ① The product comes from a good ecological environment; ② the whole product quality control is implemented; ③ the product has both "safety, high quality and nutrition," which reflects the coordination of life, resources, and environment.

11.4.4.3 Organic Aquatic Products

Organic food refers to all agricultural and sideline products and processed products that come from organic farming production systems and are produced and processed in this way, which meet international or national organic food requirements and standards, and have been certified by independent certification bodies and issued certificates. In addition to organic foods, the production materials provided by some derivative products, such as organic cosmetics, textiles, forest products, or organic foods, including biological pesticides and organic fertilizers, are collectively referred to as organic products.

11.4.4.4 Geographical Indication Aquatic Products

The geographical indication of aquatic products is an indicator that the aquatic products are derived from specific regions. The quality and related characteristics of the products mainly depend on the natural ecological environment and historical human factors, and the unique aquatic products are marked by the geographical names. And meet the following three conditions: ① the ability to supervise and manage the geographical indications and products of aquatic products; ② the ability to provide guidance services for the production, processing, and marketing of geographical indication aquatic products; ③ to have independent civil responsibility, the ability to take responsibility.

11.4.5 Market Access System

The market access system is divided into three parts: entering the market, market trading, and exiting the market, and constructing corresponding systems to carry out aquatic product quality and safety supervision. Entering the market refers to the regulation before the production of agricultural products to sales. According to the characteristics of China's agricultural production, it should mainly include three types: the production permit system, the product certification system, and the market entry inspection system. These three systems are not systems for implementing certain types of products or production objects at the same time, but complement each other, coordinate and cooperate, and provide a guarantee for the safe listing of agricultural products. The market transaction link is the core part of the agricultural product market access system and the key to maintaining the sound development mechanism of the agricultural product market. Mainly to solve the problem of product information asymmetry; secondly, counterfeit safe agricultural products, shoddy problems; and once again to prevent the secondary pollution of agricultural products in the market environment. According to the general principles of government regulation economics, it mainly uses information regulation and fair trade regulation to manage. Through the identification information, market inspection, and other systems, to effectively monitor the quality and safety of products entering the market. At the exit of the market, the government should take measures to deal with unqualified products that have passed the premarket inspection or market sampling.

11.4.6 Full-Chain Traceability and Traceability Technology

The aquatic product quality traceability system is based on the aquatic product coding rules. It establishes a traceability supervision system covering all aspects of production, processing, and circulation through bar code labeling, code encryption, and key distribution, and realizes the unification of aquatic product circulation and information flow transmission. The system mainly includes production management, transaction management, and regulatory traceability systems. The production management system mainly collects information on seedlings, water quality, and input feeding of aquaculture enterprises in the collection and breeding stage; the transaction management system collects the information on the market, merchants, quality inspection, and other aspects in the circulation stage, establishing aquaculture links and processing links. The circulation link database; the supervision and traceability system is based on the national traceability information database, and establishes a distributed traceability supervision system from the national level to the provincial level, city level, and county level. The core technology of the traceability system is the identification technology. At present, there are mainly technologies such as one-dimensional barcode, two-dimensional barcode, FRID label, NFC label, and DNA barcode.

11.5 Aquatic Product Quality and Safety Risk Management System

The modernization of the food risk management system and governance capacity is an important component of the modernization of the national governance system and governance capacity. The food safety risk management system is a systematic institutional system, which includes the target system, structural system, functional system, operational system, and tool system of governance, and each component interacts and works together. The ability of food safety risk management is formed by the integration and integration of various elements in the governance system to complement and match the elements, so that the overall function of the system is qualitatively changed, which is a comprehensive reflection of the operational efficiency of the governance system.

11.5.1 Top-Level Design of Food Safety Risk Management

In March 2013, the central government implemented a new round of reform of the food and drug regulatory system. In November 2013, the task of reforming institutional mechanisms such as food safety was put forward. In October 2014, the requirements for strengthening public security legislation and promoting the rule of law in public safety were put forward. In October 2015, a major decision was made to build a "healthy China" and implement a "food safety strategy." On October 1, 2015, the newly revised Food Safety Law, known as "the strictest in history," was officially implemented, establishing the principle of "social co-governance" for food safety risks. The report of the 19th National Congress of the Communist Party of China reaffirmed the "implementation of a food safety strategy to allow people to eat with confidence." In short, the central government has made a series of institutional arrangements for the food safety risk management system and the modernization of governance capacity. The "healthy China" and "food safety strategy" have risen to national strategies and have become a major livelihood project for the governance of the country in the new era.

11.5.2 Food Safety Rule of Law System

After the promulgation and implementation of the newly revised Food Safety Law, the revision and revision of supporting laws and regulations at the central and local levels have been successively promoted. The state and local governments have successively issued a series of supporting regulations and important normative documents. In the first half of 2016, the Standing Committee of the National People's Congress launched a law enforcement inspection of the Food Safety Law to implement the food safety strategy, promote the establishment of a sound and unified food safety supervision system, improve the governance system, improve governance, and implement the most stringent The food safety regulatory system has played an important role. Taking the opportunity of the promulgation and implementation of the new Food Safety Law, the administrative organs at all levels and the judiciary have worked together to punish the illegal and criminal activities of food safety through various effective channels and achieved remarkable results.

11.5.3 Government Food Safety Supervision System Reform

A new round of food and drug regulatory system reform was launched in March 2013. By the end of 2015, the reform of the local government food safety regulatory system had basically ended. In general, the unified food and drug safety supervision system was initially formed, but it still faces many problems, mainly due to the various forms of local food and drug safety regulatory agencies, and in some places, there has been a "frozen cake" reform. Whether local governments should set up food safety regulatory agencies separately or implement unified market supervision has become a major controversy in institutional reform.

11.5.4 Food Safety Standard System

So far, the National Health and Family Planning Commission has led the cleanup and integration of nearly 5,000 food standards, and has issued nearly 1,100 national food safety standards with the State FDA and the Ministry of Agriculture, covering all food categories. The main hazard factors are the initial establishment of a national food safety standard system in line with China's national conditions. In December 2016, the Ministry of Agriculture officially promulgated the "National Standards for Food Safety Standards for Pesticide Residues in Foods" (2016 edition), which stipulated 4,140 residue limits for 433 pesticides in 13 categories of agricultural products, which basically covered the approved use of China. Commonly used pesticides and major agricultural products are consumed by residents. However, there is still a certain gap between food safety standards and social and economic development needs. The number and indicators of food safety standards are still missing. The collaborative and sharing mechanisms of relevant industries and social institutions to participate in food safety standards in an orderly manner are still not perfect.

11.5.5 Government Food Safety Information Disclosure

The new Food Safety Law further establishes the legal status of food safety information disclosure in the form of law. The relevant state and local regulatory functions are working hard to promote the disclosure of food safety information within their respective functions and responsibilities in accordance with the functions of the institutional reform. However, there is still a big gap between the government's food safety information disclosure status and the huge social demand. The food safety supervision information disclosure, "insufficient quantity, low quality," "re-completed, light progress" has not been significantly improved. The worse the situation is at the grassroots level, the situation of government food safety information disclosure still has considerable room for improvement.

11.5.6 Government Food Safety Inspection and Testing System

As of the end of October 2015, a total of 1,054 inspection and testing institutions for food and drugs (including food, health food, medicine, cosmetics, and medical equipment, the same below) were in the National FDA System. Among them, the State FDA directly under the inspection agency, a total of 88 inspection agencies at the provincial and sub-provincial level, accounting for 6.74% of the total number of inspection and testing institutions; 361 inspection agencies at the city level, county level. The number of inspection agencies was 604, accounting for 34.25% and 57.31% of the total number of inspection and inspection institutions. According to the region, there are 368 in the eastern region, 390 in the central region, and 296 in the western region. Judging from the quantity of inspection products, the number of inspection agencies accounted for about 34.9% of the total number of the eastern regions, accounting for about 60% of the national inspection tasks. Therefore, it is very arduous to fully integrate the food inspection and testing resources of provincial, municipal, and county government departments, and strive to solve the imbalance between regions, effectively strengthen the construction of the city and county grassroots inspection and testing capabilities, and achieve resource sharing and prevent redundant construction.

11.5.7 Food Safety Risk Monitoring and Evaluation

By the end of 2016, a total of 2,656 food safety risk monitoring points had been set up across the country, covering all provinces, prefectures, and cities, and 92% of county-level administrative regions, and initially established four levels of national, provincial, prefecture-level, and county (district) levels. A three-dimensional food safety risk monitoring network is formed by the architecture. The risk monitoring variety involves 30 categories of foods, such as food, vegetables, fruits, aquatic products, and other people's daily consumption, including more than 300 indicators, and has accumulated more than 15 million monitoring data points. Basically formed a national risk monitoring plan system covering food pollution and food harmful factors monitoring and foodborne disease monitoring, including routine monitoring, special monitoring, emergency monitoring and forward-looking monitoring; basically formed to cover agricultural production, food processing, A comprehensive risk coverage monitoring system covering food contaminants and foodborne pathogens was established with comprehensive coverage and focused risk area monitoring patterns such as product circulation, catering consumption, and online shopping.

References

1 Behera, B.K., Das, A., Sarkar, D.J. et al. (2018). Polycyclic aromatic hydrocarbons (PAHs) in inland aquatic ecosystems: perils and remedies through biosensors and bioremediation. *Environmental Pollution* 241: 212–233.

2 Ruethers, T., Taki, A.C., Johnston, E.B. et al. (2018). Seafood allergy: a comprehensive review of fish and shellfish allergens. *Molecular Immunology* 100: 28–57.

3 Wu, L.H., Zhang, X.M., Wang, F. et al. (2018). Occurrence of bisphenol S in the environment and implications for human exposure: a short review. *Science of the Total Environment* 615: 87–98.

4 Wu, H.Y., Luan, Q.S., Guo, M.M. et al. (2018). Phycotoxins in scallops (*Patinopecten yessoensis*) in relation to source, composition and temporal variation of phytoplankton and cysts in North Yellow Sea, China. *Marine Pollution Bulletin* 135: 1198–1204.

5 Hong, Z., Yue, Z.J., Li, W.H. et al. (2016). The revision of aluminum-containing food additive provisions in China. *Biomedical and Environmental Sciences* 29 (6): 461–466.

6 Olmedo, P., Hernández, A.F., Pla, A. et al. (2013). Determination of essential elements (copper, manganese, selenium and zinc) in fish and shellfish samples. Risk and nutritional assessment and mercury–selenium balance. *Food and Chemical Toxicology* 62: 299–307.

7 Scott, W.C., Breed, C.S., Haddad, S.P. et al. (2019). Spatial and temporal influence of onsite wastewater treatment systems, centralized effluent discharge, and tides on aquatic hazards of nutrients, indicator bacteria, and pharmaceuticals in a coastal bayou. *Science of the Total Environment* 650: 354–364.

8 Brauge, T. and Lessard, M. (2024). Sources and contamination routes of seafood with human-pathogenic *Vibrio* spp.: a review. *Comprehensive Reviews in Food Science and Food Safety* 23.

9 Sampaio, A., Husemann, E., and Thompson, F. (2022). *Vibrio* spp.: life strategies, ecology and risks in a changing ocean. *Diversity* 14 (2): 97.

10 Cortes-Sanchez, A.D.J., Diaz-Ramirez, M., Salgado-Cruz, M.D.L.P. et al. (2020). Food safety and fish production: the case of *Staphylococcus aureus*-a review. *OnLine Journal of Biological Sciences* 20 (4): 291–306.

11 Sharp, J.H., Clements, K., Diggens, M. et al. (2021). E. coli is a poor end-product criterion for assessing the general microbial risk posed from consuming norovirus-contaminated shellfish. *Frontiers in Microbiology* 12: 608888.

12 Uzal, F.A., Henderson, E., and Asin, J. (2024). Botulism in fish: a review. *Journal of Veterinary Diagnostic Investigation* 36 (3): 312–318.

13 Ke, F. and Zhang, Q.Y. (2018). Aquatic animal viruses mediated immune evasion in their host. *Fish & Shellfish Immunology* 86: 1096–1105.

14 Gyawali, P., Fletcher, G.C., McCoubrey, D.-J. et al. (2019). Norovirus in shellfish: an overview of post-harvest treatments and their challenges. *Food Control* 99: 171–179.

15 Schlüter-Vorberg, L., Knopp, G., Cornel, P. et al. (2017). Survival, reproduction, growth, and parasite resistance of aquatic organisms exposed on-site to wastewater treated by advanced treatment processes. *Aquatic Toxicology* 186: 171–179.

16 Wu, H., Yao, J., Guo, M. et al. (2015). Distribution of marine lipophilic toxins in shellfish products collected from the Chinese market. *Marine Drugs* 13 (7): 4281–4295.

17 Munday, R. and Reeve, J. (2013). Risk assessment of shellfish toxins. *Toxins* 5 (11): 2109–2137.

18 Tsaboula, A., Menexes, G., Papadakis, E. et al. (2018). Assessment and management of pesticide pollution at a river basin level part II: optimization of pesticide monitoring networks on surface aquatic ecosystems by data analysis methods. *Science of the Total Environment* 633: 642–650.

19 Ferreira, M., Antunes, P., Costa, J. et al. (2008). Organochlorine bioaccumulation and biomarkers levels in culture and wild white seabream (*Diplodus sargus*). *Chemosphere* 73 (10): 1669–1674.

20 Tsai, Y.H. and Lein, P.J. (2021). Mechanisms of organophosphate neurotoxicity. *Current Opinion in Toxicology* 26: 49–60.

21 Santerre, C., Ingram, R., Lewis, G. et al. (2000). Organochlorines, organophosphates, and pyrethroids in channel catfish, rainbow trout, and red swamp crayfish from aquaculture facilities. *Journal of Food Science* 65 (2): 231–235.

22 Mochida, K., Ito, K., Ito, M. et al. (2018). Toxicity of the biocide polycarbamate, used for aquaculture nets, to some marine fish species. *Comparative Biochemistry and Physiology, Part C* 214: 61–67.

23 Ji, K., Rong, Y., Zhang, Z. et al. (2025). Neonicotinoids and human health: Environmental fate, toxicity mechanisms, and future directions. *Pesticide Biochemistry and Physiology* 215: 106636.

24 Wang, H., Ren, L., Yu, X. et al. (2017). Antibiotic residues in meat, milk and aquatic products in Shanghai and human exposure assessment. *Food Control* 80: 217–225.

25 Morris, D.J., Gray, A.J., Kay, J.F. et al. (2012). EU sampling strategies for the detection of veterinary drug residues in aquaculture species: are they working? *Drug Testing and Analysis* 4 (S1): 1–9.

26 Chen, J., Sun, R., Pan, C. et al. (2020). Antibiotics and food safety in aquaculture. *Journal of Agricultural and Food Chemistry* 68 (43): 11908–11919.

27 Heuer, O.E., Kruse, H., Grave, K. et al. (2009). Human health consequences of use of antimicrobial agents in aquaculture. *Clinical Infectious Diseases* 49 (8): 1248–1253.

28 Chen, P., Wang, J., Zhang, S. et al. (2024). Total syntheses of Tetrodotoxin and 9-epiTetrodotoxin. *Nature Communications* 15 (1): 679.

29 Leal, J.F. and Cristiano, M.L.S. (2021). Marine paralytic shellfish toxins: chemical properties, mode of action, newer analogues, and structure-toxicity relationship. *Natural Product Reports* 39 (1): 33–57.

30 Luis, J.G., Hélène, H., Margot, D. et al. (2022). First subcellular localization of the amnesic shellfish toxin, domoic acid, in bivalve tissues: deciphering the physiological mechanisms involved in its long-retention in the king scallop *Pecten maximus*. *Harmful Algae* 116: 102251.

31 Mudge, E.M., Miles, C.O., Ivanova, L. et al. (2023). Algal ciguatoxin identified as source of ciguatera poisoning in the Caribbean. *Chemosphere* 330: 138659.

32 Meng, J., Hong, S., Wang, T. et al. (2017). Traditional and new POPs in environments along the Bohai and Yellow Seas: an overview of China and South Korea. *Chemosphere* 169: 503–515.

33 Stancheva, M., Georgieva, S., and Makedonski, L. (2017). Polychlorinated biphenyls in fish from Black Sea, Bulgaria. *Food Control* 72: 205–210.

34 Rathna, R., Varjani, S., and Nakkeeran, E. (2018). Recent developments and prospects of dioxins and furans remediation. *Journal of Environmental Management* 223: 797–806.

35 Xu, J., Xue, Q., Zhang, M. et al. (2023). Toxic Effects and mechanisms of polybrominated diphenyl ethers. *International Journal of Molecular Sciences* 24 (17): 13437.

36 Yan, W., Xiang, L., and Sun, X. (2022). The transformation mechanism and eco-toxicity evaluation of butylated hydroxyanisole in environment. *Ecotoxicology and Environmental Safety* 231: 113179.

37 Whisel, M.A. and Rice, D.C. (2025). Tert-butylhydroquinone (TBHQ) suppresses lps- and poly (i:c)-induced raw 264.7 macrophage activation through reduced NF-κB/Type 1 interferon and enhanced antioxidant-related pathways. *Toxics* 13 (10): 883.

38 Junianto, Sandria, F., Dini, F.Z. et al. (2021). Application of food additives in the fisheries processed products development. *Asian Journal of Fisheries and Aquatic Research* 11 (1): 25–35.

39 Yang, X., Huang, C., Sanjeewa, A.K. et al. (2025). A comprehensive review of heavy metals in aquatic products and their removal techniques. *Food Control* 178: 111520.

40 Taylor, B., Ofori, F.K., Parsaeimehr, A. et al. (2025). Exploring the complexities of seafood: from benefits to contaminants. *Foods* 14 (9): 1461.

41 Li, S., Li, Y., Xie, X. et al. (2025). Unveiling in situ methylmercury production and degradation in aquaculture sediments: transformation rates, functional genes and microbial methylators. *Journal of Hazardous Materials* 489: 137659.

42 Wang, J., Lou, Y., Zhuang, X. et al. (2018). Magnetic $Pr_6O_{11}/SiO_2@Fe_3O_4$ particles as the heterogeneous catalyst for the catalytic ozonation of acetochlor: performance and aquatic toxicity. *Separation and Purification Technology* 197: 63–69.

43 Ruethers, T., Raith, M., Sharp, M.F. et al. (2018). Characterization of Ras k 1 a novel major allergen in Indian mackerel and identification of parvalbumin as the major fish allergen in 33 Asia-Pacific fish species. *Clinical & Experimental Allergy* 48 (4): 452–463.

44 Baciak, M., Sikorski, Ł., Piotrowicz-Cieślak, A.I. et al. (2016). Content of biogenic amines in *Lemna minor* (common duckweed) growing in medium contaminated with tetracycline. *Aquatic Toxicology* 180: 95–102.

45 Gu, D.C., Zou, M.J., Guo, X.X. et al. (2017). A rapid analytical and quantitative evaluation of formaldehyde in squid based on Tri-step IR and partial least squares (PLS). *Food Chemistry* 229: 458–463.

46 Lin, L., Regenstein, J.M., Lv, S. et al. (2017). An overview of gelatin derived from aquatic animals: properties and modification. *Trends in Food Science & Technology* 68: 102–112.

47 Zou, W., Yin, L., and Jin, S. (2018). Advances in rapid drug detection technology. *Journal of Pharmaceutical and Biomedical Analysis* 147: 81–88.

48 Alder, L., Greulich, K., Kempe, G. et al. (2006). Residue analysis of 500 high priority pesticides: better by GC-MS or LC-MS/MS? *Mass Spectrometry Reviews* 25 (6): 838–865.

49 Fu, Y., Li, W., and Picard, F. (2024). Assessment of matrix effect in quantitative LC-MS bioanalysis. *Bioanalysis* 16 (12): 631–634.

50 Dipietro, E.S., Bashor, M.M., Stroud, P.E. et al. (1988). Comparison of an inductively coupled plasma-atomic emission spectrometry method for the determination of calcium, magnesium, sodium, potassium, copper and zinc with atomic absorption spectroscopy and flame photometry methods. *Science of the Total Environment* 74: 249–262.

51 Ramezani, S., Mahdavi, V., Gordan, H. et al. (2022). Determination of multi-class pesticides residues of cow and human milk samples from Iran using UHPLC-MS/MS and GC-ECD: a probabilistic health risk assessment. *Environmental Research* 208: 112730.

52 Verma, N. and Bhardwaj, A. (2015). Biosensor technology for pesticides--a review. *Applied Biochemistry and Biotechnology* 175 (6): 3093–3119.

53 Xing, G., Sun, X., Li, N. et al. (2022). New advances in lateral flow immunoassay (LFI) technology for food safety detection. *Molecules* 27 (19): 6596.

54 Brainina, K.Z., Malakhova, N.A., and Stojko, N.Y. (2000). Stripping voltammetry in environmental and food analysis. *Fresenius' Journal of Analytical Chemistry* 368 (4): 307–325.

55 Wu, H., Guo, M., Tan, Z. et al. (2014). Liquid chromatography quadrupole linear ion trap mass spectrometry for multiclass screening and identification of lipophilic marine biotoxins in bivalve mollusks. *Journal of Chromatography A* 1358: 172–180.

56 Guo, Z., Zhu, Z., Huang, S. et al. (2020). Non-targeted screening of pesticides for food analysis using liquid chromatography high-resolution mass spectrometry-a review. *Food Additives & Contaminants: Part A* 37 (7): 1180–1201.

57 Raina, R. (2011). Chemical analysis of pesticides using Gc/Ms, Gc/Ms/Ms, And Lc/Ms/Ms. In: *Pesticides - Strategies for Pesticides Analysis* (ed. M. Stoytcheva), 105. InTech.

58 Boder-Pasche, S., Demir, M., Heub, S. et al. (2024). Multi-well plate lid for single-step pooling of 96 samples for high-throughput barcode-based sequencing. *Biomedical Microdevices* 26 (2): 18.

59 Govindarajan, R., Duraiyan, J., Kaliyappan, K. et al. (2012). Microarray and its applications. *Journal of Pharmacy and Bioallied Sciences* 4 (6): 310–312.

60 Liu, Y., D'Agostino, L.A., Qu, G. et al. (2019). High-resolution mass spectrometry (HRMS) methods for nontarget discovery and characterization of poly-and per-fluoroalkyl substances (PFASs) in environmental and human samples. *TrAC Trends in Analytical Chemistry* 121: 115420.

61 González-Curbelo, M.Á., Socas-Rodríguez, B., Herrera-Herrera, A.V. et al. (2015). Evolution and applications of the QuEChERS method. *TrAC Trends in Analytical Chemistry* 71: 169–185.

62 Wang, W.Q., Du, Y.G., and Chen, Y.H. (2022). Analysis of domestic and foreign standard system and research progress of detection methods for pesticide residues in aquatic products. *Journal of Food Safety and Quality* 13 (11): 3690–3696.

63 Senyuva, H.Z., Gökmen, V., and Sarikaya, E.A. (2015). Future perspectives in Orbitrap™-high-resolution mass spectrometry in food analysis: a review. *Food Additives & Contaminants: Part A* 32 (10): 1568–1606.

64 Su, H., Li, S., Jin, Y. et al. (2017). Nanomaterial-based biosensors for biological detections. *Advanced Health Care Technologies* 19–29.

Index

C

Printed in the USA/Agawam, MA
October 8, 2026

911512.002